6th International Conference on Advanced Materials, Mechanics and Structural Engineering 2019

IOP Conference Series: Materials Science and Engineering
Volume 739

Seoul, South Korea
18 - 20 October 2019

ISBN: 978-1-7138-2655-2
ISSN: 1757-8981

Printed from e-media with permission by:

Curran Associates, Inc.
57 Morehouse Lane
Red Hook, NY 12571

Some format issues inherent in the e-media version may also appear in this print version.

Institute of Physics
Dirac House, Temple Back
Bristol BS1 6BE UK

Phone: 44 1 17 929 7481
Fax: 44 1 17 920 0979

techtracking@iop.org

Additional copies of this publication are available from:

Curran Associates, Inc.
57 Morehouse Lane
Red Hook, NY 12571 USA
Phone: 845-758-0400
Fax: 845-758-2633
Email: curran@proceedings.com
Web: www.proceedings.com

TABLE OF CONTENTS

STRESS ANALYSIS OF SURROUNDING ROCK OF RECTANGULAR PIPE GALLERY IN MOUNTAIN CITY 1
Yuanqing Wang, Xiaoguang Jin, Qingting Yang

THRESHOLD DETERMINATION USING BI-HAZARD FRAGILITY CURVES FOR THE EVALUATION OF STRUCTURAL HEALTH MONITORING OF USHER TECHNOLOGY 12
M B Baylon, F A A Uy, K M S Montes, K A D Embalzado

INHOMOGENEOUS RODS IN TORSION: LONGITUDINAL FRACTURE ANALYSIS BY CONSIDERING THE ENERGY BALANCE 25
V Rizov

INHOMOGENEOUS MULTILAYERED BEAMS OF LINEARLY CHANGING WIDTH: A DELAMINATION ANALYSIS 30
V Rizov

COST IMPACT ANALYSIS ON THE DPWH GUIDELINE AND IMPLEMENTING RULES ON EARTHQUAKE RECORDING INSTRUMENTATION FOR PRIVATE BUILDINGS USING MULTI-CRITERIA DECISION ANALYSIS-ANALYTICAL HIERARCHY PROCESS (MCDA-AHP) IN THE CITY OF MANILA 35
V. C. M. Alfonso, M. B. Bagsit, J. D. Reyes, M. B. Baylon, F. A. A. Uy

RHEOLOGICAL CHARACTERISTICS OF RECLAIMED ASPHALT PAVEMENT (RAP) EVALUATION USING RECLAMITE REJUVENATING MATERIAL 42
Rindu Twidi Bethary, Bambang Sugeng Subagio

INNOVATIVE ROTARY SWAGED AL/CU LAMINATED WIRE CONDUCTORS: CHARACTERISATION OF STRUCTURE AND RESIDUAL STRESS 48
L Kuncická, R Kocich

ANALYSIS THE HYDRAULIC FRACTURING IN THE UNCONVENTIONAL SHALE RESERVOIR BASED ON FINITE DISCRETE ELEMENT METHOD 54
Zhenqian Xue, Xin Zhao

DAMAGE STATE ASSESSMENT OF A PRE-STRESS CONCRETE GIRDER (PSCG) BRIDGE USING NON LINEAR ANALYSIS FOR REAL-TIME STRUCTURAL HEALTH MONITORING 59
K M Montes, K A Embalzado, N M Sales, A J Sayno, J M Payawal, F A A Uy

CONCEPT DESIGN OF A SINGLE-AISLE AIRCRAFT SPOILER BY TOPOLOGY OPTIMIZATION 64
P Purith, C Kittikorn, J Pakpong, L Mezeix

ENGINE PERFORMANCE CHARACTERISTICS OF A COMPRESSION IGNITION ENGINE FUELLED BY TRADITIONAL DIESEL AND WASTE COOKING OIL 69
Dinh Xuan Thanh, Thanh Viet Nguyen, Dien Vu Minh, Khanh Duc Nguyen, Vinh Nguyen Duy

VARIOUS FLOW-FIELD DESIGNS FOR ENHANCING FUEL THE CELL PERFORMANCE OF PROTON EXCHANGE MEMBRANE FUEL CELLS 74
Thi Diep Tran, Duc Ha Vu, Shoudao Huang, Duy Vinh Nguyen

INTELLIGENT TRANSPORTATION SYSTEM: TRAFFIC LOAD MONITORING AND REMOTE SENSING STRUCTURAL HEALTH MONITORING OF CRITICAL BRIDGES IN METRO MANILA USING MULTI MODAL ANALYSIS 79
K M Montes, K A Embalzado, A J Sayno, L C Tani, I Marquez, F A A Uy

EFFECT OF ROAD SURFACE CONTOUR TO THE FATIGUE LIFE OF A COIL SPRING BASED ON STRAIN-LIFE APPROACH 84
T E Putra, Husaini, M Ikbal

EFFECT OF CONCRETE SURFACE CRACK FILLING CONDITIONS ON DOMINANT FREQUENCY OF SURFACE RAYLEIGH WAVE 90
Chi Hoe Liew, Foo Wei Lee, Chee Ghuan Tan, Kok Zee Kwong, Kok-Sing Lim

THE INFLUENCE OF POWDERED CRUMB RUBBER ON THE MECHANICAL PROPERTIES AND THERMAL PERFORMANCE OF LIGHTWEIGHT FOAMED CONCRETE 95
Zhi Heng Lim, Foo Wei Lee, Kim Hung Mo, Kok Zee Kwong, Ming Kun Yew

EXTRACTION OF MICROCRYSTALLINE CELLULOSE FROM TWO DIFFERENT AGRICULTURE WASTE VIA CHEMICAL TREATMENT 100
Nur Athirah Abdullah, Muhammad Hanif Sainorudin, Nilofar Asim, Masita Mohammad, Nurul Huda Abd Kadir, Zahira Yaakob

STUDY OF FAILURE ANALYSIS OF A FRACTURE CRANKSHAFT PULLEY USED ON A TRUCK ENGINE 106
Husaini, T E Putra, S Novriandika

ENCAPSULATE MATERIALS EFFECT ON ZNXCD1-XS WHITE LIGHT QUANTUM DOTS-BASED WHITE LIGHT-EMITTING DIODES 111
Y. X. Huang, S. R. Chung

THE EFFECT OF NANO CERAMIC POWDERS ON MECHANICAL AND PHYSICAL PROPERTIES OF NORMAL CONCRETE 116
Zainab Jawad Kadhim, Ahmed Raad Al-Adhadh, Dhay Ali Sabur

PHOTOBIOREACTOR SIMULATION FOR MICROALGAE CHLORELLA CULTIVATION IN PROCESS 126
Oksana A Trofimchuk, Sofia A Romanenko, Sergey B. Turanov, Alexey N. Yakovlev

EFFECTS OF BAINITE ISOTHERMAL TEMPERATURE ON MICROSTRUCTURE AND MECHANICAL PROPERTIES OF A D-TRIP STEEL 131
Xu Xin, Liu Rendong, Xu Baoyu, Yi Hongliang, Wang Guodong

NONLINEAR ELASTIC ANALYSIS OF CONCRETE BEAMS BASED ON THE SMEARED CRACK APPROACH 137
N Betancourt, C Arias, F Durán, V Arana

ADSORPTION OF DICLOFENAC PHARMACEUTICAL ONTO A CATIONIC ORGANOCLAY AND AN UNTREATED MONTMORILLONITE 144
Tiago De Oliveira, Régis Guégan

OPTICAL PROPERTIES OF CD-FREE QUANTUM DOTS-BASED FLUORESCENT FILM 150
M Y Lin, W X Lin, S R Chung

IMPROVING STABILITY OF CIS/ZNS-BASED WHITE LIGHT-EMITTING DIODES BY SILICA COATING 156
Kawinthida Moolsarn, Meng-Yen Lin, Shu-Ru Chung

STEAM INTEGRATED FLATTENING MACHINE FOR BAMBOO CULMS 161
A H Atienza, H Gutlay, P V Rodrigo, P Tamayo

MULTI-DIRECTIONAL FORCED CONVECTION KILN OVEN DRYING SYSTEM FOR BAMBOO CULMS 166
A H Atienza, R Dela Peña, R Eroles, E Fandiñola

INORGANIC INP/ZNS QUANTUM DOTS: OPTICAL PROPERTIES AND FABRICATION IN DISPLAY APPLICATION 172
Chaikit Chumchujan, Shu-Ru Chung

PERFORMANCE SEISMIC DESIGN OF THE RETROFIT OF A UNIVERSITY LIBRARY USING NON-CONVENTIONAL METHODS 177
B Chavez, C Matias, G Huaco

SYNTHESIS AND CHARACTERIZATION OF MONOCLINIC GALLIUM OXIDE NANOMATERIALS FOR HIGH-CONCENTRATION ETHANOL VAPOR DETECTION 183
H M B Dela Torre, G N C Santos

EFFECT OF ELECTRIC ARC FURNACE DUST TREATMENT ON THE PROPERTIES OF FRESH AND HARDENED MORTAR 190
R Syahyadi, A Fauzi, E Majuar, F Rizal, M Reza, Dan Fajri

MECHANICAL SPLICES FOR SEISMIC RETROFITTING OF CONCRETE STRUCTURES 195
G Huaco, J Jirsa

GRAVEL SAMPLING FOR TESTING 202
Ayad Al-Rumaithi, Mohannad Al-Sherrawi

BENDING STRENGTH OF POLYESTER COMPOSITES REINFORCED WITH STITCHED RANDOM ORIENTATION AND PLAIN WEAVE ABACA FIBER 208
M Iqbal, Y Aminanda, T Firsa, M Ali

NUMERICAL SIMULATION AND EXPERIMENT OF A MEMS ULTRASONIC CAVITATION MODEL 215
Zihong Liu, Hui You, Liangchao Li, Ronghui Lin, Ping Zhang

THERMAL SIMULATION AND SONOPORATION EXPERIMENT BASED ON A FOCUSED ULTRASONIC SYSTEM 227
Zihong Liu, Hui You, Ping Zhang, Ronghui Lin, Yongjia Chang

NUMERICAL ANALYSIS ON SEISMIC PERFORMANCE OF A NEW TYPE OF FULLY ASSEMBLED BEAM-COLUMN JOINTS 237
W Zhai, W Jiang, J H Yang, J F Yang, K Qu, L Chen

STABLE COLLOIDAL SILICA PARTICLES DOPED WITH A FLUORESCENT DYE 243
Tomiris Mulikova, Laura Khamkhash, Kanat Dukenbayev, Anara Molkenova, Timur Sh. Atabaev

EARTHQUAKE VULNERABILITY ASSESSMENT OF HIGH-RISE BUILDINGS IN SURABAYA USING RVISITS ANDROID APPLICATION 248
Wahyu Riyanto, Djoko Irawan, Tri Joko Wahyu Adi, Data Iranata, Aniendhita Rizki Amalia

A COMPARATIVE STUDY OF HYPERELASTIC CONSTITUTIVE MODELS TO CHARACTERIZE THE BEHAVIOUR OF A BIOPOLYMER MATERIAL FOR DIAPHRAGM OF BLOOD PUMP MANUFACTURING 257
L. V. Belyaev, A. V. Zhdanov, A. B. Ivanchenko, A. V. Stepenkin, D. O. Kochetov

SN-DOPED HEMATITE NANOPARTICLES FOR POTENTIAL PHOTOCATALYTIC DYE DEGRADATION ... 262
Svetlana Em, Mussa Yedigenov, Laura Khamkhash, Anara Molkenova, Timur Sh. Atabaev

ON-LINE MONITORING BY ACOUSTIC EMISSION OF BUTT JOINT STRUCTURE ON THE COMPOSITE FUSELAGE FRAME ... 267
X F Qi, Y Yang, W P Kang, Q Wang, G Zhao

EFFECT OF METHOD OF SOIL DRYING ON ATTERBERG LIMITS AND SOIL CLASSIFICATION ... 274
Ahmed Raad Al-Adhadh, Haider Kadhem Sakban, Zahraa Tawfiq Naeem

BENDING ANALYSIS OF NONLOCAL FUNCTIONALLY GRADED BEAMS ... 283
F Garbin, A Levano, R Arciniega

EVALUATION OF THE EVACUATION OF ESSENTIAL BUILDINGS: INTERACTION OF STRUCTURAL AND HUMAN BEHAVIOUR THROUGH NONLINEAR TIME-HISTORY ANALYSIS AND AGENT-BASED MODELLING ... 289
M Delgado, A Rosales, V Arana

ARTIFICIAL INTELLIGENCE OF THE WATER SUPPLY SYSTEM OF PLANT GROWING OBJECTS ... 296
D Yu Polenov

EXPLORATION AND PRACTICE OF DIGITAL FACTORY OF COMPLEX ELECTRONIC EQUIPMENT ... 301
Hu Changming, Cao Yaqi, Zhu Zhaoyu

FORECASTING CHANGE OF THE ROLLER SCREW MECHANISM EFFICIENCY ON THE BASIS OF A COMPARATIVE ANALYSIS OF COATED SURFACES ... 319
L. V. Belyaev, A. V. Zhdanov, N. S. Dovbysh, V. V. Morozov

DEVELOPMENT OF FIRE INFORMATION INTEGRATION APPLICATION PLATFORM ... 323
Ching-Shun Chen, Yu-Tsung Ho, Yi-Lin Chen, Chung-Chyi Chou

DETERMINATION OF CHANGES IN THE PARAMETERS OF ELECTRICAL MACHINES AFTER REPAIRS TO PREVENT AN UNACCEPTABLE INCREASE IN VIBRATION OF THE BUILDING STRUCTURE ... 329
A Belyi

USE OF GALACTOSE OXIDASE IN HYDRAZIDE-CAPTURING TECHNIQUE TO ISOLATE N-LINKED GLYCOALLERGENS FROM OLIVE TREE POLLEN ... 333
A. Napoli

USE OF GEOPOLYMERS IN THE TREATMENT OF WATER CONTAMINATED BY INDUSTRIAL WASTE ... 339
I Bernaudo, A Tagarelli, R Elliani, S Candamano, A Macario, P De Luca

USE OF ZINC FERRITE FOR PHOTOCATALYTIC TREATMENT OF WATER CONTAMINATED WITH ORGANIC DYE ... 345
A Cantiello, S Candamano, P De Luca

RESEARCH ON AERIAL VIEW POINT PLANNING OF DRONE BASED ON MULTI-VIEW ... 351
Zhu Yan-Yan Shi Yun

AUTOMATED PLATFORM TO THE DEPOSITION OF THIN-FILMS AND CHARACTERIZATION PROCESS IN FIBER OPTICS SENSORS 357
Adolfo Josué Rodríguez-Rodríguez, Wenceslao Eduardo Rodríguezrodríguez, Ricardo Iván Luis-Morales, Víctor Alan Cisneros-Rojas, Rocío Zamorano Báez, Jose Isidro Acosta-Morales, Luis Antonio Garcíagarza

Author Index

IOP Conf. Series: Materials Science and Engineering **739** (2020) 011001 doi:10.1088/1757-899X/739/1/011001

Preface

The 6th AMMSE 2019 is the 2019 6th International Conference on Advanced Materials, Mechanics and Structural Engineering (6th AMMSE 2019) took place in Seoul, South Korea, on October 18-20, 2019.

The conference program covered invited, oral, and poster presentations from scientists working in similar areas to establish platforms for collaborative research projects in this field. This conference will bring together leaders from industry and academia to exchange and share their experiences, present research results, explore collaborations and to spark new ideas, with the aim of developing new projects and exploiting new technology in this field.

The committee of AMMSE expresses their sincere thanks to all authors for their high-quality research papers and careful presentations. All reviewers are also thanked for their careful comments and advices. Thanks are finally given to IOP Publication as well for producing this volume.

The Organizing Committee of AMMSE 2019

Committee Chair

Prof. Mosbeh Kaloop

Incheon Disaster Prevention Research Center

Incheon National University

Conference Photograph

IOP Conf. Series: Materials Science and Engineering **739** (2020) 011001 doi:10.1088/1757-899X/739/1/011001

Sponsors

Incheon Disaster Prevention Research Center (IDPRC)

Recently, various efforts to prevent and prepare are vitally needed for prevention of Disasters and calamities. So we understand the necessity for technology of disaster and we built up the Incheon Disaster Prevention Research Center (IDPRC) in Incheon National University (INU) in 1997.

Accordingly, Incheon Disaster Prevention Research Center (IDPRC) in Incheon National University has progressed of research on the prevention of disasters and calamities through the various seminars, conference and lectures. These research could be conducted cooperation with Incheon National University (INU) in various fields structure, soil, hydraulic and environment.

Incheon Disaster Prevention Research Center (IDPRC) will try to be a leader in the disaster of industry through the various research activities and global conference.

Incheon National University

Incheon National University established a global campus by integrating two city council funded colleges into one in March 2010 and became a national university operated by the legal entity. With the start of attracting the branch of Lawrence Berkeley Research Center that is a world famous US national policy research center for education and research, Incheon National University will attract St. Petersburg University in Russia, Polymers University in Britain, and Kent University in Belgium and other foreign universities and research centers. These efforts will make INU spring to a world-wide competitive university.

As a local development leading university, Incheon National University established INU VISON 2020 and improved its competition actively through investment in selection and concentration. In compliance with the geographic feature and areal specialty, it will concentrate on the international trade, goods distribution, applied technology convergence, BNT-based life science, urban science and the sinology- based local humanity fields.

IOP Conf. Series: Materials Science and Engineering **739** (2020) 011001 doi:10.1088/1757-899X/739/1/011001

Lists of Committees

Committee name	Organizer name	Organization
Honorary Chair	Mosbeh Kaloop	Incheon National University
Conference General Chairs	I. Mansouri	Birjand University of Technology
Honorary Chair	D.K. Kim	Dong-A University
Technical Committees	J.W. Jung	Chungbuk University, South Korea
	J.W. Hu	Incheon National University
	J.K. Ahn	Incheon National University

IOP Conf. Series: Materials Science and Engineering **739** (2020) 011002 doi:10.1088/1757-899X/739/1/011002

	First name	Surename	Title	E-mail address
1	Yuanqing	Wang	Stress Analysis of Surrounding Rock of Rectangular Pipe Gallery in Mountain City	1017364393@qq.com
2	M B	Baylon	Threshold Determination Using Bi-hazard Fragility Curves for the Evaluation of Structural Health Monitoring of USHER Technology	ikingbalon@gmail.com
3	V	Rizov	Inhomogeneous Rods in Torsion: Longitudinal Fracture Analysis by Considering the Energy Balance	v_rizov_fhe@uacg.bg
4	V	Rizov	Inhomogeneous Multilayered Beams of Linearly Changing Width: a Delamination Analysis	v_rizov_fhe@uacg.bg
5	V. C. M.	Alfonso	Cost Impact Analysis on the DPWH Guideline and Implementing Rules on Earthquake Recording Instrumentation for Private Buildings u	ikingbalon@gmail.com
6	Rindu Twidi	Bethary	Rheological Characteristics of Reclaimed Asphalt Pavement (RAP) Evaluation using Reclamite Rejuvenating Material	rindutwidibethary@gmail.com
7	L L	Kunčická	Innovative rotary swaged Al/Cu laminated wire conductors: characterisation of structure and residual stress	lenka.kuncicka@vsb.cz
8	Zhenqian	Xue	Analysis the hydraulic fracturing in the unconventional shale reservoir based on Finite Discrete Element Method	hengsmith111@163.com
9	K M	Montes	Damage State Assessment of a Pre-Stress Concrete Girder (PSCG) Bridge using Non Linear Analysis for Real-time Structural Health M	katrinamaesmontes@yahoo.com
10	P	Purith	Concept Design of a Single-Aisle Aircraft Spoiler by Topology Optimization	em-purith18@outlook.com
11	Dinh Xuan	Thanh	Engine Performance Characteristics of a Compression Ignition Engine Fuelled by Traditional Diesel and Waste Cooking Oil	vinhgsyb1@gmail.com
12	Thi Diep	Tran	Various Flow-Field Designs for Enhancing Fuel the Cell Performance of Proton Exchange Membrane Fuel Cells	vinhgsyb1@gmail.com
13	K M	Montes	Intelligent Transportation System: Traffic Load Monitoring and Remote Sensing Structural Health Monitoring of Critical Bridges in Metro M	krish.angelo27@yahoo.com
14	T E	Putra	Effect of Road Surface Contour to the Fatigue Life of a Coil Spring Based on Strain-Life Approach	edi@unsyiah.ac.id
15	Chi Hoe	Liew	Effect of Concrete Surface Crack Filling Conditions on Dominant Frequency of Surface Rayleigh Wave	mmankey@1utar.my
16	Zhi Heng	Lim	The Influence of Powdered Crumb Rubber on the Mechanical Properties and Thermal Performance of Lightweight Foamed Concrete	Leon2810-lzh@hotmail.com
17	Nur Athirah	Abdullah	Extraction of Microcrystalline Cellulose from Two Different Agriculture Waste via Chemical Treatment	asimnilofar@gmail.com
18		Husaini	Study of Failure Analysis of a Fracture Crankshaft Pulley Used on a Truck Engine	husainiftm@unsyiah.ac.id
19	Y. X.	Huang	Encapsulate Materials Effect on ZnxCd1-xS White Light Quantum Dots-Based White Light-Emitting Diodes	40332205@gm.nfu.edu.tw
20	Zainab Jawad	Kadhim	The Effect ofNano Ceramic Powders On Mechanical and Physical Properties of Normal Concrete	ahmad_al_iraqi2000@yahoo.com
21	Oksana A	Trofimchuk	Photobioreactor Simulation for Microalgae Chlorella Cultivation in Process	oks.trofimchuk@gmail.com
22	Xu	Xin	Effectsof Bainite Isothermal Temperature on Microstructure and Mechanical Properties of a δ-TRIP Steel	117118419@qq.com
23	N	Betancourt	Nonlinear elastic analysis of concrete beams based on the Smeared Crack Approach	pccivara@upc.edu.pe
24	TiagoDe	Oliveira	Adsorption of Diclofenac Pharmaceutical onto a Cationic Organoclay and an Untreated Montmorillonite	regis.guegan@aoni.waseda.jp
25	M. Y.	Lin	Optical Properties of Cd-Free Quantum Dots-Based Fluorescent Film	40332202@gm.nfu.edu.tw
26	Kawinthida	Moolsarn	Improving Stability ofCIS/ZnS-Based White Light-Emitting Diodesby Silica Coating	Kawinthida_m@hotmail.com
27	A H	Atienza	Steam Integrated Flattening Machine for Bamboo Culms	ahsatienza@mcl.edu.ph
28	A H	Atienza	Multi-Directional Forced Convection Kiln Oven Drying System for Bamboo Culms	ahsatienza@mcl.edu.ph
29	Chaikit	Chumchujan	Inorganic InP/ZnS Quantum Dots:Optical Properties and Fabrication in Display Application	hibari_climax@hotmail.com
30	B	Chavez	Performance Seismic Design of the Retrofitof a University Library using Non-Conventional Methods	pccighua@upc.edu.pe
31	H M B	Dela Torre	Synthesis and Characterization ofMonoclinic Gallium Oxide Nanomaterials for High-Concentration Ethanol Vapor Detection	herbhelyne.delatorre@dlsu.edu.ph
32	R	Syahyadi	Effect of Electric Arc Furnace Dust Treatment on the Properties of Fresh and Hardened Mortar	amirfauzi@pnl.ac.id
33	G	Huaco	Mechanical Splices for Seismic Retrofitting of Concrete Structures	pccighua@upc.edu.pe
34	Al-Rumaithi	Ayad	Gravel Sampling for Testing	dr.mohannad.al-sherrawi@coeng.uobaghdad.e
35	M	Iqbal	Bending Strength of Polyester Composites Reinforced with Stitched Random Orientation and Plain Weave Abaca Fiber	mohd.iqbal@unsyiah.ac.id
36	Zihong	Liu	Numerical Simulation and Experiment of a MEMSUltrasonic Cavitation Model	751102185@qq.com
37	Zihong	Liu	ThermalSimulation andSonoporationExperiment Based on a Focused Ultrasonic System	270820424@qq.com
38	B	Thaweboon	Antimicrobial Activity of Neem Toothpaste against CariesAssociated Microorganisms	sroisiri.tha@mahidol.ac.th
39	W	Zhai	Numerical Analysis on Seismic Performance of a New Type of Fully Assembled Beam-Column Joints	815030764@qq.com
40	Tomiris	Mulikova	Stable Colloidal Silica Particles Doped with a Fluorescent Dye	timur.atabaev@nu.edu.kz
41	S	Thaweboon	Inhibitory Effect of Moringa OleiferaL. Seed Oil on Oral Bacteria Related to Periodontal Diseases and Systemic Infections	sroisiri.tha@mahidol.ac.th
42	Wahyu	Riyanto	EarthquakeVulnerability Assessmentof High-Rise Buildings in Surabaya using RViSITSAndroidApplication	wahyuriyanto303@gmail.com
43	L. V.	Belyaev	A Comparative Study ofHyperelastic Constitutive Models to Characterize theBehaviour ofaBiopolymer Material for Diaphragm ofBlood Pu	blv_vlsu@mail.ru
44	Em	Svetlana	Sn-Doped Hematite Nanoparticles for Potential Photocatalytic Dye Degradation	timur.atabaev@nu.edu.kz
45	X F	Qi	On-line Monitoring by Acoustic Emission of Butt Joint Structure on the Composite Fuselage Frame	741612110@qq.com
46	Al-Adhadh	Ahmed Raad	Effect of Method of Soil Drying On Atterberg Limits and Soil Classification	ahmad_al_iraqi2000@yahoo.com
47	F	Garbin	Bending Analysis of Nonlocal Functionally Graded Beams	roman.arciniega@upc.pe
48	M	Delgado	Evaluation of the Evacuation of Essential Buildings: Interaction of Structural and Human Behaviour through Nonlinear Time-History Analy	pccivara@upc.edu.pe
49	D Yu	Polenov	Artificial Intelligence of the Water Supply System of Plant Growing Objects	polenoff.mitya@yandex.ru
50	Changming	Hu	Exploration and Practice of Digital Factory of Complex Electronic Equipment	1169668672@qq.com
51	L. V.	Belyaev	ForecastingChangeoftheRollerScrewMechanism EfficiencyontheBasisofaComparativeAnalysisofCoated Surfaces	blv_vlsu@mail.ru
52	Ching-Shun	Chen	Development of Fire Information Integration Application Platform	dustin@mail.dyu.edu.tw
53	A.	Belyi	Determination of Changes in the Parameters of Electrical Machines after Repairs to Prevent an Unacceptable Increase in Vibration of th	belyynmstu@gmail.com
54	A.	Napoli	Use of Galactose Oxidase in Hydrazide-Capturing Technique to Isolate N-linked Glycoallergens from Olive Tree Pollen	amc.napoli@unical.it
55	I	Bernaudo	Use of Geopolymers in the Treatment of Water Contaminated by Industrial Waste	pierantonio.deluca@unical.it
56	C.	Cantiello	Use of Zinc Ferrite for Photocatalytic Treatment of Water Contaminated with Organic Dye	pierantonio.deluca@unical.it
57	Yan-yan	Zhu	Research on Aerial View Point Planning of Drone based on Multi-View	710667612@qq.com
58	R. R.	Adolfo Josué	Automated Platform to the Deposition of Thin-Films and Characterization Process in Fiber Optics Sensor	wrodriguez@docentes.uat.edu.mx

IOP Conf. Series: Materials Science and Engineering **739** (2020) 011003 doi:10.1088/1757-899X/739/1/011003

Peer review statement

All papers published in this volume of *IOP Conference Series: Materials Science and Engineering* have been peer reviewed through processes administered by the proceedings Editors. Reviews were conducted by expert referees to the professional and scientific standards expected of a proceedings journal published by IOP Publishing.

IOP Conf. Series: Materials Science and Engineering **739** (2020) 012001 doi:10.1088/1757-899X/739/1/012001

Stress Analysis of Surrounding Rock of Rectangular Pipe Gallery in Mountain City

Yuanqing Wang[1], Xiaoguang Jin[2], Qingting Yang[2]

[1]The 5th Engineering Ltd. of the 11th Engineering Bureau of China Railway, Chongqing 400037, P.R. China

[2]School of Civil Engineering, Chongqing University, Chongqing 400045, P.R. China

jinxiaoguang@cqu.edu.cn

Abstract. In this paper, the basic methods and principles of complex variable function to solve the excavation problem of caverns in single connected domain are studied. Based on the basic principle of conformal transformation and the engineering practicability, the problem of excavating rectangular section single warehouse pipe gallery in a single rock stratum is simplified as a semi-infinite plane rectangular hole problem. The rectangular hole problem is transformed into an equivalent circular hole by conformal transformation. A simplified calculation model of the extruding rectangle comprehensive pipe gallery is proposed and its surrounding rock stress calculation formula is deduced. Combined with the example of mountainous city comprehensive pipe gallery project, Theoretical calculation and numerical simulation show that the surrounding rock stress of the comprehensive pipe gallery is close, which verifies the applicability of the simplified formula. The structural design and construction design of rectangular integrated pipe gallery provide theoretical support. Comparing the stress of surrounding rock of different width pipe gallery, it is found that the stress concentration degree at the top corner of the rectangular pipe gallery increases with the increase of the width of the pipe gallery. The width of the excavation of the rectangular pipe gallery must be strictly controlled during construction.

1. Introduction

At present, the most comprehensive section form of the pipe gallery is rectangular section. Most of the construction methods are open-cut method. In some special sections, the excavation construction method is used to assist. In the mountainous city, combining the open-cut method with the under-cut method are used more in the construction of the rectangular integrated pipe gallery [1]. Wang [2] studied the feasibility and advantages of the integrated design of large tunnels and comprehensive pipe galleries; Yang and Feng [3] conducted a theoretical study on the safety distances of various pipelines in the integrated pipe gallery. Combined with the spatial interaction model, the negative power law distance attenuation curve is used to establish the safety distance model between the pipelines in the pipe gallery. The model is applied to the examples of pipe gallery engineering to verify the practicability of the model. Due to the special section form, the surrounding rock force and displacement of the rectangular integrated pipe gallery have unique characteristics, which have an important influence on the supporting form and structural arrangement [4]. Many domestic and foreign scholars have carried out in-depth research on the design and construction of the integrated pipe gallery with circular section, and given reasonable calculation methods and research methods for the stress distribution of the surrounding rock of the pipe galleries [5]. Yang, Wang et al. [6-9] used the mathematical method of complex variable function to derive the surrounding rock pressure of shallow tunnels. Huang, Wang, Wang et al. [10-12] used numerical simulation to reveal the variation of surrounding rock pressure caused by tunnel construction and give methods and parameters for numerical simulation of surrounding rock pressure. Lu[13] used the optimization method of the hybrid penalty function to transform the surrounding rock pressure into a double-pass domain by mapping the closed integral support of non-circular roadway conformally into a circular domain; Zhu and Qian[14] took the calculation and analysis of the arbitrarily shaped cavity mapping function as the starting point, combined with the elastic mechanics analysis method, the solution was solved by means of the computer program, which expands the scope and prospect of solving the underground cavern, and makes full use of the complex mapping function to solve surrounding rock pressure of underground caverns of any complex shape. Chen, Tang et al. [4] used the complex variable function to get the elastic analytical results of surrounding rock in rectangular roadways to find the influencing laws of the roadway aspect ratio and side pressure on the surrounding rock stress of rectangular-section roadways. Although these research calculation methods are relatively mature, most of them need to be calculated by computer programming, and no intuitive simplified calculation formula is given. Due to the limited computer level of on-site construction management personnel, the existing research results are not directly applicable to the project. In this paper, the solution of the surrounding rock stress of the rectangular pipe gallery is simplified and the simplified calculation formula is derived, which is convenient for the on-site construction personnel to judge the safety state of the surrounding rock during the construction process.

2. Simplified calculation model for rock pressure around pipe gallery

2.1. Simplified model

According to the characteristics of mountainous cities, the underground tunnels have strong self-supporting ability when they traverse the rock stratum with good surrounding rock. Using the conformal transformation and the elastic mechanics complex function to solve the solution principle, the surrounding rock stress can be solved by the following formula.

$$\sigma_V = \gamma H \qquad \sigma_H = \gamma\left(H + \frac{1}{2H_t}\right)\tan^2\left(45^\circ - \frac{\varphi_c}{2}\right) \tag{1}$$

The side pressure coefficient can be taken as:

$$\lambda = \frac{\sigma_H}{\sigma_V} = \left(1 + \frac{1}{2H * H_t}\right)\tan^2\left(45^\circ - \frac{\varphi_c}{2}\right) \tag{2}$$

Among them:

σ_H -horizontal pressure

σ_v -plumb pressure

$2b$ - the width of the outer contour of the rectangular integrated pipe gallery

2h - the height of the outer contour of the integrated pipe gallery

γ -the average gravity of the rock and soil

H -the buried depth of the integrated pipe gallery

H_t -the height of the section of the integrated pipe gallery

λ-the side pressure coefficient of the rock and soil body

φ_c -the calculated friction angle of the rock mass

2.2. Conformal transformation

According to the basic idea of solving the stress problem of the plane hole of elastic mechanics according to the complex variable function, the rectangular domain problem can be transformed into a circular domain by conformal transformation, which is convenient for calculation and calculation, as shown in figure 1.

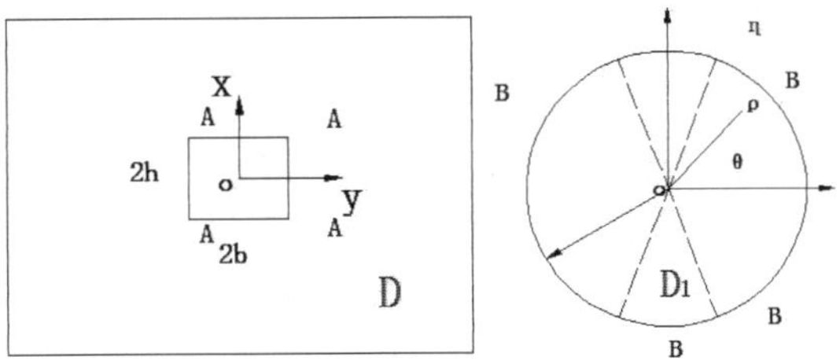

Figure 1. The conformal transformation of rectangle integrated pipeline.

Firstly, at the centroid of the rectangular integrated pipe gallery, the Cartesian coordinate system o-xy with this point as the origin is established; and the plane where the plane is located is the Z plane, and the area where the surrounding rock of the rectangular integrated pipe gallery is located is the area D. By the conformal transformation, the region D is mapped to the unit circle field with the ζ =0 as the center on the ζ plane, and as D_1.

According to the basic knowledge of the conformal transformation of the single hole problem of the elastomer, we use $\zeta=\rho e^{i\theta}$ at any point in the ζ plane. According to Schwarz-Kritofel's research method of mapping, the following general form of conformal transformation can be used:

$$\omega = K\int\left[(z-x_1)^{\frac{\alpha_1}{\pi}-1}(z-x_2)^{\frac{\alpha_2}{\pi}-1}\cdots(z-x_k)^{\frac{\alpha_k}{\pi}-1}\cdots(z-x_n)^{\frac{\alpha_n}{\pi}-1}\right]dz + C \tag{3}$$

Where α_1, x_1、 α_2, x_2 ... α_n, x_n represent the mapping angle of the polygon and the position of the mapping point respectively. From the theory of complex variable function, the rectangular outer boundary on Z-plane is mapped into the unit circle area on ζ-plane , then the following mapping relation of C=0 can be found through finding the appropriate α and x ,then through sorting and deriving.

$$z = \omega(\zeta) = \mathrm{R}\left(\frac{1}{\zeta} + c_1\zeta + c_3\zeta^3 + c_5\zeta^5 + ...\right) \tag{4}$$

Substituted into the complex function expression, the following formula can be obtained according to the mapping correspondence:

$$c_1 = \frac{b-h}{2R}\ ;\ c_3 = \frac{b+h}{2R} - 1$$

$$R = \frac{\frac{b-h}{2}\cos\theta + \frac{b+h}{2}\cos 3\theta - b}{\cos 3\theta - \cos\theta} \tag{5}$$

$$\left(\frac{b-h}{2}\cos\theta + \frac{b+h}{2}\cos 3\theta - b\right)(\sin 3\theta + \sin\theta) - \left(\frac{b-h}{2}\sin\theta + \frac{b+h}{2}\sin 3\theta - h\right)(\cos 3\theta - \cos\theta) = 0 \tag{6}$$

Where b is 1/2 the width of the pipe; h is 1/2 the height of the pipe.

Substituting the values of b and h, θ, c_1, c_3 and R can be respectively obtained, and the relevant parameters of the conformal mapping are obtained.

3. Analytical solution to the pressure of surrounding rock of pipe gallery

3.1. Complex function expression of the pressure of surrounding rock around pipe galleries

It is known from the fundamental principle of the complex variable function to solve the elastic mechanics problem that the planar problem can be represented by the complex potential functions $\varphi(z)$、$\psi(z)$ parsed everywhere in the region. Therefore, it is the main task of solving the complex analytic function that satisfies the given boundary condition that is, solving $\varphi_0(\zeta)$、$\psi_0(\zeta)$、$\varphi(\zeta)$、$\psi(\zeta)$.

First, from the Cauchy integral, if $F(\zeta)$ is continuous and resolved within the unit circle and circumference, then at any point in the circle there is:

$$F(\zeta)=\frac{1}{2\pi i}\int_\sigma \frac{F(\sigma)}{\sigma-\zeta}d\sigma \tag{7}$$

At the same time, if it is continuous and parsed outside the unit circle, then at any point outside the unit circle there is:

$$F(\infty)=\frac{1}{2\pi i}\int_\sigma \frac{F(\sigma)}{\sigma-\zeta}d\sigma \tag{8}$$

It is known by the boundary conditions:

$$\frac{1}{2\pi i}\int_\sigma\left[\varphi_0(\sigma)+\frac{\omega(\sigma)}{\omega'(\sigma)}\overline{\varphi'(\sigma)}+\overline{\psi(\sigma)}\right]\frac{d\sigma}{\sigma-\zeta}=\frac{1}{2\pi i}\int_\sigma f_0\frac{d\sigma}{\sigma-\zeta} \tag{9}$$

According to A, B which are continuous and parsed, the following expression can be got:

$$\frac{1}{2\pi i}\int_\sigma \varphi_0\frac{F(\sigma)}{\sigma-\zeta}d\sigma=\varphi_0(\zeta)$$
$$\frac{1}{2\pi i}\int_\sigma \overline{\psi_0(\sigma)}\frac{d\sigma}{\sigma-\zeta}=\overline{\psi_0(\infty)}=0 \tag{10}$$

Substituting equation (7) into equation (8) yields:

$$\varphi_0(\zeta)+\frac{1}{2\pi i}\int_\sigma \frac{\omega(\sigma)}{\omega'(\sigma)}\frac{\overline{\varphi'(\sigma)}}{\sigma-\zeta}d\sigma=\frac{1}{2\pi i}\int_\sigma f_0\frac{d\sigma}{\sigma-\zeta} \tag{11}$$

The same can be obtained:

$$\psi_0(\zeta)+\frac{1}{2\pi i}\int_\sigma \frac{\overline{\omega(\sigma)}}{\omega'(\sigma)}\frac{\varphi'(\sigma)}{\sigma-\zeta}d\sigma=\frac{1}{2\pi i}\int_\sigma \overline{f_0}\frac{d\sigma}{\sigma-\zeta} \tag{12}$$

Where $\psi_0(\zeta)=\int_\sigma \frac{\psi_0(\zeta)}{\sigma-\zeta}d\sigma$

From $\varphi_0(\zeta)=\sum_{k=1}^{+\infty} a_k\zeta^k$, it can be obtained:

$$\varphi_0'(\zeta)=\sum_{k=1}^{+\infty} ka_k\zeta^{k-1} \qquad \overline{\varphi_0'(\zeta)}=\sum_{k=1}^{+\infty} k\overline{a_k}\frac{\rho^{2k-2}}{\zeta^{k-1}} \tag{13}$$

After derivation, we can get:

$$\varphi(\zeta)=\alpha\omega(\zeta)+\varphi_0(\zeta)=\alpha R\left(\frac{1}{\zeta}+c_1\zeta+c_3\zeta^3\right)+a_1\zeta-2\alpha c_3R\zeta^3 \tag{14}$$

$$\psi(\zeta)=(\alpha_1+i\beta)\omega(\zeta)+\psi_0(\zeta)=\frac{\alpha_1R-c_3a_1}{\zeta}-2\alpha R\zeta-\alpha_1Rc_3\zeta^3-\frac{\zeta^4+c_1\zeta^2+c_3}{3c_3\zeta^4+c_1\zeta^2-1}\left(\frac{a_1}{\zeta}-6\alpha c_3R\zeta\right) \tag{15}$$

Therefore:

$$\varphi'(\zeta)=\alpha R\left(-\frac{1}{\zeta^2}+c_1+3c_3\zeta^2\right)+a_1-6\alpha c_3R\zeta^2 \tag{16}$$

$$\psi'(\zeta)=-\frac{\alpha_1R-c_3a_1}{\zeta^2}-2\alpha R-3\alpha_1Rc_3\zeta^2-\frac{2\left(a_1c_1-6\alpha c_3^2R\right)+4\left(a_1-6\alpha c_3Rc_1\right)\zeta-36\alpha c_3R\zeta^4}{3c_3\zeta^4+c_1\zeta^2-1}$$
$$+\frac{\left[a_1c_3+(a_1c_1-6\alpha c_3^2R)\zeta^2+(a_1-6\alpha c_1c_3\mathrm{R})\zeta^4-6\alpha c_3R\zeta^6\right](15c_3\zeta^4+3c_1\zeta^2-1)}{\zeta^2(3c_3\zeta^4+c_1\zeta^2-1)^2} \tag{17}$$

$$\Phi(\zeta)=\frac{\varphi'(\zeta)}{\omega'(\zeta)}=\alpha+\frac{\frac{a_1}{R}-6\alpha c_3\zeta}{-\frac{1}{\zeta^2}+c_1+3c_3\zeta^2} \tag{18}$$

$$\Phi'(\zeta)=\frac{\frac{2a_1}{R}\zeta-24\alpha c_3\zeta^3}{-1+c_1\zeta^2+3c_3\zeta^4}-\frac{\left(\frac{a_1}{R}\zeta^2-6\alpha c_3\zeta^4\right)(2c_1\zeta+12c_3\zeta^3)}{(-1+c_1\zeta^2+3c_3\zeta^4)^2} \tag{19}$$

$$\psi(\zeta)=\frac{\psi'(\zeta)}{\omega'(\zeta)}=\frac{1}{R\left(-\frac{1}{\zeta^2}+c_1+3c_3\zeta^2\right)}$$
$$\bullet\left\{\begin{array}{l}-\frac{\alpha_1R-c_3a_1}{\zeta^2}-2\alpha R-\frac{2(a_1-6\alpha c_3^2R)+(4a_1-6\alpha c_1c_3\mathrm{R})\zeta^2-36\alpha c_3R\zeta^4}{3c_3\zeta^4+c_1\zeta^2-1}\\+\frac{\left[a_1c_1+(a_1c_1-6\alpha c_3^2R)\zeta^2+(a_1-6\alpha c_1c_3R)\zeta^4-6\alpha c_3R\zeta^6\right]\left(15c_3\zeta^4+3c_1\zeta^2-1\right)}{\zeta^2(3c_3\zeta^4+c_1\zeta^2-1)^2}\end{array}\right\} \tag{20}$$

According to the method of curve coordinate expression for the stress components, there is the following formula:

$$\begin{aligned}&\sigma_\varphi+\sigma_\rho=4\mathrm{Re}\left[\Phi(\zeta)\right]\\&\sigma_\varphi-\sigma_\rho+2i\tau_{\rho\varphi}=\frac{2\zeta^2}{\rho^2\overline{\omega'(\zeta)}}\left[\overline{\omega(\zeta)}\Phi'(\zeta)+\omega'(\zeta)\psi(\zeta)\right]\end{aligned} \tag{21}$$

Then, the deduced $\Phi(\zeta)$、$\Phi'(\zeta)$、$\psi(\zeta)$ can be substituted into the above formula to obtain the surrounding rock stress of the integrated pipe gallery in polar coordinates.

3.2. Simplified calculation formula

According to the simplified calculation model of 2.1 and related derivation, combined with the actual and unsupported boundary conditions, it can be seen that at the boundary of the rectangular integrated pipe gallery chambers:

$$\rho=1,\ \zeta=e^{i\varphi},\ \sigma_\rho=\sigma_{\rho\varphi}=0 \tag{22}$$

It can be obtained from the previous derivation:

$$\begin{aligned}\sigma_\varphi &= 4\mathrm{Re}[\Phi(\zeta)]=4\mathrm{Re}\left[\alpha+\frac{s_2\alpha-6\alpha c_3e^{2i\varphi}}{-e^{-2i\varphi}+c_1+3c_3e^{2i\varphi}}\right]\\ &=4\mathrm{Re}\left[\alpha+\frac{s_2\alpha-6\alpha c_3\cos2\varphi-i6\alpha c_3\sin2\varphi}{c_1+(3c_1-1)\cos2\varphi+i(1+3c_3)\sin2\varphi}\right]\\ &=4\alpha+4\frac{(s_2\alpha-6\alpha c_3\cos2\varphi)}{[c_1+(3c_3-1)\cos2\varphi]^2+(1+3c_3)^2\sin^2 2\varphi}\end{aligned} \tag{23}$$

In the formula, $s_1=c_1+(3c_3-1)\cos2\varphi$, $s_2=2\dfrac{c_1+1+(c_1-1)\lambda}{(c_3-1)(1-\lambda)}$ 。

According to the definition in the previous formula derivation process, c_1、c_3、R are respectively obtained, and substituted into the above formula, the surrounding rock pressure of the rectangular integrated pipe gallery under polar coordinates can be obtained.

4. Calculation and comparison of examples

4.1. Theoretical calculation

Combined with the relevant pipe network specification and actual engineering, the common section size of the single-layer and single-warehouse rectangular integrated pipe gallery is around 3.5m high, with width between 3.5-6m.

For the safety of the geotechnical environment on the surface, the minimum recommended depth of the undercutting integrated pipe galleries refers to the recommended minimum value of adjacent underground pipelines and underground structures (table 1).

Table 1. Rectangular integrated pipeline and minimum spacing table of construction.

Adjacent situation \ Construction method	Open-cut construction	Under-cut construction
Integrated pipe galleries and buildings	1.0 m	outer diameter of integrated pipe galleries
Clear distance between integrated pipe galleries and underground pipelines	1.0 m	outer diameter of integrated pipe galleries
Cross vertical distance between integrated pipe galleries and underground pipelines	0.5 m	1.0 m

According to table 1, the minimum distance between the integrated pipe gallery and the building is the outer diameter of the integrated pipe gallery (generally 3~4 m). Therefore, the buried depth of the pipe gallery for theoretical analysis is 3.5m. At the same time, the commonly used section height of 3.5m was taken as the analysis section height, and the four sections with the section width of 3.5m, 4m, 5m and 6m were respectively analyzed.

The example is based on the analysis of the underground comprehensive pipe gallery project of Xincheng District, Nanchuan, Chongqing. The total length of the integrated pipe gallery is 10060 m, and it is crossed at the intersection of several main roads. The traffic flow is large. According to the design and geological survey data, the grade of the surrounding rock of the integrated pipe gallery is

IV. The geometric parameters of the pipe gallery and the mechanical parameters of the surrounding rock are listed in table 2.

Table 2. Calculation parameter table.

parameters	Parameters of pipe galleries			Parameters of rock mass		
	Width 2b /m	Height 2h /m	Buried depth H /m	$\gamma/(kn/m^2)$	$\varphi_c/°$	λ
values	3.5/4/5/6	3.5	3.5	23	40	0.22

Because the collapse of the rectangular integrated pipe gallery is mostly located at the side wall and the roof, and because the rectangular integrated pipe gallery and the model have symmetry, only the stress distribution of the surrounding rock in the range of 0° to 90° is analyzed. According to the above parameters, the theoretical values of the stresses in polar coordinates of the four rectangular sections with a height of 3.5 m and a width of 3.5 m, 4.0 m, 5 m and 6 m under unsupported conditions can be obtained, as shown in table 3.

Table 3. Analytical solution of stress of pipe gallery surrounding rock.

Width (m)	Surrounding rock pressure of the pipe galleries (kN/m^2)						
	0	15	30	θ	45	60	90
3.5	-102.31	-127.71	-204.49	-306.11	-306.11	11.89	39.3
4	-108.59	-127.71	-222.12	-339.38	-277.55	20.41	40.37
5	-120.2	-142.77	-255.11	-357.31	-215.15	30.8	41.94
6	-130.08	-155.83	-284.25	-397.97	-162.17	37.64	43.13

In order to facilitate the understanding of the trend of stress change, a coordinate system is established with φ as the angle between the horizontal plane and the centroid of the pipe in the rectangular pipe gallery. The theoretical calculation results are plotted on a graph, and the stress curve can be obtained, which is shown in figure 2.

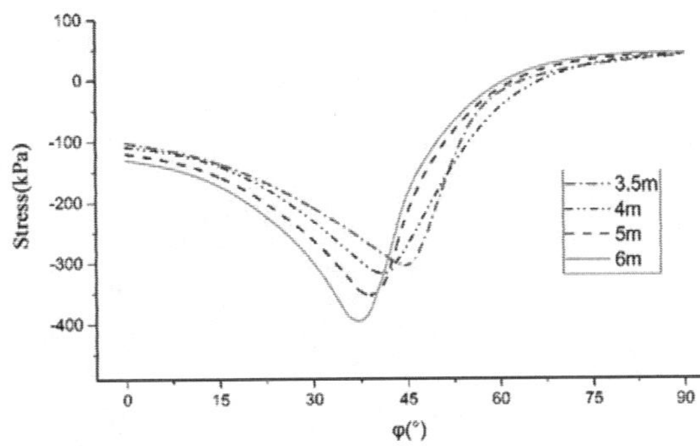

Figure 2. The stress curve of corridor chamber.

According to the curve above, most of the surrounding rock of the rectangular integrated pipe gallery is under pressure, and local tensile stress occurs at the middle of the top plate. In polar coordinates, the surrounding rock pressure of the rectangular pipe gallery is compressive stress at $\varphi=0$ (the middle of the side wall of the pipe gallery) and is the minimum pressure at the side wall. By comparing the four curves, it can be seen that as the width of the rectangular integrated pipe gallery increases, the aspect ratio of the pipe gallery decreases, and the surrounding rock stress of the pipe gallery gradually increases. Especially in the top corner of the figure, not only the stress is getting larger and larger, but also the stress curve is getting steeper. This shows that as the width of the pipe gallery increases, the compressive stress at the top corner of the pipe gallery increases and the stress concentration becomes more obvious. It can be concluded that the width of the integrated pipe gallery is an important determinant of the degree of stress concentration at the apex angle and the maximum stress.

4.2. Comparative analysis for the numerical simulation

In order to verify the rationality of the above assumptions and derivation, the surrounding rock stress of the rectangular cross-section integrated pipe gallery is taken as the research object, and the numerical simulation of the above examples is carried out by using the finite element numerical simulation software according to the theoretical calculation of the example parameters and dimensions.

Take 3B (here take the rectangular integrated pipe gallery excavation width B) to the left and the right from the excavation boundary of the cave, take 3 H_t down (here take the rectangular pipe gallery height) as the size of the model; at the same time, for the convenience of analysis and Observe, the horizontal level is in the X-X direction and the vertical direction is in the Y-Y direction. The calculation model is shown in figure 3.

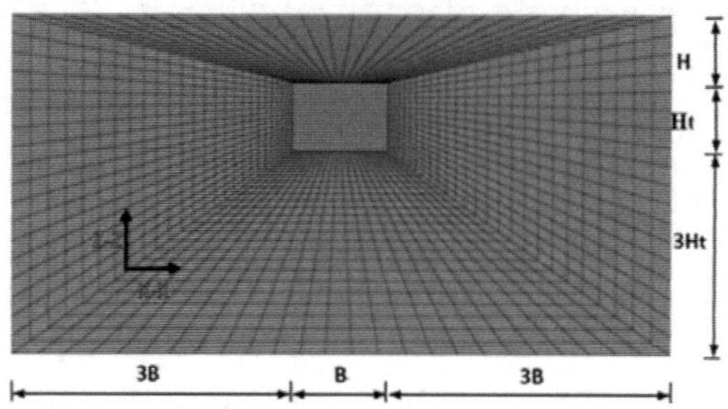

Figure 3. The computational model.

Through numerical simulation, the stress in the X-X and Y-Y directions around the pipe gallery can be obtained. According to the case where the tangential stress around the pipe gallery is zero, the numerical solution of the stress around the pipe gallery can be obtained from the two-dimensional numerical simulation of the rectangular integrated pipe gallery.

$$\sigma = \sqrt{\sigma_x^2 + \sigma_y^2 + 2\sigma_x\sigma_y \cos\langle X - X,\ Y - Y\rangle} \tag{24}$$

And $\langle X - X,\ Y - Y\rangle = \pi/2$, yields

$$\langle X - X,\ Y - Y\rangle = \pi/2 \tag{25}$$

The stress values of the two-dimensional numerical simulation of the rectangular integrated pipe porch with the section form with widths of 3.5m, 5m and 6m can be obtained from the extracted values and the above formula, as is listed in table 4.

Table 4. Analytical solution of stress of pipe gallery surrounding rock.

Angle (°)	Stress category	Stress (kN/m^2)			
		Width 3.5 m	Width 4 m	Width 5 m	Width 6 m
0.00	σ_{x-x}	-0.27	-0.43	-0.79	-0.99
	σ_{y-y}	-61.70	-63.35	-66.53	-69.54
	σ	-61.70	-63.35	-66.53	-69.55
15.00	σ_{x-x}	-12.91	-14.32	-20.92	-25.97

	σ_{y-y}	-87.61	-94.26	-99.29	-107.21
	σ	-88.56	-95.34	-101.47	-110.31
30.00	σ_{x-x}	-24.95	-25.92	-26.78	-30.29
	σ_{y-y}	-143.26	-148.97	-155.32	-158.67
	σ	-145.42	-151.21	-157.61	-161.54
θ	σ_{x-x}	-231.02	-239.42	-269.80	-330.64
	σ_{y-y}	-168.62	-182.05	-238.07	-232.74
	σ	-286.01	-300.80	-359.82	-404.34
45.00	σ_{x-x}	-231.02	-177.35	-149.57	-140.96
	σ_{y-y}	-168.62	-79.51	-68.79	-69.74
	σ	-286.01	-194.36	-164.63	-157.27
60.00	σ_{x-x}	27.61	33.21	39.72	40.95
	σ_{y-y}	7.41	10.75	9.69	14.51
	σ	28.59	34.91	40.88	43.44
90.00	σ_{x-x}	32.37	42.33	60.56	87.69
	σ_{y-y}	10.01	12.80	19.25	26.98
	σ	33.88	44.22	63.55	91.75

Comparing the numerical solution of the above table with the theoretical solution, the contrast curves of the four sections with widths of 3.5m, 4m, 5m and 6m can be obtained (figure 4).

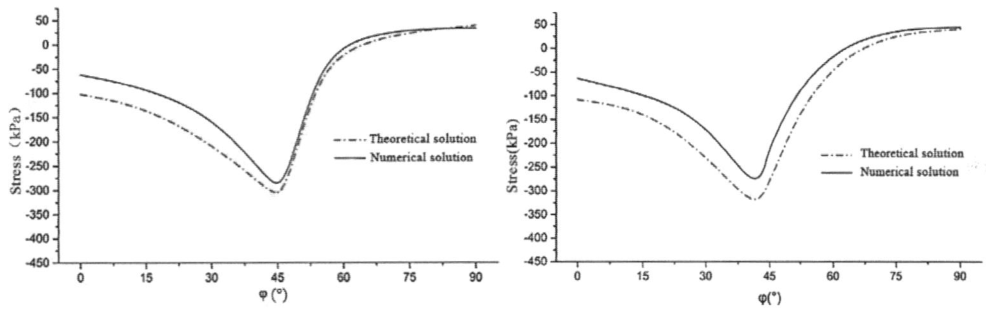

(**a**) stress comparison figure of 3.5m pipe gallery. (**b**) stress comparison figure of 4m pipe gallery.

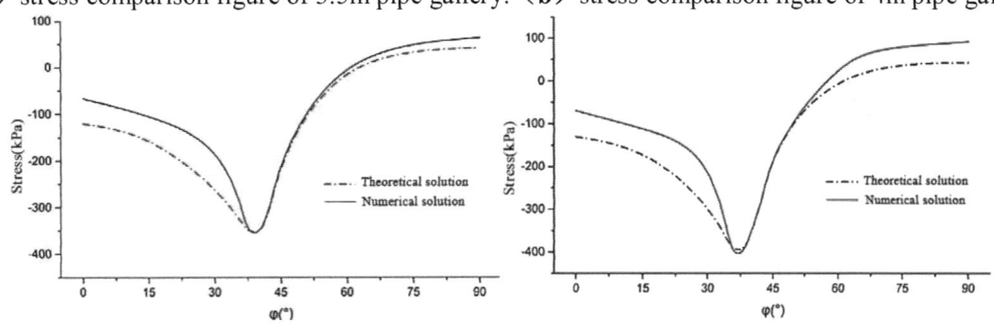

(**c**) stress comparison figure of 5m pipe gallery. (**d**) stress comparison figure of 6m pipe gallery.

Figure 4. Stress contrast diagram at the periphery of a tube Gallery.

From the comparison results of the theoretical analysis and numerical simulation of the surrounding rock stress of the rectangular integrated pipe gallery, it can be seen that the stress distribution of the theoretical calculation formula and the numerical simulation of the stress distribution of the rectangular integrated pipe gallery are generally the same. Stress concentration occurs both at the top corner of the gallery's rectangular sections. And the stress value of the same part is also close, the overall performance of the numerical calculation of the stress results is larger than the theoretical calculation results, and the main reasons for this result may have the following factors:

(1) In the theoretical calculation, the load is equivalent to the outer boundary of the model, especially the gravity load directly acts on the top of the model, which is a possible factor that causes the theoretical calculation result to be larger than the numerical calculation result;

(2)There is a certain deviation between the equivalent side pressure coefficient and the actual side pressure coefficient obtained by the formula $\lambda=\frac{\sigma_H}{\sigma_V}=\left(1+\frac{1}{2H*H_t}\right)\tan^2\left(45^\circ-\frac{\varphi_c}{2}\right)$.

Although there are some errors in the analytical solution of the pipe gallery between the theoretical formula and the numerical simulation, the formula can accurately reflect the surrounding rock stress distribution law of the rectangular integrated pipe gallery, and the error with the numerical solution is also within a certain range. And the calculated value is slightly larger than the numerical solution, which can meet the safety requirements of the project. It can be seen that the theoretical calculation formula has relatively strong engineering applicability, which can correctly calculate the stress distribution of the surrounding rock of the rectangular integrated pipe gallery, and has a reliable advantage in terms of safety and simplified calculation.

5. Conclusion

(1) From the view of the engineering application, the single-cylinder integrated pipe gallery with a rectangular section in a single rock layer is simplified into a semi-infinite planar rectangular square hole problem, and it is converted into a circular hole problem by using the conformal change, and then the basic knowledge of the complex variable function is used to derive the simplified calculation formula of the surrounding rock stress of the integrated pipe gallery and analyze the engineering examples.

(2) According to the theoretical calculation example and the selected rock mechanics parameters, a two-dimensional numerical model is established. The surrounding rock stress of the pipe gallery is compared with the theoretical calculation results to verify the applicability of the simplified calculation formula. Although the simplified calculation formula and numerical simulation results of surrounding rock stress have certain errors, the distribution trend and the overall results agree well. It is considered that the simplified calculation formula has certain engineering applicability.

(3) Based on the results of numerical simulation and theoretical calculation, comparing the surrounding rock stresses of four different widths with the width of 3.5m, 4m, 5m, 6m, it is found that the stress concentration at the top corner of the pipe gallery increases with the increase of the width. It is recommended that the width of one excavation of the rectangular pipe section must be strictly controlled during construction.

Acknowledgments

This research is supported by Chongqing Municipality Social Work and People's Livelihood Security Science and Technology Innovation Special Project (cstc2017shmsA30017).

6. References

[1] Yao S, Wang Z P 2017 Research on the construction technology of city underground pipe gallery *Green Environ. Protection Build. Mater.* **(1)** 111.

[2] Wang S S 2018 Research on integrated design of large-scale tunnel & utility tunnel *Chinese J. Underground Space Eng.* **13(4)** 170-4.

[3] Yang Q X, Feng Y N 2018 Theoretical Study on Safety Distance of Pipelines in Utility Tunnel **14(2)** 293-8.

[4] Chen K, Tang Z, Cui N X 2015 Analytical solution of rectangular roadway surrounding rock stress *J. Safety Environ.* **15(3)** 124-8.

[5] Shi G P, Zhu J H, Li B H 2014 Elastic analysis of hole-edge stress of rectangular roadway *Rock*

Soil Mechanic. **35(9)** 2587-93+2601.
[6] Yang X L, Wang Z W 2010 Limit analysis of earth pressure on shallow tunnel using nonlinear failure criterion *J. Central South Univ. Sci. Technol.* **1** 299-302.
[7] Li Y X 2012 Upper bound analysis of earth pressure and surrounding rock stability analysis for shallow tunnel *Master's thesis Central South Univ.*
[8] Yang F 2010 Investigation of shallow tunnel stability using upper bound solution of limit analysis *Doctoral Dissertation Central South Univ.*
[9] Wang Z L, Shen L F, Yao J 2010 Calculation of stress field in surrounding rocks of shallow tunnel using computational function of complex variable method *Rock Soil Mechanic.* **S1** 86-90.
[10] Huang Z P, Broch E, Lu M 2002 Cavern roof stability--mechanism of arching and stabili-zation by rockbolting *Tunnelling Underground Space Technol.* **17** 249-61.
[11] Wang Y F, Li C H, Zheng X J 2009 Research on the evolution mechanism of pressure arch of underground cavern excavated in different ways *Mining Res. Develop.* **3** 12-4.
[12] Wang Z W 2015 Characteristics of surrounding rock and calculation method of relaxation pressures of underground excavations in up-soft / low-hard jointed rock stratum. *Doctoral dissertation Beijing Jiaotong Univ.*
[13] Lu A Z 1995 New method for establishing mapping function of non-circular opening with full Support *Chinese J. Geotechnic. Eng.* **17(4)** 38-44.
[14] Zhu D Y, Qian Q H, Zhou Z S 1999 New method for calculating mapping function of opening with complex shape *Chinese J. Rock Mechanic. Eng.* **18(3)** 279-82.

IOP Conf. Series: Materials Science and Engineering **739** (2020) 012002 doi:10.1088/1757-899X/739/1/012002

Threshold Determination Using Bi-hazard Fragility Curves for the Evaluation of Structural Health Monitoring of USHER Technology

M B Baylon[1], F A A Uy[2], K M S Montes[3], K A D Embalzado[3]

[1]Senior Structural Health Engineer, USHER Technologies, Mapua University, Intramuros, Manila, Philippines

[2]Dean, School of Civil Engineering, Geosciences, Environmental Engineering, Mapua University, Intramuros, Manila, Philippines

[3]Senior Science Research Specialist 1, PHILSIMS, Mapua University, Intramuros, Manila, Philippines

embylon@usher.ph

Abstract. Structural Health Monitoring (SHM) is not new in the field of structural engineering and its application goes from the monitoring to evaluation of bridges, dams, buildings, and other similar structures. As per National Structural Code of the Philippines (NSCP) Volume 1[1-4], the Implementing Rules of earthquake recording instrumentation of buildings has been continually advancing, but not in the case of bridges and dams. The objective of this paper is to apply the established sensor-driven accelerometer developed by USHER (Universal Structural Health Evaluation and Recording) system in the evaluation of Padre Jacinto Zamora Bridge in Manila; specifically, to determine the proper installation of the developed accelerometer in the identified critical parts of the bridge. SHM addresses the problem of structural integrity assessment and help in assuring repair cost to a minimal. SHM helps quantify the strength of a structure by identifying the damage. Often evaluations are made using visual inspection and by age consideration. Most of the time, the use of this type of evaluation tend to be expensive. Repairs are not directed to the damaged component alone but the whole structure. Hence, the expenditure expands. SHM with the application of Micro-Electro-Mechanical Sensors (MEMS) will allow gathering of data that can be converted in the form of the structure's modal properties (i.e. natural frequency, mode shape, and damping ratio) [6, 7]. Structural health is then determined and be ready to compare it to the structural model simulation results. A method was developed to incorporate seismic fragility curves [5] to determine the thresholds for the evaluation of the structural health. Capacity spectrum method was utilized to derive the seismic fragility function. A Monte Carlo Simulation was used to derive the flood fragility curves.

1. Introduction
In recent times, seismic forces for engineers are inevitable in their designs, especially that of infrastructures. Structural response is utmost important to the structural engineer when he is dealing

with the earthquake hazard. In today's practice of structural engineering, data from previous earthquake event's structural response especially that of bridges are crucial in understanding the principle of performance based design (PBD) of structures. These data can be captured through recording these structural responses, e.g., displacement, velocity, and acceleration, by the system known as structural health monitoring (SHM); thus, paving the way for building instrumentation.

The Padre Jacinto Zamora Bridge is an existing concrete bridge with American Association of State Highway and Transportation Officials (AASHTO) Type IV Girders. The bridge is a reinforced concrete bridge built in the 1960's and became operational in the year 1963. Its length is about 410 meters, linking Pandacan and Paco, Manila.

The objective of this paper is to evaluate the Padre Jacinto Zamora Bridge using USHER technology. Specifically, to determine the proper installation of the developed accelerometer in the identified critical parts of the bridge and to utilize Capacity Spectrum Method (CSM) and Monte Carlo Simulation (MCS) to construct seismic fragility curves and flood fragility curves, respectively, which in turn can be used in the threshold determination for the structural health monitoring of Padre Jacinto Zamora Bridge.

2. Methodology

It can be seen in figure 1 the methodology used for this paper. Using the plans obtained from Department of Public Works and Highways (DPWH) of the Padre Jacinto Zamora Bridge, a structural model was constructed using CSiBridge software application. The structural model was used in the simulation and the installation of sensors. For the simulation, Capacity Spectrum Method was utilized in determining the performance point for each peak ground acceleration (PGA) of the capacity of the critical bridge pier. The performance point, particularly the displacement was compared to the damage levels set by Moschonas et. al. [8] in order to construct the table for the Probability of Exceedance. Using regression analysis to determine the statistical parameters of fragility function, the plot of seismic fragility curves was derived using Matlab. At a 10% probability of exceedance as per Structural Engineers Association Of California (SEAOC), the threshold values were determined at moderate and extensive damage level only. For the installation of USHER, the device was located and installed at approximately the center of mass of the floor plan, but to some extent such as accessibility and internet connectivity (wi-fi access), the device was located strategically [7]. While the device was installed and recording data, the bridge response was being archived and transmitted to the USHER portal and it is ready to download as per the certified structural engineer need for interpretation through a log-in access. The bridge response measured by USHER is then compared to the threshold derived from fragility curves; thus, the bridge is evaluated.

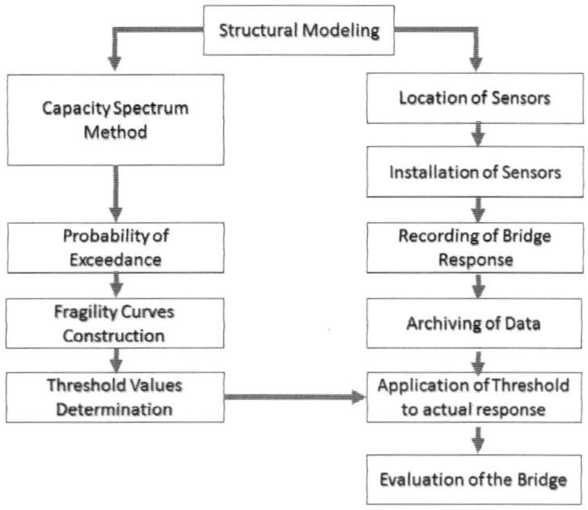

Figure 1. Methodology.

IOP Conf. Series: Materials Science and Engineering **739** (2020) 012002 doi:10.1088/1757-899X/739/1/012002

3. Results and Discussions

3.1. Modeling

The model of the bridge is shown in figure 2. In figure 3, the components of the bridge model can be seen.

Figure 2. Structural model of Father Jacinto Zamora Bridge.

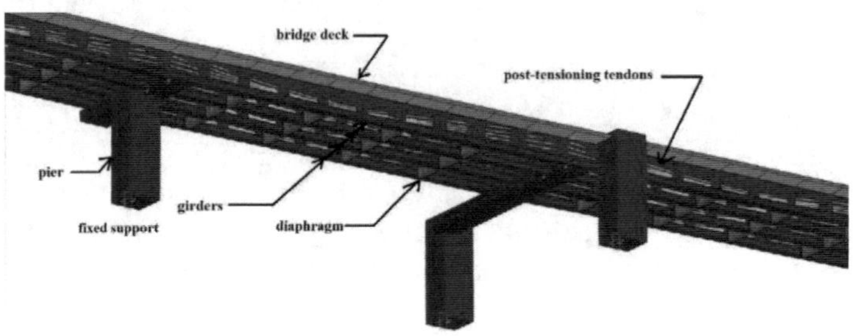

Figure 3. Some of the components of the bridge model.

3.2. Pushover Curves

The pushover curves for the three directions, X-direction, Y-direction, and Z-direction, are shown in figure 4, figure 5, and figure 6, respectively. The nonlinearity of the pushover curve declares the inelastic performance of the bridge. Thus, only the pushover curve in the Y-direction is considered for the capacity curve. Using the CSiBridge, the yield and maximum displacements were determined and it can be seen in table 1. At a base shear of 10.5 MN, the bridge pier yielded at a displacement of 60 mm, whilst at 13.7 MN, it yielded 124 mm. With these displacements, yielded (dy) and maximum (du), and substituting these values using the formulas in table 2 for the respective damage states, the damage limits (lower and upper) can be determined.

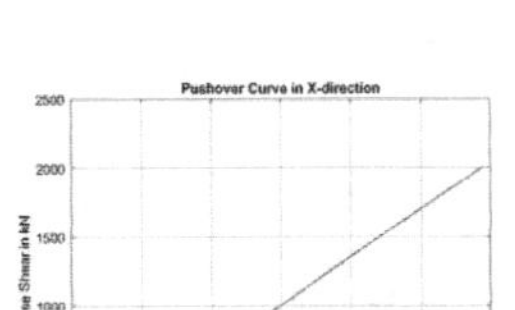

Figure 4. Pushover curve in X-direction.

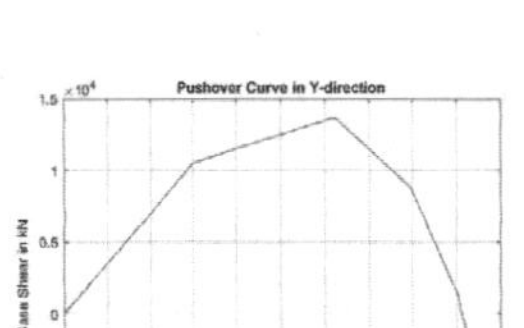

Figure 5. Pushover curve in Y-direction.

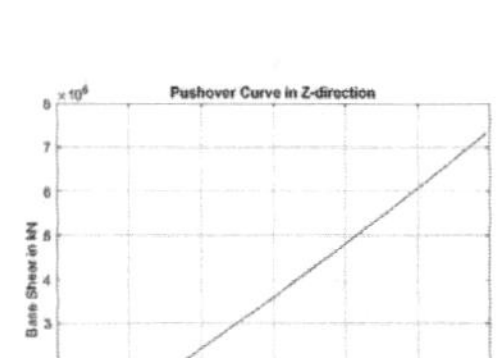

Figure 6. Pushover curve in Z-direction.

Table 1. Pushover curve of the bridge in Y-direction showing the yield and maximum displacements.

Displacement (m)	*Shear (kN)*	
0	0	
0.060	10,494.960	← yield
0.124	13,678.412	← maximum
0.160	8,713.096	
0.180	1,571.146	
0.193	-6,860.886	
0.196	-7,001.300	

Table 2. Damage state limits as per [8] page 446.

	Damage State	*Required Interventions*	*limits*	*Lower limit*	*Upper limit*
D	None	Do nothing	$\leq 0.7\ d_y$	0	0.041969
C	Minor/Slight	Inspect, adjust, patch	$> 0.7\ d_y$	0.041969	0.081376
B	Moderate	Repair components	$> \min\{1.5\ d_y, (1/3)*(d_u - d_y)\}$	0.081376	0.102796
A	Major/Extensive	Rebuild components	$> \min\{3\ d_y, (2/3)*(d_u - d_y)\}$	0.102796	0.124216
As	Complete	Rebuild bridge	$> d_u$	0.124216	∞

3.3. Capacity Spectrum Method

In table 3, the result of the performance point is shown from the simulation of the bridge using CSiBridge's feature of Capacity-Spectrum curve. In structural reliability, the performance function can be derived from the two functions: resistance and the load effect. In this paper, capacity curve is the resistance function, whilst that of response spectrum, it's the load effect. In CSM, this can be captured through the performance points. To illustrate this, one ground motion data was scaled from 0.1g to 2.0g and using an earthquake engineering software application, PRISM, these ground motion data can be transformed to response spectra. The capacity curve, which was derived from pushover curve, can be superimposed to the response spectra. This can be shown in figure 8.

Table 3. A sample tabulation of performance point summary for CSM. Based on March 3, 2006 Mindoro Earthquake.

	X-direction		*Y-direction*		*Z-direction*	
PGA in g	*V (kN)*	*D (m)*	*V (kN)*	*D (m)*	*V (kN)*	*D (m)*
0.2	8.799	0.05	10.325	0.059	7.622	0.444
0.4	17.598	0.101	20.651	0.118	22.865	0.131
0.6	35.196	0.201	41.301	0.236	30.486	0.174
0.8	43.994	0.251	51.627	0.295	45.73	0.261
1.0	52.793	0.302	61.952	0.354	53.351	0.305
1.2	61.592	0.352	72.277	0.413	60.973	0.348
1.4	70.391	0.402	82.602	0.472	76.216	0.435
1.6	87.989	0.503	103.253	0.59	83.837	0.479
1.8	96.788	0.553	113.578	0.649	91.459	0.522
2.0	105.598	0.603	123.903	0.708	106.702	0.610

IOP Conf. Series: Materials Science and Engineering **739** (2020) 012002 doi:10.1088/1757-899X/739/1/012002

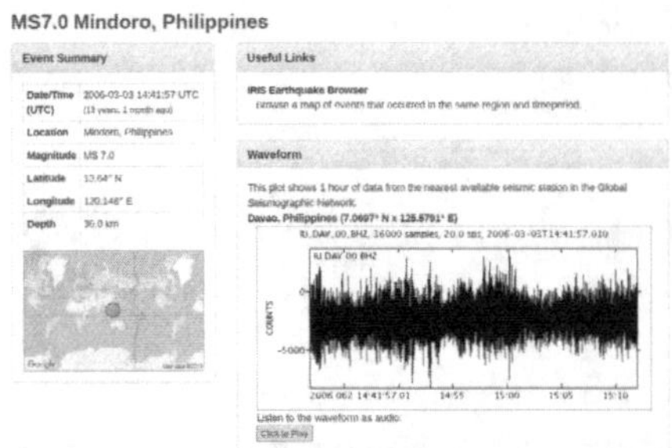

Figure 7. Pertinent data of the ground motion data used in table 3.
(Source: http://ds.iris.edu/ds/nodes/dmc/tools/event/2192624)

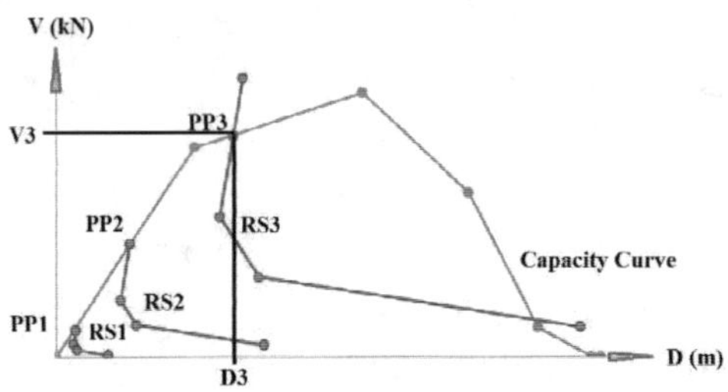

Figure 8. Illustration of the collated performance points from table 3.

All the data from different ground motion data in time history form, were collated and only the Y-direction was considered. In table 4, it summarizes the frequency for each damage state, given the peak ground acceleration (PGA) in g. From this table, the probability of occurrence can be derived as it can be seen in figure 9. It shows the occurrence of damage state present for every PGA. It can be observed that a PGA of 0.5g, the bridge has already reached the complete damage.

Table 4. Basis of the Probability of Occurrence thru frequency for each damage state in a given PGA value.

PGA	*Damage States*				
in g	*D*	*C*	*B*	*A*	*As*
0	12	0	0	0	0
0.1	2	6	0	1	3
0.2	2	3	1	1	15
0.3	0	0	1	1	10
0.4	0	0	1	1	20
0.5	0	0	0	0	12
0.6	0	0	0	0	22
0.7	0	0	0	0	12
0.8	0	0	0	0	22
0.9	0	0	0	0	12
1	0	0	0	0	22

PGA	*Damage States*				
in g	*D*	*C*	*B*	*A*	*As*
1.1	0	0	0	0	12
1.2	0	0	0	0	22
1.3	0	0	0	0	12
1.4	0	0	0	0	22
1.5	0	0	0	0	12
1.6	0	0	0	0	22
1.7	0	0	0	0	12
1.8	0	0	0	0	22
1.9	0	0	0	0	12
2	0	0	0	0	22

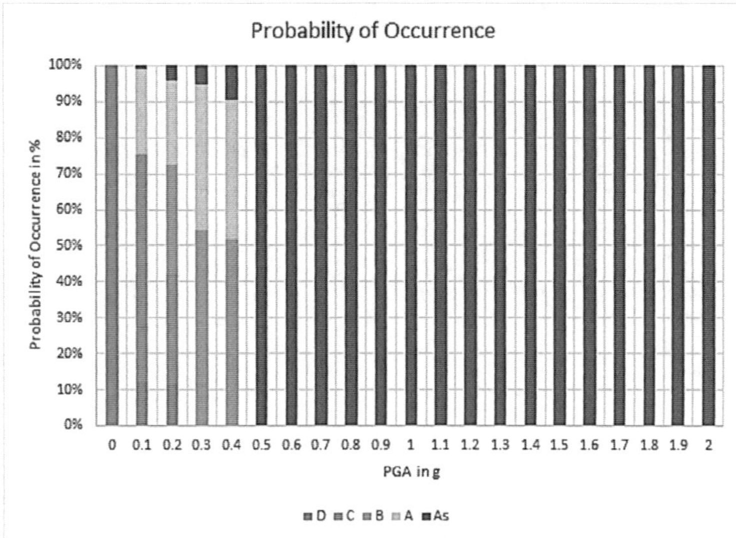

Figure 9. Column chart representation of Probability of Occurrence.

The damage ratio can then be derived from the probability of occurrence by accumulating the probability as it increases in the PGA value, as it can be shown in table 5. This damage ratio values are used as scatter points in the plot of seismic fragility curves. By plotting the natural logarithm of PGA vs. damage ratio, linear regression parameters can be determined graphically. These parameters can now be used as statistical parameters, that is, mean and standard deviation, for the lognormal function nature of the fragility curves, as in equation (1). The seismic fragility curves are then plotted using Matlab, as shown in figure 10, 11, 12, 13, and 14. The values of these fragility curves are listed in table 6. Based on the derived fragility curves, the focus in determining the threshold values for the structural health monitoring is in the Moderate damage level.

$$Pr = \Phi\left[\frac{ln(PGA)-\lambda}{\xi}\right] \tag{1}$$

Based on figure 12, the peak ground acceleration that has a probability of exceedance of 10% as per SEAOC provision, is approximately at 0.10g. It means that the bridge is deemed repairable through the bridge components which needs monitoring in the component level. The information earthquake that struck Padre Jacinto Zamora Bridge in April 8, 2017, is shown in figure 15. The superimposed threshold values where incorporated in the USHER device results, that is, the structural response, as shown in figure 16. Also, in this figure, the USHER device recorded a maximum of 2.0 m/sec velocity and a maximum displacement of 2.0 m in one of the nodes, i.e., Node 3 as per structural model of the Padre Jacinto Zamora Bridge.

Table 5. Damage Ratio.

PGA	*Damage States*				
in g	*D*	*C*	*B*	*A*	*As*
0	0.750	0	0	0	0
0.1	0.875	0.667	0	0.250	0.009
0.2	1	1	0.333	0.500	0.056
0.3	1	1	0.667	0.750	0.088
0.4	1	1	1	1	0.150
0.5	1	1	1	1	0.188
0.6	1	1	1	1	0.256
0.7	1	1	1	1	0.294
0.8	1	1	1	1	0.363
0.9	1	1	1	1	0.400
1	1	1	1	1	0.469
1.1	1	1	1	1	0.506
1.2	1	1	1	1	0.575
1.3	1	1	1	1	0.613
1.4	1	1	1	1	0.681
1.5	1	1	1	1	0.719
1.6	1	1	1	1	0.788
1.7	1	1	1	1	0.825
1.8	1	1	1	1	0.894
1.9	1	1	1	1	0.931
2	1	1	1	1	1

Table 6. Probability of exceedance values using regression analysis.

PGA	D	C	B	A	As
0	0	0	0	0	0
0.1	0.262529	0.349355	0.096104	0.154058	0.034732
0.2	0.791589	0.62023	0.338276	0.372199	0.175831
0.3	0.951411	0.761628	0.540476	0.531654	0.339343
0.4	0.988073	0.841159	0.680726	0.643232	0.481114
0.5	0.996789	0.889217	0.77495	0.72249	0.593761
0.6	0.999052	0.919944	0.838555	0.780114	0.680735
0.7	0.999696	0.940484	0.882145	0.822966	0.747407
0.8	0.999895	0.954713	0.912551	0.855487	0.798606
0.9	0.999961	0.964865	0.934138	0.88061	0.838147
1	0.999985	0.972287	0.949721	0.900324	0.868908
1.1	0.999994	0.977828	0.961143	0.916008	0.893029
1.2	0.999997	0.982041	0.969634	0.928638	0.912095
1.3	0.999999	0.985293	0.976027	0.93892	0.927283
1.4	0.999999	0.987839	0.980899	0.94737	0.939473
1.5	1	0.989856	0.984651	0.954378	0.949326
1.6	1	0.991473	0.987569	0.960234	0.957344
1.7	1	0.99278	0.989859	0.965163	0.96391
1.8	1	0.993848	0.991673	0.969341	0.969319
1.9	1	0.994727	0.993119	0.972901	0.9738
2	1	0.995455	0.994281	0.975954	0.977532

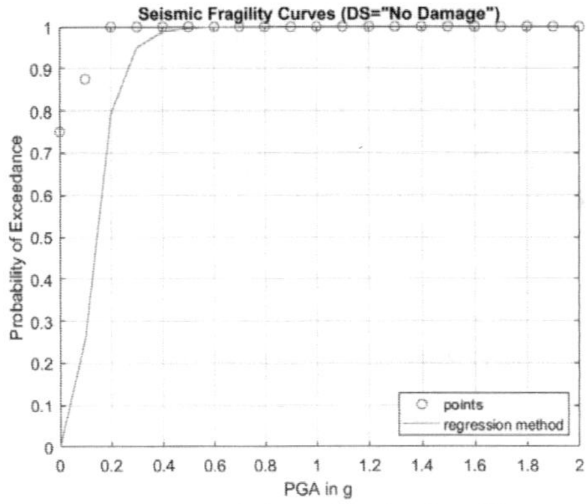

Figure 10. Plot of the seismic fragility curve for damage state of No Damage.

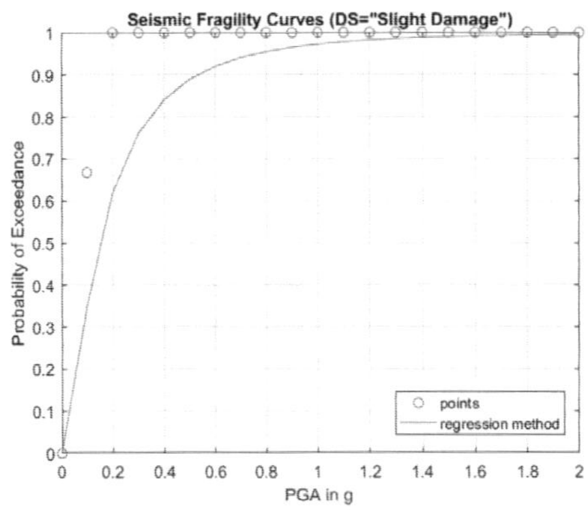

Figure 11. Plot of the seismic fragility curve for damage state of Slight Damage.

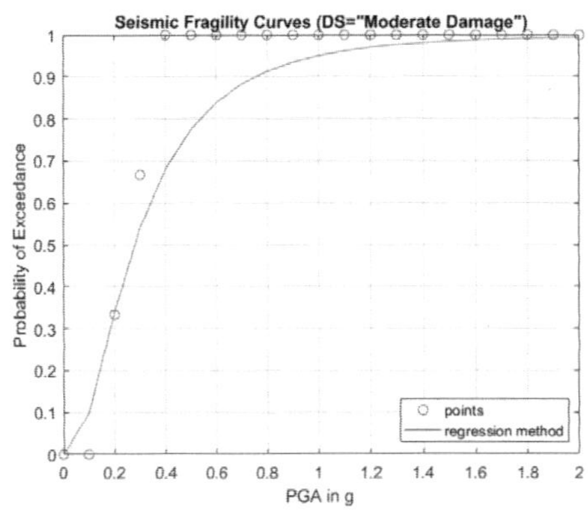

Figure 12. Plot of the seismic fragility curve for damage state of Moderate Damage.

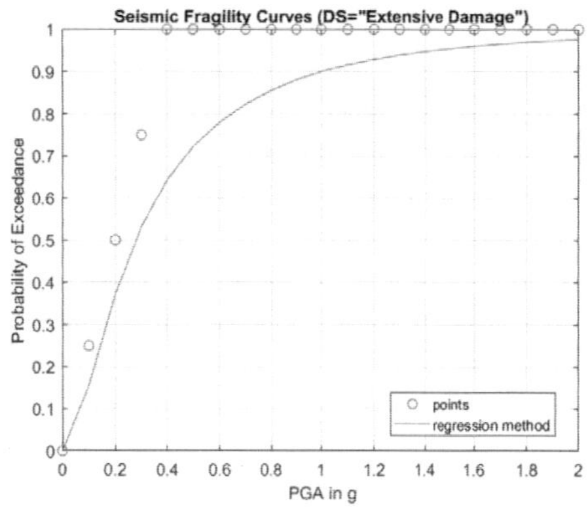

Figure 13. Plot of the seismic fragility curve for damage state of Extensive Damage.

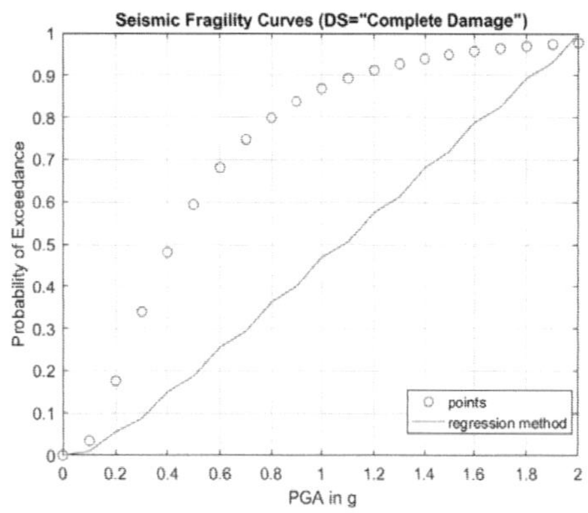

Figure 14. Plot of the seismic fragility curve for damage state of Complete Damage.

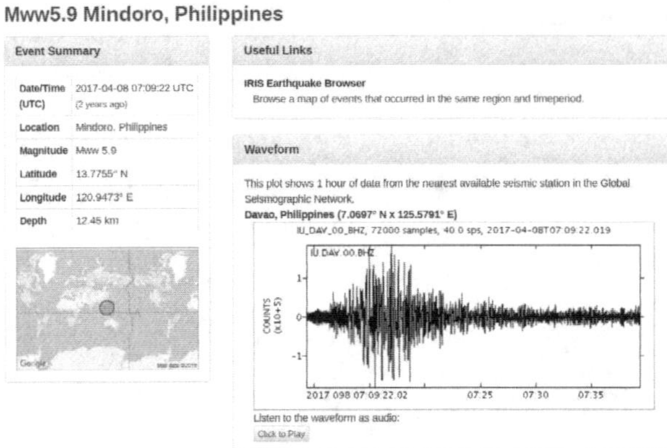

Figure 15. Information for the ground motion data in April 8, 2017.

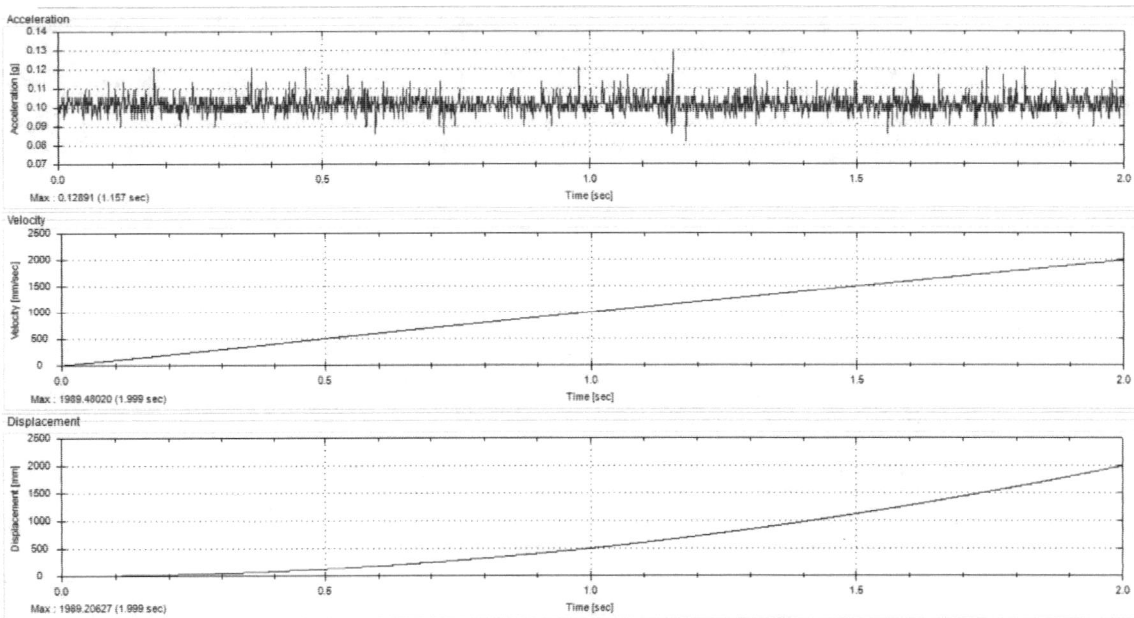

Figure 16. Structural response of the Padre Jacinto Zamora Bridge during the earthquake event last April 8, 2017 as measured by USHER in g, i.e., 0.129g.

3.4. Flood Fragility Curves by Monte Carlo Simulation

The maximum moment capacity of the columns of the piers are obtained from the column interaction diagram shown in figure 17, which can be acquired from the bridge model. This maximum moment capacity will be used in determining the damage state experienced by the columns once heavy flooding occurs.

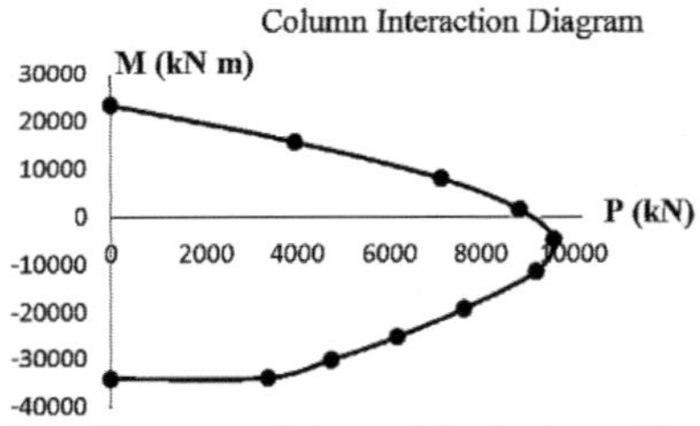

Figure 17. Column interaction diagram of the critical pier of Padre Jacinto Zamora Bridge.

The Padre Jacinto Zamora Bridge is built across the Pasig River, thus a model of the river was created using a simulation program to obtain its velocity. There were few methodologies used in the previous studies to acquire the water pressure. In the study of Kim et al. [9], the American Association of State Highway and Transportation Officials (AASHTO) and Korean Highway Bridge Design Specification (KHBDS) presented an equation for estimating the water pressure expressed in equation (2).

The damage state is a ratio of the maximum moment demand and the maximum moment capacity. Since the maximum moment capacity is already acquired from the diagram, the maximum moment demand should be identified as well. To get the maximum moment demand, the researchers established a 0 to 100 m/s range to see at what velocity will the bridge experience a collapse damage state. The water velocity is first converted into water pressure using equation (2) from the American Association of State Highway and Transportation Officials (AASHTO) and Korean Highway Bridge Design Specification (KHBDS). The water pressure takes into account the factor for the accumulated debris. Also, the pressure is multiplied to the diameter of the column to get the distributed load which will be used in the equation (3) and equation (4). The distributed load is shown in figure 18.

$$p = 5.14x10^{-4}C_dV^2 \tag{2}$$

where v is the water velocity and C_D is the drag coefficient which can be determined in table 7. In this study, two failure modes are considered: (1) lack of displacement ductility at bridge piers, and (2) pier failure.

Table 7. Drag coefficient of AASHTO and KHBDS.

Pier Type	C_D
Semicircular-nosed pier	0.7
Square-ended pier	1.4
Debris lodged against the pier	1.4
Wedged-nosed pier with nose angle 90^0	0.8

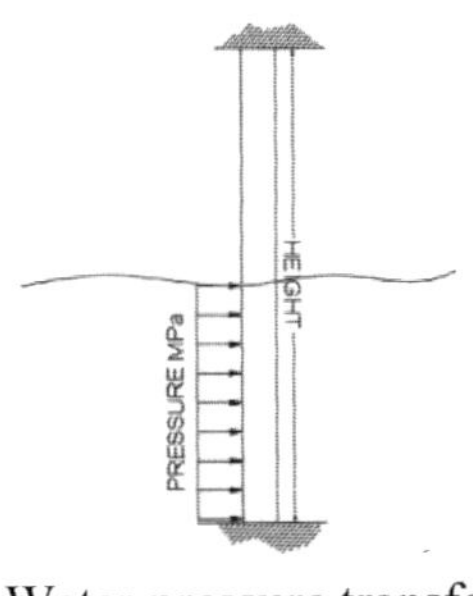

Figure 18. Water pressure transformed into distributed load applied to the fixed-end column.

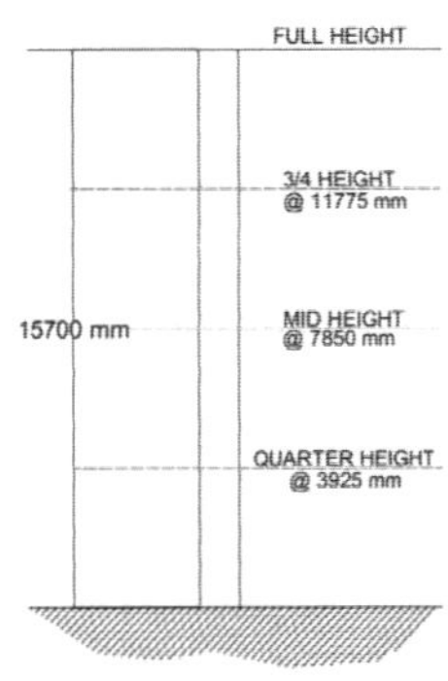

Figure 19. The flood depth classifications used in the study with their corresponding height.

The distributed load, w, will be used in the following equations to get the moment demand at the fixed ends of the column. Equation (3) and equation (4) are expressed as:

$$M_1 = \frac{wa^3(4L - 3a)}{12L^2} \tag{3}$$

$$M_2 = \frac{wa^2(6L^2 - 8aL + 3a^2)}{12L^2} \tag{4}$$

where a is the flood depth, and is the height of the column of the pier. M_1 is the moment at the top end of the column while M_2 is the moment at the bottom end. Since there are two moment demand obtained (M_1 and M_2), the greater moment will be considered as the maximum moment demand. This maximum moment demand, however, varies depending on the flood depth. In the analysis, the researchers classify the flood depth according to the height of the columns. The flood depth is divided into four classifications: (1) at quarter height, (2) at mid height, (3) at ¾ of the total height, and (4) at the full height of the column. See figure 19.

In figure 20, it is evident that the columns of Piers 1 and 2 have the same height and that the columns of Pier 3 have a shorter height. The height of the columns of Piers 1 and 2 is 15.7 meters while the columns of Pier 3 are 12.8 meters high.

Figure 20. Side elevation of the bridge model, emphasizing the height of the columns.

Since the columns have symmetrical cross-sections, only one column of the pier will be analyzed in the study. The height of that column is 15.7 meters since it is the tallest among all the columns. The analysis was conducted on four flood depth classifications, but the actual flood depths used in the computation are the mean depths, $\bar{x}$, shown in figure 21. The 20% coefficient of variance was accounted in the analysis.

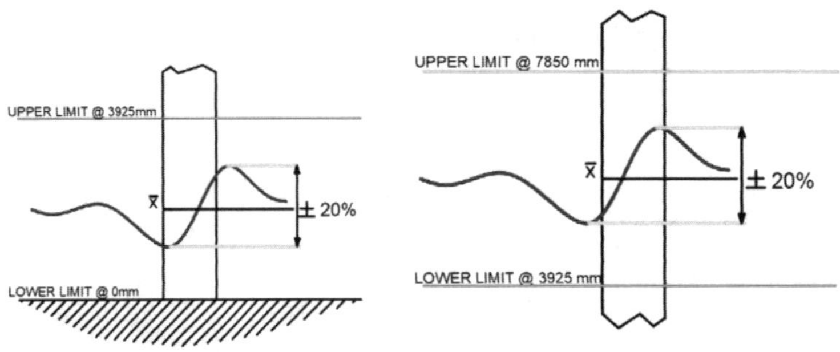

Figure 21. The mean depths at mid height and at quarter height.

Using MS Excel, 1,000 iterations were done per velocity (the 0-100m/s range) and, in those iterations, there are several trials that exhibited results wherein the moment demand exceeded the moment capacity of the column. This exceedance is the probability of failure. The equation for the probability of failure, P_f, is expressed in equation (5).

$$P_F = \frac{\Sigma n}{N} \quad (5)$$

Where N is the number of iterations and Σn is the total number of trials wherein the demand exceeded the capacity. Once the maximum moment capacity, M_C, and the maximum moment demand, M_d, are established, the damage ratio, M_D, can be obtained using equation (6).

$$M_D = \frac{M_d}{M_c} \quad (6)$$

Based on the ductility demand parameters of Dong, Frangopol, and Saydam (2013) presented in table 8, the damage state of the column per velocity is determined.

IOP Conf. Series: Materials Science and Engineering **739** (2020) 012002 doi:10.1088/1757-899X/739/1/012002

Table 8. Damage States and Ductility Demand (Dong et al., 2013).

Damage State	Ductility Demand
Minor Damage	$1 \leq MD \leq 2.9$
Moderate Damage	$2.9 \leq MD \leq 4.6$
Major Damage	$4.6 \leq MD \leq 5.0$
Collapse	$MD > 5.0$

The mean and standard deviations needed to generate the flood fragility curves are obtained once all the statistical computations are done in MS Excel. These mean and standard deviations are imported into MATLAB to generate the four flood fragility curves for each flood depth classification.

In figure 22, the bridge has a 10% probability of experiencing "Moderate Damage" at a velocity of 18 m/s when the flood depth is at full height; a velocity of 16 m/s when the flood depth is at mid height; and a velocity of 17 m/s when the it is at ¾ of the total height.

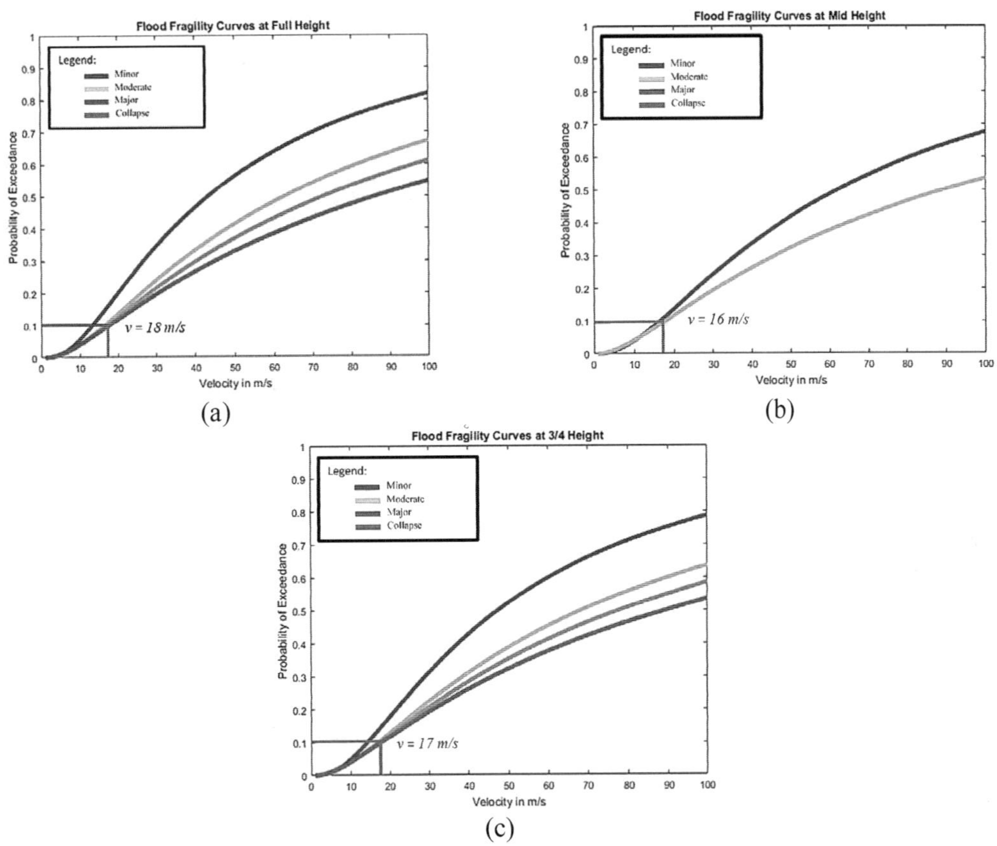

Figure 22. Flood fragility curves of Padre Jacinto Zamora at different flood depth classification.

As you can see, there is no fragility curve when the flood depth is at quarter height of the column because, within the 1,000 iterations, there were no instances that the demand exceeded the capacity. To say the least, the results generated in the iterations tell us that there is "No Damage" experienced by the bridge within the 0-100 m/s range.

4. Conclusion

Based on the findings, the threshold value for a moderate damage to Padre Jacinto Zamora Bridge is approximately at 10% of gravitational acceleration. This threshold value was surpassed by the maximum response of the structure of 0.129g when the earthquake event happened last April 8, 2017 with a magnitude of 5.9, as recorded by the Universal Structural Health Evaluation and Recording device. Hence, a thorough investigation must be done in the component level of the bridge. In terms of the flood fragility curves, at a 10% of probability of exceedance, the bridge is expected to perform in moderate damage at 18 m/s, 16 m/s, and 17 m/s at a full-height, mid-height, and 3/4 – height of the column pier flood depth.

Acknowledgments
The authors acknowledge Mapua University, Department of Public Works and Highways, Department of Science and Technology for the funding of the research where this paper is based on.

Reference

[1] Association of Structural Engineers of the Philippines. (1992) National Structural Code of the Philippines, Volume 1. Association of Structural Engineers of the Philippines (ASEP), Manila.
[2] Association of Structural Engineers of the Philippines. (2001) National Structural Code of the Philippines, Volume 1. Association of Structural Engineers of the Philippines (ASEP), Manila.
[3] Association of Structural Engineers of the Philippines. (2010) National Structural Code of the Philippines, Volume 1. Association of Structural Engineers of the Philippines (ASEP), Manila.
[4] Association of Structural Engineers of the Philippines. (2015) National Structural Code of the Philippines, Volume 1. Association of Structural Engineers of the Philippines (ASEP), Manila.
[5] Baylon M B 2017 *Developing fragility curves in seismic assessment of pier.* LAP Lambert Academic Publishing: Berlin, Germany.
[6] Dong Y, Frangopol D M, Saydam D 2014 Pre-Earthquake Multi-Objective Probabilistic Retrofit Optimization of Bridge Networks Based on Sustainability, *J. Bridge Eng.* **19(6)**.
[7] Gaviña J R, Uy F A, Carreon J P D 2017 Wireless Smart Sensor Network System Using SmartBridge Sensor Nodes for Structural Health Monitoring of Existing Concrete Bridges, *IOP Conf. Series Mater. Sci. Eng.* **216**.
[8] Moschonas I F, Kappos A J, Panetsos P, Papadopoulos V, Makarios T, Thanopoulos P 2009 Seismic fragility curves for greek bridges: methodology and case studies, *Bull. Earthquake Eng.* **7** 439-68.
[9] Kim H, Sim S-H, Lee J, Lee Y-J, Kim J-M 2017 Flood fragility analysis for bridges with multiple failure modes, Advances in Mechanical Engineering **9** **(3)** https://doi.org/10.1177/1687814017696415

IOP Conf. Series: Materials Science and Engineering **739** (2020) 012003 doi:10.1088/1757-899X/739/1/012003

Inhomogeneous Rods in Torsion: Longitudinal Fracture Analysis by Considering the Energy Balance

V Rizov[1]

[1]Professor, Department of Technical Mechanics, University of Architecture, Civil Engineering and Geodesy, 1046 - Sofia, Bulgaria

V_RIZOV_FHE@UACG.BG

Abstract. Fracture of inhomogeneous non-linear elastic rods with a longitudinal circular cylindrical crack is analyzed. Rods have circular cross-section and are loaded in torsion. General approach for analyzing the strain energy release rate is developed by considering the balance of the energy. The rods exhibit continuous material inhomogeneity in both radial and length directions. The general approach is applied to analyze the strain energy release rate for a longitudinal crack in a clamped inhomogeneous rod loaded in torsion. The strain energy release rate is derived also by considering the complementary strain energy for verification.

1. Introduction

The advance in various branches of current engineering is related to extensive use of inhomogeneous materials in load-bearing structural members and components. The basic feature of continuously inhomogeneous structural materials is the dependence of their properties on spatial coordinates (the material properties are continuous functions of spatial coordinates). The increased interest towards inhomogeneous materials is due chiefly to widespread application of functionally graded materials [1]. The main advantage of functionally graded materials is that their microstructure and properties can be formed technologically so as to meet different performance requirements in different parts of a structural member. Studying fracture behaviour of inhomogeneous materials is very important for their load-bearing structural applications. Usually, fracture of inhomogeneous (functionally graded) materials has been analyzed assuming linear-elastic behaviour [2, 3].

The present paper deals with longitudinal fracture analysis of non-linear elastic inhomogeneous rods of circular cross-section loaded in torsion. The fracture is studied in terms of the strain energy release rate. In contrast to previous publications which analyze the strain energy release rate in particular rods by using individual stress-strain relations and laws for distribution of material properties [4, 5], the aim of the present paper is to derive general solution to the strain energy release rate that is applicable for arbitrary rod configuration in torsion.

2. General approach

An inhomogeneous rod of circular cross-section of radius, R_1, is shown in figure 1. A longitudinal crack presenting a circular cylindrical surface of radius, R_2, is located in the rod as shown in figure 1. The crack length is denoted by a. The internal crack arm is treated as a rod of circular cross-section of radius, R_2, and length, a. The external crack arm is treated as a rod of ring-shaped cross-section of internal and external radiuses, R_2 and R_1, and length, a. The rod is loaded by torsion moments, T_i, as shown in figure 1. A torsion moment, T_b, is applied at the free end of the internal crack arm. The rod is in state of equilibrium under these torsion moments. The length of the rod is denoted by l. The rod exhibits continuous (smooth) material inhomogeneity in both radial and length directions. Besides, the material has non-linear elastic mechanical behaviour.

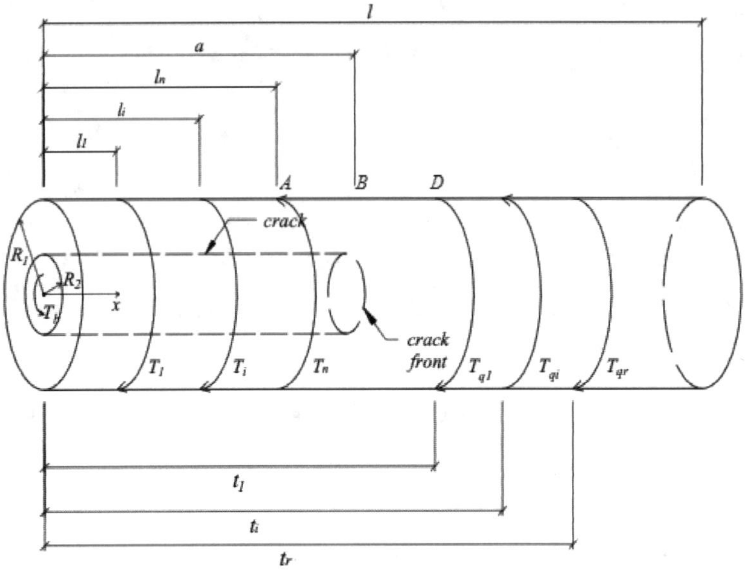

Figure 1. Geometry and loading of an inhomogeneous rod with a longitudinal circular cylindrical crack loaded in torsion.

The longitudinal fracture behaviour of the rod is studied in terms of the strain energy release rate, G. General approach for analyzing the strain energy release rate is developed by considering the balance of the energy. For this purpose, by assuming a small increase, δa, of the crack length, the strain energy release rate is derived as

$$G=\sum_{i=1}^{i=m} \frac{T_i}{2\pi R_2}\frac{\partial \omega_i}{\partial a}+\frac{T_b}{2\pi R_2}\frac{\partial \omega_b}{\partial a}-\frac{1}{2\pi R_2}\frac{\partial U}{\partial a}, \tag{1}$$

where m is the number of the torsion moments, ω_i is the angle of twist of the rod cross-section in which the i-th torsion moment is applied, ω_b is the angle of twist of the free end of the internal crack arm, U is the strain energy stored in the rod that is written as

$$U=\int_0^a\int_0^{R_2}\int_0^{2\pi} u_{0\text{int}} R dx dR d\varphi + \sum_{i=1}^{i=n}\int_{l_{i-1}}^{l_i}\int_{R_2}^{R_1}\int_0^{2\pi} u_{0ext_i} R dx dR d\varphi + \int_{l_n}^{a}\int_{R_2}^{R_1}\int_0^{2\pi} u_{0ext_{AB}} R dx dR d\varphi +$$
$$+\int_a^{t_1}\int_0^{R_1}\int_0^{2\pi} u_{0unc_{BD}} R dx dR d\varphi + \sum_{i=2}^{i=r}\int_{t_{i-1}}^{t_i}\int_0^{R_2}\int_0^{2\pi} u_{0unc_i} R dx dR d\varphi , \tag{2}$$

where $u_{0\text{int}}$ is the strain energy density in the internal crack arm, u_{0ext_i} and $u_{0ext_{AB}}$ are, respectively, the strain energy densities in the i-th portion and portion, AB, of the external crack arm, $u_{0unc_{BD}}$ and u_{0unc_i} are, respectively, the strain energy densities in portion, BD, and the i-th portion and of the un-cracked part of the rod $(a \leq l)$, n and r are, respectively, the numbers of torsion moments applied at the external crack arm and the un-cracked part of the rod, R and φ are the polar coordinates.

By applying the integrals of Maxwell-Mohr, the angles of twist, ω_i, are written as (figure 1)

$$\omega_i = \sum_{j=1}^{j=n} \int_{l_j}^{l_{j+1}} T_{ij} \frac{\gamma_{ext_j}}{R_1} dx + \int_{l_n}^{a} T_{iAB} \frac{\gamma_{ext_{AB}}}{R_1} dx + \int_{a}^{t_1} T_{iBD} \frac{\gamma_{unc_{BD}}}{R_1} dx + \sum_{k=2}^{k=r} \int_{t_{k-1}}^{t_k} T_{ik} \frac{\gamma_{unc_k}}{R_1} dx \,, \tag{3}$$

where T_{ij} and γ_{ext_j} are, respectively, the torsion moment induced by the unit loading for obtaining of ω_i and the shear strain at the periphery of the rod in the j-th portion of the external crack arm, T_{iAB} and $\gamma_{ext_{AB}}$ are, respectively, the torsion moment induced by the unit loading for obtaining of ω_i and the shear strain at the periphery of the rod in portion, AB, of the external crack arm, T_{iBD} and $\gamma_{unc_{BD}}$ are, respectively, the torsion moment induced by the unit loading for obtaining of ω_i and the shear strain at the periphery of the rod in portion, BD, of the un-cracked part of the rod, T_{ik} and γ_{unc_k} are, respectively, the torsion moment induced by the unit loading for obtaining of ω_i and the shear strain at the periphery of the rod in the k-th portion of the un-cracked part of the rod.

The integrals of Maxwell-Mohr are applied also to obtain the angle of twist of the free and of the internal crack arm (figure 1)

$$\omega_b = \int_0^a T_\beta \frac{\gamma_{\text{int}}}{R_2} dx + \int_a^{t_1} T_{BD} \frac{\gamma_{unc_{BD}}}{R_1} dx + \sum_{k=2}^{k=r} \int_{t_{k-1}}^{t_k} T_k \frac{\gamma_{unc_k}}{R_A} dx \,, \tag{4}$$

where T_β and γ_{int} are, respectively, the torsion moment induced by the unit loading for obtaining of ω_b and the shear strain at the periphery of the internal crack arm, T_{BD} is the torsion moment in portion, BD, induced by the unit loading for obtaining of ω_b, and T_k is the torsion moment in k-portion of the un-cracked part of the rod induced by the unit loading for obtaining of ω_b.

By substituting of equation (2), (3) and (4) in (1), one derives the following general expression for the strain energy release rate for the longitudinal crack in the inhomogeneous non-linear elastic rod configuration shown in figure 1:

$$G = \sum_{i=1}^{i=m} \frac{T_i}{2\pi R_2}\left(T_{iAB} \frac{\gamma_{ext_{AB}}}{R_1} - T_{iBD} \frac{\gamma_{unc_{BD}}}{R_1} \right) + \frac{T_b}{2\pi R_2}\left(T_\beta \frac{\gamma_{\text{int}}}{R_2} - T_{BD} \frac{\gamma_{unc_{BD}}}{R_1} \right) -$$

$$- \frac{1}{2\pi R_2}\left[\int_0^{R_2}\int_0^{2\pi} u_{0\text{int}} R dR d\varphi + \int_{R_2}^{R_1}\int_0^{2\pi} u_{0ext_{AB}} R dR d\varphi - \int_0^{R_1}\int_0^{2\pi} u_{0unc_{BD}} R dR d\varphi \right], \tag{5}$$

where the strain energy density, $u_{0\text{int}}$, and the distribution of shear strain, γ, in the cross-section of the internal crack arm are written as

$$u_{0\text{int}} = \int_0^\gamma \tau(\gamma) d\gamma \,, \; \gamma = \frac{\gamma_{\text{int}}}{R_2} R \,. \tag{6}$$

In equation (6), $\tau(\gamma)$ is the stress-strain relation (τ is the shear stress, γ is the shear strain), γ_{int} is the shear strain at the periphery of the internal crack arm. The following equation for equilibrium of the cross-section of the internal crack arm is used to determine γ_{int}:

$$T_b = \int_0^{R_2} \int_0^{2\pi} \tau(\gamma) R dR d\varphi \, . \tag{7}$$

After substituting of $\tau(\gamma)$ in equation (7), the equation should be solved with respect to γ_{in}. The strain energy density, $u_{0ext_{AB}}$, that participates in equation (5) is determined by the first formula in equation (6). For this purpose, $\tau(\gamma)$ is replaced with $\tau_{ext_{AB}}(\gamma_{ext})$ where $\tau_{ext_{AB}}(\gamma_{ext})$ is the distribution of the shear stresses in the cross-section of the external crack arm in portion, AB. The distribution of the shear strains, γ_{ext}, is obtained by replacing of γ, γ_{in} and R_2 with γ_{ext}, $\gamma_{ext_{AB}}$ and R_1 in the second formula in equation (7). The first formula in equation (6) is applied also to calculate the strain energy density, $u_{0unc_{BD}}$, that participates in equation (5). For this purpose, $\tau(\gamma)$ is replaced with $\tau_{unc_{BD}}(\gamma_{unc})$. Here, $\tau_{unc_{BD}}(\gamma_{unc})$ is the distribution of the shear stresses in the cross-section of the portion, BD, of the un-cracked part of the rod. The distribution of the shear strains, γ_{unc}, is obtained by the second formula in equation (6). For this purpose, γ, γ_{in} and R_2 are replaced with γ_{unc}, $\gamma_{unc_{BD}}$ and R_1, respectively.

The integration in equation (5) should be carried-out by using the MatLab computer program for particular geometry of the inhomogeneous rod, loading conditions, crack location in radial direction and laws for distribution of properties of the inhomogeneous material in the rod.

3. Numerical example

The general approach developed in the previous section of the paper is applied here to analyze the strain energy release rate for the longitudinal crack in an inhomogeneous cantilever rod that is clamped in its right-hand end.

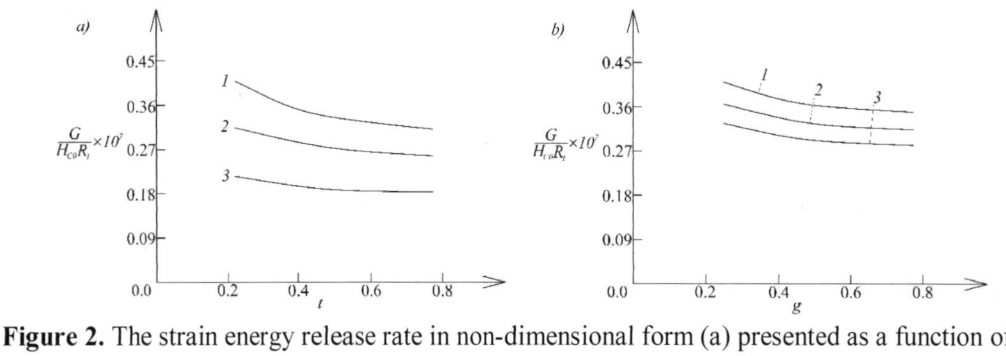

Figure 2. The strain energy release rate in non-dimensional form (a) presented as a function of t. (curve 1 – at $R_2/R_1 = 0.2$, curve 2 – at $R_2/R_1 = 0.4$, curve 3 – at $R_2/R_1 = 0.6$), and (b) presented as a function of g (curve 1 – at $a/l = 0.25$, curve 2 – at $a/l = 0.50$, curve 3 – at $a/l = 0.75$).

The rod is loaded by two torsion moments, T_1 and T_b, which are applied at the external crack arm at distance, l_1, from the free end of the rod and at the free end of the internal crack arm, respectively. A longitudinal circular cylindrical crack of radius, R_2, and length, a, is located in the rod. The non-linear elastic behaviour of the material and the distribution of material property, H, in radial and length directions are treated as

$$\tau = H\left[1 - \left(1 - \frac{\gamma}{S}\right)^f\right], \quad H = H_C e^{g\frac{R}{R_1}}, \quad H_C = H_{C0} e^{t\frac{x}{l}} \tag{8}$$

where S and f are material properties, H_C is the value of H at the longitudinal centroidal axis of the rod, g is a material property that controls the material gradient in radial direction, H_{C0} is the value of H_C at the left-hand end of the rod, t is a material property that controls the material gradient in longitudinal direction.

Calculations of the strain energy release rate are performed by applying equation (5). In order to verify equation (5), the strain energy release rate is derived also by differentiating the complementary strain energy with respect to the delamination crack area. The complementary strain energy stored in the rod is obtained by replacing of strain energy densities with complementary strain energy densities in equation (2). It should be noted that the strain energy release rate obtained by differentiating the complementary strain energy is exact match of that found by equation (5). This fact verifies the general approach developed in the present paper.

The calculated strain energy release rate is presented in non-dimensional form by using the formula $G_N = G/(H_{C0}R_1)$. It is assumed that $l = 0.200$ m, $l_1 = 0.010$ m, $R_1 = 0.005$ m, $T_1 = 4$ Nm and $T_b = 10$ Nm. The crack position in radial direction and the crack length are characterized by R_2/R_1 and a/l ratios, respectively. The influences of the material inhomogeneity in length direction and the crack location in radial direction on the longitudinal fracture are illustrated in figure 2a where the strain energy release rate is presented in non-dimensional form as a function of t at three R_2/R_1 ratios for $a/l = 0.25$, $g = 0.2$, $f = 1.2$ and $S = 0.2$. The curves in figure 2a indicate that the strain energy release rate decreases with increasing of t and R_2/R_1 ratio.

In order to evaluate the influence of the material inhomogeneity in radial direction and the crack length on the longitudinal fracture behaviour, the strain energy release rate in non-dimensional form is presented as a function of g in figure 2b at three a/l ratios. One can observe in figure 2b that the strain energy release rate decreases with increasing of g and a/l ratio.

4. Conclusions

Fracture behaviour of inhomogeneous rods with longitudinal circular cylindrical crack loaded in torsion is studied in terms of the strain energy release rate. The rods have circular cross-section and exhibit continuous (smooth) material inhomogeneity in both radial and length directions. Besides, the material has non-linear elastic mechanical behaviour. General approach for analyzing the strain energy release rate is developed by considering the energy balance. The approach is applied to analyze the strain energy release rate in a cantilever inhomogeneous rod. For verification, the strain energy release rate is derived also by considering the complementary strain energy. It is found that the strain energy release rate decreases with increasing of t and R_2/R_1 ratio. The analysis reveals that the strain energy release rate decreases also with increasing of g and a/l ratio.

Acknowledgments

The present study was carried-out with the financial support of the Research and Design Centre (CNIP) of the University of Architecture, Civil Engineering and Geodesy (UACEG), Sofia (Contract BN – 217/2019).

References

[1] Bohidar S K, Sharma R, Mishra P R 2014 *Int. J. Res.* **1** 289-301.
[2] Erdogan, F. 1995 *Comp. Eng.* **5** 753-770.
[3] Carpinteri A, Pugno N 2005 *Eng. Fracture Mechanic.* **72** 1254-67.
[4] Rizov V 2019 *Frattura ed Integrità Strutturale*, (2047) 468-481; DOI: 10.3221/IGF-ESIS.47.37
[5] Rizov V 2018 *Multidiscipline Modelling Mater. Struct.* **10** 1573-6105.

IOP Conf. Series: Materials Science and Engineering **739** (2020) 012004 doi:10.1088/1757-899X/739/1/012004

Inhomogeneous Multilayered Beams of Linearly Changing Width: a Delamination Analysis

V Rizov[1]

[1]Professor, Department of Technical Mechanics, University of Architecture, Civil Engineering and Geodesy, 1046 - Sofia, Bulgaria

V_RIZOV_FHE@UACG.BG

Abstract. Delamination behaviour of an inhomogeneous multilayered cantilever beam of linearly changing width of the cross-section along the beam length is analyzed. The material in each layer of the beam has non-linear elastic mechanical behaviour. Besides, each layer exhibits continuous material inhomogeneity along the layer height. The solution to the strain energy release rate derived by considering the complementary strain energy is valid for beams made of arbitrary number of layers of individual thickness and material properties. In order to verify the solution, the strain energy release rate is obtained also by considering the balance of the energy.

1. Introduction

Due to their high strength-to-weight and stiffness-to-weight ratios, the inhomogeneous multilayered structural members of continuously changing cross-section in the length direction are very suitable for load-bearing structural applications where the low weight is an important issue. One of the basic drawbacks of multilayered structural members and components is the high risk of separation of layers or delamination. It should be mentioned that delamination has been analyzed mainly by using the methods of linear-elastic fracture mechanics [1, 2].

Therefore, the present paper is concerned with delamination analysis of a non-linear elastic inhomogeneous multilayered cantilever beam. In contrast to previous works which are focussed on delamination analysis of multilayered beams of constant cross-section [3, 4, 5], the present paper deals with a multilayered beam of linearly changing width of the cross-section along the beam length. The delamination is analyzed in terms of the strain energy release rate by considering the complementary strain energy.

2. Solution to the strain energy release rate

A multilayered inhomogeneous cantilever beam with a delamination crack of length, a, is shown in figure 1. The beam is clamped in section, B. The length of the beam is denoted by l. The beam is made of an arbitrary number of adhesively bonded longitudinal horizontal layers of individual thickness and material properties. The cross-section of the beam is a rectangle of width, b, and height, h. The width changes along the beam length according to the following linear law:

$$b = b_0 + \frac{b_1 - b_0}{l} x_3 \tag{1}$$

where $0 \leq x_3 \leq l$. In (1), b_0 and b_1 are the widths of the cross-section at the free end of the beam and at the clamping, respectively (figure 1). The delamination is located arbitrary between layers. Thus, the heights of the lower and upper crack arms are h_1 and h_2, respectively.

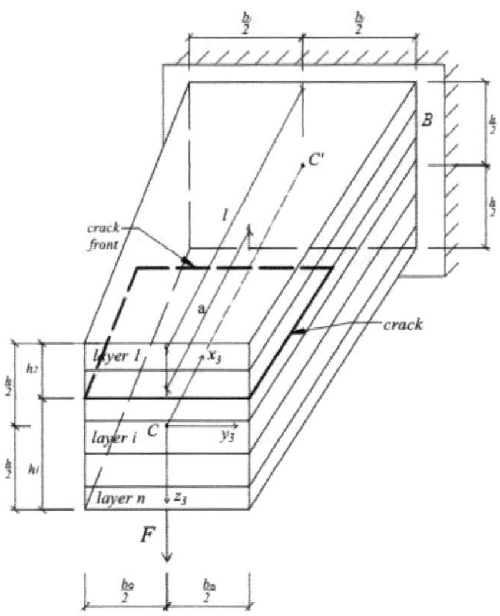

Figure 1. Geometry and loading of a multilayered inhomogeneous cantilever beam of linearly changing width along the beam length.

The beam is loaded by one vertical force, F, applied at the free end of the lower crack arm. Therefore, the upper crack arm is free of stresses. The mechanical behaviour of the material in the *i*-th layer of the beam is treated by the following non-linear stress-strain relation:

$$\sigma_i = \frac{\varepsilon}{D_i + L_i \varepsilon} \tag{2}$$

where σ_i is the normal stress, ε is the longitudinal strain, D_i and L_i are material properties. Each layer of the beam exhibits continuous material inhomogeneity in the thickness direction. The distribution of the material property, D_i, along the thickness of the *i*-th layer is described as

$$D_i = D_{gi} + \frac{D_{gi} - D_{di}}{(z_{3di} - z_{3gi})^{q_i}} \left(z_3 - z_{gi}\right)^{q_i} \tag{3}$$

where D_{gi} and D_{di} are the values of D_i in the upper and lower surfaces of the *i*-th layer, z_{3gi} and z_{3di} are the coordinates of the upper and lower surfaces, z_3 is the vertical centroidal axis of the beam cross-section, and q_i is a material property that controls the material inhomogeneity.

The delamination fracture behaviour is studied in terms of the strain energy release rate, G. For this purpose, the strain energy release rate is derived by using the formulae

$$G = \frac{dU^*}{bda}, \; U^* = U_1^* + U_2^* \tag{4}$$

where U^* is the complementary strain energy stored in the beam, da is an elementary increase of the delamination crack length, U_1^* and U_2^* are the complementary strain energies in the lower crack arm and the un-cracked part of the beam

The complementary strain energy in the lower crack arm is expressed as

$$U_1^* = \sum_{i=1}^{i=n_1} \int_0^a \int_{-\frac{b}{2}}^{\frac{b}{2}} \int_{z_{1gi}}^{z_{1di}} u_{0si}^* dx_3 dy_1 dz_1 \tag{5}$$

where n_1 is the number of layers in the lower crack arm, z_{1gi} and z_{1di} are the coordinates of the upper and lower surfaces of the i-th layer, y_1 and z_1 are the horizontal and vertical centroidal axes of the cross-section of the lower crack arm, u_{0si}^* is the complementary strain energy density in the i-th layer of the lower crack arm. The following formulae are applied to calculate u_{0si}^* :

$$u_{0si}^* = \sigma_i \varepsilon - u_{0si}, \; u_{0si} = \int_0^{\varepsilon} \sigma_i(\varepsilon) d\varepsilon, \; u_{0si}^* = \frac{\varepsilon^2}{D_i + L_i \varepsilon} - \frac{\varepsilon}{L_i} + \frac{D_i}{L_i^2} \ln\left(\varepsilon + \frac{D_i}{L_i}\right) - \frac{D_i}{L_i^2} \ln \frac{D_i}{L_i} \tag{6}$$

where

$$\varepsilon = \kappa_1 (z_1 - z_{1n}) \tag{7}$$

In (7), κ_1 is the curvature of the lower crack arm, z_{1n} is the neutral axis (it should be mentioned that the neutral axis shifts from the centroid since the beam is multilayered and inhomogeneous). The curvature and the neutral axis are determined by using the equations for equilibrium of the cross-section of the lower crack arm

$$N_1 = \sum_{i=1}^{i=n_1} \int_{-\frac{b}{2}}^{\frac{b}{2}} \int_{z_{1gi}}^{z_{1di}} \sigma_i dy_1 dz_1, \; M_1 = \sum_{i=1}^{i=n_1} \int_{-\frac{b}{2}}^{\frac{b}{2}} \int_{z_{1gi}}^{z_{1di}} \sigma_i z_1 dy_1 dz_1 \tag{8}$$

where N_1 and M_1 are the axial force and the bending moment. It is obvious that $N_1 = 0$ and $M_1 = Fx_3$ (figure 1). The distribution of D_i in the i-th layer of the lower crack arm is obtained by replacing of z_{3gi}, z_{3di} and z_3 with z_{1gi}, z_{1di} and z_1 in formula (3). After substituting of (2), (3) and (7) in (8), the equations are solved with respect to κ_1 and z_{1n} by using the MatLab computer program.

The complementary strain energy in the un-cracked beam portion is written as

$$U_2^* = \sum_{i=1}^{i=n_2} \int_a^l \int_{-\frac{b}{2}}^{\frac{b}{2}} \int_{z_{2gi}}^{z_{2di}} u_{0ri}^* dx_3 dy_2 dz_2 \tag{9}$$

where n_2 is the number of layers in the un-cracked beam portion, z_{2gi} and z_{2di} are the coordinates of the upper and lower surfaces of the i-th layer, y_2 and z_2 are the horizontal and vertical centroidal axes of the cross-section of the un-cracked beam portion, u_{0ri}^* is the complementary strain energy density in the i-th layer of the un-cracked beam portion. Formulae (6) are applied to obtain u_{0ri}^* . For this purpose, ε is replaced with ε_r where the distribution of the longitudinal strains, ε_r , along the height of the cross-section of the un-cracked beam portion is found by replacing of κ_1 and z_{1n} with κ_2 and z_{2n} in (7). The curvature, κ_2 , and the coordinate of the neutral axis, z_{2n} , are determined from equations (8). For this purpose, n_1, z_{1gi}, z_{1di}, σ_i, y_1 and z_1 are replaced with n_2, z_{2gi}, z_{2di}, σ_{ri}, y_2 and z_2 where the distribution of the normal stresses, σ_{ri}, in the *i*-th layer of the un-cracked beam portion is obtained by replacing of ε with ε_r in formula (2).

Finally, by substituting of (5), and (9) in (4), one obtains the following expression for the strain energy release rate:

$$G = \frac{1}{b}\sum_{i=1}^{i=n_1}\int_{-\frac{b}{2}}^{\frac{b}{2}}\int_{z_{1gi}}^{z_{1di}} u_{0si}^{*} dy_1 dz_1 - \frac{1}{b}\sum_{i=1}^{i=n_2}\int_{-\frac{b}{2}}^{\frac{b}{2}}\int_{z_{2gi}}^{z_{2di}} u_{0ri}^{*} dy_2 dz_2 \qquad (10)$$

The integration in (10) is performed by the MatLab computer program. It should be noted that b, u_{0si}^{*} and u_{0ri}^{*} which participate in (10) are obtained by (1) and (6) at $x_3 = a$.

In order to verify (10), the strain energy release rate is derived also by considering the balance of the energy assuming a small increase of the crack length. For this purpose, the strain energy stored in the beam is found by replacing of the complementary strain energy densities with the strain energy densities in (5) and (9). The strain energy densities are calculated by substituting of (2) in the second formula in (6). The vertical displacement of the application point of the force, F, that participates in the expression for the strain energy release rate is found by the integrals of Maxwell-Mohr. It should be noted that the strain energy release rate obtained by considering the balance of the energy is exact match of that calculated by formula (10) which is a verification of the delamination analysis developed in the present paper.

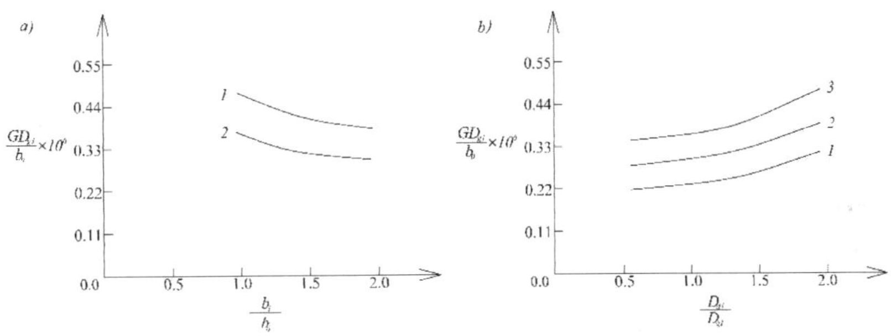

Figure 2. The strain energy release rate in non-dimensional form (a) plotted against b_1/b_0 ratio (curve 1 – for delamination crack located between layers 2 and 3, curve 2 - for delamination crack located between layers 1 and 2), and (b) plotted against D_{d1}/D_{g1} ratio (curve 1 – at $a/l = 0.3$, curve 2 – at $a/l = 0.5$, curve 3 – at $a/l = 0.7$).

The solution to the strain energy release rate (10) is applied to evaluate the influence of the linearly changing width of the cross-section along the beam length, the delamination crack location along the height of the cross-section, the material inhomogeneity in the height direction and the crack length on the delamination fracture behaviour. In order to evaluate the influence of the crack location along the height of the cross-section, two three-layered inhomogeneous cantilever beam configurations are analyzed (a beam configuration with a delamination crack between layers 2 and 3, and a beam configuration with a delamination crack located between layers 1 and 2). The thickness of each layer in both beam configurations is t. The width of the cross-section of both beam configurations changes linearly from b_0 in the free end to b_1 in the clamped end of the beam. The strain energy release rate is presented in non-dimensional form by using the formula $G_N = GD_{g1}/b_0$. It is assumed that $F = 5$ N, $l = 0.200$ m, $b_0 = 0.01$ m and $t = 0.003$ m.

The linearly changing cross-section width along the beam length is characterized by b_1 / b_0 ratio. The influences of the linearly changing cross-section width along the beam length and the crack location along the beam height on the delamination behaviour are illustrated in figure 2a where the strain energy release rate in non-dimensional form is plotted against b_1 / b_0 ratio for both three-layered beam configurations. It is assumed that $D_{d1} / D_{g1} = 2.0$, $D_{g2} / D_{g1} = 1.2$, $D_{d2} / D_{g2} = 0.6$, $D_{g3} / D_{g1} = 0.8$, $D_{d3} / D_{g3} = 0.7$, $q_1 = q_2 = q_3 = 0.7$, $L_1 / D_{g1} = 0.3$, $L_2 / D_{g2} = 0.1$ and $L_3 / D_{g3} = 0.2$. The curves in figure 2a indicate that the strain energy release rate decreases with increasing of b_1 / b_0 ratio. This finding is attributed to the increase of the beam stiffness. One can observe also in figure 2a that the strain energy release rate when the delamination crack is located between layers 2 and 3 is higher in comparison with the strain energy release rate when the delamination is between layers 1 and 2. This behaviour is due to fact that the stiffness of the lower crack arm is lower when the delamination is between layers 2 and 3.

The material inhomogeneity in layer 1 is characterized by D_{d1} / D_{g1} ratio. The delamination crack length is characterized by a / l ratio. In order to evaluate the influence of material inhomogeneity and the delamination length on the fracture, the strain energy release rate in non-dimensional form is plotted against D_{d1} / D_{g1} ratio in figure 2b at three a / l ratios. The three-layered beam configuration with a delamination crack between layers 2 and 3 is considered. It can be observed in figure 2b that the strain energy release rate increases with increasing of D_{d1} / D_{g1} ratio (this is due to the decrease of the beam stiffness). The increase of delamination crack length leads to increase of the strain energy release rate (figure 2b).

3. Conclusions

A delamination analysis of a multilayered inhomogeneous non-linear elastic cantilever beam of linearly changing width of the cross-section along the beam length is carried-out. The beam is made of arbitrary number of adhesively bonded layers of individual thickness and material properties. The material in each layer exhibits continuous material inhomogeneity in the height direction. The delamination is studied in terms of the strain energy release rate. Solution to the strain energy release rate is derived by considering the complementary strain energy stored in the beam. The strain energy release rate is obtained also by analyzing the balance of the energy for verification. Influence of such factors as the linearly changing width of the cross-section along the bean length, the crack location along the beam height, the crack length and the material inhomogeneity on the delamination fracture are evaluated by using the solution to the strain energy release rate. It is found that the strain energy release rate decreases with increasing of b_1 / b_0 ratio. The increase of D_{d1} / D_{g1} and a / l ratios lead to increase of the strain energy release rate. The analysis reveals that when the delamination crack is located between layers 2 and 3, the strain energy release rate is higher in comparison to that obtained when the delamination crack is between layers 1 and 2.

Acknowledgments

The present study was carried-out with the financial support of the Research and Design Centre (CNIP) of the University of Architecture, Civil Engineering and Geodesy (UACEG), Sofia (Contract BN – 217/2019).

References

[1] Narin J 2006 *Int. J. Fracture* **139** 267-93.
[2] Hutchinson JW, Suo Z 1992 *Adv Appl Mech* **64** 804 – 810.
[3] Rizov V 2017 *Frattura ed Integrità Strutturale* **41** 498-510.
[4] Rizov V 2017 *Archive Appl. Mechanic.* **87** 1037-48.
[5] Rizov V 2017 *Multidiscipline Modelling Mater. Struct.* **13** 434-47.

Cost Impact Analysis on the DPWH Guideline and Implementing Rules on Earthquake Recording Instrumentation for Private Buildings using Multi-Criteria Decision Analysis-Analytical Hierarchy Process (MCDA-AHP) in the City of Manila

V. C. M. Alfonso[1], M. B. Bagsit[1], J. D. Reyes[1], M. B. Baylon[2], and F. A. A. Uy[3]

[1]Undergraduate Bachelor of Science in Civil Engineering, Mapua University, Muralla Street, Intramuros, City of Manila, Philippines

[2]Senior Structural Health Engineer, USHER Techologies, Mapua University, Muralla Street, Intramuros, City of Manila, Philippines

[3]Dean, School of Civil Engineering, Geophysical, and Environmental Engineering, Muralla Street, Intramuros, City of Manila, Philippines

embylon@usher.ph

Abstract. Philippines lie in the Pacific Ring of Fire, a region where earthquakes and other seismic activities are frequent. Through the NBCDO Memorandum Circular No. 01 Series of 2015, The DPWH Guidelines and Implementing Rules on Earthquake Recording Instrumentation for buildings was implemented with the objectives of: (1) ensuring the safety of building occupants by determining the condition of buildings, and (2) collecting earthquake related data during major seismic activities. Despite being signed on March 12, 2015, the compliance rate of the buildings here in the City of Manila was low. The purpose of this study is to determine the cost impact of the said guidelines to building owners using MCDA- AHP; as well as to identify the amount needed to equip the City of Manila. The research was carried out having SHM experts (for AHP) and building owners (for MCDA) as the respondents for the corresponding parts of the study. The results show that the equipment price has the biggest impact to the building owners and their willingness to pay for ERIs is lower compared to the market.

1. Introduction

Earthquakes in the Philippines are inevitable since the country is located in the Pacific Ring of Fire. Based on the 2004 MMEIRS study, "the Big One", a 7.2 Magnitude earthquake, is expected in this lifetime to hit Metro Manila and other nearby provinces which will damage up to 1.3 Million residential structures and leave an approximate of 34,000 deaths and more than 100,000 injuries.[1]

Due to unexpected events like earthquake or typhoons the structure may suffer from damages that can lead to the failure or reduction of its structural integrity. A process called Structural Health

Monitoring (SHM) was introduced to monitor and detect the damages of a structure. Local Government Units (LGUs) and DPWH have no other way aside from visual inspection to know the extent of the damages done. A device called Earthquake Recording Instrument (ERI) was introduced as the possible solution in monitoring the structural health of the building.[2,3]

In 2015, the DPWH implemented its Guidelines, Implementing Rules, and Regulations on Earthquake Recording Instrument for buildings (DPWH IRR ERI). This guideline requires all buildings included in the categories listed to install a number of accelerographs as an additional mitigation measure and in order to safeguard lives. [4]

Department of Public Works and Highways (DPWH) adapted the DPWH IRR ERI for buildings last March 12, 2015 with the objectives of: (1) ensuring the safety of building occupants by determining the condition of buildings, and (2) collecting earthquake related data during major seismic activities. Four years after the implementation, the number of buildings that have complied with the DPWH IRR ERI in the City of Manila is still low.

The main objective of this study is to analyze the cost impact of the acquisition of ERIs to building owners. The researchers conducted this study to achieve the following specific objectives: (1) collate and assess data that represent the cost impact of the ERI installation, including the projection of the total cost needed to equip the affected buildings in the City of Manila, (2) utilize the MCDA-AHP to identify the optimum ranking of multiple criteria.

2. Methodology

The researchers gathered data and information needed to support the aim of the study. These information and data were the established facts from journals and the DPWH guidelines and implementing rules on earthquake recording instrumentation for buildings. The researchers gathered data of the buildings required by the guidelines from the office of the building official. The data gathered is a list of all buildings required and list of buildings that have not complied with the guidelines. This can be well summarized using figure 1.

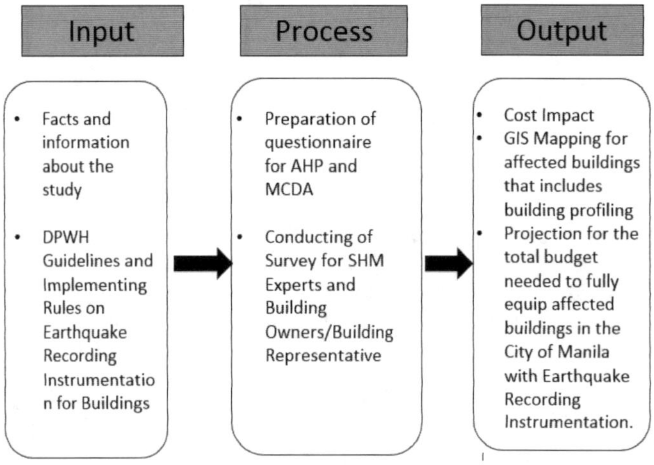

Figure 1. Research paradigm.

3. Results and Discussions

3.1. Demographics

The researchers used Cochran's formula to determine the number of respondents. The researchers used a confidence interval of 90% with a z -score value of 1.65 and a margin of error of 5%. The value of p or the percentage is 50% since it is generally used in determining the level of accuracy. Lastly, the population size was determined to be 157 as given by the Annual Inspection Division of the Office of the Building Official. Applying Cochran's formula, the researchers were able to compute for the sample size as 100.

$$n = \frac{\frac{1.65^2[0.5(1-0.5)]}{0.05^2}}{1+\frac{1.65^2[0.5(1-0.5)]}{0.05^2(157)}} \quad (1)$$

The researchers surveyed a total of 120 respondents in the City of Manila: 100 Building Owner/Representative (only 39 survey questionnaires were returned), and 20 SHM experts. The low rate of return of the questionnaires was attributed to the unwillingness of the building owners/representatives with the reason of confidentiality of their information. Figure 2 shows the composition of the sample population considering their building types. The sample size is composed of: 23 private building A, 10 private building B, and 6 private building C.

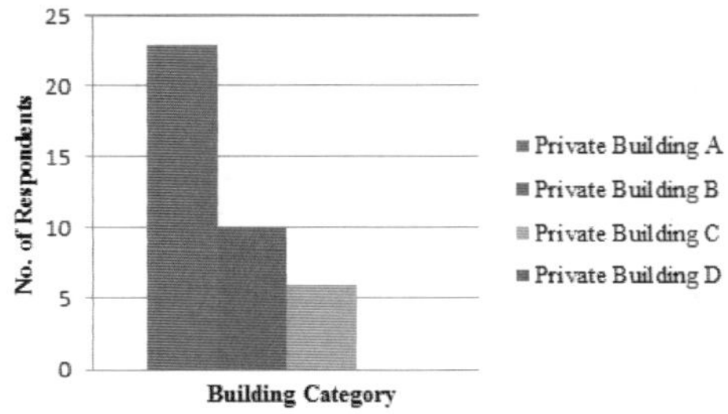

Figure 2. Population demographics regarding Building category.

The researchers also considered whether the respondents have complied with the DPWH IRR ERI or not. Only 33% or 13 respondents have complied with the said guidelines and 67% or 26 respondents have not yet complied based on the sample. This section is divided into two: (1) AHP and (2) MCDA; the two methods that the researchers used in the analysis of cost impact. All of the data presented were findings of the researchers based on the second part of the questionnaire and the AHP questionnaire.

3.2. Analytical Hierarchy Process (AHP)

AHP is used in getting the weights of each cost criterion that affects the decision making of the building owners/representatives based on the perspective of 20 experts related to structural health monitoring and ERI development from: Development of Earthquake Intensity Monitoring System (DEIMS), Universal Structural Health Evaluation and Recording System (USHER) team and the Philippine Structural Integrity Management System (PhilSIMS). The experts consider 30.66% as the weight of the equipment cost, 10.31% for the installation cost, 26.29% for the maintenance cost, and 32.74% for the operation cost.

In the perspective of the experts, the operation cost was identified to be the primary reason that affects the decision of building owners/representatives in the acquisition of ERIs, the equipment cost as the secondary criterion, maintenance cost as the third criterion to be considered, and installation cost as the last criterion to be considered.

The process in obtaining the weights of each criterion was derived using the AHP calculator by K.D. Goepel.[5] The responses of the 20 SHM experts was consolidated; the former participated in the survey. The calculated consistency ratio is 3.7% which is within the acceptable threshold; it means that the results were reliable. The error for each weight presented means that errors may happen even if the consistency ratio is below 10%. The pair-wise comparison matrix and the generated scale and weights considering the 20 SHM experts were consolidated. This shows the importance of each criterion relative to the other criteria.

3.3. Multiple Criteria Decision Analysis (MCDA)

The researchers used Multiple Criteria Decision Analysis approach in the interpretation of the responses of 39 building owners/representatives.[6] In this section, the results of each building types and the aggregate results were presented and analyzed.

There were 23 respondents for building owners/representative of private building A. The results clearly show that Equipment Cost have the biggest impact in the acquisition of ERI since it was identified to have the largest portion in the total cost of the equipment. It is followed by operation cost, maintenance cost, and installation cost consecutively.

Table 1. MCDA-AHP ranking for private building A.

Criterion	Rank
Equipment Cost	1
Operation Cost	2
Maintenance Cost	4
Installation Cost	3

Only 10 building owners have responded for the buildings in private category B. Equipment Cost also ranked first in the cost impact survey followed by operation cost, maintenance cost, and installation cost. It was identified in the result; equipment cost is the most impactful for building owners of schools and hospital.

Table 2. MCDA-AHP ranking for private building B.

Criterion	Rank
Equipment Cost	1
Operation Cost	2
Maintenance Cost	4
Installation Cost	3

There were six that responded for buildings belonging to category C. The building owners/ representatives ranked the equipment cost as first cost that affects their acquisition of ERI. Second is the operation cost; maintenance cost as third place; and installation cost as the cost that lead to least impact in the acquisition of ERIs.

Table 3. MCDA-AHP ranking for private building C.

Criterion	Rank
Equipment Cost	1
Operation Cost	2
Maintenance Cost	4
Installation Cost	3

The total number of respondents for all of the building categories is 100 however only 39 respondents or 39% of the total respondents were willing to participate in the survey. Based on the result of the cost impact survey for each building, equipment cost was always considered to have the biggest impact in the acquisition of ERI followed by the operation cost, maintenance cost and installation cost ranked consecutively.

Table 4. Aggregate MCDA-AHP ranking.

Criterion	Rank
Equipment Cost	1
Operation Cost	2
Maintenance Cost	4
Installation Cost	3

One significant finding of the research is that the willingness to pay (WTP) allocated budget of the building owners/representatives only differs on the ranking of the installation cost and maintenance cost to the results of the MCDA-AHP study. The WTP obtained were: (1) Equipment Cost- ₱593,590 (2) Operation Cost- ₱400,000; (3) Installation Cost- ₱184,615.38; and (4) Maintenance Cost- ₱120,512.82. The prices were obtained using the mean values of the WTP of the owners as presented in Appendix G. The mean values were subtracted to the preceding package to obtain the WTP of each criterion.

According to Greef & Ghoshal [7], maintenance cost is treated as one of the largest operating expense and is critical since it is needed to maintain the condition of the equipment [8]. Owners and operators see the maintenance cost only as additional cost and as a window to save money. This was proven in the findings since the allotted budget for the maintenance cost is lesser than the installation cost even though they were valued by SHM experts and the owners in reverse.

3.4. Total Cost Projection

This section discusses the cost projection in fully equipping the City of Manila with ERIs. The researchers asked the building owners/representatives about their budget or how much were they willing to pay for the installation of ERI. The researchers presented four packages in order to have a more detailed outcome.

The buildings affected by the guideline were categorized to determine the total number of ERI needed in the City of Manila based on Table 5. The number of buildings without ERI is tallied per building category and the total number of building was computed as 131. According to the Office of the Building Official of the City of Manila particularly the Annual Inspection Division the total number of private buildings required to install ERI were 157. It consists of 90 private building A, 57 private building B, and 10 private building C. Based from the data obtained only 26 private buildings or only about 17% of the total number of buildings have already complied with the DPWH IRR ERI.

To obtain the total number of ERI needed, the number of buildings without ERI is multiplied with the number of ERI needed per building category. For private building A it is required to have 3 units of ERI while for other building category only one unit is required. The total number of ERI units needed was computed as 298.

Table 5. Tally for the total number of ERIs needed for the City of Manila.

Building Category	**No. of Bldgs. in Manila w/o ERIs**	**No. of ERI req'd per bldg..**	**Total No. of ERI needed**
Private Bldg. A	79	3	237
Private Bldg. B	53	1	53
Private Bldg. C	8	1	8
Private Bldg. D	0	1	0
TOTAL	**131**		**298**

3.5. Cost Projection per Package

The researchers considered four packages in projecting the total cost needed for the City of Manila to be 100% compliant. As shown in Table 5, the total number of ERI needed in the City of Manila is 298 units.

The price per package as presented in Table 6 was computed based on the answers of the building owner/representative on the last part of the survey which is the willing to pay (WTP) part. The total cost per package was computed by multiplying the number of ERI units needed and the price per package. The following were the packages and its inclusions: (1) Package 1 - Equipment Cost, (2) Package 2 - Equipment Cost and Installation Cost, (3) Package 3 - Equipment Cost, Installation Cost and Maintenance Cost, (4) Package 4 - Equipment Cost, Installation Cost, Maintenance Cost, and Operation Cost.

The price range values of each package were computed using the frequency table. The researchers used the formula of Confidence Interval after identifying the average of the WTP of the building owners. Table 6 shows the frequency table for package 1[9].

Table 6. Frequency table for Package 1.

UL	**LL**	$\tilde{x}$	**f**
100000	200000	150000	16
200000	500000	350000	10
500000	1000000	750000	7
1000000	3000000	2000000	6

Table 7. Cost Projection for the City of Manila.

Package	Price Range	Total cost
Package 1	₱423,325 - ₱763,855	₱126.1 M - ₱227.6 M
Package 2	₱578,481 - ₱977,929	₱172.4M - ₱291.4M
Package 3	₱693,132-₱1,104,304	₱206.6 M - ₱329.1 M
Package 4	₱1,020,990-₱1,576,446	₱304.3 M - ₱469.8 M

3.6. Budget difference of building owners with and without ERIs.

The researchers divided the results of the budget allotment on the installation of ERI to those who have already installed ERIs and those who have not. On Table 6 the results show that the WTP of those who have installed ERIs was far greater compared to that of those who have not yet complied with a percentage difference ranging from 27.04%-71.04%.

WTP of the building owners/representatives presented in Table 8 is computed in the same way as presented in sample computation presented below Table 4.6 the only difference is that the samples were divided in two groups: (1) with ERI, and (2) without ERI. The percent difference was computed using the formula:

$$Percentage\ Difference = \frac{WTP_{with\ ERI} - WTP_{without\ ERI}}{\frac{WTP_{with\ ERI} + WTP_{without\ ERI}}{2}} \quad (2)$$

Table 8. Willingness to pay of building owners with and without ERIs.

	With ERI	W/o ERI	Percentage Difference
Package 1	₱542,004-₱1,265,888	₱274,716-₱602,207	65.45%-71.04%
Package 2	₱591,461-₱1,377,770	₱450,564-₱899,436	27.04%-42.01%
Package 3	₱929,044-₱1,670,956	₱474,051-₱922,103	57.76%-64.86%
Package 4	₱1,320,471-₱2,225,683	₱732,418-₱1,390,659	46.18%-57.29%

Due to unavailability of data, the researchers have only compared package 1 or the equipment cost with the market price. Maintenance cost and the other costs vary on different occasions according to technical specialists. Only Product A, a locally manufactured ERI, is within the range of WTP of the sample population.

Table 9. Price of different ERIs available in the market [10].

Brand Name	Unit Price
Product A	₱480,000
Product B	₱925,000
Product C	₱1,227,200
Product D	₱1,248,000

4. Conclusion and Recommendation

It was identified that the City of Manila has a rate of compliance of 17% (as of March 2019), a rate that is considerably low for an IRR that is in effect for four years.

Equipment cost was identified to have the most substantial impact to building owners in the compliance to the DPWH IRR ERI. It was also found out that the willingness to pay of building owners with regards to the acquisition of ERIs is lower compared to the value of ERIs available in the current market.

The units needed in order to fully equip the City of Manila with ERI shows that there is a problem in the implementation of the LGU or lack of cooperation from the building owner/representatives.

Complete compliance of the building owners to the DPWH IRR ERI for the city of Manila has a long way to go. The results show that a lot of money is needed to achieve complete compliance. The researchers have determined that the number of units needed is 298 units that costs around ₱126.1 - ₱227.6 million. [11]

Other than the cost, the researchers have identified some of the other problems relating to the non-compliance of the building owners/representative during their data gathering process. It was known that there are lack of information regarding the importance of the installation of ERIs, and the miscommunication of the Office of the Building Official to the building owners regarding the implementation. They have also pointed out that the punishments on the non-compliance were lighter compared to the cost of the ERI itself.

References

[1] National Disaster Risk Reduction and Management Council (NDRRMC), Metro Manila Disaster Risk Reduction and Management Council (MMDRRMC), Metro Manila Earthquake Contingency Plan,Oplan Metro Yakal Plus, 2015 Camp General Emilio Aguinaldo, Quezon City

[2] Concepcion R S, Cruz F R G, Uy F A A, Baltazar J M E, Carpio J N, Tolentino K G 2017 *Triaxial MEMS digital accelerometer and temperature sensor calibration techniques for structural health monitoring of reinforced concrete bridge laboratory test platform.* 2017 IEEE 9th International Conference on Humanoid, Nanotechnology, Information Technology, Communication and Control, Environment and Management (HNICEM), Manila, 1-6.

[3] Gaviña J A R, Uy F A A, Carreon J P D 2017 *Structural health monitoring of existing concrete bridges with AASHTO Type IV girder using Smartbridge Sensor Nodes.* 2017 IEEE Conference on Technologies for Sustainability (SusTech), Phoenix, AZ, 1-6.

[4] Republic of the Philippines, Department of Public Works and Highways, Office of the Secretary, 2015 Guidelines and Implementing rules on Earthquake Recording Instrumentation for Buildings, Manila, Philippines.

[5] Goepel, K.D. 2018. Implementation of an Online Software Tool for the Analytic Hierarchy Process (AHP-OS). International Journal of the Analytic Hierarchy Process, Vol. 10 Issue 3 2018, pp 469-487,

[6] Addor M L, Smutko S 2011 *Multi-Criteria Decision Analysis.* Retrieved from https://projects.ncsu.edu/nrli/decision-making/MCDA.php

[7] Greeff G, Ghoshal R 2004 *Practical E-Manufacturing and Supply Chain Management.* Newnes, 185-213.

[8] Chen J 2018 *Maintenance Expenses.* Retrieved from https://www.investopedia.com/terms/m/maintenance-expenses

[9] Feng Y., Tang H. "Confidence interval-based important-performance analysis for determining critical service attributes", 2011 2nd International Conference on Artificial Intelligence, Management Science and Electronic Commerce (AIMSEC), Dengleng, China, 2011

[10] Payawal J M 2018 USHERING A MORE RESILIENT PHILIPPINES: A Civil Engineering Innovation Story.

[11] Alfonso V C M, Bagsit M B, Reyes J D 2019 *Cost Impact Analysis on the DPWH Guideline and Implementing Rules on Earthquake Recording Instrumentation for Private Buildings using Multi-Criteria Decision Analysis-Analytical Hierarchy Process (MCDA-AHP) in the City of Manila.* Undergraduate Thesis: Mapua University.

IOP Conf. Series: Materials Science and Engineering **739** (2020) 012006 doi:10.1088/1757-899X/739/1/012006

Rheological Characteristics of Reclaimed Asphalt Pavement (RAP) Evaluation using Reclamite Rejuvenating Material

Rindu Twidi Bethary[1], Bambang Sugeng Subagio[2]

[1]Doctoral Student, Faculty of Civil and Environment Engineering, Bandung Institute of Technology, Indonesia

[2]Lecturer, Faculty of Civil and Environment Engineering, Bandung Institute of Technology, Indonesia

rindutwidibethary@gmail.com, bsugengs@si.itb.ac.id

Abstract. Using reclaimed asphalt pavement (RAP) does not only reduce costs but is also environmentally friendly. In its use, it is very important to know the empirical and rheological properties of the asphalt. The testing in this study used a Dynamic Shear Rheometer (DSR) to test asphalt pen 60/70, RAP and Reclamite rejuvenating material mixed with RAP. The results of the pen asphalt 60/70 empirical test met the Bina Marga 2010 General Specifications of Roads and Bridges, while the RAP showed the characteristics of an aging asphalt. In the rheological test based on SUPERPAVE, from the value of the performance grade (PG), it was found that Reclamite reduced the critical temperature value on RAP, so the PG value was the same as the asphalt pen 60/70. The master curve results also showed that Reclamite can reduce the RAP stiffness modulus near the value of asphalt pen 60/70 stiffness modulus. Meanwhile, as seen from the results of the black diagram, the original condition has a range of phase angle (δ) greater than the RTFOT and PAVT conditions because the asphalt has undergone aging. At the same S_{bit}, asphalt pen 60/70 has a greater (δ) than RAP, while for RAP added with Reclamite, the value is between the RAP and asphalt pen 60/70 value. This illustrates that the Reclamite rejuvenating material makes RAP more viscous.

1. Introduction

Reclaimed Asphalt Pavement (RAP) technology is a material produced by pavement scraping containing asphalt and aggregate. Based on a study conducted by asphalt association in America, the average percentage of RAP used in asphalt mixture increased by 15.6% in 2009 and increased to 20.4% in 2014, estimating the amount of RAP used in asphalt mixture is 71,900,000 tons, assuming the asphalt content in RAP is 5%. It equals 3.6 million tons of asphalt and 68 million tons of natural aggregate [7]. Reclaimed pavement performance, in addition to being economical and environmentally friendly, must also be prepared in terms of fatigue, rut, thermal resistance and good durability [2, 3].

Asphalt is a thick liquid which is a hydrocarbon compound where its performance is empirically evaluated based on its rheological properties (chemical composition and physical properties of asphalt) while mechanistically, the performance evaluation is based on its elastic modulus (G*), poisson ratio, performance grade and phase angle (δ). The asphalt rheological properties will affect the pavement performance because the rheological properties of the asphalt during production will continue to change until aging process occurs. The first significant aging of asphalt occurs in the pug mill on drum mix plant in the form of substantial rheological changes such as a decrease in penetration and an increase in viscosity which occurs during short-term mixing. Asphalt aging continues at the time of asphalt laying until it is opened for traffic. The aging process will be slower in the first two to three years until the pavement approaches density, after that, the aging rate of asphalt decreases over a longer period of time, resulting in rheological changes in asphalt [5].

The SUPERPAVE system requires the determination of asphalt rheological properties to design and evaluate good hot asphalt mix. The main rheological parameters of asphalt performance grade (PG) are dynamic modulus and phase angle at high temperatures and creep stiffness and creep relaxation levels at low temperatures. These parameters are used for the characteristics of RAP and its mix [4]. The main objective of this study is to evaluate asphalt rheology using a dynamic shear rheometer (DSR).

2. Materials and Methods

The materials used in this study were Pertamina's asphalt pen 60/70, asphalt extracted from reclaimed materials from scraping in National Road in Karawang and Reclamite rejuvenating material. The testing equipment used is a dynamic shear rheometer (DSR). The making of specimen started from testing the characteristics of asphalt pen 60/70, extracted asphalt and RAP. Then, the RAP was mixed with rejuvenating materials. In the rheological test of asphalt, there are 3 different tests: asphalt pen 60/70 as control asphalt, extracted RAP and RAP added with Reclamite using the DSR testing equipment.

3. Results and Analysis

3.1. Characteristics of asphalt pen 60/70

In this study the asphalt used was oil asphalt with 60/70 penetration from Pertamina. Some laboratory tests were carried out to determine the basic properties of the asphalt. From the results of the testing, it was found that the characteristics of the 60/70 penetration oil asphalt met the requirements for Type 1 Hard Asphalt Pen 60-70, as stated in Bina Marga 2010 General Specifications of Roads and Bridges. [6, 8].

3.2. Characteristics of RAP

Testing the characteristics of asphalt was done by separating the asphalt from RAP material with Trichloor Etyleen (TCE) liquid using an evaporator. Testing of RAP from extraction showed an asphalt content of 5.15%. Meanwhile, seen from the test of asphalt properties, there was a decrease in penetration value to 10dmm, an increase in softening point of 80°C and reduced sensitivity to elastic in the form of decreasing ductility value to 38cm. This shows that the asphalt contained in the RAP has undergone an aging process.

3.3. Asphalt Rheological Properties Using DSR

Mechanistic rheology testing (G * and δ) was carried out using a DSR equipment referring to AASHTO T315 [1]. The tests were carried out on three asphalt conditions: 1. Original, which showed the first stage of conveying and storage; 2. Conditions after RTFOT which showed the second stage of aging at the production in the asphalt mixing unit and implementation in the field; 3. Conditions after PAVT which showed the final stage of aging during the pavement service period. In this study, the temperature ranges and loading frequency were carried out based on the requirements stated in the M320 AASHTO specifications.

3.3.1. Temperature Sweep. DSR temperature sweep testing is usually carried out to verify asphalt grade in PG system, related to the mechanistic behavior of asphalt, SUPERPAVE set limits on the value of G * and δ on the rut parameter and fatigue crack parameter. For example DSR temperature sweep testing for RAP added with Reclamite can be seen on Table 1 dan Figure 1.

Table 1. DSR Testing Results for RAP added with Reclamite.

No.	Temperature (°C)	Original			RTFOT			PAVT		
		G* (kPa)	δ (°)	G*/Sin δ (kPa)	G* (kPa)	δ (°)	G*/Sin δ (kPa)	G* (kPa)	δ (°)	G*/Sin δ (kPa)
1.	18,9	-	-	-	-	-	-	$8{,}92.10^6$	41,43	13480
2.	22	-	-	-	-	-	-	$6{,}01.10^6$	44,26	8605
3.	24,9	-	-	-	-	-	-	$3{,}94\ 10^6$	47,31	5354
4.	28	-	-	-	-	-	-	$2{,}56.10^6$	50,26	3330
5.	31	-	-	-	-	-	-	$1{,}66.10^6$	53,26	2069
6.	52	8426	82,28	8,503	20620	77,48	21,13	-	-	-
7.	57,9	3801	84,66	3,817	9096	80,93	9,211	-	-	-
8.	63,9	1728	86,55	1,731	4006	83,77	4,03	-	-	-
9.	69,9	820,1	87,91	0,8206	1853	85,92	1,858	-	-	-
10.	75,9	408,8	88,74	0,4089	900,9	87,48	0,9018	-	-	-

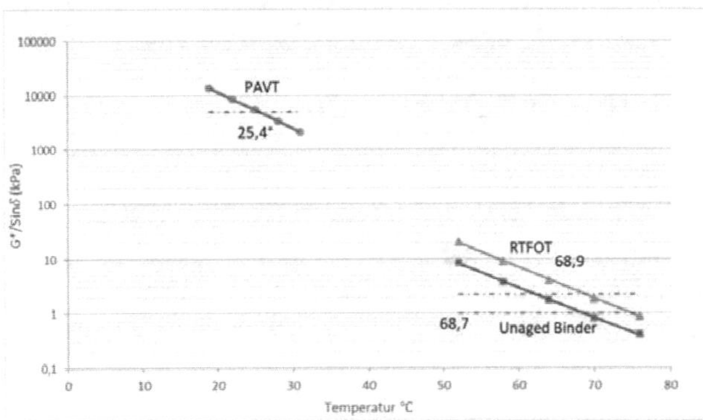

Figure 1. RAP added with Reclamite Critical Temperature Based on Rut Parameter and Fatigue Crack Parameter using DSR.

Based on these limits, a critical temperature was obtained for each asphalt condition to be used as seen in Table 2.

Table 2. Asphalt Pen 60/70, RAP and RAP added with Reclamite Performance Grade.

No.	Condition	Requirement	SUPERPAVE Criteria	Hasil Uji °C		
				Asphalt Pen 60/70	RAP	RAP + Reclamite
1	Original Binder	Min 1,00 kPa	Rutting	65,3	110,7	68,7
2	RTFOT	Min 2,20 kPa	Rutting	63,6	120,2	68,9
3	PAVT	Max 5000 kPa	Fatigue	23,7	52,8	25,4

Based on Table 2, the performance grade can be known based on the value of G * / sin δ in the original condition. For asphalt pen 60/70, the PG value is 64. RAP has a PG value of 106, while RAP added with Reclamite has a PG value of 64. The change in the value of PG on RAP added with Reclamite showed a decrease in the value of PG which resulted in the same PG value as asphalt pen 60/70. This indicates that there has been a good rejuvenation process on RAP added with Reclamite.

3.3.2. Master Curve. The relationship between Bitumen Stiffness Modulus (Sbit) and shifting factor was analyzed using a master curve which can also show the degree of asphalt sensitivity to changes in temperature and loading frequency.

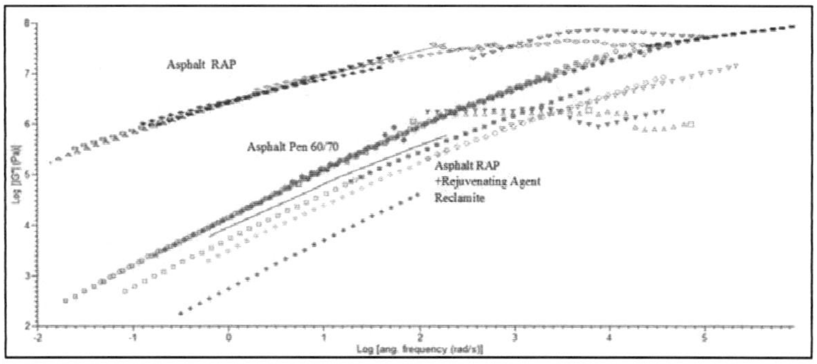

Figure 2. Master Curve of Asphalt Pen 60/70, RAP and RAP added with Reclamite.

According to figure 2 Master Curve, asphalt pen 60/70 has a gentler slope. This indicates a wider temperature range and loading time, so it is easy to apply in the field. RAP added with Reclamite also approaches the 60/70 asphalt pen master curve, which means that Reclamite can increase the temperature range and loading time when compared to RAP without any rejuvenating material because the RAP has undergone aging.

The target Sbit value can be found by using speed of 5,65 Km/H (critical point) at frequency 0,10 Hz, where the log (frequency) = (-1). Then, make a vertical line cutting the master curve. Next, make a horizontal line to get the value of G * (complex shear modulus) which will later be converted into the target Sbit value (stiffness modulus), the results can be seen in Table 3.

Table 3. Target Asbuton Stiffness Modulus (Sbit) of Asphalt Pen, RAP and RAP added with Reclamite based on the shift factor.

Asphalt Type	**Target G* (Pa)**	**Target Sbit (Pa)**
Asphalt Pen 60/70	180.300	540.900
RAP	20.100.000	60.240.000
RAP + Reclamite	73.450	220.350

The results obtained as a minimum target stiffness modulus of RAP after the addition of Reclamite becomes smaller even though it is not the same as the asphalt pen 60/70. Meanwhile, the RAP stiffness modulus is greater than the asphalt pen 60/70's because the RAP has undergone aging, so that it is stiffer.

3.4. Black Diagram

The Black Diagram illustrates the relationship between Stiffness Modulus (Sbit) of Asphalt and Phase Angle (δ) in three conditions: original, RTFOT, and PAVT. The results can be seen on figure 3 until figure 5.

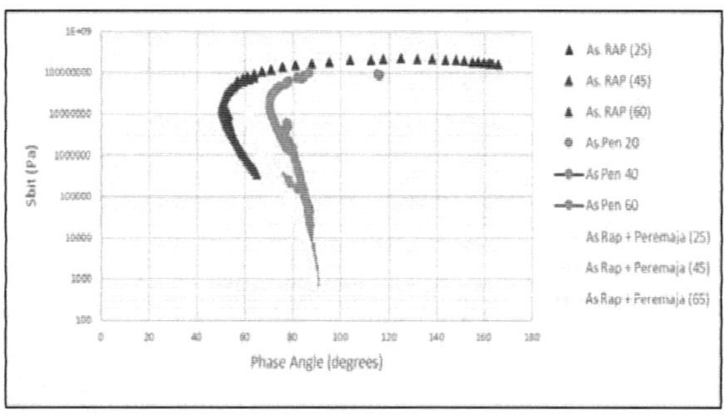

Figure 3. Black Diagram of the Original condition of RAP, Asphalt Pen 60/70 and RAP added with Reclamite.

IOP Conf. Series: Materials Science and Engineering **739** (2020) 012006 doi:10.1088/1757-899X/739/1/012006

Based on the black diagram, the original condition has a value of (δ) ranging from 40 -160. This means that the addition of Reclamite on RAP affects the Sbit value, where the Sbit value will decrease, as well as the increasing phase angle value. The phase angle increase in RAP added with Reclamite shows that the asphalt is more viscous compared to RAP without Reclamite.

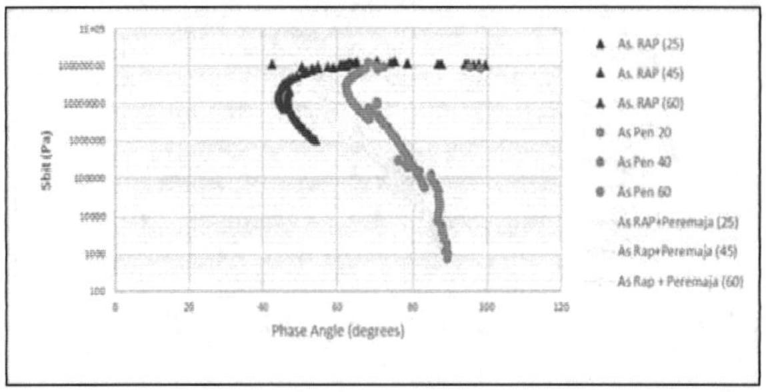

Figure 4. Black Diagram of the RTFOT condition of RAP, Asphalt Pen 60/70 and RAP added with Reclamite.

Based on the black diagram in RTFOT condition, it can be seen that there is a phase angle decrease ranging from 40 to 100. This occurs because the specimen has undergone aging. This shows that the asphalt is getting harder. Same as the original condition, when RAP is added with Reclamite, the phase angle increases and Sbit decreases.

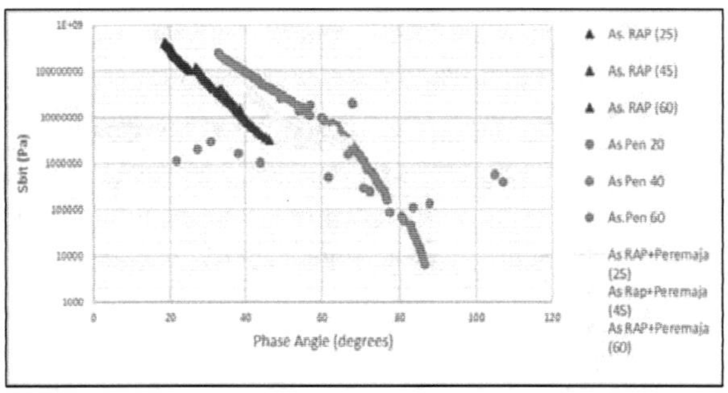

Figure 5. Black Diagram of the PAVT condition of RAP, Asphalt Pen 60/70 and RAP added with Reclamite.

On the black diagram in PAVT conditions, it can be seen that the phase angle value decreases with a range from 0 to 80. This was due to asphalt condition that continued to age. The same as the original and RTFOT conditions, when Reclamite is added to RAP, the phase angle increases and Sbit decreases.

4. Conclusion

Based on the results of the study, it can be concluded that:

1. The empirical characteristics of asphalt pen 60/70 met the requirements of the Bina Marga 2010 General Specifications of Roads and Bridges, Whereas, on RAP there was a decrease in penetration value and softening point, as well as a reduced asphalt elasticity. This shows that the RAP had undergone aging.
2. Rheological characteristics for asphalt pen 60/70, RAP and RAP added with Reclamite using DSR:

- Based on the performance grade value from the temperature sweep, it was found that the PG value of RAP before it was added with Reclamite had a PG value of 106. But after the addition of Reclamite, the PG value was the same as asphalt pen 60/70. This shows that rejuvenating material can reduce the critical temperature value on RAP.

- The master curve illustrates that asphalt pen 60/70 had a smaler asphalt stiffness modulus than RAP, but after Reclamite was added to RAP, the asphalt stiffness modulus was smaller than asphalt pen 60/70, so the addition of Reclamite can reduce the stiffness modulus of RAP.
- Black diagram, in the original condition, had a large phase angle (δ) value range, where at the same sbit, the asphalt pen 60/70 had a greater (δ) value than RAP. Meanwhile, for RAP added with Reclamite, the value was between the value of RAP and asphalt pen 60/70. This illustrates that Reclamite makes RAP more viscous. Meanwhile, the RTFOT and PAVT conditions had a smaller range of (δ) value. This is because in RTFOT and PAVT conditions, the specimen was conditioned to undergo aging, which made the asphalt more elastic than the original condition.

5. References

[1] AASHTO 2019 *Standard Method of Test for Determining the Rheological Properties of Asphalt Binder Using A Dynamic Shear Rheometer (DSR)* T315:10.
[2] Al-Qadi I L, Elseifi M, Carpenter S H 2007 *Reclaimed Asphalt Pavement- A Literature Review, Research Report Fhwa-Ict-07-001*, Illinois Center for Transportation.
[3] Asphalt Institute 2003 *Performance Graded Asphalt Binder Specification and Testing. Superpave Series* No. 1 (SP-1), 3rd Edition.
[4] Arshad H, Qiu Y J 2013 *Effect of Reclaimed Asphalt Pavement on the Properties of Asphalt Binders, The 2nd International Conference on Rehabilitation and Maintenance in Civil Engineering* Procedia Engineering **54** 840-50
[5] Chetana J, Amit P, Archana M R, Amarnath M S 2013 Determining The Rheological Properties Of Asphalt Binder Using Dynamic Shear Rheometer (DSR) For Selected Pevement Strectches. *Int. J. Res. Eng. Technol.*
[6] Direktorat J B M, Kementerian P U 2010 *General Specifications for Roads and Bridges*, Bina Marga.
[7] Kent R, Hansen P E, Audrey C, Ph.D., P.E. 2014 *Asphalt Pavement Industry Survey on Recycled Materials and Warm-Mix Asphalt Usage Information Series 138 (5th edition),* National Asphalt Pavement Association 5100 Forbes Blvd.
[8] Bethary R T, Subagio B S, Rahman H, Suaryana N 2019 Effect of recycled materials on marshall performance of hot asphalt mixture (HMA–RAP). *IOP Conf. Series: Mater. Sci. Eng.* **508**.

IOP Conf. Series: Materials Science and Engineering **739** (2020) 012007 doi:10.1088/1757-899X/739/1/012007

Innovative rotary swaged Al/Cu laminated wire conductors: characterisation of structure and residual stress

L Kunčická[1], R Kocich[1]

[1]VSB – Technical University of Ostrava, 17. listopadu 15, 70833 Ostrava-Poruba, Czech Republic

lenka.kuncicka@vsb.cz

Abstract. This study presents innovative, uniquely sequenced Al/Cu laminated composite wires having perspective usage for various applications, from transportation to electrotechnics. The composites were processed via the industrially applicable technology of rotary swaging, which was performed at 20 °C and 250 °C to provide the possibility to compare the effects of different swaging conditions on structure characteristics of the wires. The work-pieces were gradually swaged down to the final wire diameters of 5 mm, the total swaging degree for each swaged-piece was 3.6. After swaging, both the aluminium components of the swaged wires exhibited recrystallized ultra-fine-grained (UFG) structures with minor presence of residual stress, certain presence of stress was observed within the refined Cu structures. Microhardness measurements revealed occurring work hardening, especially for the 20 °C swaged composite.

1. Introduction

Research and development throughout virtually all the industrial fields has led to the emergence of various modern equipment, as well as to the introduction of a wide range of innovative materials. Heterogeneous laminates, gradient structures, and laminated and clad composites are modern materials consisting of two or more components bond at adjoining interfaces [1-4]. They are gaining increasing interest throughout various industrial and commercial branches provided by their ability to achieve enhanced properties when compared to the individual components (e.g. increased strength, better thermal and electric conductivity, enhanced corrosion resistance, decreased density, etc.). Among the most popular composite systems consisting of metallic components is the Al/Cu one [5]. Copper is used for its excellent heat and electric conductivity, while aluminium is favoured for its low density, while having reasonable strength and conductivity properties [6]. Beside other advantages, both the metals have a wide applicability, especially in the automotive and as electric conductors.

Clad composites are typically produced by welding, however, welding technologies are performed at high processing temperatures and can introduce local structure modifications and result in the formation of intermetallic phases at mutual interfaces [2, 7]. By this reason, plastic deformation methods, and especially the methods of severe plastic deformation (SPD), are advantageous for production of clad composites since they can be performed at lower temperatures and usually also at room temperature. Among the widely used and/or researched SPD methods are for example equal channel angular pressing (ECAP) [8, 9] and its modifications (TCAP [10], TCMAP [11], etc.), high pressure torsion (HPT) [12, 13], accumulative roll bonding (ARB) [14], and rotary swaging (RS) [15-17], which was used to manufacture the herein studied composites. RS is a versatile technology enabling to manufacture various axially symmetrical products via a repeated action of a set of rotating dies. The dies affect the work-piece surface with compressive radial forces by which they incrementally impose the combination of shear and compressive strains into the work-piece. The nature of the process enables to avoid large stress gradients and supports progressive grain refinement and improvement of mechanical and utility properties. The study focuses on preparation of 5 mm clad composite wires at two different swaging temperatures. The aim is to provide detailed description of the influence of RS on the development of structure and residual stress.

2. Experimental methods

The original materials the swaged composites from which were fabricated were electro-conductive CP Cu (0.015 % P, 0.002 % O, 0.002 % Zn, balance Cu) and electro-conductive CP Al (0.25 % Fe, 0.20 % Si, 0.05 % Cu, balance Al). The Al shell had the original diameter of 30 mm, while the Cu wires diameters were 3 mm each (figure 1). The composite was processed via rotary swaging (RS) at 20 °C – room temperature – and at 250 °C; the temperatures were selected based on our previous research described e.g. in studies [5, 18-20].

Figure 1. Material before swaging.

Both the composites were gradually swaged down from the original 30 mm to the final 5 mm. The total reduction ratio was calculated using equation (1), where S_0 and S_n are the cross-section areas at the input and output of the swaging dies, respectively. The reduction ratio for the 5 mm wires was 3.6.

$$\varphi = \ln\left(\frac{S_0}{S_n}\right) \quad (1)$$

The subsequent analyses focused on the influence of processing temperature on the development of substructure and residual stress. Thorough investigations of structures of the swaged composites were carried out via scanning electron microscopy (SEM). The performed examinations were done via electron backscattered diffraction (EBSD) analyses. The samples were ground on SiC papers to 2000 grit and subsequently polished electrolytically. EBSD analyses were performed on transversal sections using a JEOL FESEM 7000F equipment with an EDAX DigiView High Resolution Camera. SEM-EBSD scans were also used to provide an insight into the residual stress development. The scans were evaluated via the OXFORD Instruments [21] and ATEX [22] software. The grain size parameter was assessed as the average area of the individual grains (in μm^2). Residual stress was evaluated via analyses of internal misorientations in the scale from 0° (blue colour) to 15° (red colour).

3. Results and discussion

3.1. Structure characterization

Figures 2a to 2d depict the results of EBSD analyses of 5 mm composites after the final swaging pass to 5 mm via orientation image maps (OIM) taken to the direction of the swaging axis. The structures of both the Al and Cu components of both the swaged composites evidently exhibited restoration and significant grain refinement. However, the grains within the Cu component were larger than within the Al. This can be attributed to the intrinsic properties of Al (activation energy and stacking fault energy) [23]. Despite the fact that the imposed energy introduced recrystallization within both the Cu and Al components, the flow stress and high stacking fault energy of Al facilitate boundary migration and substructure development, which supported restoration and recrystallization within Al even during room temperature processing [24]. Also, the lower flow stress causes the Al to consume the imposed strain more easily than the Cu.

The analyses of average grain areas show that the grains within the Cu component were smaller for the 20C composite than for the 250C one (2.1 and 2.9 µm^2, respectively). Nevertheless, various shades of the colours depicting the ideal orientations within the individual grains point to substantial substructure development and formation of ultra-fine subgrains. This supposition was confirmed by the relatively high fractions of low angle grain boundaries (LAGBs) with misorientations lower than 15° pointing to the presence of subgrains [25] (46 % and 49 % for the 20C and 250C composites, respectively). The average grain areas for the Al component were 0.62 and 0.39 µm^2 for the 20C and 250C composites, respectively. As can be seen, the higher swaging temperature influenced positively the grain refinement within the Al component, but both the composites featured ultra-fine grained (UFG) Al structure. Also, due to the fact that the Al tended to be influenced by the imposed strain more intensively than Cu, recrystallization was more intense for this component and the level of substructure development, evaluated via the presence of subgrains characterized with LAGBs the volume fraction of which was 31 % and 27 % for the 20C and 250C composites, resp., was lower.

Similar conclusions were drawn in our previous study examining clad composites with the stacking sequence consisting of Cu sheath with Al wires [5, 7, 18]; the studied composites also exhibited substantial grain refinement during the last swaging pass and featured the formation of UFG structure. The Cu sheath exhibited substructure development with subgrains with the diameters of around 0.1 µm and more than 50 % of the individual grains within the 250C composite Al component had the diameters of about 0.5 µm. These findings prove that the technology of rotary swaging is suitable for production of Cu/Al clad composite wires from the viewpoint of their microstructure.

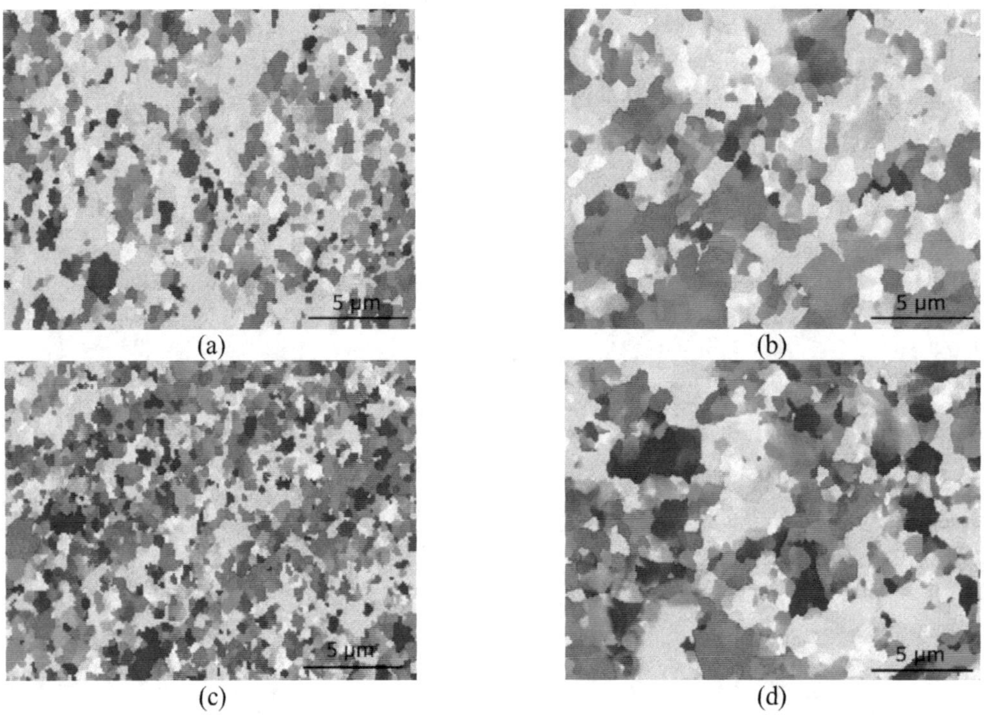

Figure 2. Orientation image maps for composites swaged at 20 °C – Al component (a), Cu component (b); swaged at 250 °C – Al component (c), Cu component (d).

3.2. Residual stress

Residual stress within a processed material can occur as a result of various causes, such as local heating during processing or inhomogeneous quenching, as well as due to non-uniform distribution of the imposed strain usually occurring when deformed under cold conditions [26]. For composites and multi-phase materials, residual stress usually develops as a result of the different properties of the individual component materials.

The analyses of residual stress were performed via evaluations of internal grains misorientations in the range from 0° to 15° (the rainbow scale from blue to red). Figures 3a to 3d show misorientations for the Al and Cu components of 20C and 250C composites, respectively. As can be seen, the fine grains within both the Al components featured only a minor presence of low misorientations, which corresponds to the above drawn conclusions about development of new stress-free grains. On the other hand, certain occurrence of internal misorientations pointing to the presence of residual stress was observed within both the Cu components; yellow and red regions within the grains point to the development of substructure and presence of low angle grain boundaries (LAGBs), the volume fractions of which were relatively high for both (Section 3.1). The findings thus correspond to the above drawn conclusion of substructure development within Cu for both the composites.

Figure 3. Maps of internal grains misorientations for composites swaged at 20 °C – Al component (a), Cu component (b); swaged at 250 °C – Al component (c), Cu component (d).

The differences in the occurrence of misorientations for the Al and Cu components can be attributed to the following factors. The above discussed intrinsic properties of Al provided this component with the ability to consume the imposed strain more readily, which resulted in a more intense recrystallization, lower grain size, and minor occurrence of residual stress when compared to Cu. Also, the lower stacking fault energy aggravated the conditions for recovery of Cu, which resulted in larger grain size and more substantial substructure development, i.e. the presence of internal grains misorientations. Mutual comparison of misorientations within the 20C and 250C composites did not reveal substantial effect of the swaging temperature on the occurrence of residual stress most probably due to the fact that the Al component exhibited recovery/recrystallization already during room temperature swaging and increase in the processing temperature only resulted in further grain refinement. For the Cu component, the temperature of 250 °C was most probably not sufficient to significantly influence the restoration processes within this component and only resulted in a slight grain growth (i.e. secondary recrystallization the necessary energy for which is significantly lower than the energy necessary for primary recrystallization ensuring grain refinement [23]).

3.3. Microhardness

The HV values for the Cu and Al original materials were 84.7 and 25.6, respectively. The results of microhardness measurements for the composites showed that the average microhardness value for the 20C composite Al component was 54.9 HV, while for the Cu component it was 107.4 HV. The values for the 250C composite components were 46.2 HV for Al and 103.8 HV for Cu. Whereas the Cu component HV values were comparable for both the composites, the Al component exhibited more substantial hardening within the 20C composite. This can be attributed to the deformation induced precipitation observed within the Al matrix of this composite; the Fe-rich precipitates within the matrix are depicted in the SEM-BSE scan in figure 4 (the figure also shows perfect bonding of both the components). The deformation induced precipitation was observed especially in the region of Al matrix adjacent to the periphery of the swaged composite due to the direct effect of the swaging dies.

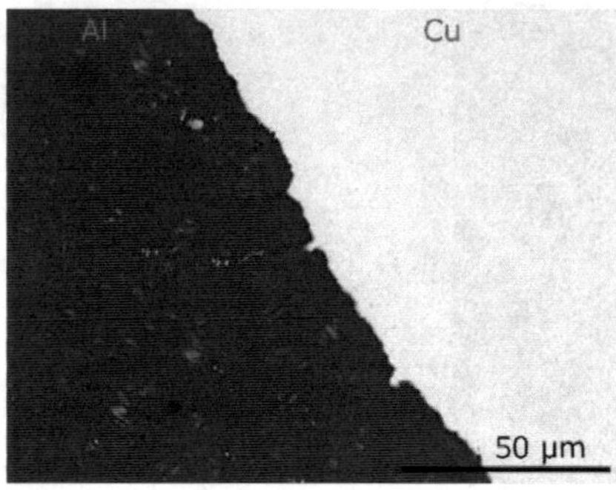

Figure 4. 20C composite Al-Cu interface 20 with evident precipitates in Al.

The comparison of the swaging temperatures shows that the higher temperature imparted lower HV values for both the components. This finding is in accordance with the results of measurements of deformation forces presented in our previous study [19]; the lower swaging temperature increased the flow stress and aggravated the plastic flow, which consequently imparted work hardening and resulted in higher HV values within both the components. The results of the herein presented HV measurements also correspond to the results of previously performed tensile and bending tests [19].

4. Conclusions

This study was focused on the deformation behaviour of Al/Cu clad composites produced using rotary swaging at two different swaging temperatures (20 °C and 250 °C). Structure analyses revealed that both the swaging temperatures resulted in the formation of UFG structures within the Al component and significant grain refinement within the Cu one. The processing temperature did not exhibit to have any substantial effect on the presence of residual stress, the occurrence of which was observed within the Cu component of both the composites. Room temperature processing resulted in higher work hardening of the Al primarily due to the deformation induced precipitation of Fe-rich particles. Suffice to say, the rotary swaging technology was successfully proven as practically applicable for fabrication of Al/Cu clad composites in the industrial scale from the viewpoint of microstructure and mechanical properties. Due to the prospective use of these composites as electro-conductive materials in the automotive, a study of electrical properties is being prepared for a subsequent publication.

Acknowledgments
The authors acknowledge the support of 19-15479S Project of Grant Agency of the Czech Republic.

References

[1] Fallah F, Taati E 2019 *Acta Mech.* **230** 2145-69.
[2] Kim I-K, Hong S I 2013 *Mater. Des.* **47** 590-98.
[3] Kunčická L, Kocich R 2018 *IOP Conf. Ser. Mater. Sci. Eng.* **369** 12029
[4] Ambekar S D, Tripathi V K 2019 *Int. J. Interact. Des. Manuf.* **13** 689-98.
[5] Kunčická L, Kocich R, Dvořák K, Macháčková A 2019 *Mater. Sci. Eng. A* **742** 743-50.
[6] Russell A, Lee K L 2005 *Structure-Property Relations in Nonferrous Metals* (John Wiley & Sons, Inc.)
[7] Kocich R, Kunčická L, Král P, Strunz P 2018 *Mater. Des.* **160** 828-35.
[8] Lukáč P, Kocich R, Greger M, Padalka O, Szaraz Z 2007 *Kov. Mater.* **45** 115-20.
[9] Suresh M, et al. 2019 *J. Alloys Compd.* **785** 972-83.
[10] Kocich R, Fiala J, Szurman I, Macháčková A, Mihola M 2011 *J. Mater. Sci.* **46** 7865-76.
[11] Kocich R, Kunčická L, Macháčková A 2014 *IOP Conf. Ser. Mater. Sci. Eng.* **63** 12006.
[12] Dobatkin S V, Gubicza J, Shangina D V, Bochvar N R, Tabachkova N Y 2015 *Mater. Lett.* **153** 5-9.
[13] Okeke U, Yilmazer H, Sato S, Boehlert C J 2019 *Mater. Sci. Eng. A* In press. doi:10.1016/J.MSEA.2019.05.102
[14] Kocich R, Macháčková A, Fojtík F 2012 *Int. J. Mech. Sci.* **64** 54-61.
[15] Kocich R, et al. 2016 *Mater. Des.* **90** 379-88.
[16] Yang Y, Nie J, Mao Q, Zhao Y 2019 *Results Phys.* **13** 102236.
[17] Gan W M, et al. 2014 *Mater. Des.* **63** 83-8.
[18] Kunčická L, Kocich R, Strunz P, Macháčková A 2018 *Mater. Lett.* **230** 88-91.
[19] Kocich R, Kunčická L, Macháčková A, Šofer M 2017 *Mater. Des.* **123** 137-46.
[20] Kocich R, Macháčková A, Kunčická L, Fojtík F 2015 *Mater. Des.* **71** 36-47.
[21] Oxford Instruments 2018 *Providing leading-edge tools for SEM, TEM & FIB – Nanoanalysis.* Available at: https://nano.oxinst.com/
[22] Beausir B, Fundenberger J J 2017 *Analysis Tools for Electron and X-ray diffraction, ATEX - software*, www.atex-software.eu
[23] Humphreys F J, Hetherly M 2004 *Recrystallization and Related Annealing Phenomena* (Elsevier Ltd.)
[24] Verlinden B, Driver J, Samajdar I, Doherty R D 2007 *Thermo-mechanical processing of metallic materials* (Elsevier)
[25] Eivani A R, Ahmadi S, Zhou J, Duszczyk J 2013 *Mater. Sci. Technol.* **29** 1297-1303.

IOP Conf. Series: Materials Science and Engineering **739** (2020) 012008 doi:10.1088/1757-899X/739/1/012008

Analysis the hydraulic fracturing in the unconventional shale reservoir based on Finite Discrete Element Method

Zhenqian Xue[1], Xin Zhao[1]

[1]China University of Petroleum (Beijing), Beijing, China

[2]DeepListen Pty Ltd, Brisbane, Australia

zhenqianxue@163.com

Abstract. Hydraulic fracturing of horizontal well is the key technology to develop unconventional resources. Simultaneous fracturing in horizontal well is the prevalent method applied in the field practice. To achieve successful and desired stimulated rock volumes and fracture networks, it is necessary to understand the influence of stress on fracture geometry. This paper proposed a 2D model to simulate the fracture propagation in simultaneous hydraulic fracture operation based on Finite Discrete Element Method (FDEM). In this model, the fracturing fluid leak-off and natural fractures are coupled. The simulation results demonstrated that the induced stress field can not only affect the fracture extended length, but also the width of fracture. The outer fractures dominate the inner fractures in growth. And the central fractures stop propagating after they reached a certain length due to the induced stress field. These simulations are meaningful for stimulation design and required spacing conditions to acquire the desired fracture lengths, proppant placement, and production rates.

1. Introduction

Hydraulic fracturing is a commonly used technique in shale gas development to enhance reservoir permeability and well efficiency [1]. Many fracturing models have been presented during the last several decades. Hydraulic fracturing is a complicated and difficult process to model because it involves the coupling of at least three processes [2, 3]: (1) mechanical deformation induced by fluid pressure on the fracture surface; (2) the fracture fluid flows within the fracture; (3) the fracture propagation under the fluid pressure. To meet the requirement of complicated hydraulic fracturing processes with branches, some multi-fractures propagation models were developed [4].

In order to determine the optimum spacing and optimum staging between fractures, a 2D model to simulate the fracture propagation in simultaneous hydraulic fracture operation based on Finite Discrete Element Method [5]. In this method, fluid flow is assumed to occur through a network generated from the same triangular mesh used for the mechanical calculations, and the flow channels are created at the interfaces between adjoining triangular finite elements [6]. The initiation and propagation of fractures in this model described by non-linear fracture mechanics. According to this theory, as the tensile strength of a material is exceeded at the tip of an opening fracture, a zone characterized by non-linear behavior, called the fracture process zone.

2. Theory and Method

FDEM is a numerical method that combines continuum mechanics principles with discrete element method to simulate interaction, deformation, and fracturing. FDEM has the ability to explicitly capture the entire deformation and fracturing process. It is a hybrid technique that combines the advantages of the FEM and DEM approaches. While the medium is undergoing elastic deformation, the behavior of intact material is explicitly modelled by FEM. As the strength of the material is exceeded fractures are initiated, giving rise to discontinuous blocks where the interaction between these blocks is captured by DEM. The FDEM approach is capable of tracking fracture initiation and propagation by applying the principles of non–linear elastic fracture mechanics. In this study, three interconnecting modules were used to simulate fracturing, namely (1) a mechanical solver which calculates the deformation of the intact rock mass as well as the initiation, propagation and interaction of fractures; (2) a cavity volume calculator then captures the changes of the cavity volume due to fracture propagation, the elastic deformation, and fluid compressibility. It also tracks the newly created wet boundaries by checking their connection with the initial source of fluid; (3) a pump model then interacts with the previous two modules to calculate fluid pressure while considering the pumping conditions.

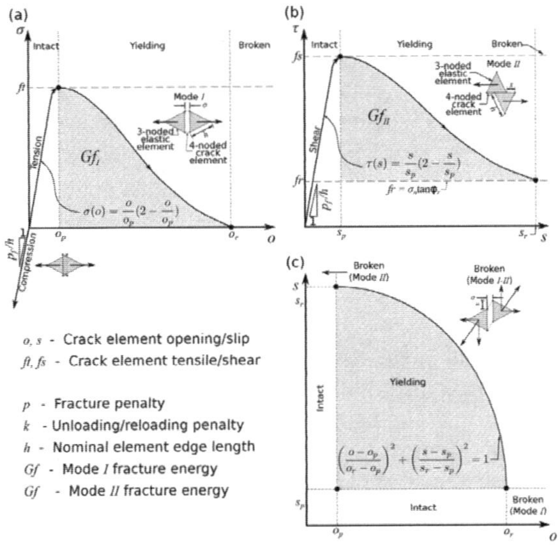

Figure 1. Fracture model implemented in FDEM: (a) cohesive model for Mode I, (b) slip weakening model for Mode II, and (c) criterion for Mixed-Mode fracture initiation.

In the FDEM, the load and element mass are treated as nodes to build the dynamic equation of a system according to Newton's second law.

$$M\ddot{x}+C\ddot{x} = F(x) \quad (1)$$

Where M and C are the mass matrix and damping matrix of all nodes, respectively, and F(x) represents the total nodal force vector which includes the node force vector $F_c(x)$ assigned by the contact force between tetrahedral elements, the node force vector F_d induced by the deformation of tetrahedral elements, the node force vector $F_j(x)$ caused by the deformation of joint elements, and the node force vector F_e (x) assigned by the external load.

The introduction of the damping matrix C consumes the kinetic energy of the system; by which we use dynamic relaxation to solve the static problem. The damping matrix is given by

$$C = \mu I \quad (2)$$

where μ is the damping coefficient, and I is a unit matrix. For a mass spring system with single degree of freedom, the critical damping coefficient is given by

$$\mu = 2h\sqrt{\rho E} \tag{3}$$

Where h is the edge length of an element, ρ is the element density, and E is the elastic modulus of the element. The system kinetic energy will be consumed at the fastest rate if the critical damping is used.

3. Results and Discussion

3.1. Model Description

A multistage fracture propagation model based on FDEM was established. The model was 180 m in the X direction (along the wellbore and parallel to the minimum horizontal principal stress), 180 m in the Y direction (perpendicular to the minimum horizontal principal stress). The target formation is a sand-rich formation, so the model was treated as pure sand, and no interlayer was built in the model. The target formation is a shale-rich formation, so the model was treated as pure shale, and no interlayer was built in the model. The input parameters in the model are listed in table 1.

Table 1. Input parameters for the fracture propagation model.

Parameter	Unit	Value
Young's modulus	$\times 10^3$ MPa	40.49
Poisson's ratio	Dimensionless	0.20
Porosity	%	2.7
Permeability	10^{-3} μm^2	0.014
Initial pore pressure	MPa	61
Vertical stress	MPa	75.55
Minimum Horizontal Principle stress	MPa	86.14
Maximum Horizontal Principle stress	MPa	95.14

3.2. Results Analysis

In this simulation, simultaneous propagation of four hydraulic fractures was considered. From the results, outer fractures tend to longer than the rest of the fractures. This is because the stress shadow on the outer fractures will be less than the fractures inside. With the propagation of outer fractures, the induced stress field would restrict the extension of inner fractures. This effect is seen when the outer fractures start to propagate more rapidly than the remaining fractures in the fracture network. The larger outer fractures exert more stress shadow on the fracture near to them, which will inhibit the growth of these fractures more than the center fracture. As the outer fractures grows large enough the stress shadow over the remaining fractures increase and completely suppress their growth after sometime. In addition, the fracture width of outer fractures is larger than that of inner fractures. The main reason account for this phenomenon was the diversion of the minimum horizontal principle stress. From the figure 2(a), when the time was 70 s, the four fractures began to initiate and propagate, and the inner two fractures did not show an obvious damage. There existed an obvious diversion of the minimum horizontal principle stress in outer fractures, while the minimum horizontal principle stress of inner fractures did not show an obvious change. When the time was 90 s (figure 2(b)), the four fractures had a clear damage. It should be noticed that the diversion of the minimum horizontal principle stress in outer fractures was much severer than that in inner fractures. What' more, the natural fractures in the model began to initiate and propagate. When the hydraulic fracture interacted with the natural fractures, the trajectory of the hydraulic fractures began to change. The main reason was that the strength of natural fracture surface was lower than that of rock matrix. When the time was 110 s and 130 s, the damage degree of inner fracture was more serious than that the time was 90 s. There was an apparent inconsistency in the propagation of fractures. The length and width of outer fractures were larger than that of inner fractures. The main reason accounted for this phenomenon was that with the propagation of fractures, the diversion of minimum horizontal principle stress was more severe. Affected by the diversion of minimum horizontal principle stress, the extension pressure of the inner fractures was much higher than the outer fractures. Under the same injection ratio and injection pressure, the net pressure in inner fractures were lower than the outer fractures, which caused the shorter fractures. In addition, the aperture of the inner fractures was also a function of net pressure. Lower net pressure contributes to a narrow fracture in the model. So an obvious “enwrapped” phenomenon occurred in this model.

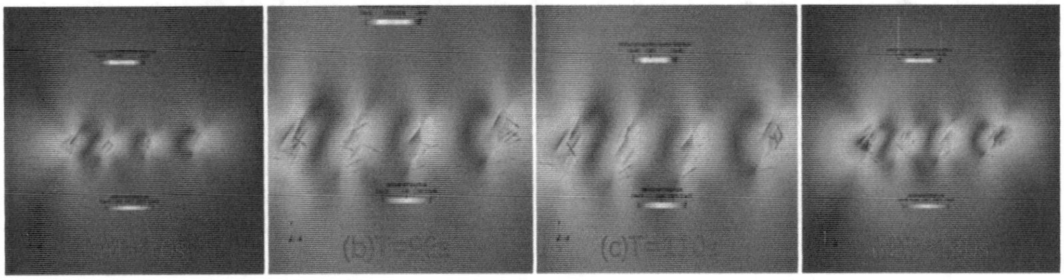

Figure 2. Fracture geometry of four fractures.

In the following discussion, fracture 1, 2, 3 and 4 represent the fractures which propagate from the left to right. As can be seen, the maximum width of fracture 1 is 0.0051m, and the average width is 0.003 m. The maximum width of fracture 2 is 0.0141m, and the average width is 0.007 m. When the width of fracture 1 decreases dramatically at x = 50 m, the width of fracture 2 and 3 rises very rapidly, and then decreased to zero when the fracture length exceeded to 80 m. Combined with figure 2, the tighter the cluster space was, the more severe the minimum horizontal principal stress diversion was. In the model, affected by fracture 1 and fracture 4, so the length of fracture 2 and fracture 3 was shorter than fracture 1 and fracture 4.

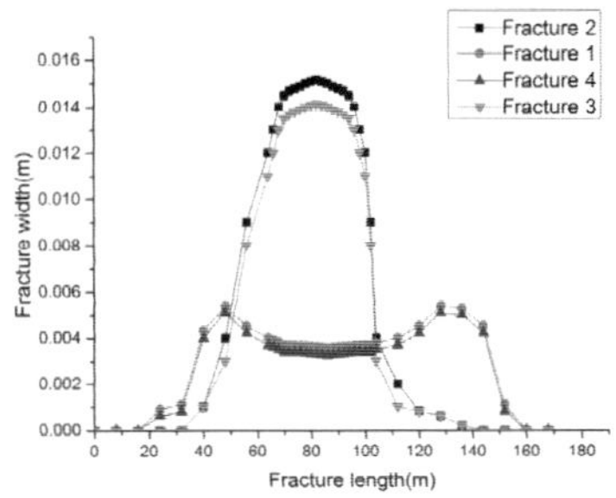

Figure 3. Fracture length versus fracture width of the four fractures.

4. Conclusion

This research was based on the 2D seepage-stress-damage coupled multi-fracture simultaneous propagation model to study the fracture propagation based on finite discrete element method with DFN. In this model, the fracture geometry, stress field redistribution was analyzed. From the simulation, the multi fracture simultaneous propagation model demonstrated that tight cluster space will insufficiently extend the middle fracture in height and length. Because of the diversion of the maximum horizontal principal stress, the induced stress field will increase the extending resistance of the middle fracture. The tighter the cluster space is, the higher the induced stress. The induced stress field can reduce the fracture propagation velocity and can even make the extension stop. Additionally, the induced stress field can also contribute to a complicated fracture network.

Reference

[1] Zhao X, et al. 2016 *Numerical study on fracture initiation and propagation of reservoirs subjected to hydraulic perforating*, **33** 544-50.
[2] Guo J, et al. 2015 Perforating cluster space optimization method of horizontal well multi-stage

fracturing in extremely thick unconventional gas reservoir. *J. Natural Gas Sci. Eng.* **26** 1648-62.
[3] Guo D l, et al. 2001 3-D Fracture Propagation Simulation and Production Prediction in Coalbed. *Appl. Mathematics Mechanic.* **22(4)** 385-93.
[4] Wang X, et al. 2018 Extended finite element simulation of fracture network propagation in formation containing frictional and cemented natural fractures. *J. Natural Gas Sci. Eng.* **50** 309-24.
[5] Munjiza A, D.R.J.O., Bicanic N 1995 A combined finite -discrete element method in transient dynamics of fracturing solids. *Eng. Computations,* **12(2)** 29.
[6] Shen B, Shi J, Barton N 2018 An approximate nonlinear modified Mohr-Coulomb shear strength criterion with critical state for intact rocks. *J. Rock Mechanic. Geotechnic. Eng.* **10(4)** 645-52.

IOP Conf. Series: Materials Science and Engineering **739** (2020) 012009 doi:10.1088/1757-899X/739/1/012009

Damage State Assessment of a Pre-Stress Concrete Girder (PSCG) Bridge using Non Linear Analysis for Real-time Structural Health Monitoring

K M Montes[1], K A Embalzado[1], N M Sales[2], A J Sayno[2], J M Payawal[3], F A A Uy[4]

[1]Science Research Specialist II, PhilSIMS, Mapua University, Intramuros, Manila, Philippines

[2]Science Research Specialist I, PhilSIMS, Mapua University, Intramuros, Manila, Philippines

[3]Senior Science Research Specialist II, DEIMS, Mapua University, Intramuros, Manila, Philippines

[4]Dean, School of Civil Engineering, Geosciences, Environmental Engineering, Mapua University, Intramuros, Manila, Philippines

katrinamaesmontes@yahoo.com

Abstract. Philippines is located in the pacific ring of fire, which indicates that seismic movement is frequently happening. West valley fault line located in the central Luzon might move due to its 400 to 500-year recurrence, and the last recorded movement was on 1958. According to MMIERS, the movement of west valley fault with 7.2 magnitude earthquake can cause one of the biggest devastation in Metro Manila. In preparation for that, this study aims to assess the vulnerability of a pre-stress bridge against seismic movement by the aid of USHER technology. The USHER system includes structural assessment, remote sensing, and real time monitoring through the portal. This study will focus on the structural assessment of the bridge, which identify the acceleration limits that the bridge can endure to interpret different level of damage. As a result, by 10% of probability of exceedance from SEAOC using the moderate damage fragility curve, the results in three directions are 0.37641g in longitudinal axis, 0.366025g in transverse axis, and 0.394498g in the vertical direction. The most critical is in the transverse direction, which can produce a maximum of 67mm displacement of the bridge in the inelastic stage. In validation through the sensor, which has a reading from the smart bridge project, the maximum acceleration recorded was 0.138g. This concludes that the bridge is safe and no retrofitting works is need.

1. Introduction

On the 22nd day of April 2019, Castillejos, Zambales was hit by a 6.1 magnitude tectonic quake that reported intensities as high as Intensity VII, described by the PHIVOLCS Earthquake Intensity Scale as destructive earthquake causing liquefaction, landslides and slight damage on well-built structure.

IOP Conf. Series: Materials Science and Engineering **739** (2020) 012009 doi:10.1088/1757-899X/739/1/012009

This kind of seismic activity has been evident within the Philippine archipelago ever since, being located within the Pacific Ring of Fire wherein collisions of tectonic plates occur. Historical records show that the Philippine archipelago experiences an average of 20 earthquakes per day at around 100 to 150 earthquakes annually, contributing to the fact that the country is the third most exposed to natural hazards among 172 countries. Earthquake hazard is a major threat to the economic development of the Philippines. The 7.2M quake in Bohol last October 2013 resulted into 2.2 billion pesos' total cost of damages, almost 67, 000 houses, 41 bridges and 18 roads damaged, and 3.2 million of people affected. [5,6]

This example of an earthquake event shows that a short period disaster can cause a huge loss in the country's economic assets and can be detrimental to thousands of lives. Earthquake hazards cannot cause as much damages only if a system or community is not exposed and vulnerable as much to these hazards. So one way to mitigate this hazard is ensuring the integrity of infrastructure in an area exposed to earthquake hazard. Among the standard practices in evaluating, the integrity of existing infrastructure is through rapid visual survey. An emerging platform is the development of structural health monitoring technologies that help in managing disaster risk. Introduction of accelerometer for structural health monitoring is now emerging, hand in hand with rapid visual inspection. [1,4,7]

For an advance monitoring process, this paper uses a sensor and a web-based portal that records and presents structural health data of selected infrastructure. The sensor utilized by the system developed through the smart bridge project and improved now by the USHER project, which enables the sensor to record three directional movements (x, y, and z). This helps to determine the different damage state of the bridge and to give an idea on the action needed to perform. This study will not cover cost analysis and foundation investigation.

2. Methodology

The bridge was a type IV AASHTO pre-stress girder. All of the piers has one column except from the middle pier, which consist of two columns. The model was illustrated using CSI Bridge software. The earthquake data came from the Kobe 1995 earthquake magnitude 6.5, Tohoku 2011 earthquake magnitude 9, and local earthquakes in the Philippines.

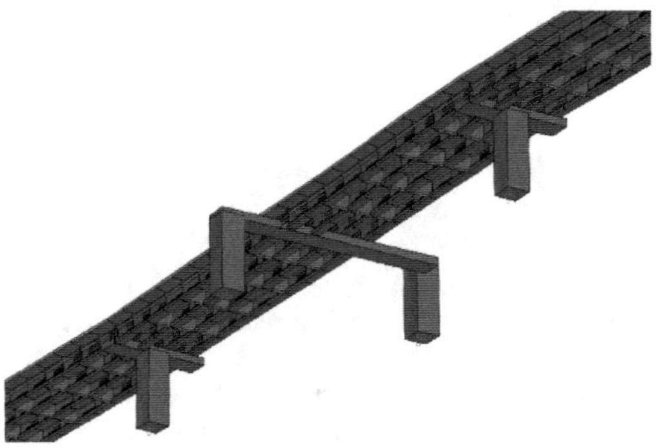

Figure 1. Bridge Model.

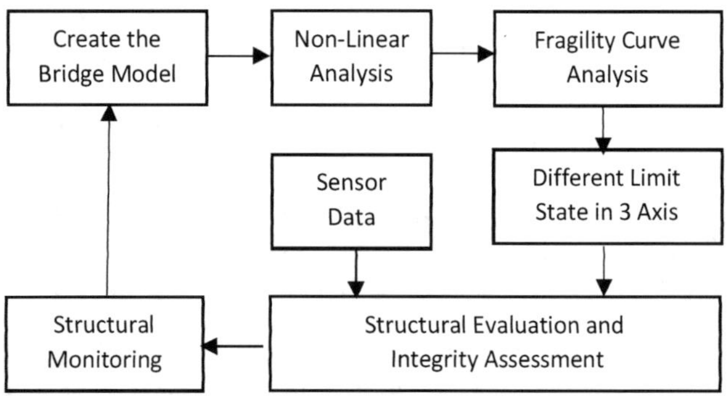

Figure 2. Conceptual Framework.

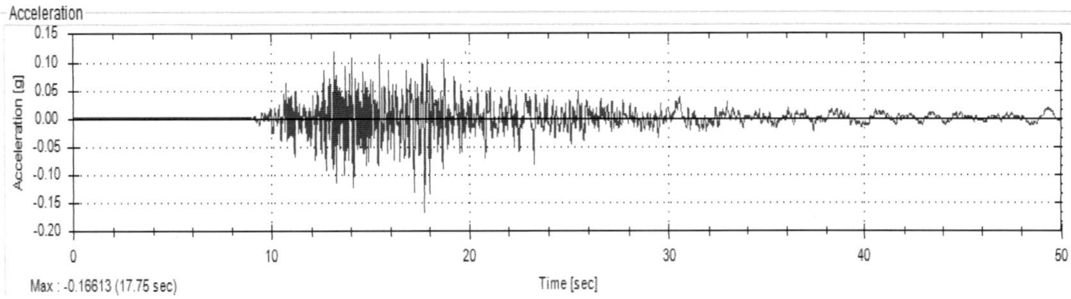

Figure 3. KOBE 1995 Earthquake Raw Values.

Push over analysis and capacity spectrum method was analyzed to get the upper and lower limit of ground acceleration to consider in levelling the damage state. These damage states were used in identifying the probability of exceedance and its corresponding peak ground acceleration through fragility curve analysis in this equation [3] :

$$\mathbf{F(a)} = \phi \left[\frac{ln\left(\frac{a}{c}\right)}{\zeta} \right] \quad (1)$$

Wherein "a" is the peak ground acceleration and ϕ [...] is the standardized normal logarithmic distribution function. Shinozuka (2001).

3. Results and Discussion

The maximum displacement in the inelastic stage of the bridge is 67.67mm which occurs at 11315.273 KN shear force, while 140.38 mm dispacement at 13415.018 KN shear force is recorded maximum in the elastic stage. This values was used in the classification of the different damage states. The maximum displacement.

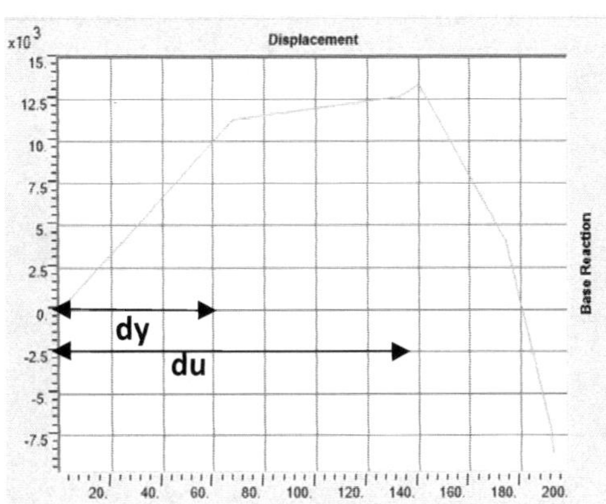

Figure 4. Push Over Curve at Longitudinal Direction.

Table 1. Push Over Curve Data.

Displacement (mm)	Base Force (KN)
67.673895	11315.273
133.033962	12708.512
140.385777	13415.018
174.155716	4057.423
191.066784	-6846.757
192.456781	-8550.317
192.462662	-8557.053
192.692406	-8569.286
192.698288	-8575.76
192.732671	-8577.613

A set of mathematical model can be use to categorized the data from the results of capacity spectrum method and push over analysis to produce the limits of different damage states.[2]

IOP Conf. Series: Materials Science and Engineering **739** (2020) 012009 doi:10.1088/1757-899X/739/1/012009

Table 2. Limits for Different Damage States.

Damage State		Lower Limit	Upper Limit
DS0	None	< 0.0473717265	0.047371727
DS1	Minor	0.047371727	0.091911189
DS2	Moderate	0.091911189	0.116148483
DS3	Major	0.116148483	0.140385777
DS4	Collapse	0.140385777	> 0.140385777

The graphs shown below are the fragility curves of different damage states at three directions produced using the data from the non linear static and dynamic analysis.

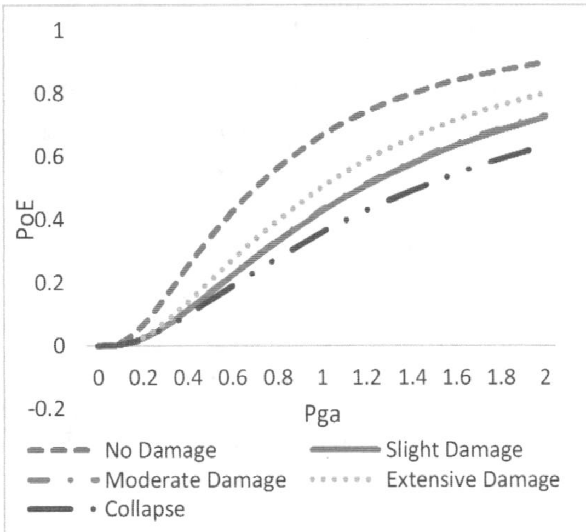

Figure 5. X- Axis Seismic Fragility Curve.

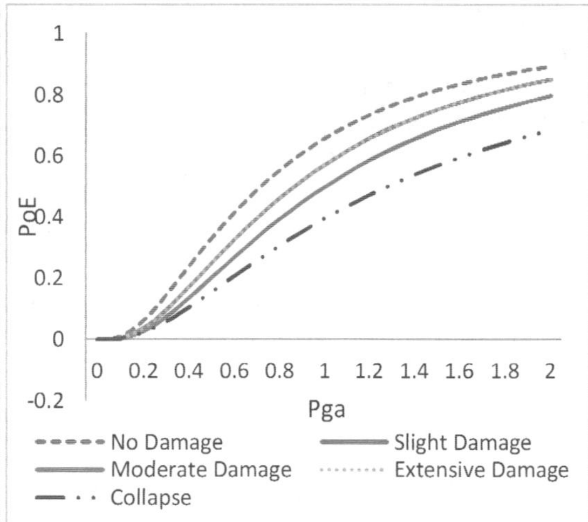

Figure 6. Y- Axis Seismic Fragility Curve.

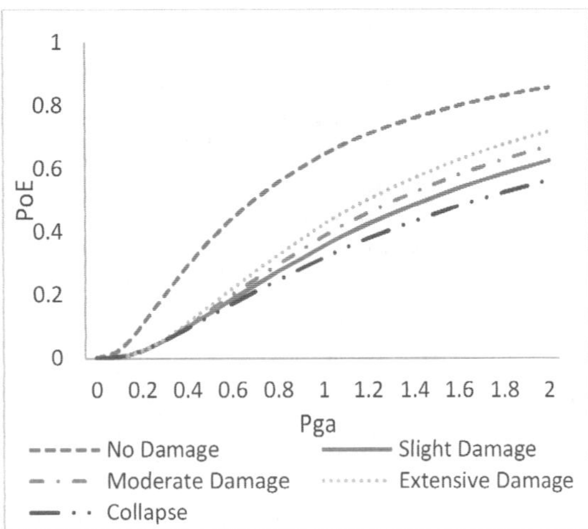

Figure 7. Z- Axis Seismic Fragility Curves.

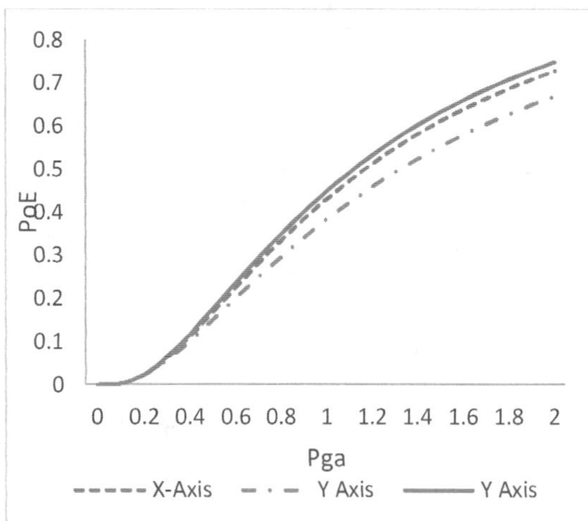

Figure 8. Moderate Damage in Three Direction.

According to the study made by SEAOC, 10% of the probability of exceedance serves as the limit state that will indicate a need for retrofitting and further structural assessment located in the moderate damage plot using the fragility curve analysis.

Table 3. Threshold in Three Direction.

	X	Y	Z
Threshold (g)	0.374641	0.366025	0.394998

4. Conclusion

The threshold obtained are 0.37641g in longitudinal axis, 0.366025g in transverse axis, and 0.394498g in the vertical direction. The weakest axis is located on the transverse direction and it indicates shear failure on the connection of bridge deck and pier. Based from the sensor reading connected on the bridge, the maximum acceleration data recorded was 0.138 g which justify that the bridge is safe and no retroffiting works to be done. The researchers recommend to investigate further the foundation and incorporate soil investigation. Analyze the foundation and how it will affect the acceleration limits.

Acknowledgements

The authors acknowledge Department of Public Works and Highways (DPWH), Department of Science Technology (DOST) PCIEERD and Mapua University for the research funding. They also acknowledge the "SMART BRIDGE" project in which the acceleration data was obtained. The authors would like to acknowledge Engr. M Baylon for his expertise in sharing his knowledge about non linear analysis and fragility curves which serves as a tool in the analysis of this paper.

References

[1] Gaviña J R, Uy F A, Carreon J P D 2017 Wireless Smart Sensor Network System Using SmartBridge Sensor Nodes for Structural Health Monitoring of Existing Concrete Bridges. *IOP Conf. Series: Mater. Sci. Eng.* **216**.

[2] Moschonas I F, Kappos A J, Panetsos P, Papadopoulos V, Makarios T, Thanopoulos P 2008 Seismic fragility curves for greek bridges: methodology and case studies. *Bull Earthquake Eng.* **7** 439-68.

[3] Shinozuka M, et al. 2001 *Statistical Analysis of Fragility Curves*. University of Southern California.

[4] 2003 the Project for Study on Improvement of Bridges through Disaster Mitigating Measures for Large Scale Earthquakes in the Republic of the Philippines.

[5] 2015 Department of Public Works and Highways. Guidelines and Implementing Rules on Earthquake Recording Instrumentation for Buildings, Manila.

[6] 2014 Disaster Risk Reduction and Management in the Philippines: Enhancing Poverty Alleviation through Disaster Reduction. The World Bank East Asia and Pacific Region Rural Development.

[7] Payawal J M G, et al. 2017 *Data Calibration of the Actual versus the Theoretical Micro Electro Mechanical Systems (MEMS) Based Accelerometer Reading through Remote Monitoring of Padre Jacinto Zamora Flyover,* IEEE Conference on Technologies for Sustainability.

Concept Design of a Single-Aisle Aircraft Spoiler by Topology Optimization

P Purith[1], C Kittikorn[2], J Pakpong[1], L Mezeix[1]

[1]Faculty of Engineering, Burapha University, 169 Long-Hard Bangsaen Road, Saensook, Muang, Chonburi 20131, Thailand

[2]Geo-Informatics and Space Technology Development Agency, Space Krenovation Park (SKP) 88 Moo 9, Thungsukhla, Sriracha, Chonburi 20230, Thailand

laurentm@eng.buu.ac.th

Abstract. Sandwich are widely used in the aerospace as flaps, spoiler, cabin floor or rotor where high flexural rigidity is required. Sandwich structures consist in a thick core material as honeycomb bonded with thin skins as carbon fibres. When the core material contains closed cells, water ingression must be considered. This water ingression may significantly increase the weight of the core and can induce damage. In this project, new spoiler design is proposed based on reinforced stiffeners in order to avoid the use of honeycomb. In a first step model is built and validated on current spoiler. Then, topology tool is used to locate the stiffeners. Finally, GFEM is performed to propose a first rough plies optimization under static analysis.

1. Introduction

Current aircraft spoiler is usually sandwich panels with Nomex honeycomb as core and carbon/epoxy as skins. Even if sandwich structures present a high ratio stiffness overweight, water ingress into honeycombs cells is one of the main concerns [1]. Redux film bonding between core and skins can be damaged and sandwich structure could completely fail. Therefore, water inside the honeycomb cells needs to be detected by Non-Destructive Tests (NDT). Different NDT methods are available for composited. For sandwich structures tap test is one of the most used even if it has several drawbacks [2]. Indeed, test quality depends of the inspector's skills and workplace noise can interfere it. In order to simplify the spoiler maintenance and control, honeycomb core could be avoided and replaced by composite stiffeners.

Design methodology can be summarized in 4 steps (figure 1). First step is the construction of the model and parameters validation, i.e. mesh, boundaries condition… Then, concept design can be proposed by using topology optimization [3].

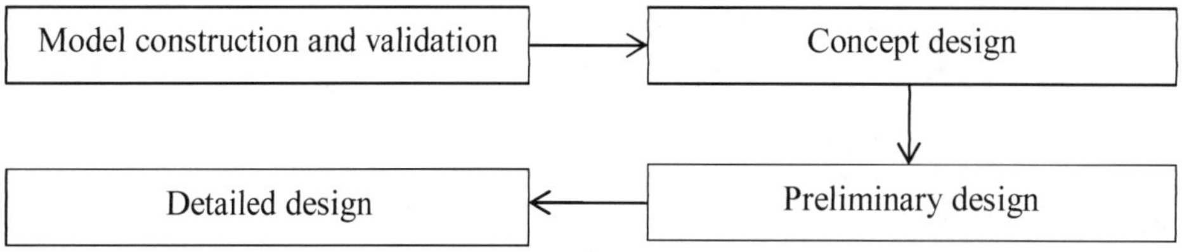

Figure 1. Design process.

In order to reach the concept design, topology density-based approach is a powerful optimization tool [4]. General definition of the topology optimization is to find an optimal distribution of materials within a specify design domain. This approach is generally formulated and solved by considering the material distribution approach. In the design a density variable (ρ) is assigned to all elements. Density variable is defined as a value that ranges from 0 to 1 during the computing. Most of the topology optimization problems based on maximum stiffness formulation or minimum global compliance. The topology optimization density-based approach optimization definitions can be written as:

$$\begin{cases} \min_{x} \quad f(\rho) \\ subject\ to \quad \begin{cases} 0 \leq \rho \leq 1 \\ state\ function\ constraint \\ Manufacturing\ constriants \end{cases} \end{cases}$$

Figure 2. Topology definition.

From the basic geometry given in the concept design step, preliminary design or Generalize Finite Element Method (GFEM) is performed. The aim is to obtain a first rough optimization by reducing the thickness of macro plies [0_n, 45_n, -45_n, 90_n]. Reaction Force (RF) between components of the structures can be also extracted. Finally, detailed design can be performed where ribs flanges, fillets... are designed. In this step ply by ply approach is used allowing lay-up optimization.

In the present paper, concept and preliminary design of spoiler will be proposed by replacing honeycomb core material by composite stiffeners. In a first step case study will be detailed. Then, model will be built and validated on current spoiler. Topology optimization will be used to determine the suitable stiffeners location and concept design will be proposed. Finally, GFEM will be performed in order to propose a first rough optimization.

2. Numerical simulation

2.1. Case study

General dimension of studied spoiler is 1.55 x 0.6 x 0.06 m^3. Spoiler is mainly used during landing and therefore drag force is the main applied load. Design load, F, is applied perpendicularly to the upper surface of the spoiler and it is given by:

$$F = \frac{1}{2} \cdot Sf \cdot Lf(\rho \cdot v^2 \cdot A \cdot C_d) \quad (1)$$

Where Cd is Drag coefficient (1.3), A the spoiler area (1 m^2), v is the landing velocity of single aisle aircraft (86.64 m/s) and ρ density of air (1.225 kg/m^3). Load and safety factor are respectively 3 and 1.5 given finally a load, F, equal to 30,790 N

2.2. Model construction and validation

Numerical simulation has been firstly validated on current type of spoiler (figure 3a). Model consists in 3D elements for the Nomex honeycomb core (48 kg/m^3) and 2D shell for unidirectional carbon (T700)/epoxy upper and lower skins. Due to the symmetry only half of the spoiler has been simulated to reduce computing time. Metallic fitting allowing spoiler rotation has been considered as rigid and has been modelled by fix boundary condition (figure 3b). Finally, load was applied on upper surface of spoiler from equation (1).

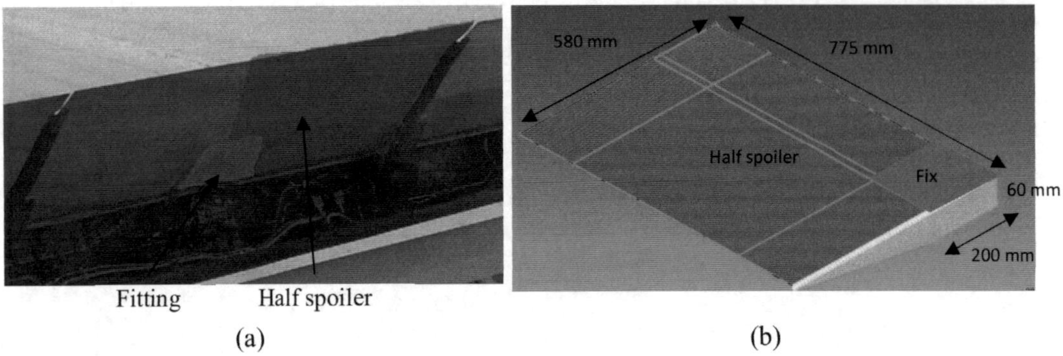

Figure 3. Half spoiler with rotation features (3a) Dimension of the spoiler (3b).

As this spoiler has been already designed and optimized to reduce the mass under aerospace design rules, the maximal compressive strain should be closed to the damage tolerance allowable, i.e. -4,000 μdef in this case study. Numerical simulation results show that the maximal compressive strain is -3,700 μdef that is closed to the allowable. Therefore, the numerical simulation model has been validated and can be used for topology optimization.

2.3. Concept design

2.3.1. Topology parameters

The objective is to replace the honeycomb core by composite stiffeners. To locate the suitable stiffeners position, topology tool was used. The current spoiler model has been used for the optimization. As upper and lower skins cannot be changed, it was considered as non-design area when the core volume have to be considered as design area. Minimizing the weight compliance has been inputted as topology objective with Volume Fraction, Vf, as constraint. Four Vf has been studied in order to identify the location of the stiffeners. Finally, extrusion function was used as manufacturing constrain to obtain only the stiffeners web.

2.3.2. Topology results

The 4 different studied volume fractions are shown in the figure 4. By analysing the evolution of the material density in function of the volume fraction, stiffeners location can be obtained. 3 main stiffeners (A-C) starts from the fix boundary condition. Moreover, a secondary stiffener is necessary (D). Finally, reinforcement is required perpendicularly to the symmetry plane (E). From conceptual step, a design can be proposed by using the stiffeners location (A-E). Another stiffener has been used to close the spoiler (figure. 5).

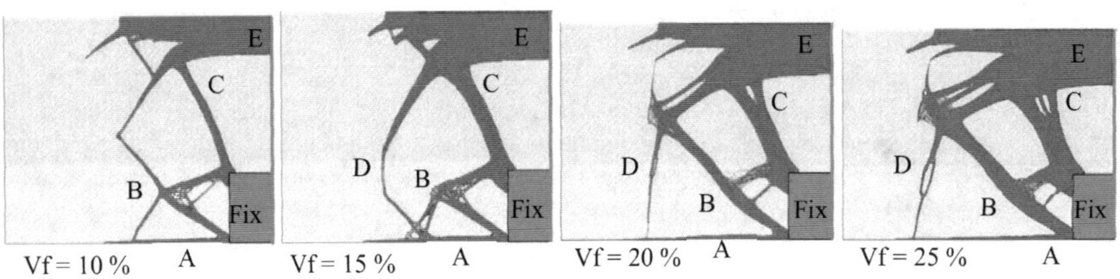

Figure 4. Topology optimization results in function of the volume fraction, Vf (Elements with density value one only).

IOP Conf. Series: Materials Science and Engineering **739** (2020) 012010 doi:10.1088/1757-899X/739/1/012010

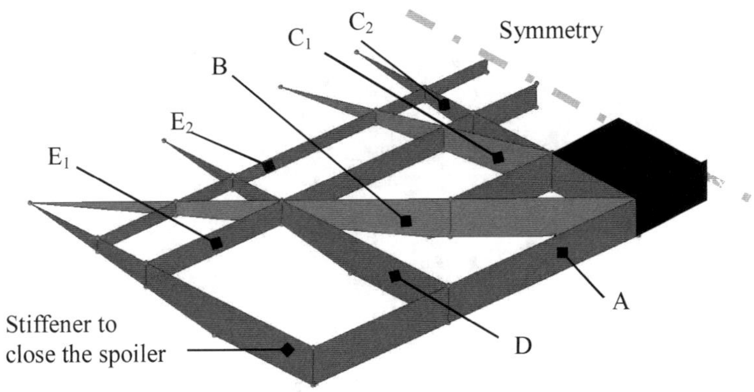

Figure 5. Conceptual design of the stiffeners by using topology tools.

3. GFEM

The aim of GFEM is to propose a preliminary design. As it is a new design, latest composite reference has been used. Unidirectional carbon/epoxy IMA/M21 was applied to the geometry obtained from the topology (figure 5). Same skin lay-up has been used as for the current spoiler. In a first step, optimization of macro plies [0_n, 45_n, -45_n, 90_n] has been performed in order to reduce the mass. First results show folding of the upper skin due to the load (figure 6a). Therefore, additional stiffeners (F) has been designed (figure 6b). Moreover, stiffeners C_1 and E_2 do not present any utility and has been therefore removed.

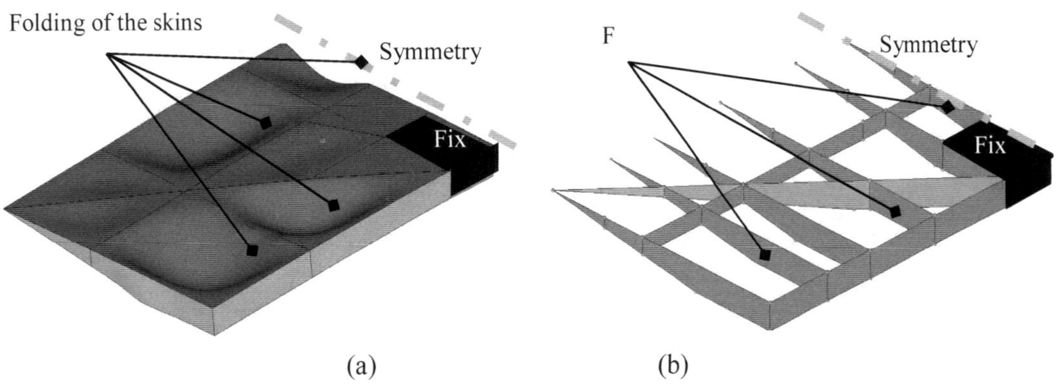

Figure 6. Folding of the skins (6a) Additional stiffeners (6b)

GFEM from optimized geometry (figure 6b) has been simulated under static analaysis. Skins lay-up are always the same as before. Different number of macro plies of stiffeners, n, has been investigated and strains have been reported (figure 7). As observed from n=10 to 2, maximum and minimum strain do not present significate evolution and are lower than allowable. Critical strain value is located on the upper skin/ stiffener C_2 edge due to the GFEM basic geometry (figure 5).

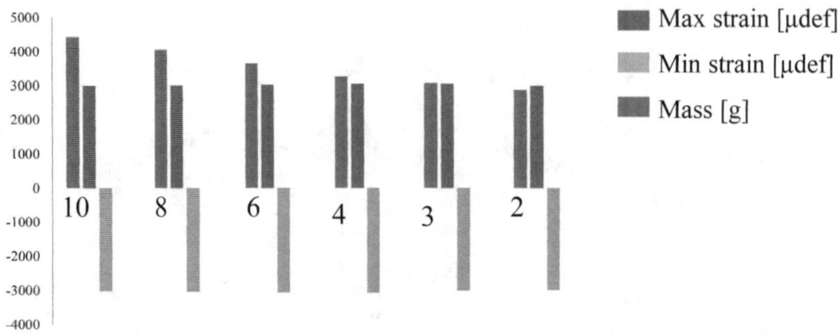

Figure 7. Summarize of the strain value each macro plies thickness.

4. Conclusion

Single aisle spoiler was studied in this paper with objective to replace Nomex honeycomb by composite stiffeners. The conclusion can be summarized as follow:

- Numerical simulation has been successfully validated on current spoiler where the minimum compressive strain is closed to the allowable.
- Stiffeners location has been determined thanks to topology optimization applied to the core material and by investigated different volume fraction.
- From topology a first CAD has been proposed
- GFEM has been performed on the CAD to reduce plies number under static analysis.

Works are in progress to extract Reaction Force between stiffeners and skins in order to design the stiffeners flange. Detailed designed will be then performed.

Acknowledgement

The authors wish to thank the Geo-Informatics and Space Technology Development Agency (GISTDA) for the financial, operational and other related matters in ensuring the progress of research.

References

[1] LaPlante G, Marble Andrew E, MacMillan B, Lee-Sullivan P, Colpitts Bruce G, Balcom Bruce J 2005 Detection of water ingress in composite sandwich structures: a magnetic resonance approach, *NDT & E Int.* **38(6)** 501-7.

[2] Gary E G, Scott L, Jeff H 1996 Electronic tap hammer for composite damage assessment In Proceedings of SPIE, *Int. Soc. Optical Eng.* **2945**.

[3] Grihon S, Krog L, Bassir D 2009 Numerical Optimization applied to structure sizing at AIRBUS: A multi-step process, *Int. J. Simulation Multidisciplinary Design Optimization,* **3** 432-42.

[4] Verbart A, van Keulen F, Langelaar M 2015 *Topology Optimization with Stress Constraints* Ipskamp Drukkers TU Delft, 2-14.

[5] Bouvet, C. and S. Rivallant 2016 Damage tolerance of composite structures under low-velocity impact. *Dynamic Deformation, Damage and Fracture in Composite Materials and Structures*: 7-33.

IOP Conf. Series: Materials Science and Engineering **739** (2020) 012011 doi:10.1088/1757-899X/739/1/012011

Engine Performance Characteristics of a Compression Ignition Engine Fuelled by Traditional Diesel and Waste Cooking Oil

Dinh Xuan Thanh[1], Thanh Viet Nguyen[2], Dien Vu Minh[1, 2], Khanh Duc Nguyen[2], Vinh Nguyen Duy[3]*

[1]Department of Automobile Engineering, Hanoi University of Industry, Km 13, 32 Road Minh Khai - Tu Liem – Hanoi, Viet Nam

[2]Department of Internal Combustion Engine, School of Transportation Engineering, Hanoi University of Science and Technology, 1 Dai Co Viet Road, Hanoi, Viet Nam

[3]Faculty of Vehicle and Energy Engineering, Phenikaa University, Yen Nghia, Ha-Dong District, Hanoi, Viet Nam

Abstract. This research presents a comparison of the test engine's performance characteristics fuelled by diesel and waste cooking oil synthetic diesel (WCOSD). Commercial diesel (CD) fuel used in this research was commercial diesel in the market; meanwhile, WCO was manufactured from cooking waste product using Catalytic Cracking Method. The synthesizing process known as Catalytic Cracking Method at pilot scale level using non-food oil such as waste cooking oil to manufacture biodiesel fuel oil in place of diesel with a stable energy output. The results of this research showed that it can be concluded that the engine torque with WCO drop from 1.9 Nm to 5.4 Nm at all speed while the BSFCs drop at almost speed at full load condition compared to that of CD. The BTEs of the engine fuelled with WCOSD are higher than that of the engine fuelled with CD at almost engine speed at full load. The result of this research is the foundation for using WCO in worldwide application.

1. Introduction

Recently, environmental pollution and energy crisis has been one of the most serious problems facing humanity and other life forms on our planet today. And thus, it is necessary to find out the solution to replace the traditional energy sources as mentioned in [1-5]. Biodiesel is a fuel that is produced from renewable biological sources and equivalent to diesel. Biodiesel is the common name for all kinds of ester-based oxygenated fuels derived from renewable biological sources. It can be produced from animal fats and organic oils [6]. Chemically, biodiesel refers to mixtures of mono alkyl-esters of long chain fatty acids produced from animal fats or vegetable oils. It is normally manufactured by trans-esterification of an animal fat or vegetable oil with a monohydric alcohol, such as ethanol or methanol in the presence of catalyst to produce glycerin and esters (biodiesel). Generally, methanol is favored for trans-esterification [6-7]. The purpose of the trans-esterification process is to reduce the viscosity of the animal fat or vegetable oil. Trans-esterification is basically a sequential reaction. Firstly, triglycerides are reacted to di-glycerides, then di-glycerides are subsequently reacted to mono-glycerides and finally mono-glycerides are reacted to fatty acid esters. Reaction environment such as reaction temperature and time, catalysts, free fatty acid and water content in animal fats or vegetable oils and the molar ratio of alcohol to glycerides are the main factors affecting trans-esterification [6]. Most of the biodiesel made today is produced with the base catalyzed reaction for some reasons: It get high efficiency (98%) with minimal reaction time and side reactions; Pressure and temperature are low; Construction materials are simple.

The combustion process of engine can be improved as the engine is fuelled with biodiesel because of their oxygen content. The oxygen content of biodiesel also reduces its oxidation potential. The combustion efficiency of biodiesel increases in comparison to that of petroleum-based diesel. Biodiesel involves average 11% oxygen by weight and biodiesel usually contains no sulfur. When CI engine is used with biodiesel its life can be extended since biodiesel is more lubricating than petroleum-based diesel. Lubricant properties of biodiesel are better than those of petroleum-based diesel.

A. Abu-Jrai et al [14] has tested blends of treated waste cooking oil and commercial diesel on a naturally aspirated, water-cooled, four stroke, direct-injection, compression-ignition engine to investigate exhaust emission characteristics and combustion characteristics of the engine. They observed that the total combustion duration of B50 was expanded in comparison with that of commercial diesel fuel. They also saw that the combustion was advanced at the all examined engine loads and BSFC for B50 slightly increases in comparison with conventional diesel. In terms of emissions, they also showed that the NOx emissions concentration of the engine fuelled with B50 were increased by 37%, 29%, and 22%, but the smoke emission was dropped by 42%, 31%, and 30% in comparison with conventional diesel at 25%, 50%, and 75% loads respectively. CO2 emissions increased nearly linearly as the load increases.

In 2012, S. Padalkar [9] analyzed effects of waste fried oil methyl esters on exhaust emission and performance of an engine. They observed that the emission characteristics of engine fuelled with different waste fried oil methyl ester blend improved significantly. For example, CO emissions were decreased with different blends of 21–45% of fried oil methyl esters; for blend B50, CO concentration was lower by 21% than that of conventional diesel. Particulate matter percentage was dropped by 23–47% as fried oil methyl esters quantity in the blends increase. Sulfur dioxide emission was dropped by 50% to 100% for different waste fried oil methyl ester blend. However, NOx emission concentration was increased by 4–10% as fried oil methyl esters concentration in the blend increase. In another research, H. An et al [10] investigated effects of biodiesel derived from waste cooking oils on emission characteristics, combustion characteristics and performance of the test engine. They concluded that the use of biodiesel/blend fuels lead to a higher BSFC, particularly at partial load conditions and low engine speed. For instance, the BSFC of the engine running with biodiesel increased by 34.4% and 42% in comparison with conventional diesel at 1200 rpm and 800 rpm respectively and at 25% engine load and the opposite trend was found at 50% and 100% load. For emission characteristics, the major emissions such as HC and NO_x slightly reduced for biodiesel in comparison with conventional diesel. With respect to combustion process, ignition delay is slightly shorter and peak heat release rate is lower with the use of biodiesel. But at low engine speed trend was opposite, disclosing that the low engine speed affect significantly on the emission formation and combustion processes in the engine.

Although there have been many studies as listed above that investigated the use ability of waste cooking oil on compression-ignition engine; however, these studies used biodiesel produced from waste cooking oil throughout the transesterification reaction directly waste cooking oil to fuel for the engine. In addition, these studies only focused on testing at some operating engine points, such as speed characteristic at full load or load characteristic at rated engine speed. For these reasons, the objective of this study is to produce biodiesel from waste cooking oil and investigate the compression-ignition engine performance and its combustion characteristics fueled by WCO corresponding to all main operating engine points.

2. Testing equipment, experiment setup and test procedure

In this study, waste cooking oil synthetic diesel produced from waste cooking oil successfully manufactured in this research was used and compared with conventional diesel fuel when they were used run on internal combustion engine. The physical and chemical properties of WCOSD and CD fuels are shown in Table 1. The test engine is a four stroke, single-cylinder, water-cooled, naturally aspirated, direct injection diesel engine, a KUBOTA model RT140.

To conduct the specific engine speed, the experiments were conducted with the single cylinder diesel engine corresponding to various engine speeds. In the experiment, the engine was coupled with a 40-kW regenerative dynamometer. The cylinder pressures were recorded by a high-speed data acquisition system including two high accuracy AVL piezoelectric pressure transducers, a Crank Angle Encoder model Kistler 2613B - GU12P and a combustion analyzer model DEWE-5000-CA-SE. The fuel consumption was determined by FC2210 Advanced Fuel Measurement Device. In order to measure the air flow rate, TH01-40 air consumption meter was set up on the intake manifold. In addition, six different digital thermocouples were set up to measure the exhaust gas, engine oil, coolant inlet and outlet, fuel, and air inlet temperature.

Table 1. Chemical and physical properties of tested fuels.

Properties		WCO	CD
Composition (%)	C, H, N	77.14, 14.6, 0.078	81.17, 15.29, 0.067
Density (kg/m³)		820.289	827.485
Heating Value (MJ/kg)		44.245	44.864
Flash Point (°C)		93.5	81.5
Viscosity (mm²/s)		2.92	3.74

3. Results and discussion

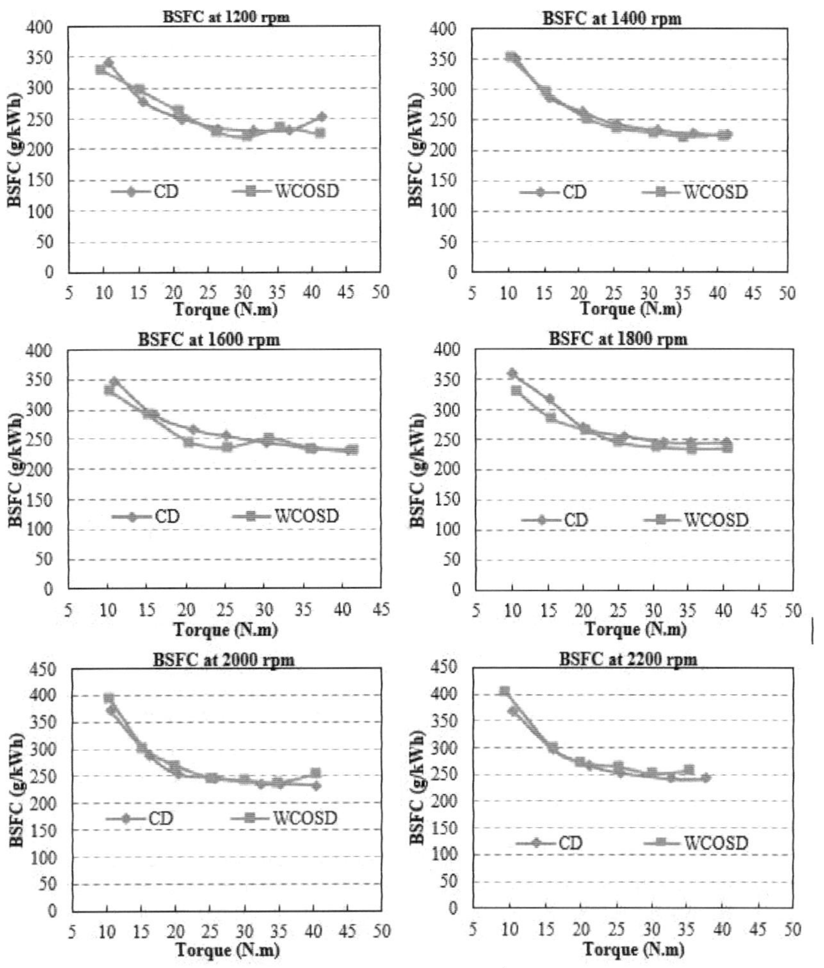

Figure 1. Brake specific consumption at part load different engine speeds.

Figure 1 shows the experiment results of brake specific fuel consumptions (BSFCs) of the test engine fuelled by two different fuels WCO and CD at part load condition and seven different engine speeds.

From the results, it can be seen that the BSFCs of CD is almost higher than that of WCO. At 1200 rpm, when the engine toques are 15 Nm, 20 Nm, and 35 Nm, the BSFCs of the engine running with WCOSD are respectively 19.6 g/kWh, 13.3 g/kWh and 5.6 g/kWh higher than that of the engine running with CD. Whereas at 10 Nm, 25 Nm, 30 Nm and 40 Nm, the BSFCs of the engine running with WCOSD decrease 13.5 g/kWh, 5.5 g/kWh, 8.8 g/kWh and 26.2 g/kWh respectively.

In comparison with CD at 1400 rpm and part load condition, it can observe that, the BSFCs of the engine running with WCOSD drop at 5/7 test point, particularly, 11.7 g/kWh at 20 Nm, 8.0 g/kWh at 25 Nm, 5.3 g/kWh at 30 Nm, 6.9 g/kWh at 35 Nm and 0.5 g/kWh at 40 Nm; and increase at the other points, particularly, 0.7 g/kWh and 7.8 g/kWh. At 1600 rpm, the BSFCs of the engine running with WCOSD drop 17.0 g/kWh, 0.2 g/kWh, 20.6 g/kWh and 18.8 g/kWh at 10, 15, 20 and 25 Nm respectively. Whereas they increase 6.2 g/kWh, 0.9 g/kWh and 2.2 g/kWh at 30, 35 and 40 Nm. The BSFCs of the engine running with WCOSD drop at all test points at 1800 rpm, particularly 31.4, 31.2, 4.6, 10.4, 7.8, 9.0 and 8.9 g/kWh at load from 10 to 40 Nm.

At 1600 rpm, the BSFCs of the engine running with WCOSD drop 17.0 g/kWh, 0.2 g/kWh, 20.6 g/kWh and 18.8 g/kWh at 10, 15, 20 and 25 Nm respectively. Whereas they increase 6.2 g/kWh, 0.9 g/kWh and 2.2 g/kWh at 30, 35 and 40 Nm.

The BSFCs of the engine running with WCOSD drop at all test points at 1800 rpm, particularly 31.4, 31.2, 4.6, 10.4, 7.8, 9.0 and 8.9 g/kWh at load from 10 to 40 Nm. However at 2000 rpm and 2200, the BSFCs of the engine running with WCOSD increase at almost part load conditions, with the increase amount from 1.6 to 35.6 g/kWh.

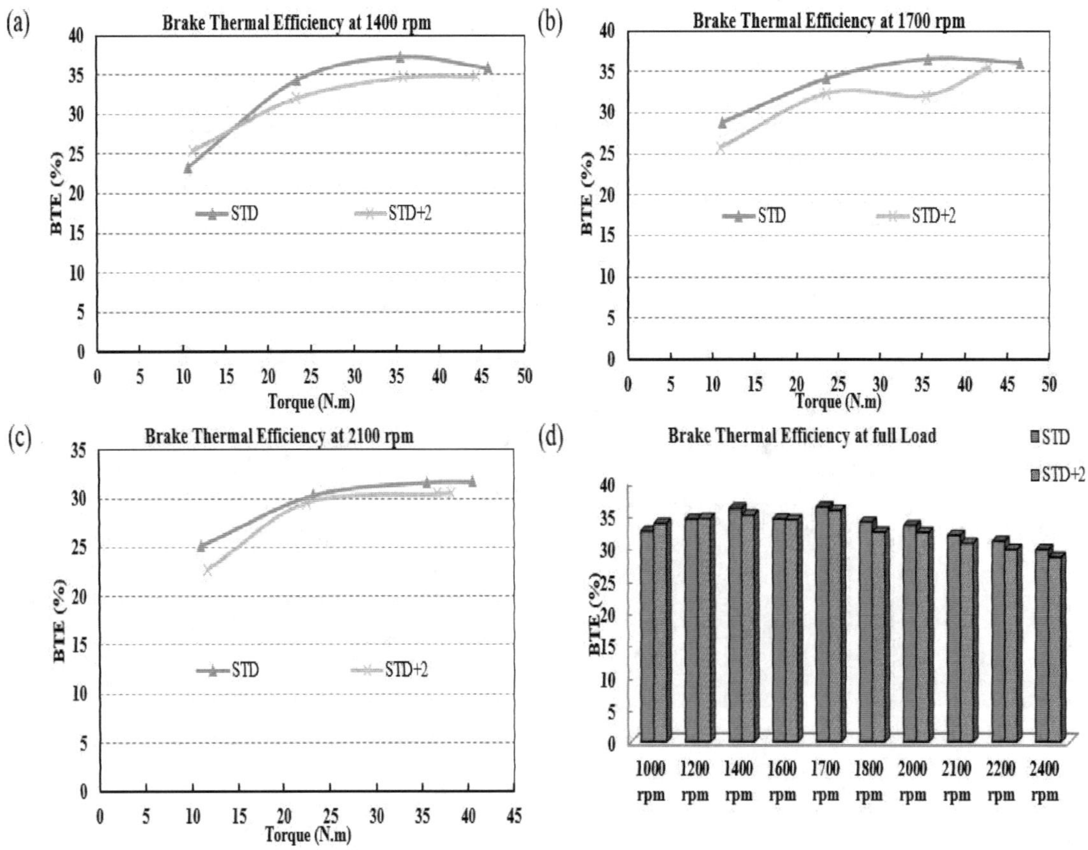

Figure 2. Brake Thermal Efficiency of the engine running with WCOSD for standard (STD) and retard injection timings (STD+2) at (a) 1400 rpm, (b) 1700 rpm, (c) 2100 rpm and (d) full load.

The results referring to the brake thermal efficiency (BTE) of the engine running with WCOSD for standard and retard injection timings, at 1400 rpm, 1700 rpm and 200 rpm constant speeds and full load, are shown in Figure 2.

It may be readily seen that the BTEs of the engine run by WCOSD drop than at almost of all engine operating conditions when injection timing is retarded 2^oCA.

It can be seen from the experiment results that when injection timing is retarded 2^0CA, the BTEs of the engine running with WCOSD at 1400 rpm decrease 2.3% and 2.7% at 23 Nm, 35 Nm respectively. The BTEs drop 3.1%, 1.8% and 4.5% at 11 Nm, 23 Nm and 35 Nm and 1700 rpm as well. At 2100 rpm the BTEs of the engine with standard injection timing is 2.4%, 0.8% and 1.2% higher than that of the engine with 2CA retarded injection timing at 11 Nm, 23 Nm and 35 Nm respectively as shown in Figure c.

It can be seen from Figure 2d that the BTEs drop at almost all speeds at full load condition when the injection timing of the engine is changed from STD to STD+2. The BTEs drop 1.1%, 0.1%, 0.6%, 1.5%, 1.1%, 1.2%, 1.3% and 1.2% at 1400 rpm,1600 rpm, 1700 rpm, 1800 rpm, 2000 rpm, 2100 rpm, 2200 rpm, and 2400 rpm respectively. However The BTEs increase at 1000 rpm and 1200 rpm.

4. Conclusion

In this paper, catalyst cracking biodiesel from waste cooking oil was used. The first target of this study is to evaluate the influence of using WCOSD on the performance and emissions of a conventional direct-injection single cylinder compression-ignition engine, a KUBOTA model RT140, comparing with commercial diesel fuel as base line. The results indicate that in comparison with CD at full load condition the engine torque of the engine running with WCOSD drops from 1.9 Nm to 5.4 Nm at all speeds while the BSFCs drop at almost speeds, thus the BTEs of the engine fuelled with WCOSD are higher at almost engine speed.

Reference

[1] Duc K N, Tien H N, Duy V N 2018 A Study of Operating Characteristics of Old-Generation Diesel Engines Retrofitted with Turbochargers, *Arab. J. Sci. Eng.* **43** 4443-52.

[2] Anh T, Duy V, Thi H, Xa H 2018 Experimental Investigation on Establishing the HCCI Process Fueled by N-Heptane in a Direct Injection Diesel Engine at Different Compression Ratios, *Sustainability,* **10(11)** 3878.

[3] Duc K N, Tien H N, Duy V N 2018 Performance enhancement and emission reduction of used motorcycles using flexible fuel technology, *J. Energy Inst.* **91** 145-52.

[4] Duc K N, Duy V N 2018 Study on performance enhancement and emission reduction of used fuel-injected motorcycles using bi-fuel gasoline-LPG Energy, *Sustain. Dev.* **43** 60-7.

[5] Nguyen Duc K, Nguyen Duy V, Hoang-Dinh L, Nguyen Viet T, Le-Anh T 2019 Performance and emission characteristics of a port fuel injected, spark ignition engine fueled by compressed natural gas, *Sustain. Energy Technol. Assess.* **31** 383-9.

[6] Demirbas A 2008 *Biodiesel: A Realistic Fuel Alternative for Diesel Engines* (Springer).

[7] Moser B R 2009 Biodiesel production, properties, and feedstocks, *In Vitro Dev. Biol.-Plant,* **45(3)** 229-66.

[8] Balasaheb J, Padalkar A S 2012 Performance and emission analysis of a compression ignition Engine operated on waste fried oil methyl esters, *Appl. Energy* **90** 68-72.

[9] Kannan G R, Karvembu R, Anand R 2011 Effect of metal-based additive on performance emission and combustion characteristics of diesel engine fuelled with biodiesel, *Appl. Energy* **88** 3694-703.

[10] An H, Yang W M, Maghbouli A, Li J, Chou S K, Chua K J 2013 Performance, combustion and emission characteristics of biodiesel derived from waste cooking oils, *Appl. Energy* **112** 493-9.

IOP Conf. Series: Materials Science and Engineering **739** (2020) 012012 doi:10.1088/1757-899X/739/1/012012

Various Flow-Field Designs for Enhancing Fuel the Cell Performance of Proton Exchange Membrane Fuel Cells

Thi Diep Tran[1,2], Duc Ha Vu[1,2], Shoudao Huang[1], Duy Vinh Nguyen[3]

[1]College of Electrical and Information Engineering, Hunan University, Hunan, P.R. China

[2]Faculty of Electrical Engineering, Sao Do University, Hai Duong, Vietnam

[3]Faculty of Vehicle and Energy Engineering, Phenikaa University, Yen Nghia, Ha-Dong District, Hanoi, Viet Nam

Abstract. The flow-field design is an essential key in the operation of fuel cells, it conducts main functions such as contributing reactants to the membrane electrolyte assembly (MEA) over gas diffusion layers, conductive part, clamping MEA as well as water management & thermal management and so on. As a result, it is necessary to optimize the flow-field design of fuel cells in order to enhance fuel cell operation characteristics. This paper shows numerical analyses of fuel cell characteristics based on 4 configurations of flow-field in order to find out the best design for enhancing fuel cell performance. The results showed that the different design of flow-field for the anode side and the cathode side can contribute to enhancing fuel cell operation characteristics due to their difference in water formation and discharge. Indeed, the fuel cell configuration with the serpentine flow field in anode side and the pin flow field in the cathode side is the best design because it leads to reducing the water flooding in the cathode side and drying in the anode side. The maximum power density of configuration 1-4 is 0.515, 0.528, 0.585, respectively. It means an improvement of 13.6% percentage can be observed. The research results of this research are the foundation of optimizing the design of the flow-field design.

1. Introduction

In recent several decades, proton exchange membrane fuel cell (PEMFC) has been developed to consider as one of the best solutions to cope with the depletion of energy resources and air pollution due to its fewer emissions and high efficiency compared to the traditional energy sources [1]. A fuel cell working as a battery transfers chemical energy into electrical energy; however, it is continually fed with fuel and thus it can solve the restriction of battery-related to the low energy storage.

As clearly shown in figure 1, a PEMFC includes some main parts: a polymer electrolyte membrane in the centre of the fuel cell, an anode and a cathode, catalyst layers, gas diffusion layers and bipolar plates containing the flow-fields. In the operation of PEMFC, hydrogen enters the fuel cell at the anode side, and thus it is oxidized into two protons and electrons. This is an electrochemical reaction process, and only ions can pass through the membrane but the electrons cannot due to the special design of electrolyte. In the case of the cathode side, oxygen enters the fuel cell and reacts with the hydrogen ions and the electrons to produce water. Otherwise, the free electrons pass through a connected wire from the anode to the cathode generating the electric current.

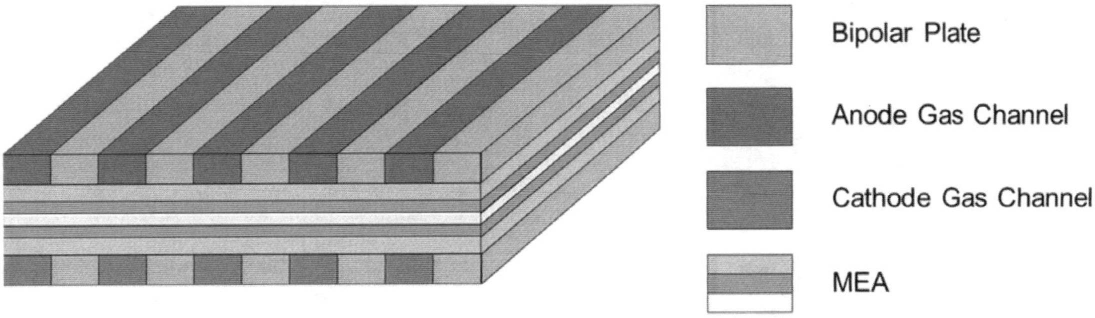

Figure 1. Schematic of a computational domain for PEMFC simulation.

In the fuel cell operation, the flow-field design leads to hydrogen and oxygen through gas diffusion layers to the MEA due to its special design. In addition, it also plays an important role in the water & thermal management; consequently, the flow-field pattern selection is very important due to the affection of its shape, size and pattern on the fuel cell operation characteristics. Many previous studies have focused on developing the novel flow-field design to improve fuel cell performance by both numerical simulation and experiment. D.H. Jeon et. al performed computational fluid dynamics (CFD) simulations based on 4 configurations of single, double, cyclic and symmetric channel patterns to compare their effect on the fuel cell performance. This research showed that the flow-field of the double channel has a higher performance of about 3% and the most uniform current density distribution compared to the other configurations with high inlet humidity. However, corresponding to the low inlet humidity, the difference in performance and current density distribution among four configurations is trivial [1]. In another similar research topic [2], the configurations of single serpentine, parallel, interdigitated and pin flow-field were performed to evaluate the electrochemical reaction behaviours, water formation as well as water discharge of the reference fuel cell. This research showed that corresponding to the same identical operating conditions, the single serpentine flow-field design shows better characteristics compared to those of the interdigitated flow-field. In addition, the worst mass transfer characteristics can be observed in case of the pin and the single parallel flow-field show due to their membrane flooding and drying phenomena.

As mentioned in many previous studies [3-8], the serpentine flow-field is the good design and has been considered in worldwide applications because this design contributes to the equal distribution of supply gases as well as formed water. The design of serpentine is also complicated with many types of design conducted by scientists. A serpentine flow field added sub-channels (SFSC) can improve fuel cell performance as the results shown in [8]. Indeed, an enhancement performance of approximately 18% in comparison with the conventional serpentine flow-field (CSFF) design can be observed by using SFSC design. Furthermore, when flexible applying SFSC for the cathode side and CSFF for the anode side, the fuel cell performance can be enhanced up to 24% because this design overcomes the water flooding in the cathode side and drying in the anode side as shown in [7].

The results in [7] suggested that the flow-field should be designed differently between the anode and cathode side since water is formed in the cathode side and transferred to the anode side. From the ideas of previous results, this research aims to find out the best configuration for the anode and the cathode side. As a result, we performed four configurations based on serpentine flow-field (CF1), parallel flow-field (CF2), interdigitated flow-field (CF3) and pin flow-field (CF4),. The numerical configurations are as follows: Configuration 1 in which CF1 is used at both the anode and the cathode, configuration 2 in which a CF1 and a CF2 are used at the anode and the cathode, respectively, configuration 3 in which a CF1 and a CF3 are respectively used at the anode and the cathode, and configuration 4 in which a CF1 and a CF4 are used at the anode and the cathode, respectively. Figure 2 shows the assembly of 4 configurations using for the simulation process in this research.

IOP Conf. Series: Materials Science and Engineering **739** (2020) 012012 doi:10.1088/1757-899X/739/1/012012

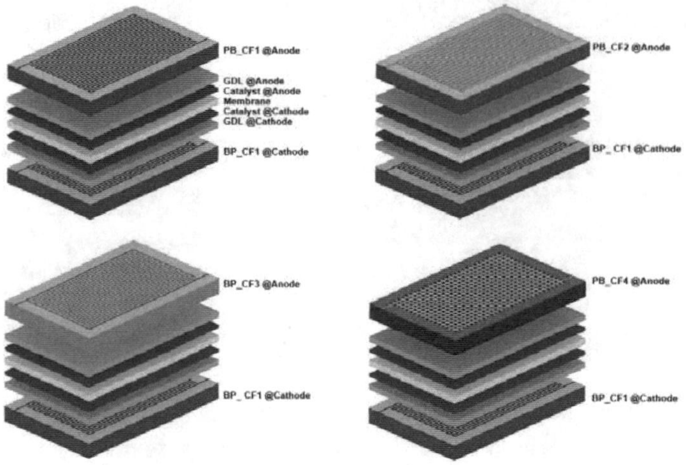

Figure 2. Schematics of the computational domains: (a) configuration 1, (b) configuration 2, (c) Configuration 3 and (d) configuration 42. Testing equipment, experiment setup and test procedure.

2. Model development

In this study, the numerical work is based on a steady-state, multi-phase phenomena, and three-dimensional mass transfer model, including heat transfer aspects of a PEM fuel-cell fuel cell. The Version 14.5 ANSYS program was used to simulate the fuel cell simulation. Figure 2 illustrates based on the computational domain listed in figure 1. The Bipolar palates used in this simulation are based on the serpentine (SPP), the parallel (PRP), the interdigitated (IDP) and the pin flow field (PIP) of the BP as depicted in figure 2. The equations used in this work, conservation of mass, Navier-Stokes equations, species transport equations, energy equation, and water phase change mode, and so on. Table 1 shows the parameters of the test fuel cell used in this research.

Table 1. Properties and Paramenters in this Numerical.

Parameters	**Value**	**Unit**
Porosity after compressed	70	%
GDL thermal conductivity	0.31	W/m·K
MEA thickness	100	μm
Membrane thermal conductivity	0.15	W/m·K
Dry membrane density	2.0	g/cm^3
Dry membrane equivalent weight	1050	g/mol
Cathode exchange current density	0.05	A/cm^2
Cathode transfer coefficient	0.5	
Anode exchange current density	0.2	A/cm^2

3. Results and Discussion

The polarization and power density curves corresponding with 4 configurations of the fuel cell are shown in figure 3.

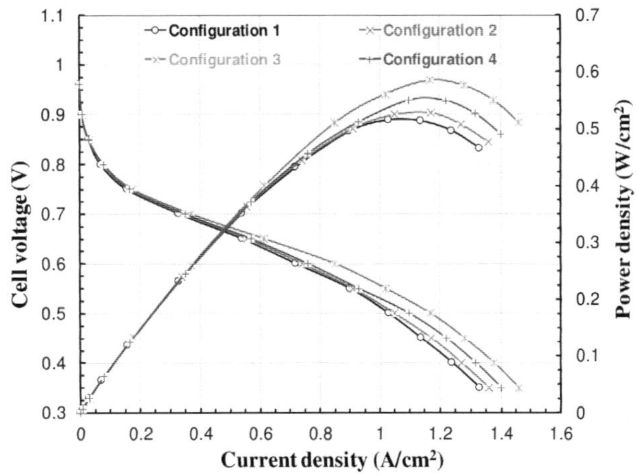

Figure 3. The polarization and power density curves of four different flow field patterns.

Figure 3 represents the average polarization and power density curves at the membrane electrolyte surface of configurations 1-4 corresponding the maximum values of power densities are 0.515 (W/cm^2), 0.525 (W/cm^2),0.55 (W/cm^2), 0.585 (W/cm^2), respectively. Generally, the power densities among four configurations are different from each other when compared with the same current density point. Indeed, the power density of configuration 1 retrofitted SPP flow-field for both the anode and the cathode side is lower than other configurations, it suggested that the fuel cell performance can be enhanced by flexible applying the different flow-field for the anode and the cathode bipolar plate. The reason for this phenomenon is the difference in the reaction process in two sides. At the membrane cathode surface, oxygen is reduced and forms water as a product. Afterwards, water moves along the gas channels as well as in the direction perpendicular to the anode side by the electro-osmotic drag. Increasing current density results in an increase of water formation in the cathode side since back diffusion is not enough to build up for the electro-osmotic drag in drying at the anode side, and thus the anode drying shows faster progress than the moving water velocity. As a result, the water flooding can be observed in the cathode side, meanwhile, the membrane drying may happen in the anode side. The serpentine flow-field is the good design for equally leading the supplied gases throughout the catalyst and also supports for the water discharge in both the anode and the cathode side. This can cause the membrane drying in another side which reduces the fuel cell performance. The PRP, IDP and PIP configuration can solve the problem because they show a lower water discharge process than SPP design as mentioned in [2]. Therefore, the fuel cell performances of configurations 2-3 are higher than those of configuration 1 as shown in figure 3.

4. Conclusions

This research shows the numerical optimization of flow field pattern based on the various configuration of flow-field of SPP, PRP, IDP and PIP using ANSYS software. The results show that with the same active area and operating conditions are identical, the configuration strictly affects the fuel cell performance. In addition, due to the difference in the operating characteristics between the anode and the cathode side, the flow-field of the anode and the cathode side should be designed differently in order to reduce the anode drying in the anode side and the water flowing in the cathode side.

References

[1] Jeon D H, Greenway S, Shimpalee S, Van Zee J W 2008 The effect of serpentine flow-field designs on PEM fuel cell performance, *Int. J. Hydrogen Energy* **33** 1052-66.

[2] Lee B, Park K, Kim H 2013 Numerical Optimization of Flow Field Pattern by Mass Transfer and Electrochemical Reaction Characteristics in Proton Exchange Membrane Fuel Cells, *Int. J. Electrochem. Sci.* **8** 219-34.

[3] Tran T D, Huang S, Vu D H, Duy V N 2018 Effects of gas channel design on water management and on the performance of polymer electrolyte membrane fuel cells: A review, *Int. J. Electrochem. Sci.* **13** 10480-95.
[4] Duy V N, Lee J K, Park K W, Kim H M 2015 Enhancement of PEM fuel cell performance by flow control, *Mater. Sci. Forum* **804** 75-8.
[5] Duy V N, Kim K, Lee J, Ahn J, Park S, Kim T, Kim H M 2015 Parametric simulations of optimum flow-field configuration for efficient proton exchange membrane fuel cell, *Int. J. Electrochem. Sci.* **10** 5842-61.
[6] Vinh N D, Kim H-M 2016 Comparison of numerical and experimental studies for flow-field optimization based on under-rib convection in polymer electrolyte membrane fuel cells, *Energies* **9(10)** 1-17.
[7] Duy V N, Lee J, Kim K, Ahn J, Park S, Kim T, Kim H-M 2015 Dynamic simulations of under-rib convection-driven flow-field configurations and comparison with experiment in polymer electrolyte membrane fuel cells, *J. Power Sources* **293** 447-57.
[8] Park K, Kim H M, Choi K S 2013 Numerical and experimental verification of the polymer electrolyte fuel cell performances enhanced by under-rib convection, *Fuel Cells* **13** 927-34.

IOP Conf. Series: Materials Science and Engineering **739** (2020) 012013 doi:10.1088/1757-899X/739/1/012013

Intelligent Transportation System: Traffic Load Monitoring and Remote Sensing Structural Health Monitoring of Critical Bridges in Metro Manila using Multi Modal Analysis

K M Montes[1], K A Embalzado[1], A J Sayno[2], L C Tani[3], I Marquez[4], F A A Uy[5]

[1]Science Research Specialist II, PhilSIMS, Mapua University, Intramuros, Manila, Philippines

[2]Science Research Specialist I, DEIMS, Mapua University, Intramuros, Manila, Philippines

[3]Project Assistant, PhilSIMS, Mapua University, Intramuros, Manila, Philippines

[4]Senior Science Research Specialist II, PhilSIMS, Mapua University, Intramuros, Manila, Philippines

[5]Dean, School of Civil Engineering, Geosciences, Environmental Engineering, Mapua University, Intramuros, Manila, Philippines

katrinamaesmontes@yahoo.com

Abstract. Bridge load limit monitoring is a major problem in the Philippines. Since this was not implemented strictly, the structural integrity of the bridges was affected hence it resulted to frequent maintenance, retrofitting of strucutral members and collapse. The integrity of the bridge continously decreases through time once the maximum load limit was experienced by the structure. Therefore, this study aims to emphasize the importance of traffic load monitoring in conducting structural health assessment of bridges aside from earthquake and mass loads. The 7.7% increase in the flexure stress of the bridge deck indicates that truck loading has a significant effect on the bridge. In addition for this, variation on the truck volume for peak hours and non peak hours will correlate to the reading of the smart sensor attached on the bridge for structural health monitoring.

1. Introduction

One of a pre stress concrete bridge in metro manila was closed due to a visible major cracks on the bridge deck and deformation on girders. The bridge was constructed 50 years ago and included on the list of critical bridges to be retrofitted by the governent agency two years ago, but due to expected traffic the repair was delayed. Also, the bridge load limit was not strictly supervised and 6,000 trucks passes daily. Fourty more bridges in Metro Manila, considered as critical bridges are scheduled to be retrofitted. [1]

Numerous incidents of bridge failure due to overloading happens just like in Davao city, a bridge in Calinan riverside fails because of a 10 tons truck pass despite of the 5 tons bridge load limit.

Considering those incidents, it is really importat to consider the traffic loads in assessing the structural integrity of a bridge.

This study focused on the significance of traffic load monitoring for structural health monitoring of a certain bridge in Navotas, Philippines. Through the use of accelerometer for the monitoring of the frequency of the bridge with influence of the truck loads for traffic load monitoring. Theoretical model and values of the stresses are included in this paper, but the actual reading of the sensor will not be disclosed at this moment. [4,5]

Figure 1. Cracks on Bridge Deck.

Figure 2. Actual Bridge Footage

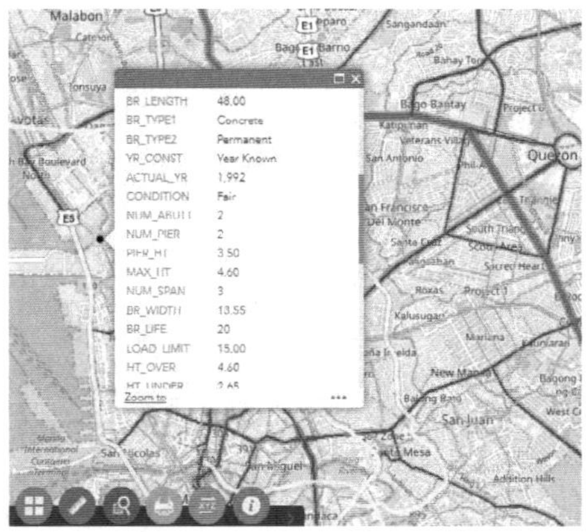

Figure 3. DPWH Bridge Inventory

The figure above shows the actual bridge footage and the properties came from the dpwh bridge inventory. The actual load limit of the bridge is 15 Tons, and the main vehicles passing are cargotrucks because it is near the port area.

According to the Map given by the Philippine Vulcanology and Seismology, the bridge is located in the part of Metro Manila which has high susceptibility of liquefaction but this paper does not include the analysis of the foundation and scouring. The bridge was located near the port area and the foundations of the piers are submerged in a river. This will open to a series of studies which will include multi hazards analysis affecting this bridge. [2]

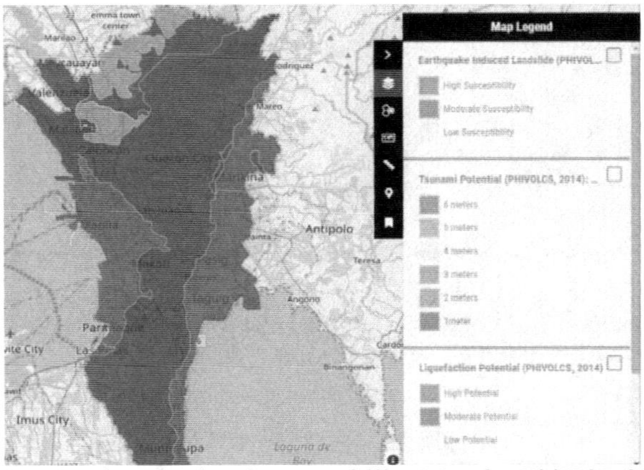

Figure 4. Liquefaction Susceptability Map by PhiVOLCS.

2. Methodology

The bridge consist of three lanes and three span concrete deck, two pier and one column each, and the abutments are integral at the support. The bridge model was done using STAAD bentley structural software. The supports are roller at the supports while pin at the piers. The load cases considered are dead load, lane load, and truck load. Single and Double HS 20-44 are considered and enveloped by 1 meter increment. [3]

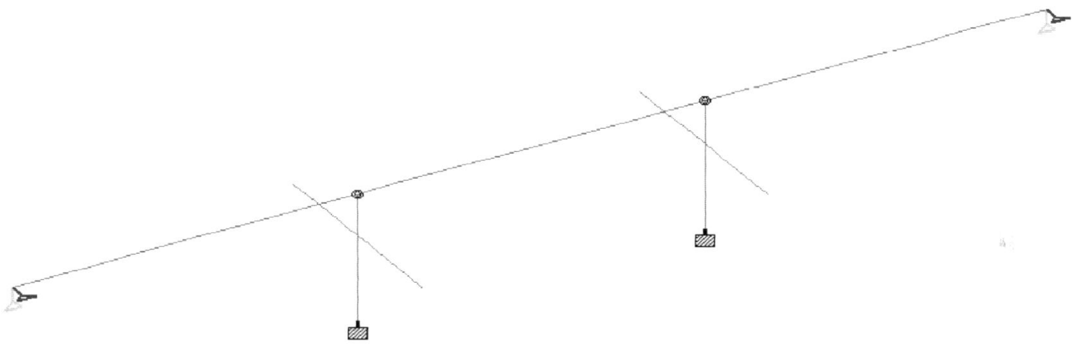

Figure 5. Bridge Model.

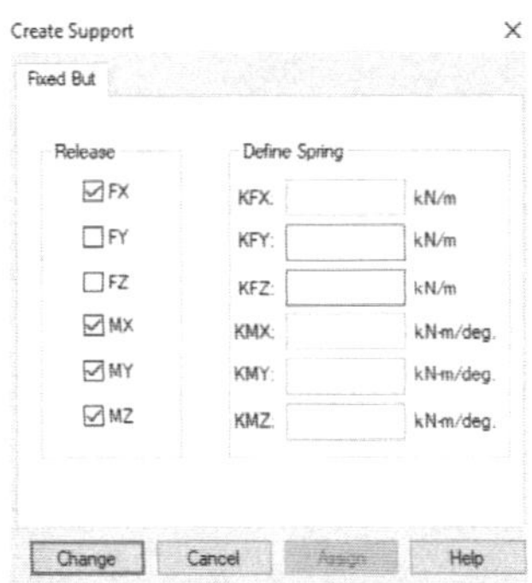

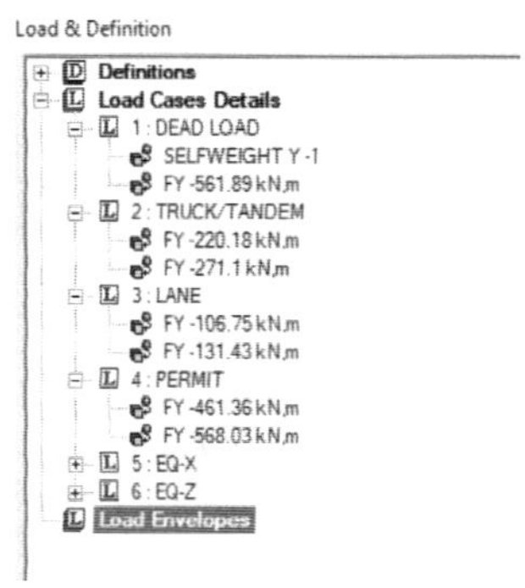

Figure 6. Supports at the Piers **Figure 7.** Support at the Abutment **Figure 8.** Loads and Definitions

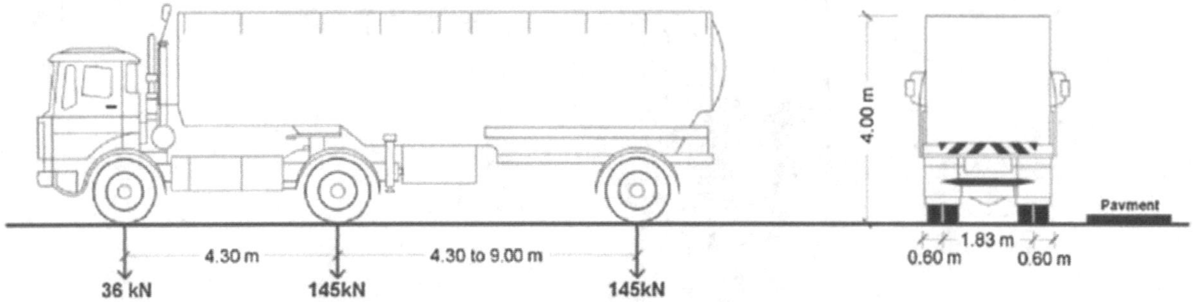

Figure 9. AASHTO HS 20-44 Truck Loading

3. Results and Discussion

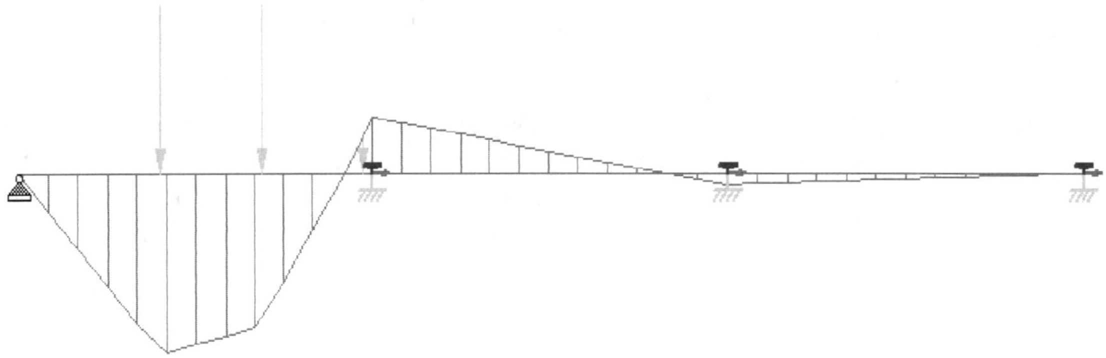

Figure 10. Forces and Stresses with Truck Loadings

The maximum bending moment, 705 KN-m occurs at the first span wherein the truck's front axle is located 0.3 meter near the first pier. While the maximum bending moment without considering the truck loading is 653 KN-m. There's an increase of 7.7% in the moment experienced by the bridge.

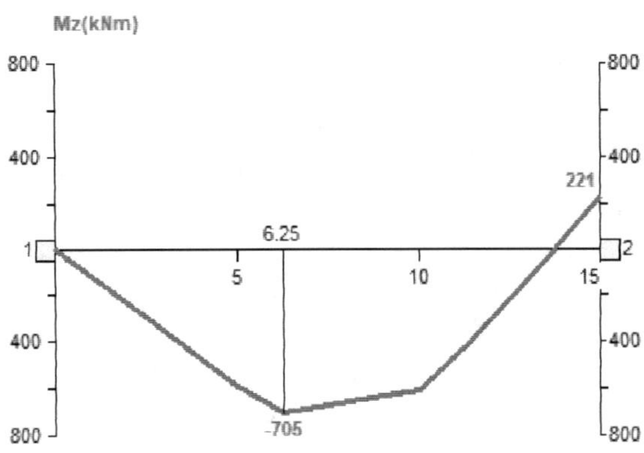

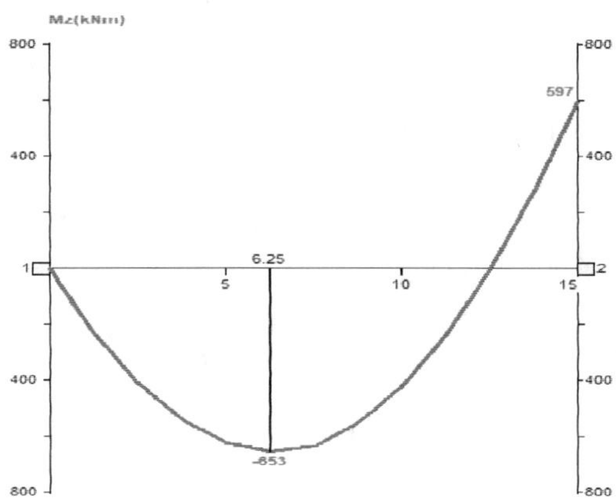

Figure 11. Moment Diagram with Truck Load **Figure 12.** Bending Moment Diagram with Dead Load

4. Conclusion

The increase by 7.7% in the maximum bending moment significantly evaluates the importance of considering the truck loading in analyzing the structural health monitoring of a bridge. Aside from earthquake loadings, live load overloading affects the integrity of the bridge in the the long run. The load limit must be strictly followed to avoid bridge failures and unexpected repairs. CCTV will be installed on the approach of the bridge and traffic calculation will be done. The installation and readings of the accelerograph will be the continuation of this study. The different data analysis in peak hours and non peak hours will also be considered.

Acknowledgements
The authors would like to acknowledge the Dpeartment of Public Works and Highways, Mapua University, and Department of Science and Technology- PCIEERD for the funding of this research project.

References

[1] Department of Public Works and Highways Bridge Inventory http://www.dpwh.gov.ph/DPWH/gis/
[2] Philippine Volcanology and Seismology, Susceptibility Map of Metropolitan Manila https://gisweb.phivolcs.dost.gov.ph/phivolcs_hazardmaps/
[3] AASHTO 2002 *Standard Specifications for Highway Bridges 14th ed* (Washington: Association General Office) p 412
[4] Gaviña J R, Uy F A, Carreon J P D 2017 Wireless Smart Sensor Network System Using SmartBridge Sensor Nodes for Structural Health Monitoring of Existing Concrete Bridges. *IOP Conf. Series: Mater. Sci. Eng.* 216.
[5] Payawal J M G, et al. 2017 *Data Calibration of the Actual versus the Theoretical Micro Electro Mechanical Systems (MEMS) Based Accelerometer Reading through Remote Monitoring of Padre Jacinto Zamora Flyover,* IEEE Conference on Technologies for Sustainability.

IOP Conf. Series: Materials Science and Engineering **739** (2020) 012014 doi:10.1088/1757-899X/739/1/012014

Effect of Road Surface Contour to the Fatigue Life of a Coil Spring Based on Strain-Life Approach

T E Putra[1], Husaini[1], M Ikbal[1]

[1]Department of Mechanical Engineering, Syiah Kuala University, Darussalam 23111, Banda Aceh, Indonesia

edi@unsyiah.ac.id

Abstract. This study aims to determine the effect of road surface contours to the fatigue life of an automotive coil spring. A strain gauge was affixed at the critical point based on the stress distributions. The vehicle passed smooth and rough roads with the speeds above 30 km/h and less than 20 km/h, respectively. The strain signals obtained were analyzed using the Coffin-Manson, Morrow, and Smith-Watson-Topper models. According to the analysis, when the vehicle passed the rough road, the coil spring received a higher stress leading to a shorter fatigue life, of 1,284 cycles to failure. This value was 526 % lower compared to the smooth road because it consisted of bumps and holes.

1. Introduction

Fatigue failure of a vehicle is affected by the physical condition of road surfaces. This is because its stability is reliant on the road surface and tire grinding [1]. Dynamic friction generates vibrations, which increases just as a vehicle moves on a rough road surface at high speeds. Repeated vibration damages vehicle components.

The coil spring is one of the major components of a vehicle which aims at reducing vibrations [2, 3]. However, it often fails owing to the repeated existence of residual stresses [4-6]. The coil spring is expected to not only be able to withstand vibrations, but also loads due to vehicle maneuvers, such as speed, braking or bending while in motion [2, 3]. It must also be able to reduce, withstand, and absorb impact twisting, and fatigue loads. Therefore, the aim of this study is to predict the fatigue life of a coil spring when a vehicle accelerates on smooth and rough road surfaces.

2. Materials and method

Chemical composition test is used to analyze and determine the compositions inherent in a material. Its size must be 1 - 10 cm (figure 1), to enable easy grinding and free from contaminants which agitates the testing results. The sample was inserted into PDA 7000 YS and was injected with an argon fluidized electrode for about 1 minute.

Figure 1. A chemical composition testing sample.

The finite element analysis (FEA) works by dividing the component into small elements. Modeling is very important in FEA therefore a computer aided design is used to produce a model that is in accordance with the original. The efficiency of the FEA is dependent on the pre-processing stage, where the boundary conditions and the load are inserted. In the current study, tetrahedron was used for the meshing [7] and produced 10,999 nodes and 3,710 elements. Considering the vehicle and passenger's weight was 1,045 kg and 420 kg, respectively, with the total weight divided into four by assuming the load was evenly distributed to all wheels, in such a way that the load was 366.25 kg or 3,591.7 N. The load was administered at the bottom of the coil spring while the upper section was fixed support.

von Mises stress [8] is analyzed and acquired from hoop, longitudinal, and radial directions known as the three-dimensional main stress relationship. The yield citerion develops when the equivalent stress σ_e is equal to the yield stress σ_{ys}, which can be defined by:

$$\sigma_e = \frac{1}{\sqrt{2}}\left[\left(\sigma_x - \sigma_y\right)^2 + \left(\sigma_y - \sigma_z\right)^2 + \left(\sigma_z - \sigma_x\right)^2 + 6\left(\tau_{xy}^2 + \tau_{yz}^2 + \tau_{zx}^2\right)\right]^{1/2} \quad (1)$$

where σ_x, σ_y, σ_z are the normal stresses in the direction of the *x*, *y* and *z*-axes, while τ_{xy}, τ_{yz}, τ_{zx} are the shear stresses.

The strain signal measurement was carried out by attaching a 3-mm strain gauge with a resistance of 350 Ω at the coil spring. Furthermore, it was installed at a higher stress distribution area therefore the strain gauge easily detected the signals. The strain signal measurement must have a sampling frequency greater than 400 Hz [9]. Therefore, the frequency of 500 Hz used, was adequate to discover and store all destructive load cycles on the coil spring. After installing the equipment, the vehicle was driven on the smooth and rough road surfaces with its speed above 30 km/h and less than 20 km/h. Figure 2 shows the strain gauge installation.

Figure 2. Installation of strain gauge.

Fatigue signals are generally characterized by statistical parameters which are used to monitor patterns and classify random signals. Parameters often used to observe strain signals include mean, standard deviation (SD), root-mean square (r.m.s.), and kurtosis.

For signals F_j with a spesific number of data *n*, the mean value $\bar{x}$ is calculated by:

$$\bar{x} = \frac{1}{n}\sum_{j=1}^{n} Fj \tag{2}$$

SD determines how data are distributed in a sample. It is also attributed to the number of average variability in a data set. It value for numbers F_j above thirty can be stated by:

$$SD = \sqrt{\frac{1}{n}\sum_{j=1}^{n}(F_j - \bar{x})^2} \tag{3}$$

r.m.s. helps to estimate the total amount of energy in a discrete data F_j. This is expressed by:

$$r.m.s = \sqrt{\frac{1}{n}\sum_{j=1}^{n}F_j^2} \tag{4}$$

Kurtosis is a statistical parameter sensitive to spikes. The equation used to estimate kurtosis K in a discrete data F_j is:

$$K = \frac{1}{n(SD)^4}\sum_{j=1}^{n}(F_j - \bar{x})^4 \tag{5}$$

The strain-life technique accuratelly predicts fatigue life. This takes into account plastic events that occur in certain areas. It is quite good to use, does not require high costs or exhaustive crack observation provided failure occurs, and it is applied to small components. Elastic and plastic strain is combined to produce a relationship between the strain and the fatigue life, called the Coffin-Manson model [10,11] and this can be expressed by:

$$\varepsilon = \frac{\sigma'_f}{E}\left(2N_f\right)^b + \varepsilon'_f\left(2N_f\right)^c \tag{6}$$

where ε is the strain amplitude, σ'_f is the fatigue strength coefficient, E is the material modulus of elasticity, N_f is the number of cycles to failure for a particular stress range and mean, b is the fatigue strength exponent, ε'_f is the fatigue ductility coefficient and c is the fatigue ductility exponent.

Mean stress is an important factor in the analysis of fatigue and this affects its low and high cycles. Its effect can be investigated using the Morrow model [12], expressed by:

$$\varepsilon = \frac{\sigma'_f - \sigma_{mean}}{E}(2N_f)^b + \varepsilon'_f(2N_f)^c \tag{7}$$

where σ_{mean} is the normal mean stress. Furthermore, K.N. Smith, P. Watson, and T.H. Topper recomended the SWT parameter [13], a mean stress correction model mathematically defined as:

$$\sigma_{max}\varepsilon = \frac{\sigma'^2_f}{E}(2N_f)^{2b} + \sigma'_f\varepsilon'_f(2N_f)^{b+c} \tag{8}$$

where σ_{max} is the maximum stress.

Fatigue damage is determined by the addition of normal and shear energy of the peak-valley in each loading. It can be influenced by variations in amplitude and frequency. Its damage for each loading cycle D_i is:

$$D_i = \frac{1}{N_f} \tag{9}$$

Palmgren-Miner rule [14,15] was used to determine the cumulative fatigue damage, as follows:

$$D = \sum \left(\frac{n_i}{N_f} \right) \quad (10)$$

3. Results and Discussion

Analysis from the chemical composition testing results shows that the material for the coil spring consited of 0.64 % carbon (C), 0.16 % silicon (Si), 0.77 % manganese (Mn), 0.76 % chromium (Cr), 0.18 % molybdenum (Mo), 0.15 % vanadium (V), 0.01 % sulfur (S) and 0.003 % phosphor (P). Therefore, it can be concluded that the coil spring was made from the SAE 5160 carbon steel [16], used in the automotive industries. Table 1 presents its mechanical properties.

Table 1. The mechanical propertis of the SAE5160 carbon steel [17].

Properties	Values
Ultimate tensile strength, S_u (MPa)	1,584
Material modulus of elasticy, E (GPa)	207
Yield strength [MPa]	1,487
Fatigue strength coefficient, σ'_f (MPa)	2,063
Fatigue strength exponent, b	-0.08
Fatigue ductility exponent, c	-1.05
Fatigue ductility coefficient, ε'_f	9.56
Cyclic strain-hardening exponent	0.05
Cyclic strength coefficient (MPa)	1,940
Poison ratio	0.27

The stress concentrations are available in several color contours as shown in figure 3. The area with the highest stress concentration is represented by red, orange, yellow, green and blue. The red area has a maximum von Mises stress of 0.997 GPa. With reference to table 1, the maximum stress was low in comparison to the yield strength of the SAE 5160 carbon steel, which was 1.487 GPa. This illustrated that the load did not collapse the component. The surface area of the inner spring was smaller than the outside so that the critical point occurred in the inside. This was identical to other research results [18-21].

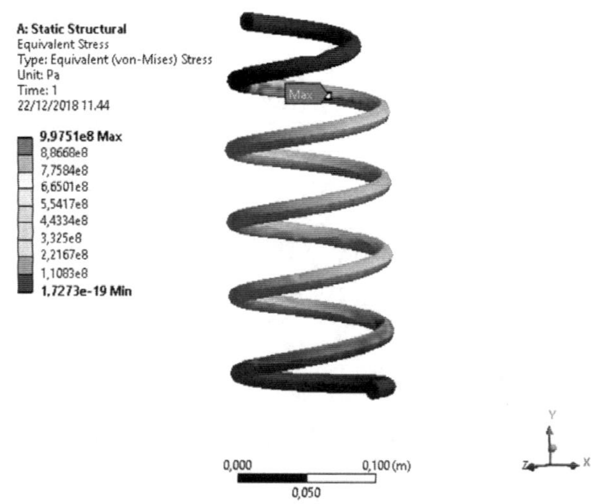

Figure 3. Stress distribitions of the coil spring.

30,000 data gave a total length of 60 seconds as shown in figure 4. All strain signals have non-zero-mean values, which were -684 με and -658 με for the smooth and rough road surfaces, respectively. The negative mean value indicated that the coil spring received a compressive load. The smooth and rough roads provided the SD of 20 με and 22 με, while for the r.m.s., their values were 685 με and 658 με, respectively. Both signals were non-stationary as displayed on the kurtosis above 3 [22,23], which was 4.1 and 5.8.

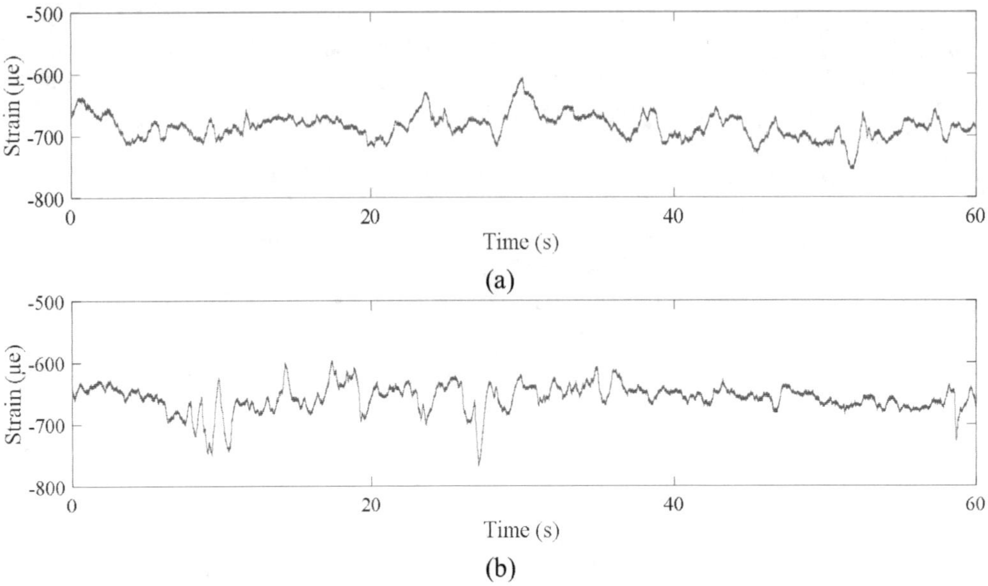

Figure 4. Strain signals: (a) smooth, (b) rough.

Based on the Coffin-Manson model, the fatigue damage for the smooth and rough roads was 2.9E-2 damage per block and 4.9E-2 damage per block, respectively. Using the Morrow model, it was 2.2E-2 damage per block and 4.1E-2 damage per block. For the SWT model, it was 1E-2 damage per block and 2.5E-2 damage per block. A higher damage caused a decrease of the fatigue life in the coil spring. Therefore, the rough road provided a shorter fatigue life, which was 1,284 cycles to failure with a higher vertical load contour, while the smooth road gave 19,060 cycle to failure.

4. Conclusion

This study aims to determine the effect of road surface contours on the fatigue life of a vehicle coil spring. According the results, it can be concluded that just as a vehicle plies a rough road surface, the coil spring received a higher stress contributing to a shorter fatigue life. It also accelerated the useful life of the coil spring by 526 %.

5. Acknowledgements

The authors would like to express their gratitude to Universitas Syiah Kuala for financial support for this research through the grant No. 55/UN11.2/PP/SP3/2019 and 63/UN11.2/PP/PNBP/SP3/2019.

6. References

[1] Jin L Q, Ling M, Yue W 2017 *PLoS ONE* **12**.
[2] Ansari A A, Jain K K 2016 *Global J. Eng. Sci. Res.* **3**.
[3] Kumhar V 2016 *VSRD Int. J. Mechanic. Civil, Automobile Product. Eng.* **6**.
[4] Cruz M L S, Carrillo J, de Almeida S F M 2016 *Latin American J. Solids Struct.* **13** 435-55.
[5] Mahalov M S, Blumenstein V Y 2016 *IOP Conf. Series: Mater. Sci. Eng.* **126**.
[6] Yanxun X, Da T, Mingxi D, Yunze L, Changjun L, Fuzhen X 2018 *Metals* **8**.

[7] Mulia T M, Kadam S J, Kengar V S 2012 *Int. J. Mechanic. Industrial Eng.* **2** 74-7.
[8] Von Mises R 1913 *Mathematisch-Physikalische Klasse 1* 582-92.
[9] Ilic S 2006 *Methodology of Evaluation of In-Service Load Applied to the Output Shafts of Automatic Transmissions* Ph.D. Thesis The University of New South Wales.
[10] Coffin Jr. L F 1954 *Transactions of the ASME* **76** 931-50.
[11] Manson S S 1965 *Experimental Mechanics* **5** 193-226.
[12] Morrow J 1968 *Fatigue Design Handbook* Society of Automotive Engineers.
[13] Smith K N, Watson P, Topper T H 1970 *J. Mater. JMLSA* **5** 767-78.
[14] Palmgren A 1924 *Zeitschrift VDI* **68** 339-41.
[15] Miner M A 1945 *J. Appl. Mechanic.* **67** A159-64.
[16] ASTM A322-91 2001 *Standard Specification for Steel Bars, Alloy, Standard Grades* ASTM International.
[17] nCode 2018 *GlyphWorks* nCode International, Ltd.
[18] Zhu Y, Wang Y, Huang Y 2014 *Case Studies in Engineering Failure Analysis* **2** 169-73.
[19] Putra T E, Abdullah S, Schramm D, Nuawi M Z, Bruckmann T 2015 *Mechanical Systems and Signal Processing* **60-61** 485-97.
[20] Putra T E, Abdullah S, Schramm D, Nuawi M Z, Bruckmann T 2017 *Mechanical Systems and Signal Processing* **90** 1-14.
[21] Putra T E, Abdullah S, Schramm D, Nuawi M Z, Bruckmann T 2017 *Mechanical Systems and Signal Processing* **94** 432-47.
[22] Chen X 2014 *Engineering Structures* **74** 145-56.
[23] Kihm F, Ferguson N S, Antonic J 2015 *Procedia Engineering* **133** 698-713.

IOP Conf. Series: Materials Science and Engineering **739** (2020) 012015 doi:10.1088/1757-899X/739/1/012015

Effect of Concrete Surface Crack Filling Conditions on Dominant Frequency of Surface Rayleigh Wave

Chi Hoe Liew[1], Foo Wei Lee[1], Chee Ghuan Tan[2], Kok Zee Kwong[1], Kok-Sing Lim[3]

[1]Lee Kong Chian Faculty of Engineering and Science, Universiti Tunku Abdul Rahman, Jalan Sungai Long, Bandar Sungai Long, Cheras, 43000 Kajang, Selangor, Malaysia

[2]Department of Civil Engineering, University of Malaya, 50603 Kuala Lumpur, Malaysia

[3]Photonics Research Centre, University of Malaya, 50603 Kuala Lumpur, Malaysia

mmankey@1utar.my, leefw@utar.edu.my, tancg@um.edu.my, kwongkz@utar.edu.my, kslim@um.edu.my

Abstract. Elastic Surface Rayleigh Wave (R-wave) based non-destructive test (NDT) techniques have been studied broadly for concrete surface crack assessment in the past few decades. However, the effect of environment factors on the accuracy of R-wave NDT results are often being neglected. Sand and dust are ones of them that easily accumulate within the surface cracks of concrete structures, which potentially affect the characteristics of R-waves propagation when it travels through the concrete medium that containing cracks. This study is aimed to investigate the effect of concrete surface crack filling conditions to the dominant frequency of R-wave. Numerical simulations were conducted to study the changes of R-wave waveform when it propagated through a cracked concrete model with two various filling conditions, namely fine sand and charcoal powder. Fast Fourier Transform (FFT) for post-signal-processing (PSP) of raw R-wave signals to determine its dominant frequency. The results of numerical simulations were then justified by experimental measurements. It was found that the dominant frequency of R-wave is independent to the volume and the type of filling materials, but sensitive to the surface discontinuity of the concrete.

1. Introduction

Concrete surface crack is a common defect that can be found on civil concrete structures such as dam, bridge and most of the commercial and residential buildings. This defect is one of the indications of the degradation of concrete structures [1]. With the presence of the crack, the durability of the concrete structures may be reduced dramatically by allowing the penetration of water and other aggressive agents [2]. Therefore, concrete surface cracks detection and evaluation became a crucial step for maintaining the safety of concrete structures as early inspection allows preventive measures to be conducted to prevent damage and possible failure of the concrete structures [3]. In order to assess the concrete surface crack without further damaging the existing concrete structures, non-destructive test (NDT) is developed to fulfil this particular task.

Elastic wave is one of the popular NDT methods that have been utilised for concrete defects assessment. Recently, elastic surface Rayleigh wave (R-wave) based NDT method is broadly studied for evaluating concrete surface crack [4-8]. However, past researches which utilising R-wave in characterising concrete surface discontinuities characterisation are seldom contemplate the environmental factors that could possibly affect R-wave propagation behaviour. In an actual situation, surface cracks on concrete structures are having a certain possibility to be exposed to the natural environment that is dusty and contains moisture. Sand and dust will start to accumulate inside the surface crack after some period of time. At this moment, the applicability of NDT in the crack assessment will be doubted. The doubts can be eliminated by clearing the accumulated substances within the crack but the process is troublesome as the accessibility is low within the crack.

Saltation of sand bring fine sand particles into the surface cracks concrete structures. Besides, soot that containing mainly black carbon, which emitted from motor vehicles or originated from incomplete combustion of fuel, will eventually accumulate within the concrete surface cracks as well. For this reason, fine sand and fine charcoal powder are used in this study as the filling materials within the concrete surface crack. It is aimed to determine the effect of the proposed filling materials to the dominant frequency of R-wave.

2. Numerical Simulation

A multi-channel R-wave assessment method was adopted in this study. A commercially available software, Wave 2000 by CyberLogic Incorporation was used to simulate motions of elastic software for a series of designed numerical simulation cases. The software solves the two dimensional (2D) acoustic (elastic) wave equation based on a method of finite difference and simulating received waveforms under a variety of spatial and temporal acoustic interrogations, give as equation 1 [9]:

$$\rho \frac{\partial^2 w}{\partial^2 t} = \left[\mu + \eta \frac{\partial}{\partial t}\right] \nabla^2 w + \left[\lambda + \mu + \emptyset \frac{\partial}{\partial t} + \frac{\eta}{3} \frac{\partial}{\partial t}\right] \nabla(\nabla . w) \quad (1)$$

where ρ is material density, λ is the first Lame constant, μ is the second Lame constant, η is shear viscosity, $\emptyset$ is bulk viscosity, ∇ is the gradient of operator, $\nabla.$ is the divergence operator, ∂ is the partial differential operator, t is the time and w is a two dimensional column vector.

2.1. Parameters of Numerical Simulations

Table 1 shows the materials properties for simulation work while, table 2 presents the parameters of interest which includes the type of filling materials, volume of crack to be filled and the frequency of source excitation.

Table 1. Material properties in numerical simulations.

Parameter	Concrete	Sand	Charcoal
First Lame constant, λ [MPa]	11100	0.482	1.381
Second Lame constant, μ [MPa]	16600	0.482	0.615
Density, ρ [kg/m^3]	2375	1212	344

Table 2. Properties of simulated crack.

Type of filling materials	Volume of crack to be filled [%]	Frequency of source excitation, f[kHz]
Charcoal powder	20, 40, 60, 80 and 100	20, 40, 60, 80 and 100
Fine Sand	20, 40, 60, 80 and 100	20, 40, 60, 80 and 100

Figure 1 (a) depicts the filled cracked simulation model with a dimension of 500 mm (width) x 100 mm (depth) in size. The material and geometry of the concrete medium were set to constant throughout the simulation works. Four sensors linear array was deployed in this study, where two of the sensors were located before and after the crack.

(a) (b)

Figure 1. (a) Schematic Diagram of Concrete Simulation Model with a Vertical Surface Crack with Filling Material (Side View) and (b) the actual set up.

3. Experimental Measurement

The concrete specimen in this study was cast with a water-to-cement ratio of 0.51, elastic modulus of 25 GPa, density of 2375 kg/m^3 and aggregate with a maximum size of 20 mm. The artificial surface crack was created by installing a 5 mm thick polystyrene board at the middle of the concrete specimen vertically, with a depth of 50 mm.

3.1. Set up of laboratory experiment.

This study adopts a four-sensor array measurement as shown in figure 1 (b), which also known as linear measurement array. Impaction of steel balls was implemented by the same operator in order to maintain the consistencies in the generation of stress waves. Different size of steel ball impactors allows generation of R-waves with varying dominant frequencies. The sizes of the steel balls used in this study were 1.0 cm, 1.5 cm, 2.0 cm and 2.5 cm. It is attempted to investigate the changes of dominant frequencies of R-wave when propagated through the surface crack filled with different materials and different volume.

4. Results

The arrival of the Rayleigh wave component can be identified after the P-wave component (the first disturbance), followed by strong bursts in the positive phases [10, 11]. By obeying this principle, R-waves were extracted from each set of raw recorded waveforms. The extracted R-wave waveforms were then processed through a series of Fourier transform using equation 2 to determine its frequency components (MATLAB®):

$$F_{k+1} = \sum_{j=0}^{n-1} \omega^{jk} A_{j+1} \quad (2)$$

Where $\omega = e^{-2\pi i/n}$ is one of the n complex roots of unity and i is the imaginary unit. A is the amplitude of recorded signals, whereas F is the computed Fast Fourier transform (FFT) of the signals. The indices j and k ranged from 0 to n-1.

4.1. Simulation results

Figure 2 displays the simulated dominant frequencies of all four receivers for the cases of sound concrete (plain concrete without absence of defect or crack), control sample (plain concrete with presence of crack), fully sand filled control sample and fully charcoal powder filled control sample with the excitation frequencies are from 20 kHz to 100 kHz, at 20 kHz increment. Several phenomena can be observed from results of laboratory experiments. {1} Dominant frequencies retrieved from each receivers in simulation cases of sound concrete remain almost constant, due to the absence of inhomogeneity and discontinuity in the concrete model that could possibly alter the propagation behaviour of R-wave; {2} Dominant frequencies recorded by Receiver 3 of each cases show a significance difference (as much as 50%) when compared to Receiver 1, 2 and 4. This indicates that R-wave dominant frequency is sensitive to the presence of flaws within a concrete, which also proved that R-wave is feasible for concrete surface crack evaluation; {3} The value of dominant frequencies of Receiver 3 is shifted to a higher value when compared to ones collected from sound concrete cases with excitation frequencies from 20 kHz to 60 kHz, but vice-versa for simulation cases with 80 kHz and 100 kHz; {4} The presence of accumulated filling materials within the concrete surface crack is not affecting the R-wave propagation behaviour. As can be seen from the insignificant changes between the dominant frequencies obtained from all the cases (less than 0.3%). Sand and charcoal powder may not possess the shear properties that required to create a bridging effect for R-wave to travel through the concrete surface crack, as R-wave propagation is strongly depends on the shear properties of the medium [12]. In addition, results analysed from 20% to 80% sand filled and charcoal filled cases are not plotted as the results are similar to the 100% filled cases, which exhibit an insensitivity towards the propagation behaviour of R-wave.

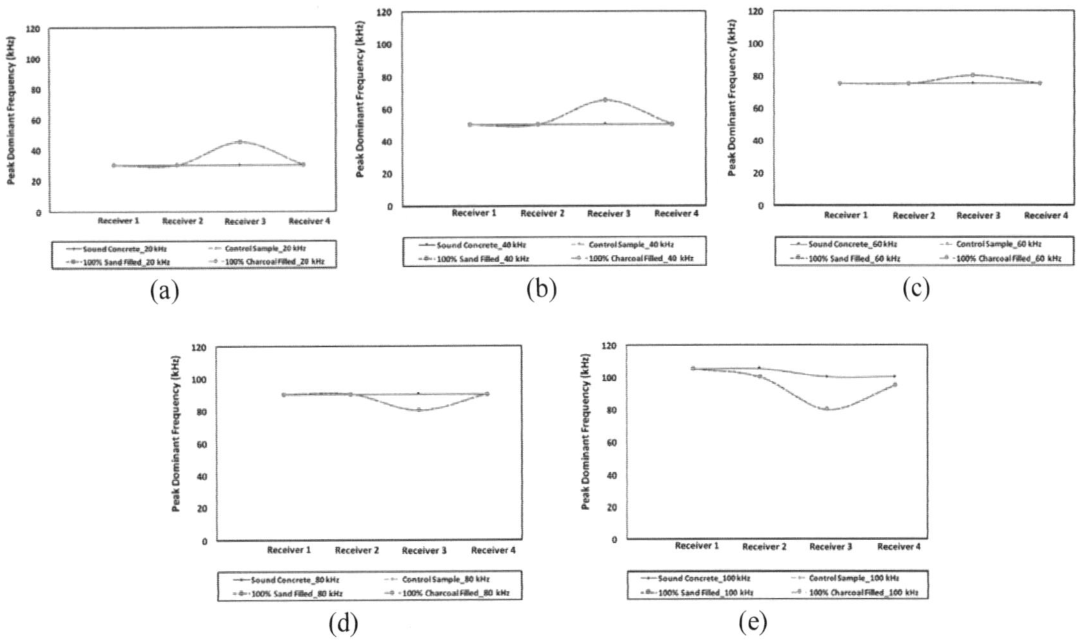

Figure 2. Collected dominant frequencies from each respective receiver for simulation cases of sound concrete, control sample and control sample fully filled by sand and charcoal powder, with the excitation frequencies of (a) 20 kHz, (b) 40 kHz, (c) 60 kHz, (d) 80 kHz and (e) 100 kHz.

4.2. Experimental verification

The experimental results justified the findings obtained from the simulation that the R-wave dominant frequency is one of the parameters that sensitive to concrete surface crack. However, it's the filling conditions do not affect the behaviour of R-waves propagation. An average of variation of 4.1% is reported when compared both of the filled and control sample cases. However, instead of Receiver 3 alone being affected as mentioned in simulation results, the dominant frequencies retrieved from Receiver 2 and 3 are both influenced by the presence of the concrete surface crack. This is due to the raw waveform collected from laboratory experiments are mixed with the reflective wave from the surface of the vertical crack. The effort to task for filtering out the reflective wave is quite challenging when compare with the waveforms recorded from numerical simulations. Further investigation should focus on this signal processing process to enhance the efficiency of the work.

5. Conclusion

It can be concluded that dominant frequency of R-wave is independent regardless the volume and the type of filling materials, sensitive to the surface discontinuity of the concrete. A maximum of 0.3% and average of 4.1% difference are reported between both filled materials and control sample cases in numerical simulations and laboratory experiments. For this reason, R-wave based NDT is practical to be utilised in concrete surface crack assessment without the needs of considering the effects of filling conditions within the crack.

Acknowledgement
This work was supported under the University Research Funding (IPSR/RMC/UTARRF/2017-C2/L02).

References

[1] Yamaguchi T, Hashimoto S 2010 Fast crack detection method for large-size concrete surface images using percolation-based image processing. *Machine Vision Appl.* **21(5)** 797-809.

[2] Sham F C, Chen N, Long L 2008 Surface crack detection by flash thermography on concrete surface. *Insight-Non-Destructive Testing Condition Monitor.* **50(5)** 240-3.

[3] Dhital D, Lee J R 2012 A fully non-contact ultrasonic propagation imaging system for closed surface crack evaluation. *Experimental Mechanic.* **52(8)** 1111-22.

[4] Zerwer A, Polak M A, Santamarina J C 2005 Detection of surface breaking cracks in concrete members using Rayleigh waves. *J. Environ. Eng. Geophysic.* **10(3)** 295-306.

[5] Aggelis D G, Shiotani T 2007 Repair evaluation of concrete cracks using surface and through-transmission wave measurements. *Cement Concrete Compos.* **29(9)** 700-11.

[6] Shin S W, Zhu J, Min J, Popovics J S 2008 Crack depth estimation in concrete using energy transmission of surface waves. *ACI Mater. J.* **105(5)** 510.

[7] Aggelis D G, Shiotani T, Polyzos D 2009 Characterization of surface crack depth and repair evaluation using Rayleigh waves. *Cement Concrete Compos.* **31(1)** 77-83.

[8] Ghosh D, Beniwal S, Ganguli A, Mukherjee A 2016 Near Surface Defect Detection in a Concrete Slab Using Ultrasonic Rayleigh Wave-A Numerical Study. In *ISSS National Conf. MEMS, Smart Mater. Struct. Syst.*

[9] Wave 2000, CyberLogic, Inc., New York, http://www.cyberlogic.org.

[10] Qixian L, Bungey J H 1996 Using compression wave ultrasonic transducers to measure the velocity of surface waves and hence determine dynamic modulus of elasticity for concrete. *Construct. build. Mater.* **10(4)** 237-42.

[11] Sansalone M J, Streett W B 1997 Impact-echo. Nondestructive evaluation of concrete and masonry. *TRID* 339.

[12] Chai H K, Momoki S, Aggelis D G, Shiotani T 2010 Characterization of Deep Surface-Opening Cracks in Concrete: Feasibility of Impact-Generated Rayleigh-Waves. *ACI Mater. J.* **107(3)** 305-11.

IOP Conf. Series: Materials Science and Engineering **739** (2020) 012016 doi:10.1088/1757-899X/739/1/012016

The Influence of Powdered Crumb Rubber on the Mechanical Properties and Thermal Performance of Lightweight Foamed Concrete

Zhi Heng Lim[1], Foo Wei Lee[1], Kim Hung Mo[2], Kok Zee Kwong[1], Ming Kun Yew[1]

[1]Lee Kong Chian Faculty of Engineering and Science, Universiti Tunku Abdul Rahman, Jalan Sungai Long, Bandar Sungai Long, Cheras, 43000 Kajang, Selangor, Malaysia.

[2]Department of Civil Engineering, Faculty of Engineering, University of Malaya, Jalan Universiti, 50603 Kuala Lumpur, Malaysia.

Abstract. The introduction of powdered crumb rubber into the lightweight foamed concrete influenced various mechanical and physical behaviour of the concrete. The objective of this study is to investigate the influence of 15 %, 30 % and 45 % crumb rubber, in replacement of sand on the mechanical properties and thermal performance of lightweight foamed concrete. Compressive strength, flexural strength and thermal conductivity tests were conducted to evaluate the performance of crumb rubber lightweight foamed concrete. Laboratory results indicated that the strength properties of crumb rubber lightweight foamed concrete were weaker than those of conventional lightweight foamed concrete with a similar density. The compressive strength and flexural strength were reduced from 14.02 MPa to 10.66 MPa and 4.92 MPa to 3.06 MPa respectively, as the crumb rubber proportion was increased from 0 % to 45 %. However, it is worth mentioning that the optimum incorporation of 45 % crumb rubber in lightweight foamed concrete enhanced the thermal insulating performance by lowering the heat conductivity of the concrete from 1.1863 $W.K^{-1}m^{-}$ to 0.9733 $W.K^{-1}m^{-1}$.

1. Introduction

Increasing piles of waste tires have been one of the infamous environmental issues in modernized countries. According to Environmental Protection Agency (EPA), the accumulation of dumped tires at various landfill sites can pose severe health and safety hazards such as the habitation of rodent and mosquitos which will facilitate the spread of contagious diseases, increased risk of uncontrolled fires and production of toxic chemicals contaminating soil and vegetation. The amount of scrap tire generated has been skyrocketing as it is estimated to reach 1.2 billion tonnes annually by 2030 [1]. This is considered as one of the worldwide environmental challenges because waste tires are not biodegradable even after a long period of biological treatment. Burning of such wastes will induce serious air pollution involving the emission of poisonous gas to both human and environment. Thus, one of the effective solutions is the reuse of tire rubber as a partial replacement of aggregate in cement-based materials, known as rubberized concrete.

The aim of this study is to investigate the potential of crumb rubber to be adopted as one of the useful concrete components to enhance thermal properties of the concrete, and at the same time, minimize the existing environmental issue. In addition, another objective of this study is to examine the influence of crumb rubber on the mechanical properties and thermal performance of lightweight foamed concrete with the density of 1800 kg/m^3. The crumb rubber will be adopted to substitute fine aggregate in concrete at different proportions (15 %, 30 % and 45 %).

2. Experimental program

2.1. Mix proportion

The mix proportion of the control specimen (CR0) was set at 1.00:1.00:0.50 (cement: sand: water) and it contains no crumb rubber. The foam generated was introduced to the specimen until the fresh density of the fresh concrete dropped to 1800 kg/m^3. In the case of crumb rubber lightweight foamed concrete, powdered crumb rubber, as illustrated in figure 1, was utilized to replace fine aggregates at 15 % (CR15), 30 % (CR30) and 45 % (CR45) by volume fraction. The overall mix proportions of all specimen sets are summarized in table 1.

Figure 1. Crumb rubber particles with the maximum size of 40 mesh.

Table 1. The mix proportions of various specimen sets.

Designation	Weight per unit volume (kg/m^3)				
	Water	Cement	Sand	Crumb Rubber	Foam
CR0	357.45	714.90	714.90	0	12.83
CR15	370.85	741.70	630.45	45.47	11.62
CR30	385.25	770.50	539.35	94.46	10.31
CR45	400.9	801.80	440.99	147.44	8.92

2.2. Casting of the specimen

During the casting process, the concrete components including cement, sand and crumb rubber were dry-mixed in a ball mixer for 5 minutes. Water was then poured into the mixer for wet-mixing for another 5 minutes. After the concrete mix had attained homogenous condition, foam was introduced into the mix to lower the density of the fresh concrete to approximately 1800 kg/m^3. Eventually, the fresh concrete was cast and compacted into various moulds for subsequent laboratory tests.

2.3. Laboratory tests of the specimen

Three tests were conducted to accomplish the objectives of the study namely, compressive strength test, flexural strength test as well as steady-state heat flux measurement thermal transmission test.

As specified in the standard BS EN 12390-3, the concrete cube with dimension of 100 mm x 100 mm x 100 mm was fabricated as the compressive strength test specimen. The axial compressive load was applied on the specimen at the loading rate of 0.02 mm/s. The peak load attained prior to the ultimate failure of the specimen was taken as the compressive strength of the specimen.

In accordance with the requirements specified in the standard ASTM C496, the concrete rectangular specimens with the dimension of 160 mm x 40 mm x 40 mm were cast. The axial load of the flexural strength machine was pre-set to be 1.2 mm/min. The maximum load attained by each specimen was taken as the flexural strength of the specimen.

The thermal conductivity test was carried out based on the requirements stated in the ASTM C332 standard. The concrete cube specimens with the dimension of 50 mm x 50 mm x 50 mm were oven dried for 24 hours, and then cooled to room temperature prior to their placement in the steady-state heat flux machine. The cold plate with a temperature of 25 °C was placed at the top of the specimen while the hot plate with a temperature of 40 °C was placed at the bottom of the sample. The heat transfers and heat change between the two plates were recorded automatically by the data logger in every minute for one day. The specimens were covered by rubber insulators to minimize the effect of transfer of heat from the external surrounding.

3. Experimental results

3.1. Compressive strength test

Both table 2 illustrate the relationship between the compressive strength of concrete and the different crumb rubber proportions. It is clearly shown that the compressive strength drops as the crumb rubber ratio in the concrete increases. Both 7 days and 28 days' compressive strength of the concrete depicts a similar trend. One notable finding is that the 28 days' compressive strength of the lightweight foamed concrete only reduces slightly as 15 % of crumb rubber is used to replace the sand, which is from 14.02 MPa to 13.89 MPa (0.93 %). Beyond the crumb rubber proportion of 15 %, the compressive strength begins to decline vigorously. For instance, the 28 days' compressive strength of the lightweight foamed concrete descends to 10.66 MPa, upon the incorporation of 45 % of crumb rubber into the concrete. The reduction percentage is relatively significant and reported to be23.97 %.

It is found that the primary reason causing the decreasing of the compressive strength is the relatively low strength of the rubber particles [2]. As the crumb rubber proportion of the concrete increases in the concrete, the properties of the crumb rubber increasing tend to dominate the properties of the concrete resulting in the loss of strength. Besides, the decline in compressive strength of the concrete is generally due to the lack of adhesion between the cement paste and the smooth crumb rubber particles [3]. During the loading stage, cracks will emerge and develop quickly around the rubber particles which results in rapid rupture of concrete. In addition, the fine nature of crumb rubber also leads to the generation of voids in the hardened concrete. Benazzouk et al. [4] also reported similar behaviour as packing of crumb rubber particles becomes difficult at high substitution level due to the generation of voids. It was observed that reduction in the concrete compressive strength was also partially due to the lower stiffness of crumb rubber particles as compared to the original fine aggregate.

Table 2. The 7 days and 28 days compressive strengths of concrete cube specimens with different crumb rubber proportions.

Designation	7 days		28 days	
	Compressive strength (MPa)	Reduction percentage as opposed to the control sample (%)	Compressive strength (MPa)	Reduction percentage as opposed to the control sample (%)
CR0	13.00	-	14.02	-
CR15	13.29	-2.23	13.89	0.93
CR30	10.23	21.31	12.66	9.70
CR45	7.37	43.31	10.66	23.97

3.2. Flexural strength test

The relationship between the flexural strength of the concrete and the crumb rubber proportion is presented in table 3. Both 7 days and 28 days' flexural strengths of the concrete behave in the same manner as the crumb rubber ratio increases. However, the reduction in flexural strength is more obvious and drastic as compared to that in compressive strength. For example, the matured flexural strength (28 days) of the concrete descends from 4.92 MPa to 3.96 MPa, which is equivalent to 37.80 % of reduction, when 45 % of the fine aggregates are substituted by the crumb rubber.

The reduction in flexural strength is primarily due to inadequate bonding between the cement paste and the crumb rubber particles. Gupta et al. [5] claimed that the decrease in flexural strength relies on the size and shape of the crumb rubber. The smooth texture of the crumb rubber will accelerate the propagation of cracks during the loading phase and the overall resistance against the applied force will be reduced.

Table 3. The flexural strength of concrete rectangular specimen with different crumb rubber proportions.

Designation	7 days		28 days	
	Flexural strength (MPa)	Reduction percentage as opposed to the control sample (%)	Flexural strength (MPa)	Reduction percentage as opposed to the control sample (%)
CR0	4.85	-	4.92	-
CR15	4.69	3.30	4.37	11.18
CR30	3.88	20.00	3.82	22.36
CR45	3.03	37.53	3.06	37.80

3.3. Thermal conductivity test

Table 4 shows the variation in thermal conductivity of the concrete as the crumb rubber proportion increases. The final result was obtained by calculating the average value of thermal conductivity, which was generated and recorded once in an hour, and tested for 24 hours. A worth mentioning finding is that the thermal conductivity drops as the crumb rubber proportion in the concrete increases, and the decreasing trend is approximately linear with the change in crumb rubber proportion. The thermal conductivity of the concrete is decreased from 1.1863 $W.K^{-1}.m^{-1}$ to 0.9733 $W.K^{-1}.m^{-1}$, which is equivalent to 17.95 % of reduction.

The decline in thermal conductivity is most probably due to the insulating properties of crumb rubber particles in the concrete. Since crumb rubber essentially has a better insulating effect than sand, the employment of crumb rubber in replacement of sand simply implies that that insulating ability will be carried over to the resultant concrete as well. Another reason is that the rubber particles possess non-polar nature which tend to entrap air in their surface. Air entrapped in the porosity is not a good medium for the heat to be transferred.

Table 4. The thermal conductivities of concrete specimens with different crumb rubber proportions.

Designation	Thermal conductivity ($W.k^{-1}.m^{-1}$)	Reduction percentage as opposed to the control sample (%)
CR0	1.1863	0%
CR15	1.0737	9.49%

CR30	0.9969	15.97%
CR45	0.9733	17.95%

4. Conclusion

This paper examines the use of crumb rubber as a substitute of fine aggregate at different proportions through experimental study. Several conclusions can be drawn based on the findings obtained through a series of laboratory tests.

The incorporation of crumb rubber in the lightweight foamed concrete will induce reduction in the strength properties of concrete, both compressive strength and flexural strength. However, the reduction in compressive strength is almost negligible (0.93 %) if the substitution of sand by crumb rubber does not exceed 15 %. For higher crumb rubber proportion, the compressive strength declines at a greater rate. The reduction in compressive strength when 45 % crumb rubber is introduced to the lightweight foamed concrete is reported as 23.97 %. In term of flexural strength of the concrete, a strong reduction effect by crumb rubber proportion can be spotted. At 45 % crumb rubber proportion, the flexural strength is reduced by 37.80 % as compared to the one without crumb rubber.

However, crumb rubber lightweight foamed concrete exhibits superior thermal properties than plain lightweight foamed concrete. This is proved by the decrease in thermal conductivity, from 1.1863 $W.K^{-1}.m^{-1}$ to 0.9733 $W.K^{-1}.m^{-1}$ (17.95 %) as 45 % of crumb rubber is introduced to the plain lightweight foamed concrete.

Acknowledgement

The authors would like to express their gratitude to Malaysian Rubber Export Promotion Council (MREPC) for financially support this research (8041-000), and also Universiti Tunku Abdul Rahman (UTAR) for providing facilities and resources for the research to progress smoothly.

Reference

[1] Mohammed B S, Adamu M, Shafiq N 2017 A review on the effect of crumb rubber on the properties of rubbercrete *IJCIET* **8** 599.

[2] Atahan A O, Yücel A Ö 2012 Crumb rubber in concrete: static and dynamic evaluation *Constr. Build. Mater.* **36** 617-22.

[3] Bisht K, Ramana P 2017 Evaluation of mechanical and durability properties of crumb rubber concrete *Constr. Build. Mater.* **155** 811-7.

[4] Benazzouk A, Douzane O, Langlet T, Mezreb K, Roucoult J M, Quéneudec M 2007 Physico-mechanical properties and water absorption of cement composite containing shredded rubber wastes *Cem. Concr. Compos.* **29(10)** 732-40.

[5] Gupta T, Chaudhary S, Sharma R K 2014 Assessment of mechanical and durability properties of concrete containing waste rubber tire as fine aggregate *Constr. Build. Mater.* **73** 562-74.

Extraction of Microcrystalline Cellulose from Two Different Agriculture Waste via Chemical Treatment

Nur Athirah Abdullah[1], Muhammad Hanif Sainorudin[1], Nilofar Asim[1], Masita Mohammad[1], Nurul Huda Abd Kadir[2], Zahira Yaakob[3]

[1]Solar Energy Research Institute, Universiti Kebangsaan Malaysia, 43600 Bangi, Selangor, Malaysia

[2]School of Fundamental Sciences, Universiti Malaysia Terengganu, 21030 Kuala Nerus Terengganu, Malaysia

[3]Departments of Chemical and Process Engineering, Faculty of Engineering and Built Environment, Universiti Kebangsaan Malaysia, 43600, Bangi, Selangor, Malaysia

masita@ukm.edu.myalam

Abstract. Microcrystalline cellulose (MCC) was successfully extracted from coconut husk fibre (CHF) and pineapple leaf fibres (PLF) using alkali and bleaching treatments. The extracted cellulose from the chemical treatments were characterized using X-ray diffraction (XRD), thermogravimetric analysis (TGA) and scanning electron microscopy (SEM). XRD analysis revealed that the extracted microcellulose from CHF and PLF using chemical treatments have high crystallinity. The thermal stability analysis of the extracted microcrystalline celluloses was investigated revealed that the CHF and PLF lose their lignin and hemicellulose during chemical treatments. Morphological investigation using SEM shows that chemical treatment could remove the layer of extractives from surface of CHF and PLF.

1. Introduction

Agricultural waste is known as one of the rich resources in Malaysia. Most of these agricultural wastes have high quantities of organic and inorganic materials which could produce various useful products. Utilizing these agricultural wastes as resource materials in different application not only provide us cheap and sustainable resource materials but also could tackle the waste disposal issue which making big pollution for environment [1]. In last few decades, more research is focusing on the use of cellulosic waste as filler and this study field has grown rapidly. Many researchers studied production of crystalline cellulose from agriculture waste such as coconut husk fibres, soy hulls, pineapple leaves, sugarcane bagasse, cornstalks, banana rachis, mulberry bark, straw, soybean pods [2]. Due to its biodegradability, availability, renewability, and environmental friendliness, natural fibres or lignocellulose fibres has attracted much attention which are being used as reinforcing components for thermoplastic and thermoset matrices. Coconut coir or coconut husk fibre is the natural fibres that has good thermal and low specific weight while benefit from its low cost and friendly processing [3]. However, its application affected by its limitation for high temperature process conditions and its degradability towards humidity [4]. pineapple leaf fibre contains natural fibres that displays high specific strength and rigidity. Pineapple leaf fibre is abundantly available, low cost, and very hygroscopic. The mechanical properties of pineapple leaf fibres can be associated with high cellulose content and low microfibrillar angle [5].

Natural fibres mainly contain of cellulose, hemicellulose and lignin as their main content. The most dominant reinforcing phase in plant structures called cellulose. cellulose is one of the most abundant biopolymers on earth available in plant-based materials such as wood, cotton and any other plants. It is relatively non-toxic, biodegradable and renewable materials which consider one of the most popular materials for polymer reinforcement in new research fields [6]. It can be processed into whisker-like microfibrils, microcrystalline cellulose (MCC), nanocrystalline cellulose (NCC) and many more. Extraction of cellulose commonly involve chemical treatment which include alkali treatment and bleaching, mechanical treatment and chemo-mechanical treatment [7].

To improve the compatibility of natural fibres, various methods have been studied. To modify the fibre surface, both physically and chemically methods were used. Most of the methods have been used for improving the properties of natural fibres such as surface geometry, dirt removal, fibre strength, and the interaction between the fibre and the matrix are chemical treatments [8]. Alkali treatment is the common chemical treatment method which used to extract most of hemicellulose, lignin, waxes and oils and make fibre surface more rough to reduce fibre aggregation [9]. Studies on alkali treatment of the agricultural waste to extract microcrystalline cellulose from natural fibres has been done by many researches using different agriculture wastes [10–12]. However, there isn't enough research on comparing the extraction of microcrystalline cellulose from different agriculture waste using the same treatment methods. Hence, the main objective of this study was to study the extracted microcrystalline celluloses from the coconut husk fibre (CHF) and pineapple leaf fibre (PLF) using the alkali treatment and bleaching method for both wastes.

2. Materials and methods

2.1. Materials

The CHF and PLF were collected from local agriculture farm, Sabak Bernam, Malaysia. Both wastes were washed and cut into small pieces before dring in an oven at temperature 35°C. The dried samples then were grind using grinder machine until become powder. Samples then washed again repeatedly with distilled water and dried in a temperature-controlled oven at 60°C for 18 h. The dried samples were kept in a desiccator before analysis.

Toluene, ethanol, potassium hydroxide acetic acid and sodium chlorite were purchased from R & M Chemicals Ltd., United Kingdom (reagent grade) and were used directly without any purification.

2.2. Preparation of microcrystalline cellulose from CHF and PLF

2.2.1. Dewaxing. The ground samples were dewaxed under reflux using toluene and ethanol with the ratio 2:1 for 6 hours. The solid was then filtered and washed several times using distilled water.

2.2.2. Bleaching process. The dewaxed fibres were sequentially delignified with acidified sodium chlorite (acetic acid was used to reach pH 4) at 70°C for 1 hour and this treatment were repeated thrice until the treated samples became white. Then it was filtered to remove water-soluble components and washed with distilled water until the pH was neutral.

2.2.3 Alkaline Treatment. The alkali treatment was performed to extract the cellulose by removing lignin and hemicellulose from the coconut husk fibre (CHF) and pineapple leaf fibre (PLF). The residue from bleaching process was then treated using potassium hydroxide (6 wt. %) overnight at room temperature. After filtration, the treated samples were washed carefully with distilled water until the pH was neutral and dried in an oven for 16 h at 40 °C. The dried cellulose was stored for further use.

3. Results and Discussion

The prepared samples were characterized by X-ray Diffraction (XRD) analysis, Thermogravimetric Analysis (TGA), and Scanning Electron Microscopy (SEM). Figure 1. present the raw coconut husk fibre (CHF) and pineapple leaf fibre (PLF) powder and the extracted microcrystalline cellulose from them.

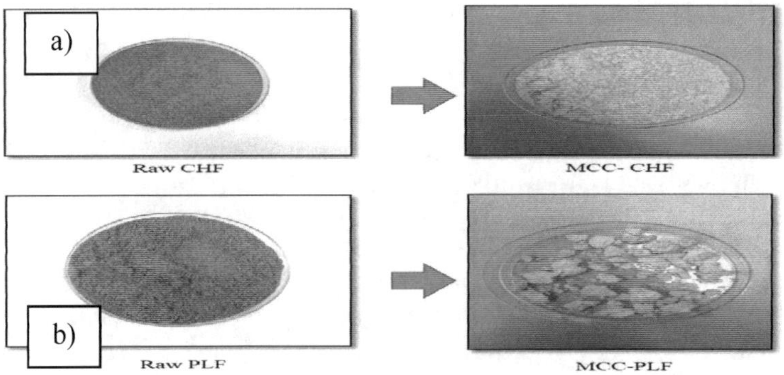

Figure 1. (a) coconut husk fibre powder and its extracted Microcrystalline cellulose (b) pineapple leaf fibre powder and its extracted Microcrystalline cellulose.

3.1. X-ray Diffraction (XRD) Analysis

X-ray diffraction analysis was conducted to analyze the crystallinity of prepared samples. The crystallinity index of the prepared samples were calculated based on Segal *et al.*, (1959) [13] method using equation 1:

$$C_I (\%) = I_{cry} - I_{ams} / I_{cry} \text{ x } 100 \quad (1)$$

Where, C_I is the crystallinity index (%), I_{cry} is the maximum intensity at a 2θ, and I_{ams} is the intensity of the amorphous material at an angle in the valley between the peaks. The Scherrer formula was used to calculate the crystallite size (table 1.). Figure 2 depicted the X-Ray diffractograms of prepared samples. Cellulose has a crystalline structure due to its hydrogen bonding interactions and Van der Waals forces between their adjacent molecules[14]. From the XRD of MCC-CHF and MCC-PLF, a sharp peak is set which is indicate that crystalline in nature at 2θ = 22.80° and 2θ = 22.76° respectively and both peaks are broad which conclude that as amorphous nature.

Table 1. The crystallinity index and crystallite size of MCC-CHF and MCC-PLF.

Samples	Crystallinity Index (%)	Crystallite size (nm)
MCC-CHF	71.8	4.47
MCC-PLF	68.2	4.14

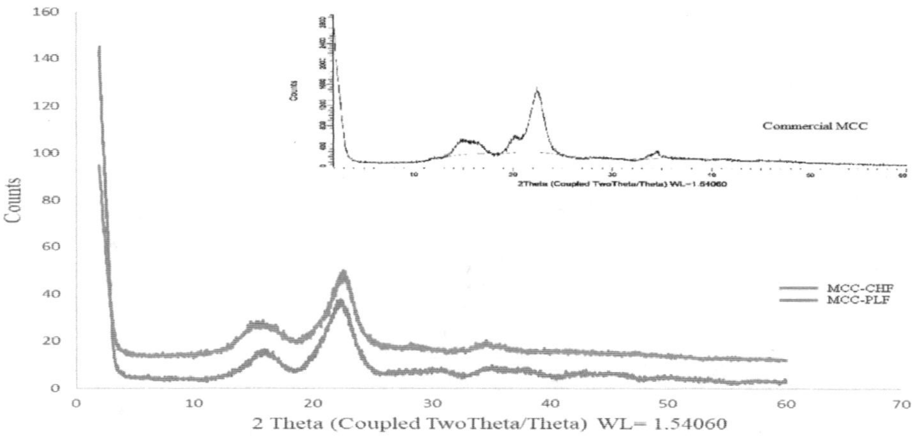

Figure 2. X- ray diffractograms of MCC-CHF, MCC-PLF and commercial MCC.

As shown in table 1, both MCC-CHF and MCC-PLF show a partial crystalline and amorphous nature. The crystallinity index of MCC-CHF (71.8%) is higher than MCC-PLF (68.2%). It is known that the crystallinity of cellulose on natural fibres can be affected by using chemical treatment.

3.2. Thermogravimetric Analysis (TGA)

TGA was performed by using Perkin Elmer Thermal Analyser (Waltham, MA, USA). This analysis was performed to evaluate the thermal stability of both fibres, MCC-CHF and MCC-PLF. The thermo-stability property of the sample prepared was analyzed by using about 6 mg of sample that were used from room temperature to 600°C with a heating rate of 10°C/min under nitrogen (30 ml/min). Figure 3 shows the TGA and derivative thermogravimetry (DTG) curves for prepared samples. According to figure 3 there is significant weight loss for both MCC-CHF and MCC-PLF within temperature 60 ºC to 100ºC due to the removal of bound water in the cellulose [15].The graphs also show the removal of lignin and hemicellulose within 210ºC - 500ºC for both samples. The fastest weight change was observed within 220ºC- 340 ºC.

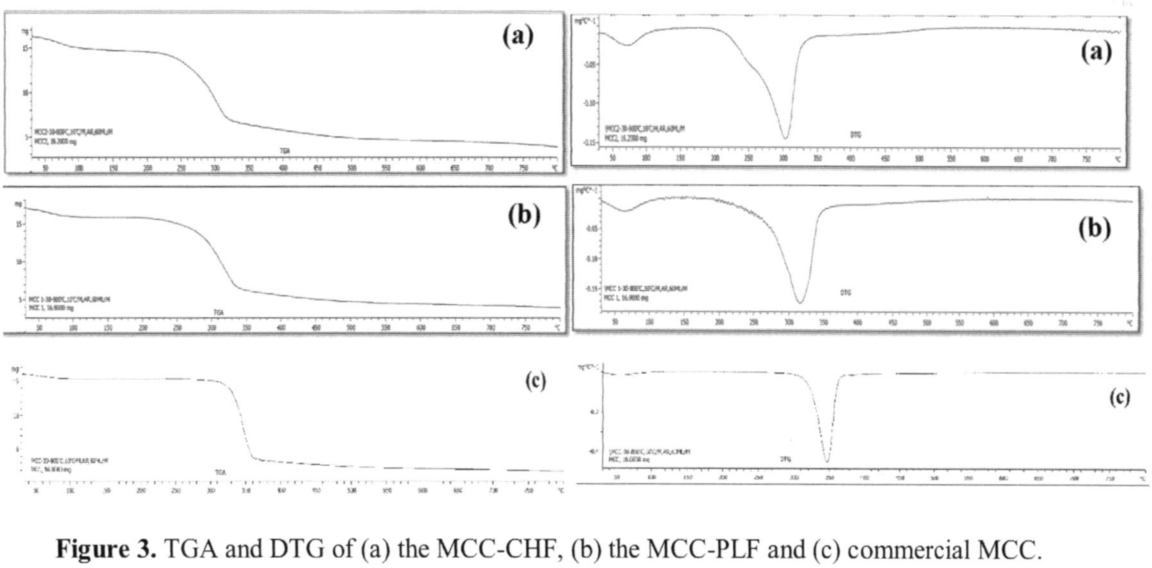

Figure 3. TGA and DTG of (a) the MCC-CHF, (b) the MCC-PLF and (c) commercial MCC.

3.3. Scanning Electron Microscopy (SEM)
Surface morphology of the fibres was studied by JEOL 175 Model JSM-6390 LV scanning electron microscope. The samples were coated with gold in a plasma sputtering apparatus. Figure 4 shows the SEM micrographs of extracted MCC-CHF and MCC-PLF respectively.

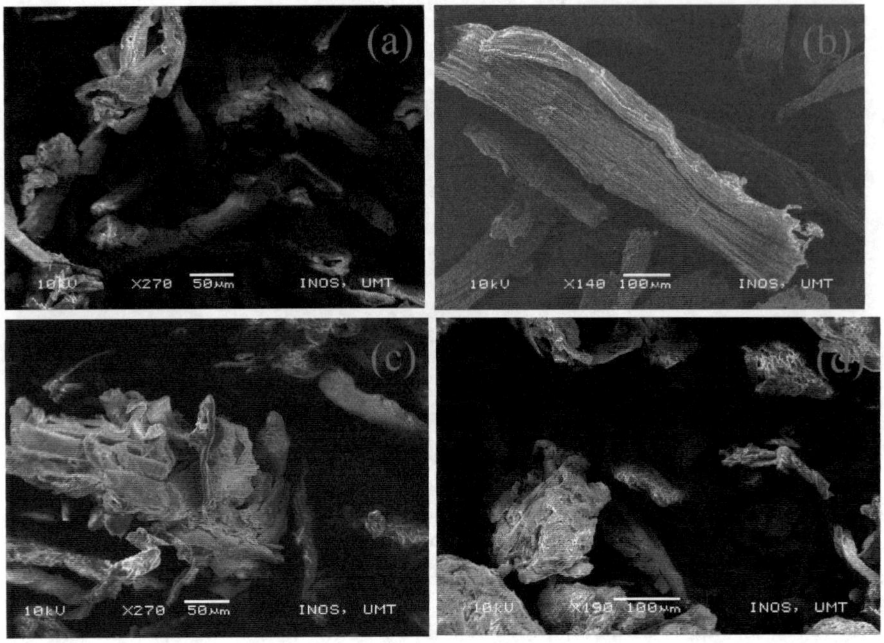

Figure 4. SEM images from MCC-CHF (a) and (b); MCC-PLF (c) and (d).

The agglomeration of fibrse could attribute to the incomplete removal of water in extracted samples. Meanwhile, the roughness of the surface is resulted from partially removal of hemicellulose, lignin, pectin, wax from the samples surfaces by chemical treatments [16].

4. Conclusion
In this study, microcrystalline cellulose successfully extracted from coconut husk fibre (CHF) and pineapple leaf fibres (PLF) using same chemical treatment methods which include alkali and bleaching treatment. The extracted microcrystalline celluloses exhibit rough surface because of partially removal of hemicellulose, lignin, pectin, wax from the sample's surfaces. The extracted microcellulose from coconut husk fibre (MCC-CHF) showed higher crystallinity index compared to extracted microcellulose from pineapple leaf fibres (MCC-PLF) which attributed to the various effect of chemical treatment on the fibers with different composition and microstructures.

Acknowledgement
The work was financially supported by FRGS/1/2014/TK04/UKM/03/2 from Centre of Research and Instrument Management (CRIM) and Ministry of Higher Education Malaysia.

References
[1] Nuruddin M, Chowdhury A, Haque S A, Rahman M, Farhad S F, Sarwar Jahan A Q 2011 Extraction and Characterization of Cellulose Microfibrils from Agricultural Wastes In An Integrated Biorefinery Initiative. *Cellulose Chem. Technol.* **45** 347-54.
[2] Johar N, Ahmad I, Dufresne A 2012 Extraction, preparation and characterization of cellulose fibres and nanocrystals from rice husk. *Ind. Crops. Prod.* **37** 93-9.

[3] Mir S S, Hasan S M N, Hossain M J, Hasan M 2012 Chemical modification effect on the mechanical properties of coir fiber. *Eng. J.* **16(2)** 73-83.

[4] Sabri M, Hafiz F, Shahril K, et al. 2013 Effects of Compatibilizer on Mechanical Properties and Water Absorption of Coconut Fiber Filled Polypropylene Composite. *Adv. Mater. Res.* **795** 313-7.

[5] Cherian B M, Leão A L, de Souza S F, et al. 2010 Isolation of nanocellulose from pineapple leaf fibres by steam explosion. *Carbohydr. Polym.* **81(3)** 720-5.

[6] Khatoon N, Ramezani O, Kermanian H 2012 Production of Nanocrystalline Cellulose from Sugarcane Bagasse. 4th Int Conf Nanostructures 12-4.

[7] Dufresne A 2013 Nanocellulose: A new ageless bionanomaterial. *Mater. Today* **16(6)** 220-7.

[8] Arsyad M, Wardana I N G, Pratikto, Irawan Y S 2015 The morphology of coconut fiber surface under chemical treatment. *Rev. Mater.* **20(1)** 169-77.

[9] Carvalho K C C, Mulinari D R, Voorwald H J C, Cioffi M O H 2010 Chemical modification effect on the mechanical properties of hips/ coconut fiber composites. *BioResources* **5(2)** 1143-55.

[10] Daifullah A A M, Girgis B S, Gad H M H 2003 Utilization of agro-residues (rice husk) in small waste water treatment plans. *Mater. Lett.* **57(11)** 1723-31.

[11] Ludueña L, Fasce D, Alvarez V A, Stefani P M 2011 Nanocellulose from rice husk following alkaline treatment to remove silica. *BioResources* **6(2)** 1440-53.

[12] Deepa B, Abraham E, Koshy R R, et al. 2015 Extraction and characterization of cellulose nanofibers from banana plant. *Handb. Polym. Nanocompos. Process Perform Appl. Vol. C Polym. Nanocompos. Cellul. Nanoparticles* 65-80.

[13] Segal L, Creely J J, Martin A E, Conrad C M 1959 An Empirical Method for Estimating the Degree of Crystallinity of Native Cellulose using X-ray Diffractometer. *Textile Res. J.* **29(10)** 786-94.

[14] Zhang Y H, Lynd L R 2004 Toward an aggregated understanding of enzymatic hydrolysis of cellulose: Noncomplexed cellulase systems. *Biotechnol. Bioeng.* **88(7)** 797-824.

[15] Julie Chandra C S, George N, Narayanankutty S K 2016 Isolation and characterization of cellulose nanofibrils from arecanut husk fibre. *Carbohydr. Polym.* **142** 158-66.

[16] Le Troedec M, Sedan D, Peyratout C, et al. 2008 Influence of various chemical treatments on the composition and structure of hemp fibres. *Compos. Part A Appl. Sci. Manuf.* **39(3)** 514-22.

Study of Failure Analysis of a Fracture Crankshaft Pulley Used on a Truck Engine

Husaini[1], T E Putra[1], S Novriandika[1]

[1]Laboratory of Computational Mechanics, Mechanical Engineering Department, Universitas Syiah Kuala, Darussalam – Banda Aceh, 23111, Indonesia

husainiftm@unsyiah.ac.id

Abstract. A pulley is a mechanism consisting of a wheel, with a groove between its two edges, which is fastened to a shaft. The belt usually passes along the groove of the pulley to give it the power of movement and rotation. In December 2016 there was a case of failure of the crankshaft pulley in a truck. This study aims to investigate the causes of fracture. Therefore, it was necessary to perform a failure analysis. Investigation methods included visual examination, scanning electron microscope analysis, chemical analysis of the material and mechanical tests. A finite element analysis (FEA) was also performed to quantify the stress intensity factor *(K_I)* distribution in the crankshaft pulley that is near the crack tip. From the finite element analysis, it was found that K_I was greater than the fracture toughness *(K_{IC})* of the material. Therefore, the propagation of crack occurs from the initial crack until final failure.

1. Introduction

The failure analysis has always been a source of interest for the technical and scientific community as a way through to collection and analysis of data, determine the cause and prevent mechanical failure [1]. Mechanical failure can thus be defined as any change in size, shape of a structure, machine part that renders it incapable of satisfactorily performing his intended function [2]. Generally, failure to a product still occurs frequently due to incidents and other factors [3]. Many causes of fracture failure in a pulley in a truck are related with fatigue and fracture, improper fastening of support [4] or high stresses, friction and fretting [5]. Some methodologies to prevent the fracture failure of crankshaft pulley implies a better knowledge of mechanical design [6], stress concentration factors [7] and crack initiation and propagation mechanisms [8]. Failures due to incidents usually occur because of loads that exceed the strength of a component, for example, shock loads due to collisions, overloads, mixed loads between tensile loads and bending loads [9, 10]. Moreover, many automotive components can be associated with failure due to fatigue during their operations [11].

The crankshaft pulley of a truck was damaged on 20 October 2016. Therefore, this condition caused the engine of the truck to malfunction. Consequently, it is clear that cracks are a serious problem, and efforts must be made to overcome and prevent crack propagation [12, 13]. Furthermore, the purpose of this study was to investigate the causes of failure in a crankshaft pulley.

2. Methodology

Pulley is made of AISI 1045 material (table 1) as will be explained further in section 3.2. The engine component that was used was the crankshaft pulley of a truck. The pulley was located on the engine, as shown in figure 1 [14].

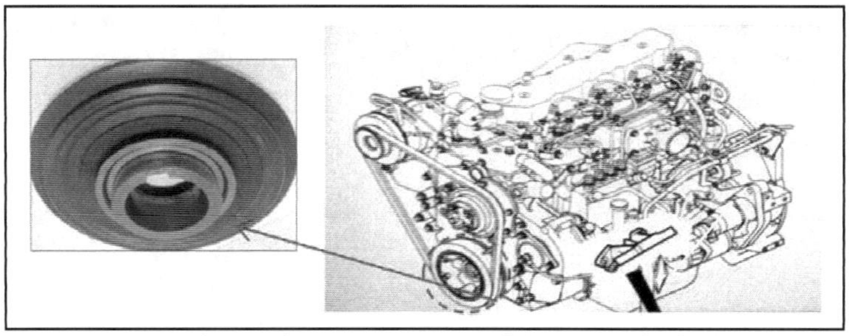

Figure 1. The engine components of the crankshaft pulleys.

To determine the shape, appearance, and location of the component that experienced macro failure were investigated by visual observations . The macroscopic observations were carried out on the component as shown in figure 1.

The identification of the chemical composition is carried out using optical emission spectroscopy (OES) to determine what chemical elements were quantitatively present in the crankshaft. The instrument used in this experiment was the PDA-700 spectrometer. Moreover, the fracture surface was observed by means of a scanning electron microscope (SEM) [15]. Hereafter, the testing of hardness with the Rockwell B method (with a loading of 100 Kgf) is aimed at determining the hardness of a material by means its resistance to specimens in the form of steel balls or diamond cones, which are exerted on the surface of the test material [16].

A *Stress Intensity Factor* (K_I) analysis *using the finite element method (FEM)* was used to determine the operating conditions of the component. The maximum stress intensity factor (K_I) was used to determine whether the cracks in the crankshaft pulley had propagated or not.

3. Results and Discussion

3.1. Visual Observations

The visual observations of the failure in the crankshaft pulley in figure 2 show the initial location of the cracks, and cracks in the visible area of the keyway. Then, figure 3 shows the surface fractures that occurred when the cracks began to appear in the parts. In this case, the area of stress concentration in the pulley was at the keyway because this was the area with the smallest radius and sharp angles.

3.2. Chemical Composition

The results of *chemical composition* **test** indicated that the crankshaft pulley material met the standard of the AISI 1045, as shown in table 1. The standard of the AISI 1045 [16] with regard to mechanical properties can be seen in table 2. The fracture toughness, K_{IC} had a value of 41 MPa√m [17].

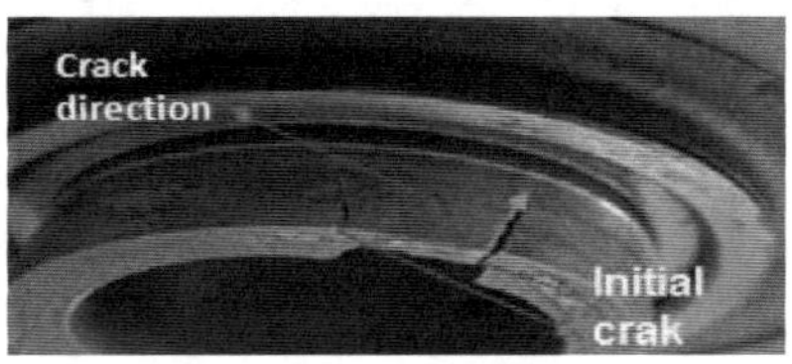

Figure 2. Initial crack on the crankshaft pulley.

Figure 3. Fracture surface.

3.3. Fracture Surface Observation Results Using SEM

The next stage, after the visual observations, was a fractographic examination using a Digital SLR camera. The results obtained were finally able to show the fracture surface pattern of the pulley. The observations this time, using a Digital SLR camera, referred to the predictions at the time of the visual observations, namely the crack initiation area, crack propagation, and also the final fracture. Then, to find out more about the pattern that occurred in the initial area of the fault, observations were made at a magnification of 41x. The results can be seen in figure 4 (c), which shows the initial cracks that occurred and the propagation of the cracks in two directions, namely, upwards and sideways. There were also chevron lines in the image.

Table 1. Comparison of results of chemical composition testing and AISI 1045.

Type	Composition (wt %)				
	Fe	C	Mn	S	P
The test results using Spectroscopy	89.8	0.45	0.773	0.0336	0.201
AISI 1045	98.4	0.43-0.5	0.60-0.90	0.04 (max.)	0.05 (max.)

Table 2. Mechanical properties of AISI 1045 [16].

No	Properties	Value
1	Tensile strength (σ_B)	585 MPa
2	Poisons Ratio (υ)	0.29
3	Yield strength (σ_{ys})	450 MPa
4	Young's Modulus, E	200 GPa
5	Hardness, Rockwell B	84 HRB
6	Density, ρ	7.87 Kg/m^3
7	Shear modulus (μ)	80 GPa
8	Shear Stress (τ_a)	146 MPa
9	Fracture Toughness, K_{IC}	41 MPa.m$^{1/2}$

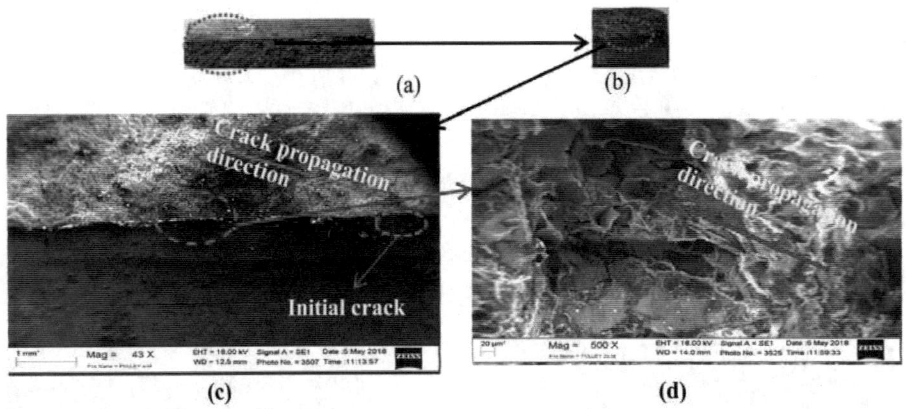

Figure 4. (a) Cut Part of the Pulley for SEM observation. (b) Specimens for Fractographic testing. (c) Initial crack location. (d) Initial Cracks Performed 500X Magnification.

Then, the next area to be observed was at the tip of the fracture surface area. For these observations to be made, a magnification of 43x was sufficient as it was clear that there was a fracture surface pattern in this area, as shown in figure 5.

3.4. Hardness Test Results

In a failure analysis research, it was necessary to test the hardness of the pulley pieces that failed, as shown in figure 6. Tests carried out using the Rockwell B hardness test equipment. The hardness test curve of the crankshaft pulley resulted from 9 hardness testing points using a load of 100 kgf.

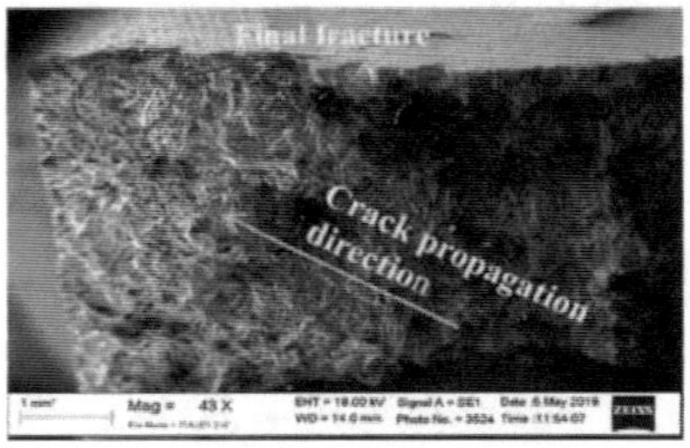

Figure 5. Fracture surface pattern.

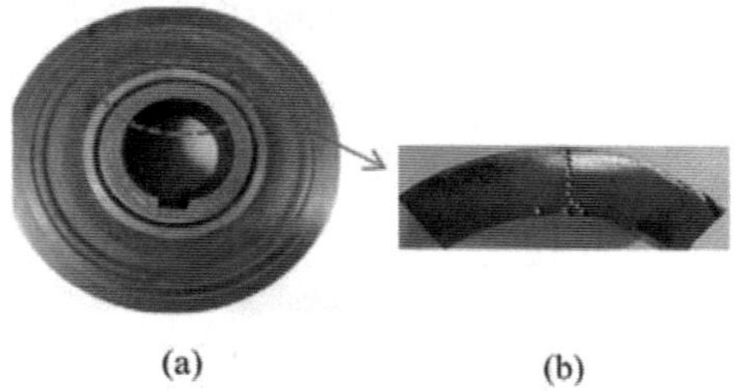

Figure 6. a) Parts of Pulleys Cut for hardness testing. b) Hardness Testing Specimen.

From the data obtained through the hardness test, as shown in figure 7 the average value of the hardness of the perpendicular cross-section was 88.6 HRB. The hardness of the inner area of the pulley was lower than that of the outer area of the pulley.

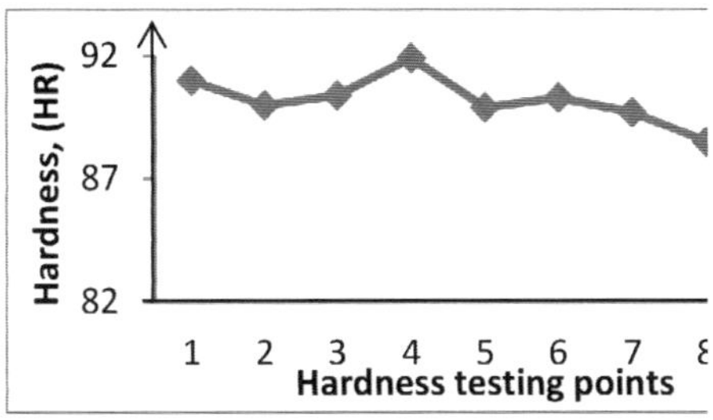

Figure 7. Hardness Testing Curve.

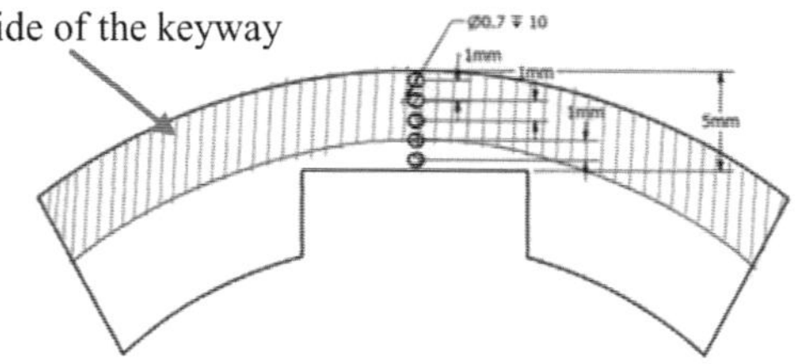

Figure 8. Sketch the hardness testing point around the keyway.

Figure 8 shows the keyway section, where the tightly-shaded part of the material with a thickness of 4 mm was harder than the part that was not shaded, while the remaining thickness of about 1 mm had a lower material strength than the shaded part, and it was this weaker part that caused the crack.

3.5. Stress Intensity Factor (SIF) Analysis Using the Finite Element Method

The results of the FEA on the crankshaft pulley showed the conditions that existed when a torsion moment force of 339876.9 N.mm was exerted. The geometry of the crankshaft pulley was made using software as can be seen in figures. 9(a), and (b)). Furthermore, a crack was made on the crankshaft pulley as shown in figure 10 so that simulations could be carried out to find the stress intensity factor (K_I) [18-20]. figure 10(a) shows the geometry of the crank axle and the given crack in the area of the groove, while figure 10(b) shows the enlargement of the crack area. The crack that was made had both a width and depth of 1 mm.

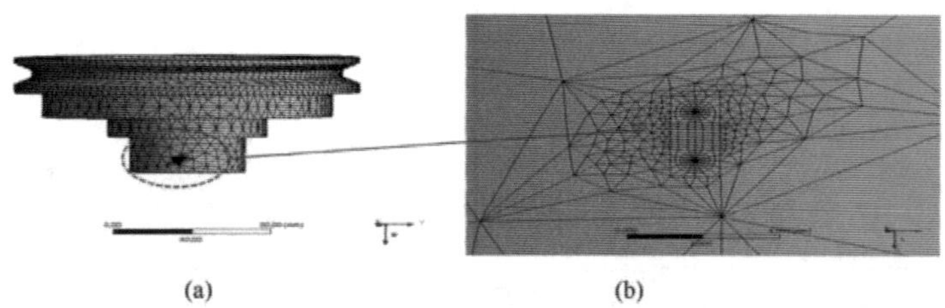

Figure 9. Geometry Crankshaft Pulleys after meshing.

The finite element analysis to find the stress intensity factor (K_I) near the crack tip on the crankshaft pulley by modelling the crack length and depth of 1.0 mm in the keyway area is shown in figure 9. A stress intensity factor (K_I) of 88 MPa.$m^{1/2}$ was obtained from the simulation results larger than the fracture toughness (K_{IC}) of the material was about 41 MPa.$m^{1/2}$ (as shown in table 2) . Therefore, this result indicated that crack propagation occurred on the crankshaft pulleys until final failure.

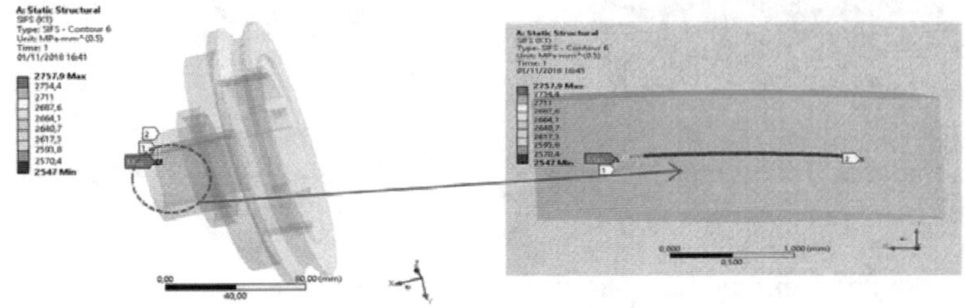

(a) (b)

Figure 10. (a) Analysys of SIF (K_I). (b) Magnification around crack tip.

4. Conclusions

According to the results, it can be concluded that the initial crack occurred because of the stress concentration in the sharp-angled keyway. Moreover, from the FEA results K_I that occurred near the crack tip was greater than the fracture touhness, K_{IC}. This also caused the crack propagation in the groove of the crankshaft pulley.

Acknowledgement
The authors would like to express their gratitude to Universitas Syiah Kuala for financial support for research through the Professor's research grant No. 03/UN11.2/PP/PNBP/SP3/2019.

References

[1] Becker W T, Shipley R J 2002 ASM Handbook Failure Analysis and Prevention **11** Ohio.
[2] Collins J A *Fail. Mater. in Mech. Design : Anal. Predic. Prevent.* **193** 2nd ed Wiley New York
[3] Husaini, Ali N, Riantoni R, Putra T E 2019 *IOP Conf. Ser. Mater. Sci. Eng.* **541** 012046.
[4] T E Putra, Husaini, Ali N, Zulfikar 2019 *IOP Conf. Ser. Mater. Sci. Eng.* **523** 012067.
[5] Husaini K, Kishimoto M, Hanji, Notomi M 2016 ARPN *J. Eng. Appl. Sci.* **11** 885.
[6] Khurmi R S, Gupta J K 2005 *Mach. Des.* (Eurasia Publishing House Ltd New Delhi) 110055.
[7] Noda N A, Takase Y, Monda K 1997 *Int. J. Fatigue* **19** 75.
[8] M de Freitas, Reis L, M da Fonte, Li B 2011 *Eng. Fract. Mech.* **78** 826.
[9] Husaini, Saputra E, Husni, Putra T E 2019 *IOP Conf. Ser. Mater. Sci. Eng.* **523** 012073.
[10] Husaini, Zuhaimi 2016 *Int. J. Tech.* **3** 456.
[11] Husaini, Putra T E, Ali N 2018 *Int. J. Automotive Mech. Eng.* **15** 5251.
[12] Putra T E, Husaini, Ali N, Zulfikar 2019 *IOP Conf. Ser. Mater. Sci. Eng.* **523** 012067.
[13] Akhyar H, Husaini 2016 *Int. J. Metalcasting* **10** 452-6.
[14] HINO DUTRO 2002 *Workshop Manual*: Engine W04D-J & S05C-B Serial WU300 340 XZU342 (*in Indonesia*) PT. HINO INDONESIA *MANUFACTURING.*
[15] Zyl V G, Al-Sahli A 2013 *Case Studies Eng. Failure Anal.* **3** 144.
[16] ASTM E18-15 2015 Standard Test Method for Rockwell Hardness of Metallic Materials *USA.*
[17] Samoila A 2011 *Thesis at the Mech. Eng. Dept.* **69** Faculty of Technology and Science, Karlstads University Press Sweden
[18] Husaini, Kishimoto K, Notomi M 2004 *Key Eng. Mater.* Trans Tech Publications Ltd. **261** 711.
[19] Husaini, Kishimoto K, Notomi M, Shibuya T 2001 *Fatigue Fract. Engng. Mater. Struct.* **24** 895.
[20] Husaini, Kishimoto K 2001 *Proc. SPIE - Int. Soc. Opt. Eng.* **4317** 111.

Samoila A 2011 *Liftning for lifting of the locomotive with bogies,* Karlstads University Press
Degree project at the Mechanical Engineering program

Encapsulate Materials Effect on $Zn_xCd_{1-x}S$ White Light Quantum Dots-Based White Light-Emitting Diodes

Y. X. Huang[1], S. R. Chung[1]

[1]No. 64, Wenhua Rd., Huwei Township, Yunlin County 63201, Taiwan (R.O.C.)

srchung@nfu.edu.tw

Abstract. In this study, colloidal ternary $Zn_xCd_{1-x}S$ white light quantum dots (WQDs) were prepared, which have the characteristics of broad emission window. We used epoxy, styrene ethylene butylene styrene block copolymer (SEBS), and silicone as encapsulate materials to pack the device. The results show that the $Zn_{0.5}$ and $Zn_{0.8}$ have band edge and surface state emission. The quantum yield (QY) and particle size of $Zn_{0.5}$ and $Zn_{0.8}$ WQDs is 19, 32 % and 4.5±0.5 and 3.1±0.5 nm, respectively. The white light-emitting diodes (WLEDs) were obtained by mixing WQDs with different resins, and pumping by UV chip. The luminous efficacy is 6.7 and 8.5 lm/W for $Zn_{0.5}$ and $Zn_{0.8}$-based WLED by using epoxy resin as encapsulate material. Moreover, the long period test can maintain 201 and 184 h for $Zn_{0.5}$ and $Zn_{0.8}$-based WLED, respectively. The results of this study present that the device encapsulated by epoxy resin can improve the luminous efficacy and stability of device. This outstanding finding may contribute to the future development of solid-state lighting (SSL).

1. Introduction

Quantum dots (QDs) have high quantum yield (QY), wavelength tunability and broad excitation band [1], so they can be applied to many applications, such as solid state lighting (SSL) [2, 3] and display [4]. In order to improve the disadvantage of color rendering and self-absorption effects of commercially available white light emitting diodes (WLEDs) [5, 6], the white light quantum dots (WQDs)-based WLEDs were developed. The WQDs can cover the entire visible light range because of its broad emission window, and preparation of high color rendering WLEDs [7]. However, the low luminous efficacy and low stability of device limits the application in solid state lighting (SSL). The luminous efficacy and stability of WLEDs are affected by the transparency of the encapsulate materials. The compatibility between encapsulates and QDs is influence the performance of WLED. Chung et al. reported that the Zn0.5 and Zn0.8 WQDs were used silicone resin to pack WLED, the luminous efficacy of Zn0.5 and Zn0.8 are 1.0 and 4.1 lm/W [8], while used polyurethane Acrylic (PUA) UV resin to pack WLED, the luminous efficacy of Zn0.8 increases to 11.9 lm/W, and the long period test can maintain for 10 h [9]. In this study, three types of encapsulate materias are selected to form WLEDs, and the performances of WLEDs are compared.

2. Experimental

2.1. Chemicals

Cadmium oxide (CdO, 99.99 %) was purchased from Alfa Aesar. Stearic acid (SA, 99 %), zinc oxide (ZnO, 99.99 %), Sulfur (S, 99.99 %), 1-octadecene (ODE, 90 %), hexadecylamine (HDA, 90 %), and trioctylphosphine oxide (TOPO, 90 %) were purchased from Sigma-Aldrich. Epoxy was purchased from Everwide Chemical Co., Ltd., SEBS block copolymer was purchased from Phon Tech Industrial Company, and silicone and UV chip provided by Lextar Electronics Corporation. The standard dye, rhodamine 101, used in quantum yield (QY) measurement, was also purchased from Aldrich. Hexane (99.7 %) and methanol (99 %) gotten from Macron Chemicals. All chemicals were used as received without purification.

2.2. Synthesis of $Zn_xCd_{1-x}S$ quantum dots
0.15 mmol of CdO and 0.15 mmol of ZnO for $Zn_{0.5}$, 0.06 mmol of CdO and 0.24 mmol of ZnO for $Zn_{0.8}$ and SA were placed in the three-necked flask and then heated to 230 °C under Ar flow. CdO, ZnO and SA powder was dissolved until the solution transfer to optical transparent with light yellow color. Meanwhile, the solution was cooled down to room temperature. HDA and TOPO were added into above solution and stirred together under Ar at room temperature, and then reheated up to 320 °C. S powder dissolved in ODE injected rapidly into the reaction flask and reacted at 290 °C to growth $Zn_{0.5}$ and $Zn_{0.8}$ QDs. After reacted for 60 min, the solution was cooled down to room temperature to stop the reaction. Samples were purified by adding methanol, and then dispersed in hexane for further measurement.

2.3. Preparation of white light-emitting diode
The $Zn_{0.5}$ and $Zn_{0.8}$ QDs blended with epoxy, SEBS, and silicone gel under different weight ratios, and then putted above mixture into 413 nm UV chip and cured at room temperature. The performance of WLEDs such as CIE coordinates, correlated color temperature (CCT, K), color rendering index (CRI) and luminous efficacy measured by integrating sphere. The samples were named as $Zn_{0.5}$-$LED_{y,z}$ and $Zn_{0.8}$-$LED_{y,z}$. The y means the type of gel, and z means the weight percentage of QDs.

*2.4. Characterization *
The optical properties of samples measured with UV-vis spectrometer (UV-vis, Hitachi UH-5300 spectrometer) and fluorescence spectrophotometer (FL, Hitachi F-7000). Relative QY of samples was determined by comparing the peak area under the curve of FL emission for the $Zn_{0.5}$ and $Zn_{0.8}$ QDs with that of fluorescent dye (rhodamine 101 in ethanol). Transmission electron microscope (TEM, JEOL JEM-2010) was used to analyze the morphologies and size distribution of samples. The LED measuring system (ISUZU Optics, SLM-6Z) is used to measure the performance of devices.

3. Results and Discussion

3.1. FL spectra and TEM images of $Zn_xCd_{1-x}S$ QD
The absorption and emission spectra of $Zn_{0.5}$ and $Zn_{0.8}$ QD shows in figure 1. We can find that the $Zn_{0.5}$ and Zn0.8 QD have very broad emission window, which combines with band edge and surface state emission [8]. The band edge emission peak and QY of Zn0.5 and Zn0.8 QDs are 429, 409 nm, and 19, 32 %, respectively. Moreover, band edge emission is blue shift with increasing Zn content due to the larger band gap of ZnS. The photographs of Zn0.5 and Zn0.8 QD stored in the hexane under UV light is also shown in figure 1. The light color is warm and cold white light for Zn0.5 and Zn0.8 QD, respectively. This result is consistent with that reported by Siao and Chen et al. [8-11] In the TEM images we can find that the morphology of Zn0.5 and Zn0.8 is spherical with 4.5±0.5 and 3.1±0.5 nm

and the size distribution is uniform.

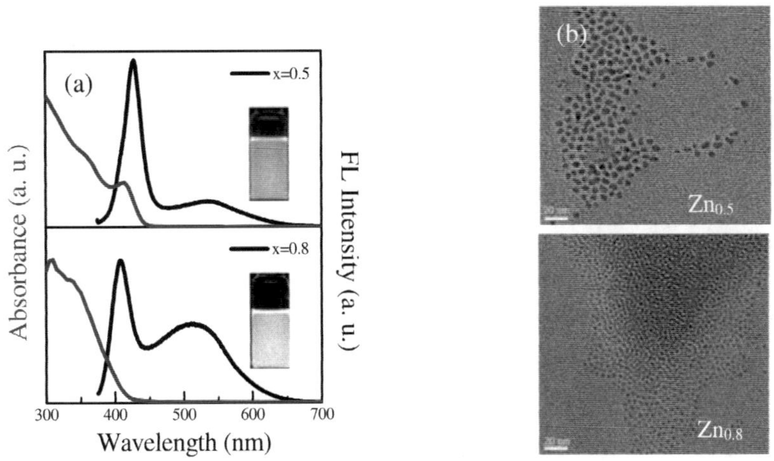

Figure 1. (a) Absorption and emission spectra and (b) TEM image of samples. Insert photograph in (a) shows the $Zn_{0.5}$ and $Zn_{0.8}$ QDs under UV irradiation.

3.2. White light-emitting diodes

The $Zn_{0.5}$ and $Zn_{0.8}$ used as light conversion materials to form WLED. The EL spectra and CIE coordinates of WLED show in figure 2, and the light transmittance of encapsulates show in figure 3. We can find that the EL intensity of Zn_x-LED_{epoxy} is higher than that of Zn_x-LED_{SEBS} and Zn_x-$LED_{silicone}$ significantly, and the CIE coordinates close to the position of warm white light. We can find that the transmittance of epoxy and silicone film is higher than 80 % in visible light range, while that of SEBS is lower 80 %. The luminous efficacy of $Zn_{0.5}$-$LED_{epoxy,50}$, $Zn_{0.5}$-$LED_{SEBS,80}$ and $Zn_{0.5}$-$LED_{silicone,67}$ are 6.7, 5.2 and 5.9 lm/W, and that of for $Zn_{0.8}$-$LED_{epoxy,33}$, $Zn_{0.8}$-$LED_{SEBS,50}$ and $Zn_{0.8}$-$LED_{silicone,33}$ are 8.5, 5.6 and 8.0 lm/W, respectively. The results are summarized in table 1. The stability of Zn_x-$LED_{y,z}$ is shown in figure 4. The long period test of $Zn_{0.5}$-$LED_{epoxy,50}$, $Zn_{0.5}$-$LED_{SEBS,80}$ and $Zn_{0.5}$-$LED_{silicone,67}$ can maintain 201, 28 and 4 h, and the reliability for $Zn_{0.8}$-$LED_{epoxy,33}$, $Zn_{0.8}$-$LED_{SEBS,50}$ and $Zn_{0.8}$-$LED_{silicone,33}$ are 184, 161 and 44 h. We can find that the luminous efficacy and reliability of Zn_x-LED_{epoxy} is better than that of Zn_x-LED_{SEBS} and Zn_x-$LED_{silicone}$, owing to the light transmittance of the epoxy is the highest, and the compatibility with QDs is the best.

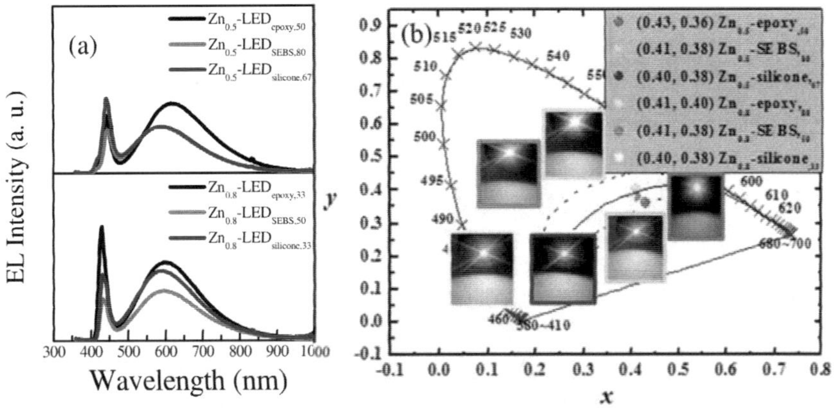

Figure 2. The EL spectra (a) and CIE coordinates (b) of Zn_x-$LED_{y,z}$. Insert photographs show the devices under 20 mA injection current.

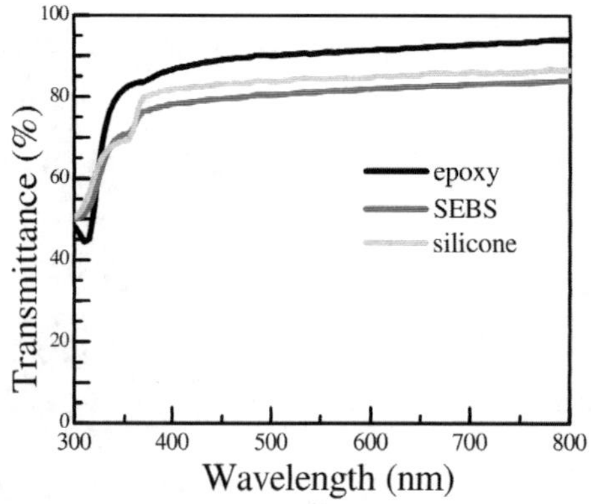

Figure 3. The transmittance of epoxy, SEBS, and silicone.

Table 1. The properties of Zn_x-$LED_{y,z}$.

QD	gel	CIE (x, y)	CCT (K)	CRI	Efficacy (lm/W)
	$epoxy_{50}$	(0.43, 0.36)	2760	85	6.7
$Zn_{0.5}$	$SEBS_{80}$	(0.41, 0.38)	3776	86	5.2
	$silicone_{67}$	(0.40, 0.38)	4508	90	5.9
	$epoxy_{33}$	(0.40, 0.35)	3379	83	8.5
$Zn_{0.8}$	$SEBS_{50}$	(0.41, 0.38)	3392	83	5.6
	$silicone_{33}$	(0.40, 0.38)	3593	83	8.0

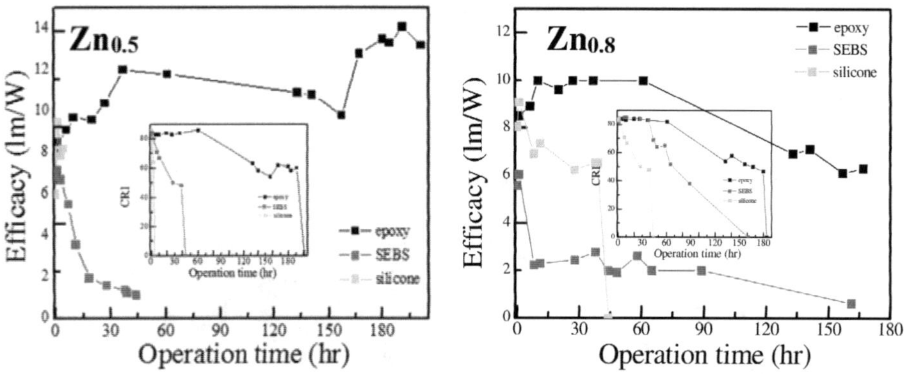

Figure 4. Stability of Zn_x-$LED_{y,z}$ for long period of working time under 20 mA. Insert diagram shows the CRI variation of devices.

4. Conclusion

In conclusion, we present a facile thermal pyrolysis organometallic route to synthesize white light Zn_x WQDs, and used epoxy, SEBS and silicone mixed WQDs to pack WLED. We can find that $Zn_{0.5}$ and $Zn_{0.8}$ QD have broad emission window, and QY is 19 and 32 %, respectively. The morphology of $Zn_{0.5}$ and $Zn_{0.8}$ QD is spherical. Using epoxy as encapsulate material, the $Zn_{0.8}$-$LED_{epoxy,33}$ device have the best luminous efficacy of 8.5 lm/W and operating time for 184 h. This study shows that the high light transmittance of the encapsulate material and the good compatibility of the QDs enhance the luminescent properties of the WLED.

Acknowledgement
This work was supported by the Ministry of Science and Technology of Taiwan under Contract nos. 107-2221-E-150-034 and 108-2221-E-150-016.

References

[1] Chung S R 2016 Phosphors, Up Conversion Nano Particles, Quantum Dots and Their Application **2** 419-29.
[2] Thejokalyani N, Dhoble S J 2014 *Renewable Sustainable Energy Rev.* **32** 448-67.
[3] Crawford M H 2009 *IEEE J. Selec. Topic Quant. Elect.* **15** 1028-37.
[4] Luo Z, Chen Y, Wu S T 2013 *Optics Express* **21** 26269-84.
[5] Gao J, Dane J 2003 *Appl. Phys. Lett.* **83** 3027-9.
[6] Wang Z L, Liang H B, Zhou L Y, Wang Gong M L, Su Q 2008 *J. Lumin.* **128** 147-54.
[7] Su Y S, Chung S R 2016 *Proc. SPIE* 9954.
[8] Chen H S, Wang K W, Chen S S, Chung S R 2013 *Opt. Lett.* **38** 2080-2.
[9] Siao C B, Wang K W, Chen H S, Su Y S, Chung S R 2016 *Opt. Mater. Express* **6** 749-58.
[10] Siao C B, Chung S R, Wang K W 2016 *Proc. SPIE* 9954.
[11] Chen H S, Wang K W, Chung S R 2011 *Proc. IEEE* 445-8.

The Effect of Nano Ceramic Powders On Mechanical and Physical Properties of Normal Concrete

Zainab Jawad Kadhim[1], Ahmed Raad Al-Adhadh[2], Dhay Ali sabur[3]

[1]Department of Civil Engineering, Al-Mustaqbal University College, Babylon, Iraq
[2]Department of Civil Engineering, Engineering College, Al-Muthanna University, Al-Muthanna, Iraq
[3]Department of Physics Science, Al-Mustaqbal University College, Babylon, Iraq

ahmad_al_iraqi2000@yahoo.com

Abstract. Concrete is one of the most important materials used in construction. Properties of concrete can be improved using additives like nano-ceramics which acts as reinforced phase to raise the mechanical properties of concrete. Effect of nano-powder on some properties of concrete has been determined. In this study, two concrete mixes have been prepared, first one, by adding different contents of nano–ZrO_2, second using nano – Al_2O_3 powder. Properties of concrete have been studied through the compressive strength, Splitting tensile strength, workability and water absorption. Results showed that the slump reduced from 103 mm to 62 mm when 1.1% nano-zirconia added, the same behavior observed in mix that contain nano alumina powder. The compressive strength increases when nano-zironia added, the maximum value of compressive strength obtained when 0.7% nano-zirconia added (45.3 MPa at 28 days, 48.3 MPa at 60 days and 52.3 MPa at 90 days). Moreover, splitting tensile strength was improved with increasing nano-zirconia content and the maximum values were at 0.7% of nano-zirconia, (5.2 MPa at 28 days ,6.6 MPa at 60 days and 8.2 MPa at 90 days), water absorption decreased when nano-zirconia percentage increased (0.12 % at 28 days and 0.1 % at 60 and 90 days). In this research, it is also observable that nano alumina powder has more influence for enhancing concrete properties than zirconia powder. Compressive strength and splitting tensile strength of concrete mix with nano alumina increased highly with increasing content of Al_2O_3 and reach its optimum value at 1.1 % of Al_2O_3 in NA6. Alumina powder has a high ability to diminish pores size and to increase density of concrete mixture; so, low permeability, low water absorption concrete produced.

1. Introduction

Concrete is the most important ingredient used in construction. Mechanical properties and durability of concrete is an essential issue nowadays. Properties of this concrete can improve by adding additives like nano-ceramic powders, which acts as reinforced phase to improve the properties of concrete.

Nano particles are the simple building block in nanotechnology science and are consist of up to thousands of atoms of 1 – 100 nm. Nano-technologies are developing with high rate, due to unique applications and using of nano-technology; there is interest in studying the effect of nano–ceramic powders on characterizations of hardened cement paste, cement mortar and concrete. There are few researches focused on utilization of nano powders with concrete and other construction materials. Shekariand and Razzaghi [1] investigated the effect of nano particles on performance and properties of high strength concrete. The nano materials he used to be nano ZrO_2, nano Fe_3O_4, nano TiO_2, nano Al_2O_3. Results reveal that Nano powders can be very successful in improvement of mechanical properties and durability of concrete. Moreover, the Nano-Al_2O_3 contribution on enhancements the properties were more than other nano particles. Ali Nazari [2] studied the effect of nano alumina particles on mechanical properties of blended concrete, it has been concluded that the partial replacement of cement with nano alumina powder improved mechanical properties of concrete but decreased its workability. Konstantin Sobolev [3] reviewed the beneficial effects of the nano-particles for the enhancement of concrete properties and performance. Ali Nazari [4] has studied the compressive strength and workability of concrete by partial replacement of cement with nano Fe_2O_3 particles, the results showed that strength of concrete has been improved with incorporation of nano Fe_2O_3 particles up to maximum substitution percent of 2%. The workability of fresh concrete decreased with increasing nano particles content. Hui Li [5] studied the characterizations of mortar mixed with nano particles. Results of this study showed that the compressive and flexural strength of mortar mixed with nano particles at curing age of 7 days and 28 days were higher than that of plain cement mortar.

In this research, the influence of Nano – ZrO_2 and Nano – Al_2O_3 powders on properties of concrete through the compressive strength, splitting tensile strength, workability and water absorption experimentally investigated. For this purpose, different concentration of nano-ZrO_2 and Al_2O_3 such as 0.1, 0.3, 0.5, 0.7, 0.9 and 1% (wt. %) as partial replacement of cement has been added to concrete mix.

2. Materials and primary test methods

The physical and chemical properties of the materials used in this study are specified in this section. The materials used in this include ordinary Portland cement, coarse aggregate, fine aggregate, Nano – Powders and water. *A subsection*
Some text.

2.1. Cement

Ordinary Portland cement has is used in this study. Tables (1 and 2) show the physical and chemical properties of cement used in this study according to Iraqi Specification No.5 / 1984.

Table 1. Physical properties of O.P.C (according to IQS No. 5/1984).

Physical Properties	Test Results	IQS No. 5/1984
Setting Time (Vicat's Method)	103 235	**$\geq$ 45 min** **$\leq$ 600 min**
Fineness (Blaine Method), m^2/kg	310	**$\geq$ 230 m^2/Kg**
Compressive Strength, MPa **3 days** **7 days**	 18.7 26.9	 **$\geq$ 15 MPa** **$\geq$ 23 MPa**

Table 2. Chemical composition of O.P.C (according to IQS No. 5/1984).

Compound	Chemical Name	Test Results %	IQS No. 5/1984
Lime	CaO %	60.14	-
Silica	SiO_2 %	19.77	-
Alumina	Al_2O_3 %	4.38	-
Iron Oxide	Fe_2O_3 %	5.20	-
Sulphate	SO_3 %	2.15	$\leq$ 2.8
Magnesia	MgO %	1.71	$\leq$ 5
Free Lime	Free CaO %	1.12	-
Loss On Ignition	L.O.I. %	3.19	$\leq$ 4
Insoluble Residue	I.R. %	1.23	1.5
Lime Saturation	L.S.F.	0.95	0.66 – 1.02
Main Compounds	**Percent by weight of cement**		

Tricalcium Silicate	C_3S	57.47	-
Dicalcium Silicate	C_2S	12.55	-
Tricalcium Aluminates	C_3A	6.06	-
TetracalciumAlumino Ferrite	C_4AF	10.83	-

2.2. Coarse aggregate (Gravel)

Aggregate of 20 mm maximum size and 2.65 specific gravity were used in this study. It was brought from AL- Nibai quarry. Grading of aggregate has been determined regarding the Iraqi Standard No. 45/1984. Physical and chemical Properties of coarse aggregate are shown in table 3. Figure 1 shows the Grain size distribution curve of the coarse aggregate.

2.3. Fine aggregate (Sand)

Natural sand used with maximum size of 5 mm. It was brought from Al–Ukhaider quarry. Table 4 shows the chemical and physical properties of sand which conforms to Iraqi Specification No.45/1984. Figure 2 demonstrates. Grain-size distribution curve of sand.

Table 3. Physical and Chemical Properties of Coarse Aggregate (according to IQS No.45/1984).

Physical Properties	Test Results	IQS No. 45/1984 for Zone II
Specific Gravity	2.65	-
Sulfate Content	0.10 %	≤ 0.1 %
Absorption	0.5 %	-
Clay Content	0.14 %	≤ 3 %

Table 4. Chemical and Physical Properties of the sand (according IQS No. 45/1984).

Properties	Test Results	IQS No. 45/1984 for Zone II
Specific Gravity	2.60	-
(SO_3) %	0.28	≤ 0.5 %
Absorption %	0.75	-
Clay Content %	3.67	≤ 5 %

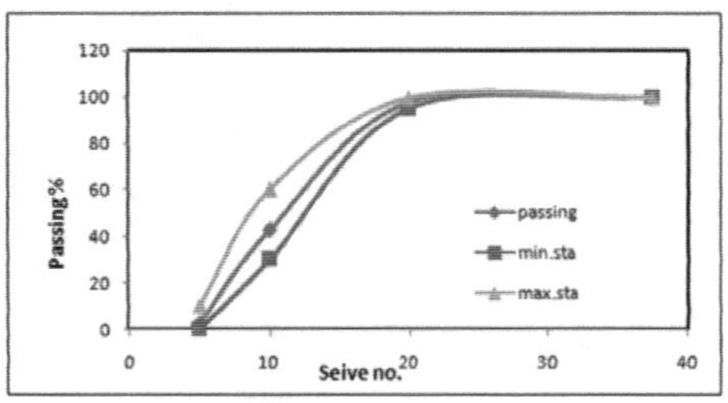

Figure 1. Grain-size distribution curve of Coarse Aggregate.

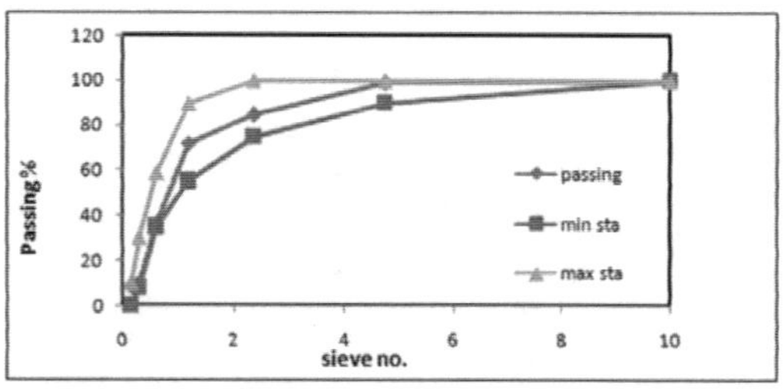

Figure 2. Grain-size distribution curve of Fine Aggregate.

2.4. Nano–Powders

Two types of nano-materials are used in this study, Nano–Zirconia and Nano–Alumina. These materials are used as a form of nano powder.

2.4.1. Nano–Zirconia powder. Nano Zirconia powder manufactured by (SIGMA Company) has been used in this study. Table 5 illustrates the purity of this powder.

2.4.2. Nano–Alumina powder. In this study, fine, very pure alumina (> 99 %) was used. Manufacturing company tested the properties of powder and the results of tests are shown in table 6.

Table 5. Chemical composition of zirconia nano-powder.

ZrO2	Al	Pb	Fe
> 99.9%	< 0.07%	< 0.02%	< 0.01%

Table 6. Characters of alumina nano-powder.

Phase	Purity %	Average Size (nm)	Surface area (cm^2/g)	Density Cm^3/g
γ	99.95	< 170	13.5	0.4 – 0.5

2.5. Mixing water
In this study, portable water was used for mixing and curing.

3. Concrete Mixture
Series of NZ and NA mixtures were prepared using various percentages of nano powders. Content of Nano ceramic powders within each sample was (0, 0.1%, 0.3%, 0.5%, 0.7%, 0.9% and 1.1% by weight of cement) for both alumina and zirconia powder as shown in table 7.

Table 7. Mix proportions of concrete specimens.

Zirconia Samples				Alumina Samples			
Samples	**Nano Zirconia %**	**W/C**	**Cement Kg/m^3**	**Samples**	**Nano Alumina %**	**W/C**	**Cement Kg/m^3**
C0	0	0.5	580	**C0**	0	0.5	580
NZ1	0.1	0.5	579.42	**NA1**	0.1	0.5	579.42
NZ2	0.3	0.5	578.26	**NA2**	0.3	0.5	578.26
NZ3	0.5	0.5	577.1	**NA3**	0.5	0.5	577.1
NZ4	0.7	0.5	575.94	**NA4**	0.7	0.5	575.94
NZ5	0.9	0.5	574.78	**NA5**	0.9	0.5	574.78
NZ6	1.1	0.5	573.62	**NA6**	1.1	0.5	573.62

4. Methodology
To study the effect of nano-ceramic powders on properties of concrete, concrete samples were prepared with selective mix proportions of 1:1.5:2.5 by weight and water / cement ratio of 0.5. Fresh concrete Specimens with dimensions of (150*150*150 mm) cubes were prepared in order to evaluate the compressive strength. Moreover, samples with dimensions of (300*150 mm cylinder) are prepared for testing the splitting tensile strength. All specimens were cured for 28, 60 and 90 days using portable water at room temperatures.

4.1. Compressive strength test
Compressive strength tests were carried out according to B.S 1881 part 116: 1983 [6]. All cubes were tested using a hydraulic compressive machine of 2000kN (ELE digital testing machine). All specimens were cured in water until testing age is up. The compressive strength value was taken to be an average value of three tests.

4.2. Splitting Tensile Strength
The splitting tensile strength was resolute according to the procedure outlined in BS 1881 part 117:1983 [7]. Cylinders were cast, molded and cured in a similar way as the cubes. Each value was average of three specimens for each age.

4.3. Water absorption
This test was performing on three samples of each mixture according to BS 1881 part 122:1983 [8].

4.4. Workability
Slump, compacting factor and V-B test were performed on fresh concrete according to BS 1881 part 2:1970 [9].

5. Results and Discussion
Al the tests results are presented in this section. The results contain the effect of nano ceramic powder on the workability of concrete mix, compressive and tensile strength, and finally water absorption results.

5.1. Workability Results
Workability of control fresh concrete (without additives) and of these samples containing different percentages of Nano Zirconia and Nano Alumina powders are shown in table 8. Workability of Control concrete (C0), NZ and NA series concrete are shown in figure. 3. Results demonstrate that control concrete samples have high slump value (110mm). It can be clearly noticed that workability decreases as the nano ceramic powder, Nano Zirconia and Nano Alumina, increases. This might be relevant to the concept that explains how fine materials increase bonding between cement paste and aggregate to provide adequate workability without segregation [4]. On the other hand, increasing of nano powders concentration leads to high surface area mixture and consequently increased water requirements which necessary for wetting the surface of cement particles [2, 3], for this reason, slump values decreased with increasing nano powders content as water-cement ratio is constant which is 0.5.

Compacting factor and V-B value are within acceptable range, indicating that fresh concrete was flow easily and at the same time liberated from segregation as indicated in table 10.

Table 8. Workability of concrete specimens.

Zirconia Samples				Alumina Samples			
Samples	Slump (mm)	Compacting Factor %	V – B test (second)	Samples	Slump (mm)	Compacting Factor %	V – B test (second)
C0	110	0.78	10.2	**C0**	110	0.78	10.2
NZ1	101	0.84	8.5	**NA1**	106	0.84	9.5
NZ2	100	0.86	8.1	**NA2**	102	0.87	8.8
NZ3	95	0.89	7.5	**NA3**	98	0.92	8.4
NZ4	93	0.94	7.3	**NA4**	95	0.95	8.1
NZ5	89	0.90	7.1	**NA5**	90	0.96	7.7
NZ6	62	0.85	7.0	**NA6**	65	0.96	7.3

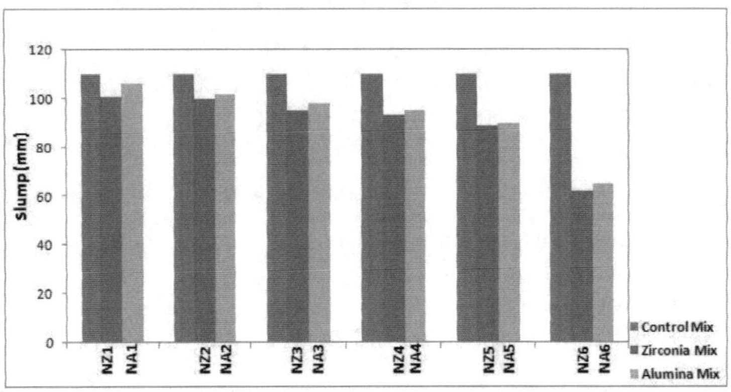

Figure 3. Slump of investigated specimens.

5.2. Strength Results

5.2.1. Compressive strength. Compressive strength of control concrete and samples that contain Nano powders after 28, 60 and 90 days are indicated in tables 9 and 10. Figure 4 illustrates that compressive strength of concrete samples that contain nano zirconia and nano alumina powders at age of 28 days. Compressive strength of NZ series increases as the percentage of Nano ZrO_2 powder increases up to 0.7% and then it starts to fall down slightly with (0.9 and 1.1 %) but still higher than that of control samples. Therefore, the optimum percentage ZrO_2 to reach maximum compressive strength of concrete is 0.7%. On the other hand, For NA series, compressive strength increases as the content of Al_2O_3 increases and there is no optimum percentage of Al_2O_3 to reach maximum compressive strength conversely to the effect of ZrO_2.

Figure 5 and 6 reveal the compressive strength of NZ and NA series at curing age of 60 and 90 days respectively. At age of 60 and 90 days NZ and NA series follow the same behavior that at 28 days. As observed from figure 4, 5 and 6, Nano alumina powder more effective in enhancing the strength of normal concrete than nano zirconia powder.

High enhancement in compressive strength has been attributed to high activity of Nano powder which accelerate the hydration reactions of cement and lead to form large quantities of hydration products (high gel volume) and consequence, consuming of $Ca(OH)_2$ that formed at early ages of the hydration process [10]. Also, Nano powder responsible about increasing the packing density to cement paste due to its tiny particles size, which act as filler materials, and then decrease the porosity and pores volume in hydrated cement paste. Maximum strength of concrete samples is related with workability and this can only obtain when the fresh concrete has adequate degree of workability because of ability on self-compacting [11-13].

Table 9. The effect of nano Zirconia on the compressive and tensile strength of concrete.

Zirconia Samples	Compressive Strength (MPa)			Splitting Tensile Strength (MPa)		
	28 days	60days	90days	28 days	60days	90days
C0	33.3	38.5	41.3	4.1	5.5	6.2
NZ1	41.3	45.4	47.7	4.3	6.1	6.5
NZ2	42.1	45.8	48.4	5.1	6.5	7.5
NZ3	43.8	46.6	49.2	5.6	6.8	7.6
NZ4	45.3	48.3	52.3	6.5	7.3	8.2
NZ5	42.5	47.5	49.4	6.1	6.8	7.8
NZ6	42.1	47.1	48.6	5.7	6.3	7.4

Table 10. The effect of nano Alumina on the compressive and tensile strength of concrete.

Alumina Samples	Compressive Strength (MPa)			Splitting Tensile Strength (MPa)		
	28 days	60days	90days	28 days	60days	90days
C0	33.3	38.5	41.3	4.1	5.5	6.2
NA1	42.5	45.7	48.5	4.8	6.7	6.8
NA2	43.6	46.2	49.1	5.6	7.2	7.5
NA3	44.2	46.8	49.6	6.2	7.7	8.3
NA4	46.2	48.7	52.4	6.8	8.1	8.8
NA5	47.1	49.2	52.8	7.1	8.7	9.2
NA6	47.7	49.6	53.2	8.2	9.3	9.7

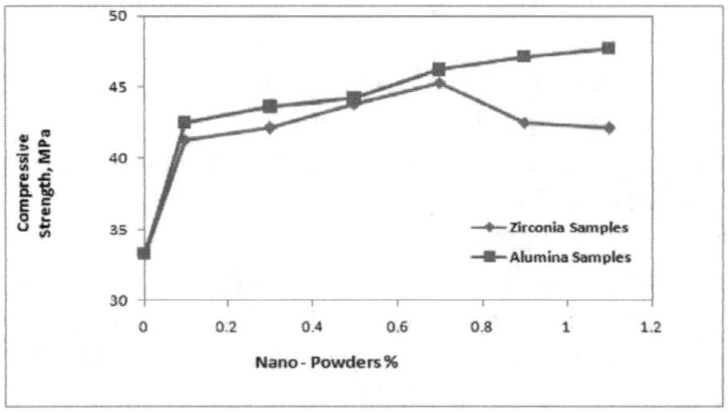

Figure 4. Compressive strength for concrete samples with Nano Al_2O_3 and ZrO_2 at 28 days.

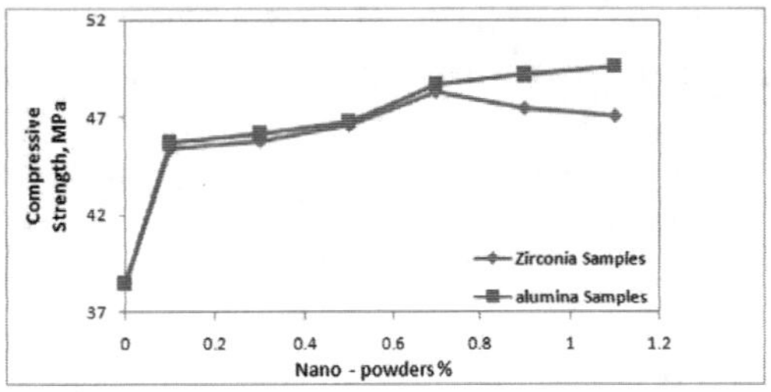

Figure 5. Compressive strength for concrete with Nano Al_2O_3 and ZrO_2 at 60 days.

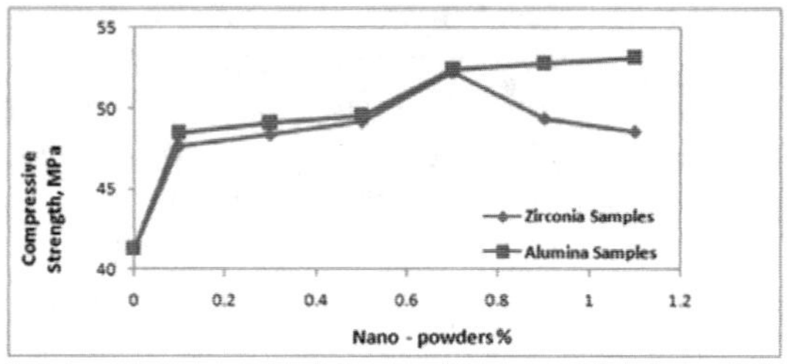

Figure 6. Compressive strength for concrete samples with Nano Al_2O_3 and ZrO_2 at 90 days.

5.2.2. Splitting tensile strength. Tensile strength of control concrete and samples that contain Nano powders after 28, 60 and 90 days are demonstrated in table 9 and 10. Figure 7 shows splitting tensile strength for concrete mixes that contain nano zirconia and nano alumina powders at age of 28 days. Tensile strength follows same behavior as compressive strength. Splitting tensile strength developed in specimens containing Nano ZrO_2powder higher than that of control specimen and increased with increasing Nano ZrO_2 powder concentration up to 0.7% and then it falls down slightly with (0.9 and 1.1 %) at all ages but still higher than that of control samples. Figure 8 and 9 illustrate the tensile strength of NZ and NA series at curing age of 60 and 90 days respectively.

At all ages and for each case, tensile strength of NA series follow the same behavior as compressive strength. It can be seen that the effect of nano alumina powder in enhancement of splitting tensile strength of concrete is more than the effect of nano zirconia powder. High enhancement in mechanical properties achieved by nano alumina powder might be attributed to the reaction of nano alumina with hydrated lime to form hydrated calcium aluminate (one of hydration products), which consequently fills the capillary pores between hydrated cement paste and aggregate. This may lead to increase the densification of interfacial zones between concrete components [12, 13]; so, the density increases and both compressive and splitting tensile strengths increases higher than that in case of using nano zirconia powder.

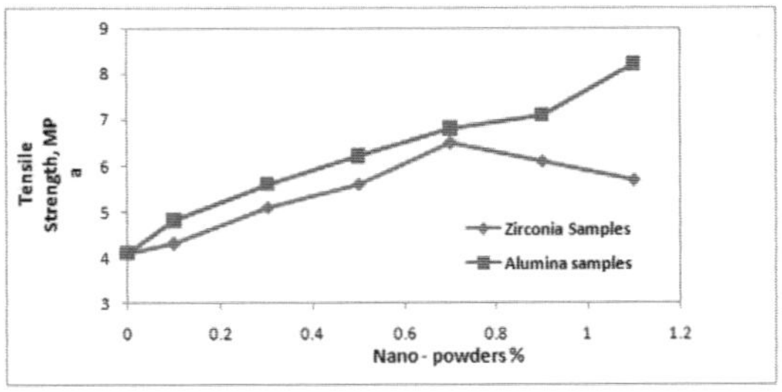

Figure 7. Splitting tensile strength for concrete Samples with Nano Al_2O_3 and ZrO_2 at 28 days.

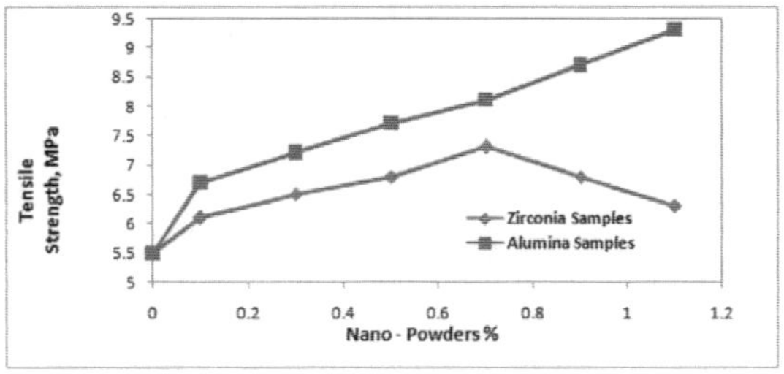

Figure 8. Splitting tensile strength for concrete with Nano Al2O3 and ZrO2 at 60 days.

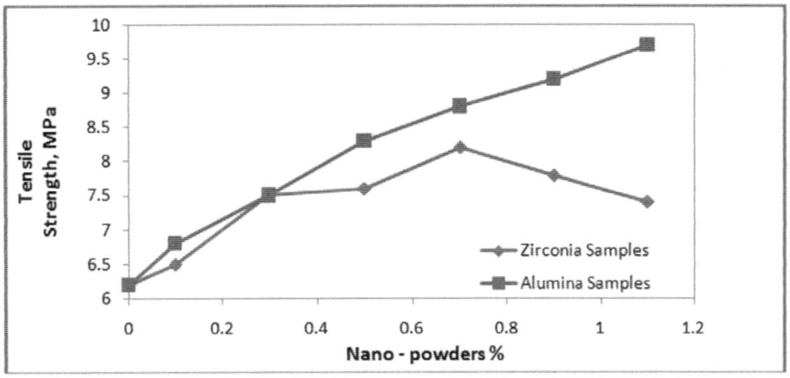

Figure 9. Splitting tensile strength for concrete samples with Nano Al_2O_3 and ZrO_2 at 90 days.

5.3. Water absorption result

Values of water absorption (in percent) for control concrete and samples that contain Nano powder at age of 28, 60 and 90 days are indicated in table 11. As illustrated in figure 10, 11 and 12, the water absorption percent decrease with increasing the concentration of nano powder at all ages and for all specimens. This behavior attributed to decreasing the volume of pores and spacing between particles in cement paste after addition of Nano powder [10, 11].

Table 11. The effect of nano powder on the concrete water absorption.

Zirconia samples				Alumina Samples			
Samples	28 days	60days	90days	Samples	28 days	60days	90days
C0	0.48	0.42	0.38	**C0**	0.48	0.42	0.38
NZ1	0.42	0.35	0.29	**NA1**	0.39	0.34	0.28
NZ2	0.38	0.30	0.25	**NA2**	0.30	0.25	0.22
NZ3	0.32	0.28	0.21	**NA3**	0.28	0.23	0.19
NZ4	0.17	0.15	0.12	**NA4**	0.19	0.14	0.12
NZ5	0.14	0.13	0.09	**NA5**	0.13	0.11	0.08
NZ6	0.11	0.09	0.07	**NA6**	0.11	0.06	0.04

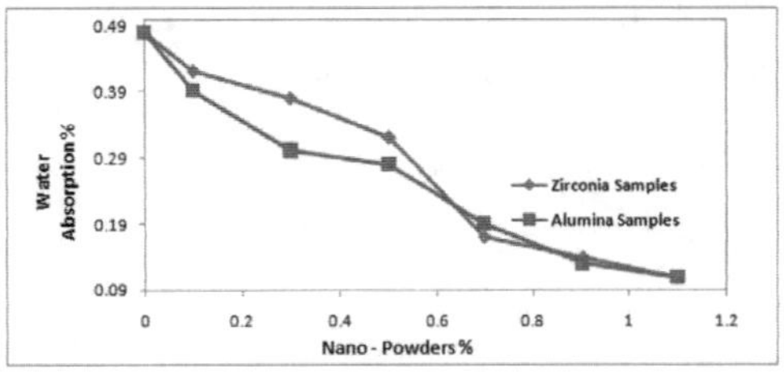

Figure 10. Water absorption for concrete Nano Al_2O_3 and ZrO_2 at 28 days.

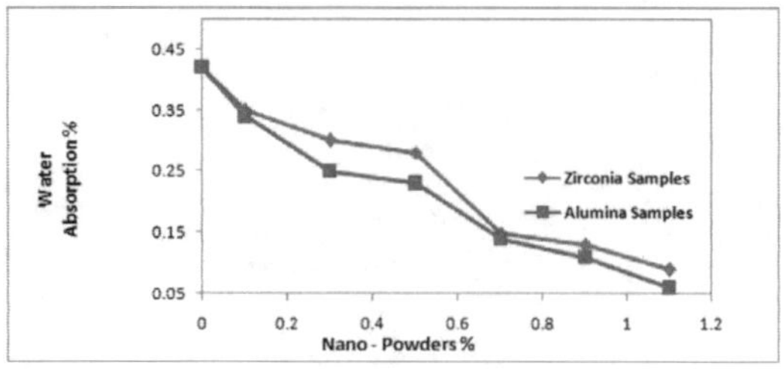

Figure 11. Water absorption for concrete samples with with Nano Al2O3 and ZrO2 at 60 days.

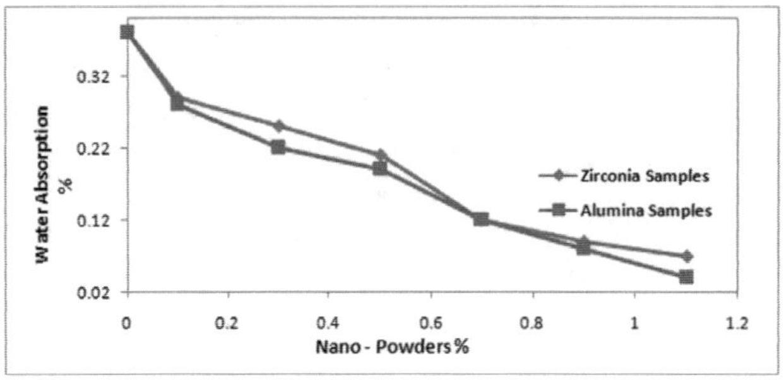

Figure 12. Water absorption for concrete samples with Nano Al_2O_3 and ZrO_2 at 90 days.

6. Conclusions

- Results show that all of concrete samples, which contain nano powder, have significantly higher splitting tensile strength and compressive strength, less workability and water absorption as compare to that of concrete without nano ceramic powder.
- Slump of fresh concrete decreased from 110 mm to 62 mm when zirconia content increased from 0.0 % to 1.1 % in NZ series, while in NA series it is decreased from 110 mm to 65 mm, this attributed to high surface area of nano powders which cause increasing water requirements at constant W/C.
- Compressive strength of NZ series increased with increasing Nano ZrO_2 powder percent up to 0.7% in NZ4 of 52.3 MPa, and then it is decreased with (0.9 and 1.1 %) in NZ5 and NZ6.
- Splitting tensile strength developed in specimens containing Nano ZrO_2 powder higher than that of control specimen and increased with increasing Nano ZrO_2 powder concentration up to 0.7% in NZ4, which equal to 8.2 MPa and then it is decreased slightly with 0.9 and 1.1 % at all ages but still higher than that of control samples.
- Optimum percentage of ZrO_2 to reach maximum compressive strength and splitting tensile strength of concrete is 0.7%.
- For NA series that contain nano alumina powder, compressive strength and tensile strength increases linearly with increasing content of Al_2O_3 and have no optimum percentage to reach maximum value of strength.
- Tensile strength of NA series followed the same behavior as compressive strength.
- Low pores volume and small spacing between particles in cement paste after addition of Nano powder result in low permeability, low water absorption concrete mixes for both NZ and NA series.

Acknowledgements
This study was established by laboratory of Civil Engineering Department at Al-Mustaqbal University College. Words of appreciation are presented to the staff of the laboratory and the authors are grateful for this support.

References

[1] Shekari A, Razzaghi M 2011 Influence of Nano Particles on Durability and Mechanical Properties of High Performance Concrete, *Proced. Eng.* **14** 3036-41.
[2] Nazari A, et al. 2010 Influence of Al_2O_3 Nano-Particles On the Compressive Strength and Workability of Blended Concrete, *J. American Sci.* **6(5)** 6-9.
[3] Sobolev K, Florence S, Ismael F 2012 The Use of Nano Particles Admixtures to Improve the Properties of Concrete, *12th Int. Conf. Recent Adv. Concre. Technol. Sustainab. Iss.*455-69.
[4] Nazari A, et al. 2010 Benefits of Fe_2O_3 Nano Particles in Concrete Mixing Matrix, *J. American Sci.* **6(4)** 102-6.
[5] Li H., Xiao H, Yuan J and Ou J 2004 Microstructure of Cement Mortar with Nano-Particles *Compos. Part B: Eng.* **35(2)** 185-9.
[6] BS 1881 1983 Method for Determination of Compressive Strength of Concrete Cubes; British Standard Institution (London) Part 116
[7] BS 1881 1983 Method for Determination of Splitting Tensile Strength of Concrete; British Standard Institution (London) Part 117
[8] BS 1881 1983Method for Determination of Water Absorption of Concrete; British Standard Institution (London) Part 122
[9] BS 1881 *1970* Method for determination of fresh concrete workability; British Standard Institution (London) Part 2
[10] Ye Q, Zhang Z, Kong D, Chen R 2007 Influence of Nano-SiO2 Addition on Properties of Hardened Cement Paste as Compared with Silica Fume, *Constr. Build. Mater.* **21(3)** 539-45.
[11] Neville A M 1995 Properties of Concrete, 4th and final edition *Pearson Edu. Limit.*
[12] Shaikh F, Hosan A 2019 Effect of Nano Alumina on Compressive Strength and Microstructure of High Volume Slag and slag–fly ash bended paste, *Front. Mater.* **6(90)**.
[13] Saba J, Seyed K 2014 Effect of Nanon–Alumina (N–Al) and Nano silica (NS) As Admixtures On Concrete Behavior, *Int. Conf. Adv. Agricult. Biolog. Enviro. Sci.*

IOP Conf. Series: Materials Science and Engineering **739** (2020) 012021 doi:10.1088/1757-899X/739/1/012021

Photobioreactor Simulation for Microalgae Chlorella Cultivation in Process

Oksana A Trofimchuk[1], Sofia A Romanenko[1], Sergey B. Turanov[1], Alexey N. Yakovlev[1]

[1]National Research Tomsk Polytechnic University, Lenin Avenue 30, Tomsk, 634050, Russia

Abstract. Three-dimensional simulation of light fields in the photobioreactor for Chlorella cultivation was performed. It was shown, that light flux reduced due to the increase of optical density and growth of microalgae cells concentration. According to the fact that light intensity is an important factor, affecting growth rate and development of microalgae cells, process automation and light flux increase in dependence to Chlorella cells gain are needed to be introduced [1-3].

1. Introduction

As the microbiology has developed to the industrial level, lot of data has been cumulated, allowing use of additional intense lighting for microalgae cultivation. At the start of microalgae operations, the influence of light flux on Chlorella investigations were conducted. The dose-effect, the relation between the consumed by the object light dose and the level of cultivation, was the result. Following that, two logically defined solutions were created – orienting cells uniformly against the light source or raising light intensity of the irradiation source.

The problems have become far more challenging with the transition to industrial and semi-industrial levels of desired products derivation in light-dependent technical process. It is complicated to expose microalgae suspension to radiation equally in large volume reservoirs and at the same time avoid photoinhibition of the cells in close proximity to the light source. This implies that radiance sources should be strictly specified and located to maximally expose the entire volume of the suspension and remain energy-efficient.

2. Materials and Methods

Information gain about the characteristics of existing bioreactor models was accomplished before starting simulation. The desired light intensity for microalgae should be at the average rate of 14 klx [4, 5].

The photobioreactor simulation consists of components with properties as following:

1. Solid-state light source (100 pcs) – light-emitting diode XLamp XT-E White. Radiators are placed separately at the distance of 44 mm (from one exposed edge to another). Overall dimensions 3.45×3.45 mm [6].
2. The reservoir is a cylindrical glass tank. Reservoir if 68 cm in height, 20 cm in diameter. Glass properties: Rho = 6%; transmission capacity = 70%; undulation = 0%; reflection effect = 90%.
3. Top cover. Conjectured material for construction – polystyrene: Rho = 90%; transmission capacity = 0%; undulation = 60%; reflection effect = 20%.

4. Reflector. Material – tinfoil, properties set in accordance with measurements: Rho = 80%; transmission capacity = 0%; undulation = 0%; reflection effect = 80%.

5. Suspension simulation, which fills the reservoir to the height of 50 cm, has following properties:

- Properties of the medium at the beginning of cultivation process. Suspension imitation with transmission capacity = 80% was added to the simulation with optical density 0.098 of the initial rate.
- Properties of the medium at the end of cultivation process. Suspension imitation with transmission capacity = 58% was added to the simulation with optical density 0.233 of the initial rate.

General configuration of the simulation is depicted in the figure 1.

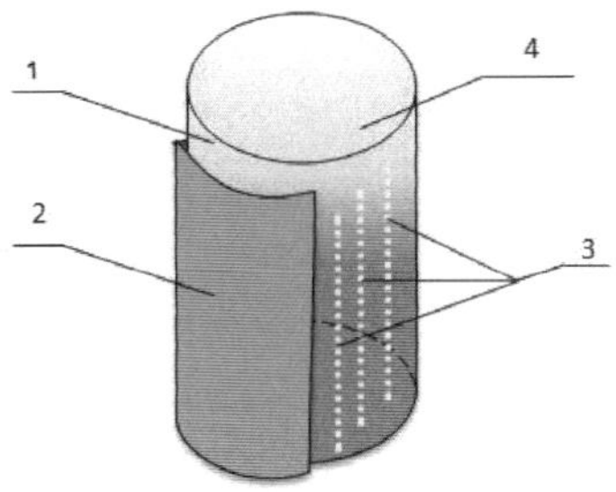

Figure 1. Photobioreactor simulation: 1–reservoir; 2–reflector; 3–LED lines; 4–top cover.

Three-dimensional simulation of the cultivation process was conducted in software application Dialux [7].

Dialux enables to take into consideration colour, reflection index and surface texture during lighting calculation. As a result of data processing complete general 3D elevation of the reservoir and graphics image of light distribution on the surface could be obtained. The software also allows viewing the light intensity distribution with help of fictitious colour.

In present work the simulation, which represented two stages of cultivation, was created. The two stages were: initial–when microalgae concentration was low; final–after 12 hours when the cultivation was over and Chlorella concentration has risen. Simulation without suspension have been created as well.

The final step was to calculate the lost factor while analyzing the results of two stages simulations not including medium agitation.

With the use of fictitious colour in Dialux, light intensity distribution on the surface inside the reservoir could be observed. Figure 2a illustrates light distribution at the beginning of the process, figure 2b–after 12 hours of cultivation, figure 2c–without suspension.

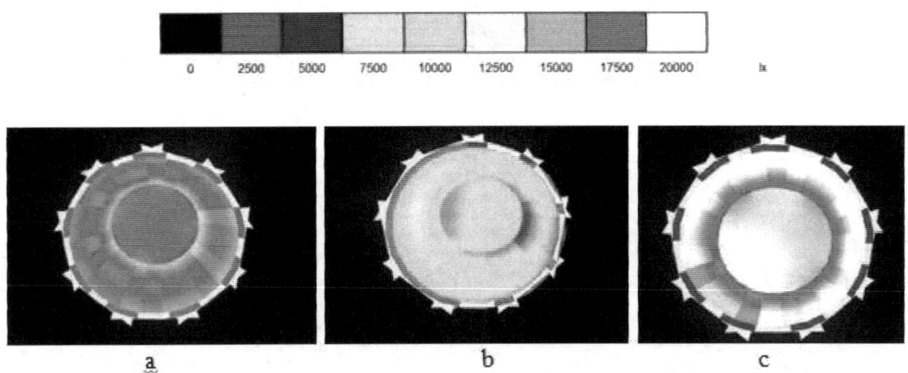

Figure 2. Light intensity distribution using fictitious colour (plan view): a-beginning of the cultivation process; b-after 12 hours; c-without suspension.

Figure 3. shows estimated points in the inline directions of the two cultivation stages.

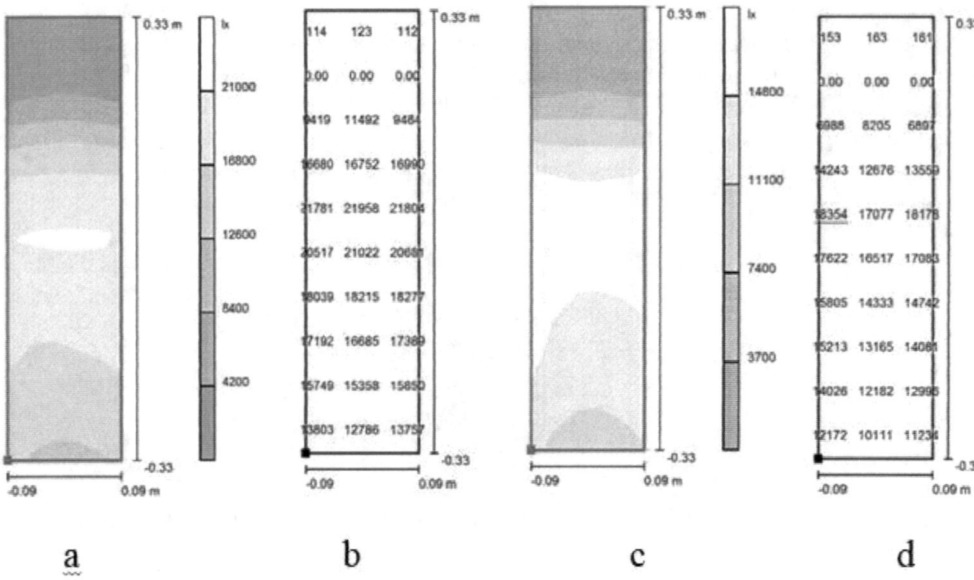

Figure 3. Photosynthetic photon flux density (PPFD, μmol/cm^2) of luminescent tubes (Philips, TLD 18 W) at the surface of *C. hybridum* plants *in vitro* cultured: control group–achromatic (white) light WL, group–WL and red light WL+RL, group 2–WL and blue light WL+BL, group 3–WL and green light WL+GL.

Lost factor was calculated by using target exposed points, which were located at the same positions for both simulations. Figure 4 depicts estimated points order; the results are arranged at the table 1.

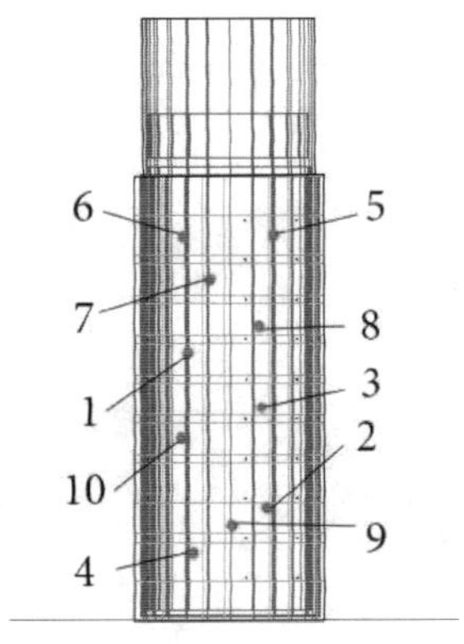

Figure 4. Target points placement.

Table 1. Values in target points, simulating microalgae suspension, at the beginning ant the end of cultivation process.

Target point	Light intensity without suspension (lx)	Light intensity at the beginning (lx)	Light intensity after 12 hours (lx)	Light intensity reduction after 12 hours (%)
1	24338	16712	8907	46,7
2	26024	18158	12352	32,0
3	25813	17724	8318	53,1
4	23645	16388	3917	76,1
5	24491	17056	14008	17,9
6	13108	7386	4082	44,7
7	15476	9437	3871	59,0
8	23610	16024	8043	49,8
9	24916	17126	9453	44,8
10	28859	20120	4661	76,8

3. Results
The simulation showed the decrease of light flux by an average 50.1% after 12 hours of microalgae cultivation in comparison to initial characteristics. The cause of this result is microalgae Chlorella cells growth. Thus concentration rate raises and transmittance of the suspension reduces. This means the necessity of lighting adjustment, light flux augmentation throughout the microalgae cultivation particularly. By applying fully-automated light control system desired light intensity distribution in the reservoir could be maintained and energy efficiency of the photobioreactor could be increased.

4. Conclusion
Nowadays, microalgae are widely used in various spheres: agriculture, food industry, medicine and esthetics, as wastewater treatment, oxygen and biofuel manufacturing. However, cultivation of light-dependent microorganisms is associated with high energy cost for lighting. Applying LED light sources with appropriate spectrum for maximum microalgae cells gain and revealing the relation between light intensity and optical density of microalgae suspension could significantly reduce energy consumption of industrial photobioreactors.

Acknowledgments
The work was supported by the Ministry of Education and Science of the Russian Federation under the State Program in the field of scientific investigations №13.3647.2017/PC.

References
[1] Trofimchuk O A, et al. 2019 The influence of PAR irradiance on yield growth of Chlorella microalgae, *IOP Conf. Series: Mater. Sci. Eng.* **510** 012017.
[2] Pattanaik A, Sukla L B, Pradhan D 2018 Effect of LED Lights on the Growth of Microalgae, *Inglomayor* **14** 17-24.
[3] Malapascua J R, Ranglová K, Masojídek J 2019 Photosynthesis and growth kinetics of chlorella vulgaris R-117 cultured in an internally LED-illuminated photobioreactor, *Photosynt.* **57** 103-12.
[4] Meshcheryakova Yu V, Nagornov S A 2012 Cultivation of Microalgae Chlorella for Biofuel Production, *Prob. Contemp. Sci. Pract. Vernadsky Uni.* **43** 33-6.
[5] Official page OOO "Bashkirskaya ptitsa" [Web log post] Retrieved September 10, 2013 from http://башптица.рф/

[6] XLamp XT-E White [Web log post] Retrieved June 15, 2019 from https://www.cree.com/led-components/products/xlamp-leds-discrete/xlamp-xt-e-white
[7] Official page DIALux Help: calculation and engineering [Web log post] Retrieved June 15, 2019 from http://www.dialux-help.ru/.

IOP Conf. Series: Materials Science and Engineering **739** (2020) 012022 doi:10.1088/1757-899X/739/1/012022

Effects of Bainite Isothermal Temperature on Microstructure and Mechanical Properties of a δ-TRIP Steel

Xu Xin[1, 2], Liu Rendong[2], Xu Baoyu[1], Yi Hongliang[1], Wang Guodong[1]

[1]The State Key Laboratory of Rolling and Automation, Northeastern University, Shenyang110819, P.R. China
[2]Angang Steel Co., Ltd. Technology Centre, Anshan114009, P.R. China

ag_xuxin@126.com

Abstract. In this work, the microstructure, properties and retained austenite characteristics of a δ-TRIP steel containing 3.0% Al at different isothermal temperatures in the bainite region were researched by means of scanning electron microscopy, tensile test and XRD. The evolution of austenite and the effect on microstructure and mechanical properties were investigated. The results show that the high Al element in δ-TRIP steel can effectively inhibit the precipitation of carbides in austenite and obtain more stable austenite which can lead to an excellent combination of elongation and tensile strength. When the isothermal temperature is 400 °C, the comprehensive mechanical properties are the best, the tensile strength reaches 800 MPa, and the elongation exceeds 40%.

1. Introduction

With the requirements of energy saving, environmental protection and safety in the automotive industry, lightweight body has become an inevitable choice [1], and the steel industry has also been developed to produce a new generation of automotive steel with high strength and excellent forming performance [2].

At present, major steel companies are committed to the development of the third generation of automotive steel, the representative steels are mainly quenching and partitioning(Q&P) steel, medium manganese steel and δ-TRIP steel [3-5]. Among them, δ-TRIP steel, as a typical third-generation automobile steel, promotes the formation of δ-ferrite by adding Al element, and can also hinder the precipitation of carbides, thereby forming stable retained austenite, achieving a perfect match between strength and plasticity [6]. Therefore, δ-TRIP steel has broad application prospects. In this paper, a new type of δ-TRIP steel with tensile strength of over 800 MPa and total elongation of more than 40% has been designed. The effects of bainite isothermal temperature on the microstructure and properties of δ-TRIP steel is studied.

2. Material and procedure

In this paper, the experimental δ-TRIP steel was casted into an 80 kg ingot using a vacuum induction melting furnace and rough rolled to a cross-sectioned dimension of 35 mm × 150 mm× 300 mm square slabs. The chemical composition of the steel in wt.% was Fe-0.40C-0.21Si-1.01Mn-2.89Al. The slabs were completely austenitizing at 1200 °C for 2 hours and then hot rolled into a 3.0 mm sheet by seven passes on a reversing hot rolling mill, and the final rolling temperature was set at 850 °C, then air cooled to room temperature. The hot-rolled plates were pickled by 10% hydrochloric acid and then cold rolled to 1.5mm.

A rectangular specimen of 220 mm × 120 mm × 1.5 mm was cut along the rolling direction on the cold-rolled sheet, and the heat treatment process of specimens was performed on a continuous annealing simulation machine. In order to study the effects of bainite isothermal temperature on microstructure and properties of δ-TRIP steel, three as-cold-rolled specimens were kept at the critical zone temperature (820°C) for 5 min, then cooled to the bainite transformation zone with a 20 °C/s rate, which involved isothermally holding at 375、400、425°C for 10min, respectively. The annealing process is shown in figure 1.

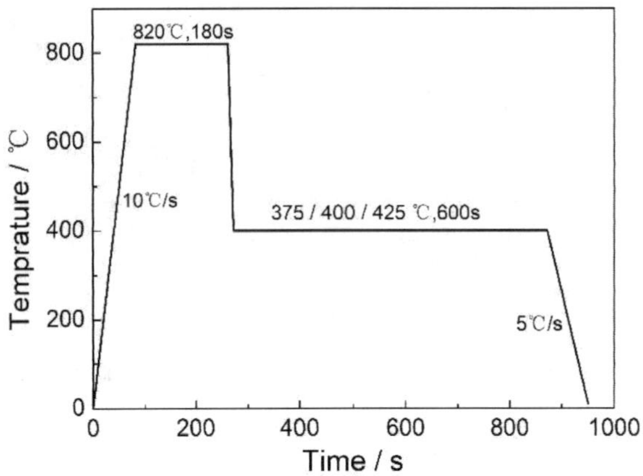

Figure 1. Schematic of annealing process of the experimental δ-TRIP steel.

The annealed samples were processed into tensile specimens with a gauge length of 50 mm according to the national standard GB/T 228.1-2010, and the prepared tensile specimens were subjected to tensile tests using the Zwick Z100 material test system. Microstructure of specimens was examined by SUPRA 55 scanning electron microscope (SEM). The D5000 X-ray diffractometer was used to obtain the diffraction pattern; the formula of retained austenite content was as follows [7]:

$$V_r = {}^{1.4I_r}\!\big/_{(I_\alpha + 1.4I_r)} \tag{1}$$

where V_r is the volume fraction of retained austenite; I_r is the average integrated intensity of the austenite at {200}, {220}, and {311} plane diffraction peaks; I_α is the average integrated intensity of ferrite at {200},{211} plane diffraction peaks. The carbon content of retained austenite is calculated by the following formula [7]:

$$W_c = {}^{(a_r - 3.578)}\!\big/_{0.033} \tag{2}$$

where W_c is the mass fraction of carbon in retained austenite, %; a_r is the lattice constant of retained austenite.

3. Results and discussion

3.1. Effect of bainite isothermal temperature on microstructure

The microstructure of the experimental δ-TRIP steel samples at different isothermal temperatures in the bainite region is shown in figure 2. It can be seen from figure 2 that the microstructure of the three specimens is relatively fine, mainly composed of ferrite, bainite and retained austenite. Since the annealing temperature and soaking time of the samples are identical, the proportion of ferrite should theoretically be consistent [8]. With the bainite isothermal temperature increasing, the ratio of bainite and retained austenite changes accordingly. When the bainite isothermal temperature is at 375 °C, due to the lower temperature, some austenite may transform into martensite after the bainite transformation, and the retained austenite content in the sample is low, so the tensile strength is highest. When the bainite isothermal temperature is raised to 400 °C and 425 °C, most of the austenite in the sample is transformed into bainite.

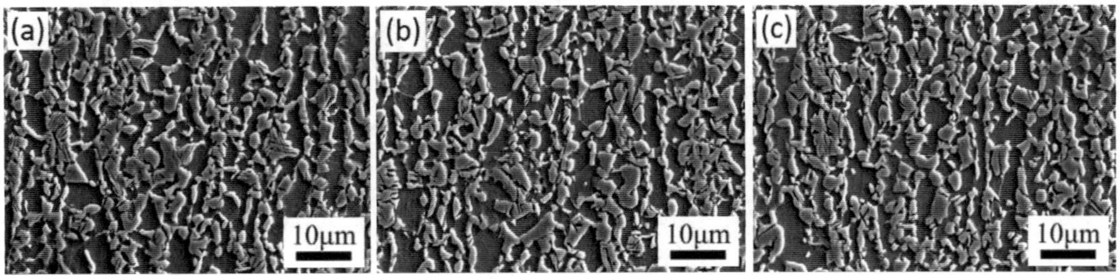

Figure 2. SEM micrographs of δ-TRIP specimens at different bainitic isothermal temperatures (a) 375°C; (b) 400°C; (c) 425°C.

When the bainite isothermal temperature is at 375 °C, the Carbon diffusion is limited, resulting in insufficient Carbon enrichment in austenite, and the stability of retained austenite is deteriorated. Some of the austenite is transformed into martensite in the subsequent cooling process after bainite isothermal transformation. Finally, less retained austenite is obtained at room temperature. When the bainite isothermal temperature is raised to 400 °C, the rate of Carbon diffusion from ferrite to austenite is accelerated. During the bainite isothermal treatment, the austenite completes carbon enrichment, which increases the stability and content of retained austenite [9]. When the bainite isothermal temperature is raised to 425 °C, the austenite is more easily transformed into bainite, so that the amount of retained austenite finally obtained is slightly lower than that at 400 °C.

3.2. Effect of bainite isothermal temperature on mechanical properties

Figure 3 shows the mechanical properties of the experimental steel at different isothermal temperatures in bainite regions. It can be seen that the isothermal temperature in the bainite region has a great effect on the mechanical properties of the experimental steel [10]. When the isothermal temperature in the bainite region is 375 °C, the yield strength of the specimen is 544 MPa, the tensile strength is up to 831 MPa, and the elongation of the specimen is 33.1%. As the bainite isothermal temperature increases, the tensile strength of the specimen decreases while the elongation increases. When the bainite isothermal temperature is at 400 °C, the yield strength and tensile strength of the specimen are 550 MPa and 794 MPa respectively, and the elongation is significantly improved, reaching 40.6%. When the isothermal temperature continues to rise to 425 °C, the yield strength and tensile strength change little, 543 MPa and 781 MPa respectively, and the elongation rate is also reduced to 37.9%. It can be seen that when the bainite isothermal temperature is at 400 °C, the comprehensive mechanical properties of the sample are the best, and the product of strength and elongation reaches a maximum value of 32.2 GPa·%.

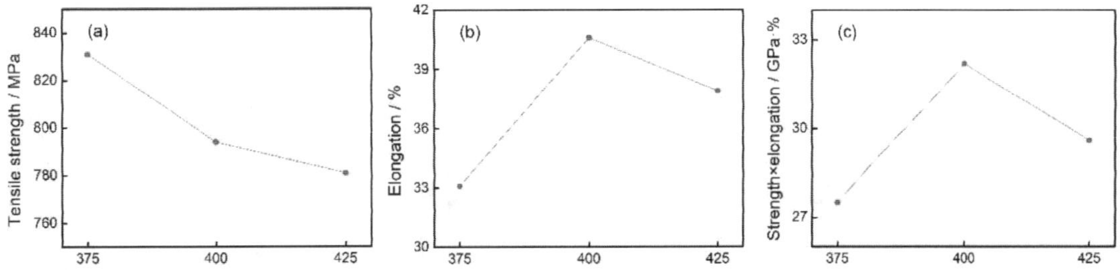

Figure 3. Mechanical properties of δ-TRIP specimens at different bainitic isothermal temperatures

(a) tensile strength; (b) elongation; (c) product of strength and elongation.

In δ-TRIP steel, ferrite is softer than bainite [11], when the experimental steel is subjected to stress during the stretching process, the yield first occurs in the ferrite matrix [12]. Although the bainite isothermal temperature rises to 400 °C and 425 °C, the yield strength of δ-TRIP specimens doesn't change much, because the ferrite ratio was theoretically the same. For δ-TRIP steel, the volume fraction and stability of retained austenite in the specimen will greatly affect the mechanical properties.

3.3. Effect of isothermal temperature on austenite characteristics

Table 1 shows the volume fraction and carbon content of retained austenite in the specimens at different bainite isothermal temperatures calculated from the relevant data. It can be seen from table 1 that when the isothermal temperature in the bainite region is 400 °C, the volume fraction of retained austenite reaches a maximum value of 15.9%. In δ-TRIP steel, the high content of Al plays an important role in the stabilization of austenite, thus providing a condition for retained austenite content increasing [13].

Table 1. Volume fraction and carbon content of retained austenite of δ-TRIP specimens under different bainitic isothermal temperatures.

Bainitic isothermal temperature / °C	Retained austenite amount / vol.%	Carbon content of retained austenite / wt.%
375	11.6	1.97
400	15.9	2.03
425	13.2	2.12

Even the isothermal temperature of 375 °C is higher than the Ms point. Due to the larger phase change driving force, a large amount of austenite transforms, and the volume fraction of retained austenite at room temperature is small. With the bainite isothermal temperature increases, the driving force of bainite transformation decreases, the carbon diffusion accelerates, the carbon content of retained austenite increases, and the stability of retained austenite increases.

Figure 4 shows the gradient of retained austenite content distribution in different regions of the specimens after stretching. It can be seen that much of the austenite does undergo martensitic transformation during deformation, this part of the austenite transformation not only improves the plasticity of the steel, but also increases the strength of the steel [14]. At the same time, some of the austenite is retained even at a distance less than about 5mm away from the point of fracture, it indicates that there is much relatively stable retained austenite in the specimens.

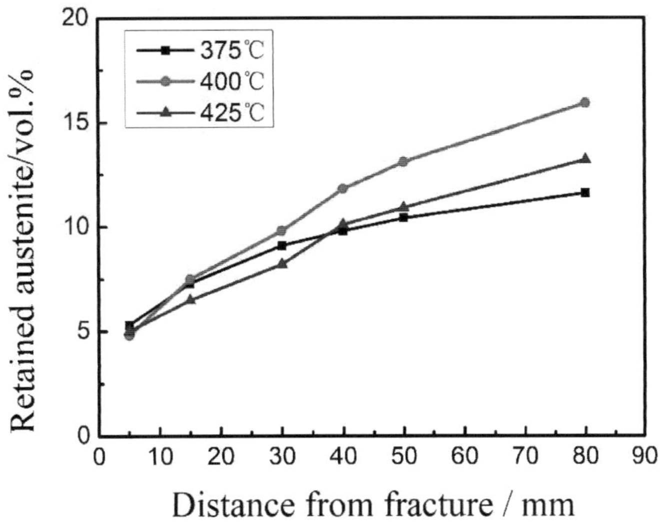

Figure 4. Retained austenite in different regions after fracture in tensile specimen.

4. Conclusions

This study examined the retained austenite stability of a δ-TRIP steel with different bainitic isothermal temperatures. The main findings of the investigation are as follows:

(1) Aluminum is a strong austenite stabilizing element, which is beneficial to the formation of retained austenite, thereby increasing the retained austenite content at room temperature.

(2) With the increase of bainite isothermal temperature, the proportion of ferrite in the microstructure of δ-TRIP steel is basically unchanged, and the ratio of bainite and retained austenite changes, which determines the different mechanical properties of TRIP steel;

(3) In δ-TRIP steel, retained austenite is the key to product of strength and elongation. When the bainite isothermal temperature is at 400°C, the volume fraction of retained austenite is the highest, reaching 15.9%, and the product of strength and elongation exceeds 32.2GPa%.

References

[1] Wang X J, Chen Y S 2013 Discuss the Applications of the Automotive Lightweight Materials in China Briefly, *Adv. Mater. Res.* **621** 148-52.

[2] Bhagat A N, et al. 2010 Development of Cold-Rolled High-Strength Formable Steel for Automotive Applications Advanced, *Mater. Manufac. Proces.* **25(1-3)** 202-5.

[3] Ying C, Han S, Li X, et al. 2018 Effect of shearing clearance on formability of sheared edge of the third-generation automotive medium-Mn steel with metastable austenite, *J. Mater. Proces. Technol.* **259** 216-27.

[4] Bouaziz O, Zurob H, Huang M 2013 Driving Force and Logic of Development of Advanced High Strength Steels for Automotive Applications, *Steel Res. Int.* **84(10)** 937-47.

[5] Li X G, Zhao A M, Zheng H H, et al. 2016 Study of the Microstructure and Mechanical Properties of a Medium Manganese Quenching and Partitioning (Q&P) Steel, *Mater. Sci. For.* **850** 659-63.

[6] Shiri S G, Jahromi S A. J, Palizdar Y, et al. 2016 Unexpected Effect of Nb Addition as a Microalloying Element on Mechanical Properties of δ-TRIP Steels, *J. Iron Steel Res. Int.* **23(9)** 988-96.

[7] Dyson D J, Holmes B 1970 Effect of alloying additions on the lattice parameter of austenite, *J. Iron Steel Inst.* **208** 469-74.

[8] Park D B, Lee J W, Lee Y S, et al. 2008 Effects of the annealing temperature and time on the microstructural evolution and corresponding the mechanical properties of cold-drawn steel wires, *Metals Mater. Int.* **14(1)** 59-64.

[9] Hu H J, Xu G, Wang L, et al. 2015 Effect of ausforming on the stability of retained austenite in a C-Mn-Si bainitic steel, *Metals Mater. Int.* **21(5)** 929-35.

[10] Zeng S W, Zhao Z Z, Zhao A M, et al. 2014 Effect of isothermal bainite transformation temperature on microstructure and mechanical properties of TRIP800 steel containing vanadium, *Transac. Mater. Heat Treatm* **35(1)** 120-4.

[11] Yi H L, Lee K Y, Bhadeshia H K D H 2011 Mechanical stabilisation of retained austenite in δ-TRIP steel, *Mater. Sci. Eng. A* **528(18)** 5900-3.

[12] Cheng L, Cai Q W, Dong E T, et al. 2016 Influence of ferrite matrix and precipitation status on the mechanical properties of low carbon low alloy steel during high temperature tension, *Mater. Sci. Eng. A* **678** 1-9.

[13] Kang S, Moor E D, Speer J G 2015 Retained Austenite Stabilization Through Solute Partitioning During Intercritical Annealing in C-, Mn-, Al-, Si-, and Cr-Alloyed Steels, *Metallurgical Mater. Transact. A*, **46(3)** 1005-11.

[14] Ul'Yanin E A, Sorokina N A, Georgieva I Y, 1976 Properties of stainless Cr-Mn steels with a difference in the stability of austenite during deformation, *Metal Sci. Heat Treat.* **18(9)** 799-802.

IOP Conf. Series: Materials Science and Engineering **739** (2020) 012023 doi:10.1088/1757-899X/739/1/012023

Nonlinear elastic analysis of concrete beams based on the Smeared Crack Approach

N Betancourt[1], C Arias[1], F Durán[1], V Arana[1]

[1]Department of Civil Engineering, Universidad Peruana de Ciencias Aplicadas (UPC), Surco, Lima, Peru.

Abstract. In the present study, an analysis of plain and reinforced concrete beams under monotonic loading was made based on the Fixed Smeared Crack approach. The objectives of this research were to analyze the nonlinear behavior of the selected cases of analysis and to propose an alternative and simple model for the analysis of beams under service loadings, by means of Committee 435 of the American Concrete Institute. A brittle model for concrete and a linear-elastic behavior for steel reinforcement bars were considered. Results are presented through force-displacement curves and the sequence of cracking propagation. Also, a comparison of calculated instantaneous deflections of simply supported beams was made between the proposed model and other researches. It was verified that the proposed algorithm can predict adequately the cracking process and the deflections of beams subjected to service loadings, taking into account experimental results from other authors.

1. Introduction

Concrete structures undergo a high nonlinear behavior after cracking because the tensile strength decays and develops an anisotropic behavior [1]. Likewise, when the formation of micro-cracks entails in an important failure zone, the use of linear elastic models for structural analysis is insufficient [2]. Moreover, several authors have proposed complex models and methods that can simulate the behavior of concrete structures [3-6].

These cracking models can be categorized into two groups. The first of them, the Discrete Crack Approach, which evaluates cracking propagation as a physical phenomenon [6]. That is to say, it generates a geometrical discontinuity in the finite element mesh. This methodology was the first implemented for the analysis of cracking [7]. However, one disadvantage of this method is that it requires a constant recalculation of nodal connectivity between the elements [6]. Therefore, the computational cost of the method is high, because it employs sophisticated algorithms for the re meshing along the cracking process [8].

On the other hand, the Smeared Crack Approach, proposed originally by Rashid [3], evaluates cracking in a continuum along finite elements, in which the propagation of crack implies a localized modification of the constitutive equation [9]. In that regard, after the formation of a crack in an element, the material changes from an isotropic behavior to an orthotropic behavior [6]. Compared with the Discrete Crack Approach, this method follows the same principles as the finite element method based on displacements because the continuity of the displacement field is guaranteed in the system [10]. Likewise, since the simulation of cracking is in the finite element and not in their boundaries the direction of the cracks is not restricted [6].

In the analysis of beams under service loading, the ACI 435 Committee [11] mentions that the evaluation of deflections is as important as the calculation of the ultimate strength. Therefore, the Branson model [12] is suggested for the calculation of instantaneous elastic deflections in reinforced concrete beams, by means of the use of an effective inertia that takes into account the cracked inertia and the real inertia of the section and the cracking moment. Considering these parameters, limits for deflections are imposed for the design of reinforced concrete beams.

In this context, the present research deals with the analysis of unreinforced and reinforced concrete beams under monotonic loading for the evaluation of cracking, the ultimate strength and the deflections produced under service loading. The Smeared Crack Approach is used and the fragile model is selected for the evaluation of cracking, considering the fracture initiation perpendicular to the principal tensile stress when the stresses are greater than the ultimate tensile strength of the material. A linear elastic model for reinforcement bars is used. Finally, the obtained results are compared with other authors [12-15].

2. Method

In the present investigation the Smeared Crack Approach is used because it can evaluate cracking by altering the constitutive equation of the material [6].

This approach can be categorized into three groups: fixed, rotational and fixed multidirectional [6]. The principal difference between these lies the evaluation of the crack direction during the fracture process. While in the fixed model the direction of the crack remains invariant along the computational process, in the rotational model the crack direction may change in function of the principal stresses. On the other hand, a multidirectional model considers the direction of the crack by means of an angle of retention [8]. This approach is proposed with a combination of a failure criterion, the level of stress transfer after cracking and the softening model employed for the material which can be fragile, linear, multilinear and nonlinear [15]. This study employs a fragile softening model because it is capable to evaluate the structural degradation generated by cracking [6].

For the failure criterion, only the stresses of the cracks in the mode of fracture I are evaluated. This hypothesis holds true if compressive stresses are low [17]. Therefore, is optimal for the evaluation of beams, because the failure of the element is caused by the tensile stress produced.

For the evaluation of cracking is essential to differentiate the uncracked concrete from cracked concrete. For the first one it's assumed a linear elastic and isotropic behavior, governed by the following constitutive relation:

$$\boldsymbol{D}^{co} = \frac{E}{1-\upsilon^2}\begin{bmatrix} 1 & \upsilon & 0 \\ \upsilon & 1 & 0 \\ 0 & 0 & \frac{1-\upsilon}{2} \end{bmatrix} \quad (1)$$

Where E is the Young's Module of concrete and υ is the Poisson's module. In regard of the constitutive relation for the cracked material it's assumed that a cracked element lost all his mechanical properties, therefore its Young's Modulus is reduced to a negligible value in order to ensure mathematical stability during the computational process.

3. Results

3.1. Simple supported unreinforced notched concrete beam - Kormeling & Reinhardt [19]

First, to validate the results obtained from the proposed model they are compared with other computational model results obtained by Sena, et al. [13], which compared the results from their computational model with numerous constitutive relationships and the experimental results by Kormeling y Reinhardt [19]. The following figure shows the mesh implemented for the beam. It considers four-node quadrilateral isoparametric elements. The problem has a notch at the middle of the beam. Dimensions are expressed in millimeters. The beam has a width of 100 mm.

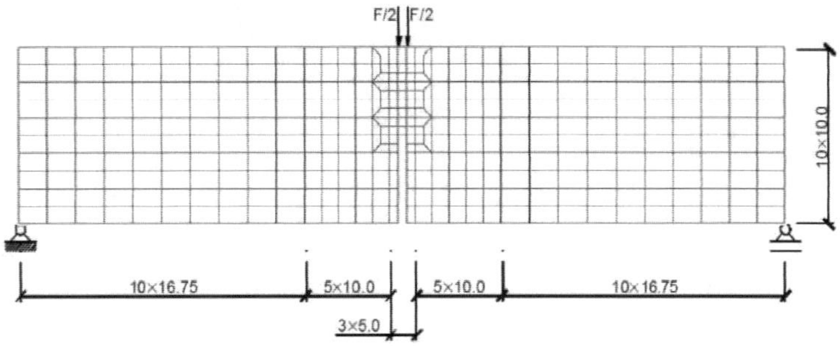

Figure 1. Proposed mesh.

For the analysis of the model a concrete density of 2400 kg/m3 was considered as well as a Poisson's module of 0.20 and an elasticity module of 20000.0 Mpa. For the parameters necessary to model crack it was considered an ultimate tensile strength of 2.4 Mpa.

For the solution of the softening problem, a displacement control in the point of loading was implemented as proposed by de Borst, Crisfield, Remmers y Verhoosel [20]. Thus, under the application of a vertical deflection of -1.00 mm the following force-deflection curve was obtained.

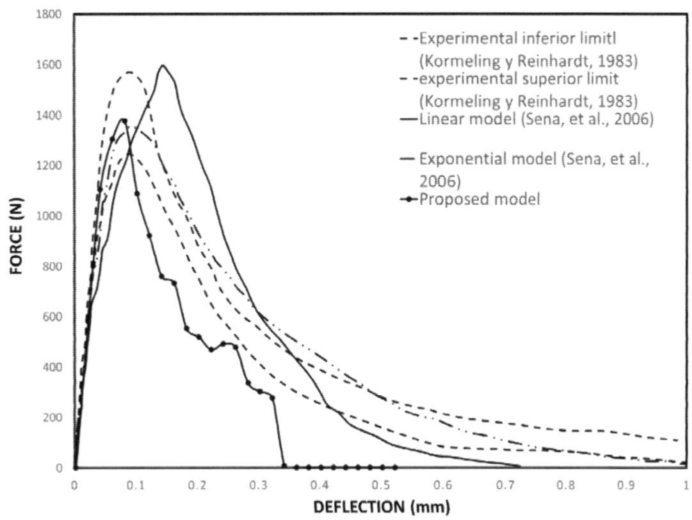

Figure 2. Force vs deflection curve at middle spam of the beam.

The results show that the model can predict the maximum bearing capacity of the beam, as the calculated value is in the range of the experimental results of 1200N and 1600N approximately. Further, the progressive degradation of the bearing capacity of the beam is illustrated. However, this section of the curve decays abruptly because of the fragile constitutive relationship implemented.

Then, the progression of cracking is presented. The first group of crack appears when a displacement of -0.0225 mm is applied with a load of 586.39 N. On the other hand, the maximum bearing capacity of the beam was found under a deflection of -0.0823 mm, under a load of 1.376 kN approximately. It's showed that the peak force is within the experimental result range. This crack state can be observed in figure 3.

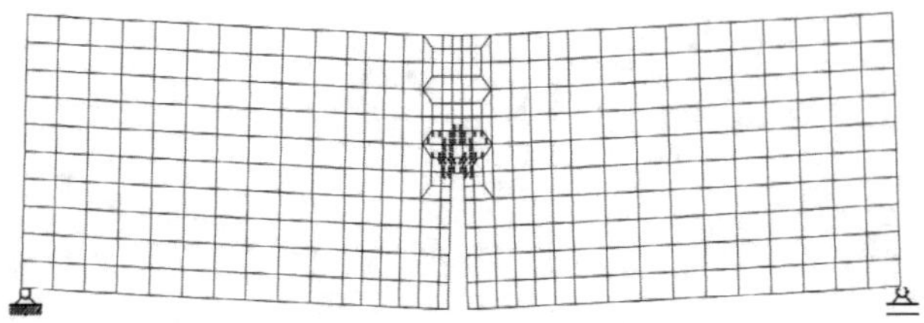

Figure 3. Fractured system under its maximum resistant load.

3.2. Simple supported reinforced concrete beam - Walraven [21]

Secondly, the reinforced beam experimentaly tested by Walraven [21] was analysed. Numeric result obtained by de Borst y Nauta [14] were compared. The beam studied has a clear span of 0.45m and a width of 0.20 m. The reinforcement was composed of a single 20 mm diameter bar and two 14 mm diameter bars, both of them placed at 30 mm from the base of the beam. The dimensions, in meters, of the beam to be evaluated are presented in figure 4. Dimensions are presented in meters.

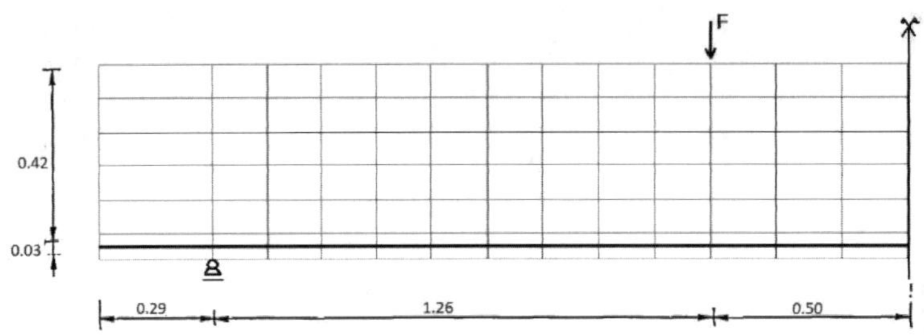

Figure 4. Case of study [14].

An elastic module of 28000.0 Mpa, a Poisson's module of 0.20 and an ultimate tensile strength of 2.5 Mpa were used. The reinforcement was assumed lineal elastic with an elastic module of 210000 Mpa.

For the solution of the nonlinear, a force control was applied for an ultimate load of 80 kN, the following force vs deflection diagrams were obtained, for the mesh 1, proposed by de Borst y Nauta [14], and for the mesh 2, more refined, proposed by the present investigation. In the figure 5 the results of the force vs deflection diagram at the center of the beam are showed.

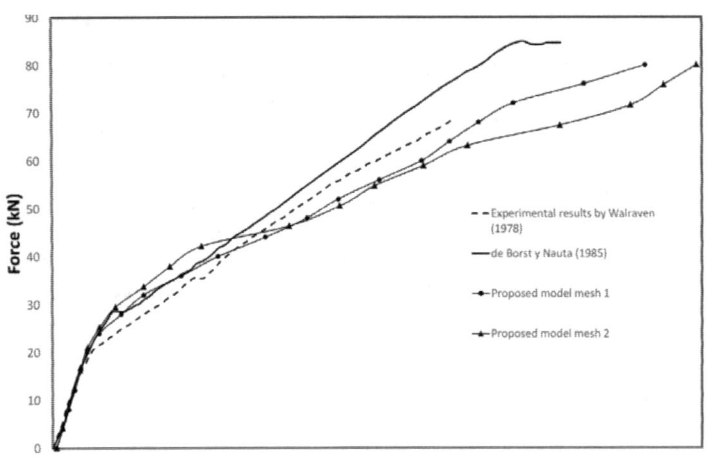

Figure 5. Force vs deflection curve at middle of the beam.

It's showed that the results obtained follow the same tendency as the results obtained by other authors. It is observed that the result has an almost linear behavior until a vertical deflection of approximately 0.8 mm and a force of 17 kN is reached. From this point, the beam loses his bearing capacity at the middle, because of the severity of the crack propagation. After surpassing this point, only the reinforcement provides stiffness to this part of the element. However, because the reinforcement its assumed linear elastic, the curve follows almost a linear tendency.

For the proposed mesh, figure 6 shows the cracked system when the firs fracture forms. It develops when the beam has reached a vertical deflection of -0.5382 mm and a force of -14.01 kN.

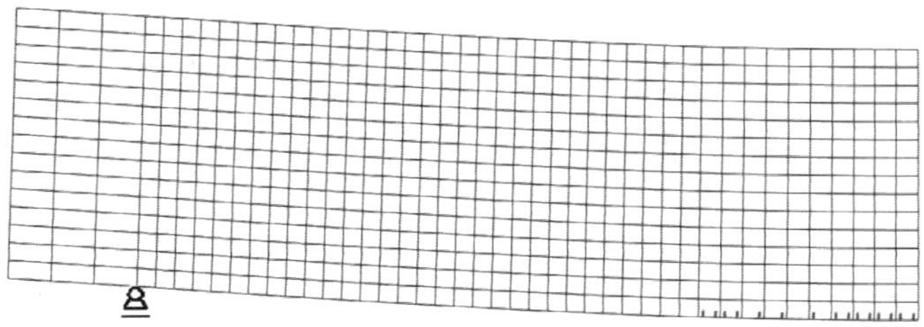

Figure 6. First cracks developed with the proposed mesh.

3.3. Unreinforced and reinforced concrete beams under service loadings - Branson [12] and Washa & Fluck [16]

For the evaluation of instant deflections, the ACI Committe 435 [11] suggests the use of the method proposed by Branson [12]. An effective inertia was proposed for the calculation of deflections. Moreover, a linear elastic analysis is performed for a unidimensional beam, considering the effective inertia for the computation of the deflection under service loading [12].

For the validation of the method, Branson [12] compared experimental results of tested beams and the experiments developed by Washa y Fluck [16] with the results from his calculation. A few of the analyzed beams were selected with the objective of comparing the result of the method proposed by [12] with the model proposed by this investigation. The properties of the selected beams can be found in [12]. The mesh employed for the analysis was proposed under guidelines for solving nonlinear problems for reinforced concrete by the Finite Element Method [22].

It is observed that the percentage of error obtained in the analysis in comparison with the experimental results is acceptable when compared with the results obtained by the recommendations of the ACI [11]. In the table 1 the calculated deflections by the proposed investigation and the Branson method [12] are showed.

Table 1. Comparative results.

N°	Reference	Beam	Exp Deflection	Newmark's Deflection	Effective inertia Deflection	Prop Model Deflection	Expvs Newmark's	Exps vs Effective inertia	Exp vs Prop Model
			mm	mm	mm	mm	%	%	%
1	[12]	SB-1	1.04	1.27	1.27	1.22	18%	18%	15%
2	[12]	SB-3	3.89	5.16	5.23	3.65	25%	26%	6%
3	[16]	A1,A4	13.46	15.49	15.75	13.94	13%	15%	3%
4	[16]	B1,B4	23.37	25.15	25.15	26.53	7%	7%	12%
5	[16]	C1,C4	40.13	44.45	44.45	46.30	10%	10%	13%
6	[16]	D1,D4	11.94	16.00	16.00	15.82	25%	25%	25%
7	[16]	B2,B5	24.89	25.91	25.91	26.08	4%	4%	5%
8	[16]	C2,C5	43.43	44.70	44.96	45.49	3%	3%	5%
9	[16]	D2,D5	14.22	16.26	16.51	16.19	13%	14%	12%
10	[16]	A3,A6	17.02	16.76	16.76	17.02	2%	2%	0%
11	[16]	B3,B6	26.42	25.91	25.91	26.90	2%	2%	2%
12	[16]	C3,C6	47.75	46.48	46.23	45.74	3%	3%	4%
13	[16]	D3,D6	17.78	16.76	16.76	16.49	6%	6%	8%

It is observed that the proposed model can analyze the propagation of cracks under incremental loading. Likewise, on top of calculating instant deflections and evaluate designs code criteria's, it is possible to verify the damage due to cracking in the element.

4. Conclusions

The results obtained by the computational model proposed in the present investigation were satisfactory by comparing force-deflection curves proposed by other authors. In this regard, the analysis showed that the model is capable of calculate the maximum bearing capacity of reinforced and reinforced concrete beams and present diagrams of the propagation of cracks, i.e. it can predict the localization and direction of the crack and the progressive failure of elements. On the other hand, the proposed model is capable of acceptably predicting the deflection of reinforced concrete beams under service loading. Moreover, its showed that the method used for the proposed model can predict cracking patterns under service loading. Because of this, it can be possible to stablish more design criteria's to control the damage produced by cracks.

However, due to the fact the constitutive relation employed by the proposed model is fragile, the evaluation of the degradation of the bearing capacity of the beams is overestimated. This disadvantage can be overcome by the implementation of more complex constitutive relations. It was observed that results were mesh dependent which can also be prevented by the implementation of more complex constitutive relations. This represent an oportunity of development of the presented model. Finally, numerical instability was observed during the development of the proposed model which are of common occurrence in softening materials. For this reason, more complex solution procedures like the Modified Riks Method enhanced with a local constrain equation or the Generalized Displacement Control Method are recommended for this type of problems.

References

[1] Bažant Z P 1976 Instability, Ductility and Size Effect in Strain-Softening Concrete, *Eng. Mech. Division*, **16(2)** 331-44.
[2] Bažant Z P 1992 Fracture Mechanics of Concrete Structures Part , I State-of-Art Report, **140** 17-30.
[3] Rashid Y 1968 Ultimate strength analysis of prestressed concrete pressure vessels, *Nuclear Eng. Desg.* **7(4)** 334-44.
[4] Hillerborg A, Modéer M, Petersson P E 1976 Analysis of crack formation and crack growth in concrete by means of fracture mechanics and finite elements, *Cem. Concr. Res.* **6(6)** 773-81.
[5] Bazant Z P, Oh B H 1983 Crack band theory for fracture of concrete, *Mat. Struct.* **16(3)** 155-77.
[6] Rots J 1968 Computational Modeling of Concrete Fracture, Ph.D. dissertation, Delft, Netherlands.
[7] ACI 446.3R-97, 1997 Finite Element Analysis of Fracture in Concrete Structures: State-of-the-Art, *ACI Comm.* **3R-97** 33.
[8] Edalat-Behbahani A, Barros J A O, Ventura-Gouveia A 2017 Three dimensional plastic-damage multidirectional fixed smeared crack approach for modelling concrete structures, *Int. J. Solids Struct.* **115-116** 104-25.
[9] Menin R C G, Trautwein L M, Bittencourt T N 2009 Smeared Crack Models for Reinforced Concrete Beams by Finite Element Method, *IBRACON Struct. Mater.* **2(2)** 166-200.
[10] Rots J G, Blaauwendraad J 1989 Crack Models for Concrete: Discrete or Smeared? Fixed, Multi-Directional or Rotating?, *Heron*, **34(1)** 1-59.
[11] ACI 435R-95 1995 Control of Deflection in Concrete Structures, *ACI Comm.* **R-95**.
[12] Branson D E 1963 Instantaneous and Time-Dependent Deflections of Simple and Continous Reinforced Concrete Beams, *HPR Publication* **7(1)** 1-78.
[13] Sena Cruz J, Barros J, Azevedo Á 2006 Elasto-plastic multi-fixed smeared crack model for concrete, *Braga, Portugal, Dec.*
[14] de Borst R., Nauta P 1985 Non-orthogonal cracks in a smeared finite element model, *Eng. Comput.* **2(1)** 35-46.
[15] Menin R C G, Trautwein L M, Bittencourt T N 2009 Smeared Crack Models for Reinforced

Concrete Beams by Finite Element Method Modelos de Fissuração Distribuída em Vigas de Concreto Armado pelo Método dos Elementos Finitos, *IBRACON Struct. Mater.* **2(2)** 166-200.

[16] Washa G W, Fluck P G 1952 The Effect of Compressive Reinforcement on the Plastic Flow of Reinforced Concrete Beams, *ACI J.* **9(8)** 89-108.

[17] Kupfer H, Hilsdofr H K, Rusch H 1969 Behaviour of concrete under biaxial stresses, *J. Am. Concrete Inst.* **66(8)** 656-66.

[18] Rots J G 1989 Smeared crack approach. Fracture Mechanics of Concrete Structures: From Theory to Applications, *RILEM Report. Editor L. Elfgren. London: Chapman and Hall*, 138-46.

[19] Reinhardt H A, Kormeling H W 1983 Determination of the fracture energy of normal concrete and epoxy modified concrete, The Netherlands.

[20] De Borst R, Crisfield M A, Remmers J J C, Verhoosel C V 2012 Non-Linear Finite Element Analysis of Solids and Structures, *2nd ed. Chichester, Wiley*.

[21] Walraven J C 1978 The influence of depth on the shear strength of lightweight concrete beams without shear reinforcement, Report 5-78-4, Stevin Laboratory, Delft University of Technology, Delft.

[22] Rijkswaterstaat Ministry of Infrastructure and the Environment 2017 Guidelines for Nonlinear Finite Element Analysis of Concrete Structures, p. 45.

IOP Conf. Series: Materials Science and Engineering **739** (2020) 012024 doi:10.1088/1757-899X/739/1/012024

Adsorption of Diclofenac Pharmaceutical onto a Cationic Organoclay and an Untreated Montmorillonite

Tiago De Oliveira[1], **Régis Guégan**[1, 2]

[1]Institut des Sciences de la Terre d'Orléans, UMR 7327, CNRS-Université d'Orléans, 1A Rue de la Férollerie, 45071 Orléans Cedex 2, France
[2]Faculty of Science and Engineering, (Global Center for Science and Engineering), Waseda University, 3-4-1 Okubo, Shinjuku-ku, Tokyo 169-8555, Japan

regis.guegan@univ-orleans.fr, regis.guegan@aoni.waseda.jp

Abstract. Among pharmaceutical products (PPs) recalcitrant to water treatments, diclofenac shows a high toxicity and remains at high concentration in natural aquatic environments. The aim of this study concerns the understanding of the adsorption mechanism of this anionic PP onto an organoclay prepared with a long-alkyl chains cationic surfactant for various experimental pH and temperature conditions. The experimental data obtained by a set of complementary techniques (X-ray diffraction, elemental analyses, gas chromatography coupled with mass spectrometry, and Fourier transform infrared spectroscopy) and the use of Langmuir, Freundlich and Dubinin-Radushkevish equation models, reveal that organoclays show a good affinity to diclofenac while the chemical nature of surfactant appears to play a minor role. This study stresses out the application of organoclays for the adsorption of a recalcitrant PP in numerous aquatic compartments that can be used as a complement with activated carbon for waste water treatment for instance.

1. Introduction

Pharmaceutical products (PPs) are non-biodegraded substances which are more and more commonly found in sewage and surface water [1]. Although being present at low concentration, these substances have been recognized to drive to toxic effects on several ecosystems. Among PPs recalcitrant to water treatments, diclofenac, a nonsteroidal anti-inflammatory substance, shows a high toxicity [2] and remains at high concentration in the environment [3] due to its particular resistance to the current treatment [4]. For this purpose, new techniques including adsorbents have to be developed for the adsorption of PPs [5-12]. Adsorption represents an effective and easy way to remove both inorganic and organic micro-pollutants limiting the formation of pharmaceutical residues with unknown effect in the environment. Clay minerals were for a long time known for their outstanding adsorption properties. However, although being used in water treatment factory, these layered materials turn out to be ineffective for the sequestration of persistent nonpolar hydrophobic contaminants and/or negatively charged organic species such as diclofenac. The chemical modification of clay mineral via the use of cationic surfactants, like the benzyl dimethyl tetradecyl ammonium (BDTA), for the preparation of organoclays generate a hydrophobic environment which enhance the adsorption for emerging micro-pollutants [5-12]. Indeed, the intercalation of cationic surfactants switches the chemical nature of the starting layered material from hydrophilic to hydrophobic. Moreover, the resulting organoclay composite layered materials show an opening of their interlayer space that significantly improve the adsorption of numerous organic compounds [7]. The aim of this research work was to study the adsorption of diclofenac onto a cationic organoclay for water remediation.

2. Materials and Methods

2.1. Adsorbents

Wyoming sodium montmorillonite (Mt), obtained from the Source Clay Minerals Repository, University of Missouri (Columbia, MO) was used in this study as a starting material. The structural formula can be expressed as: $(Ca_{0,12}\ Na_{0,32}\ K_{0,05})\ [Al_{3,01}\ Fe(III)_{0,41}\ Mn_{0,01}\ Mg_{0,54}\ Ti_{0,02}]\ [Si_{7,98}\ Al_{0,02}]\ O_{20}(OH)_4$. Mt was fractioned to < 2 µm by gravity sedimentation, purified and Na^+ exchanged. This Mt clay mineral shows a cation exchange capacity (CEC) of 92.4 meq per 100 g clay. The benzyl dimethyl tetradecyl ammonium (BDTA) cationic surfactant was supplied from Sigma Aldrich Chemical. Aqueous solutions of BDTA with Millipore deionized water were prepared at ambient temperature for concentration at 4 times the cation exchange capacity (CEC) of Mt clay. The surfactant solution was poured in aqueous Mt dispersions before being stirred at 300 rpm for 24 h. At the end of this procedure, the solution was centrifuged at 3000 rpm for 20 min and the supernatant was removed. The resulting organoclay was dried at 100°C for 24 h before analysis and used as sorbents.

2.2. Experimental Techniques

The concentration of diclofenac before and after being in contact with the organoclay was obtained by GC/MS analysis using Trace GC Ultra chromatograph and TSQ Quantum XLS mass spectrometer (Thermo Scientific). Fourier transform infrared (FTIR) measurements in the range 650-4000 cm^{-1}, were recorded using a Thermo Nicolet 6700 FT spectrometer equipped with a Deuterated Triglycine Sulfate (DTGS) detector and a Nicolet Continuum microscope. The powder samples were spread over a NaCl window of the microscope. The analyzed sample area was a square of side 100 µm chosen under the microscope 15X Infinity Reflechromat objective. The analyses were performed in transmission mode and each spectrum corresponded to the average of 256 scans collected at 2 cm^{-1} resolutions.

The structural changes of the layered materials (organoclays and diclofenac/organoclay complexes) were determined by X-ray diffraction. X-ray diffraction patterns were recorded in a conventional θ-θ Bragg-Brentano configuration by using a Thermo Electron ARL'XTRA diffractometer equipped with a Cu anode (CuK_α = 1.5418 Å) coupled with a Si(Li) solid detector. The diffractograms on dry samples (100°C for 24 h) were performed between 1 and 24° (2θ) with an angular and time steps of 0.04° and 10 s, respectively.

2.3. Adsorption of diclofenac

Diclofenac (2-[(2,6-dichlorophenyl) amino] benzeneacetic acid) purchased from Sigma Aldrich Chemical, and assumed to have a purity > 98 % was used in its sodium salt form which shows a solubility better than 10 g L^{-1}. A batch adsorption experiment of diclofenac onto organoclays was conducted with a solution at 1 g L^{-1}. The solid to liquid ratio was kept constant, where 100 mg of organoclays was used for 50 mL of diclofenac aqueous solutions in centrifuge tubes. Sample was shaken on a rotary shaker at 50 rpm during 24 h in order to reach the equilibrium final concentrations and was centrifuged at 5000 rpm for 25 min. Both supernatants and $BDTA_4$-Mt organoclay after contact with the diclofenac were removed and analysed through GC/MS. The amount of adsorbed diclofenac was calculated by the difference between the initial and equilibrium final concentrations. Before their XRD and FTIR characterizations, the entire resulting organoclay and Mt after adsorption of diclofenac were dried at 100°C for 48 h for both XRD and FTIR characterizations.

3. Results and Discussion

3.1. Adsorbent properties

X-Ray diffraction patterns of the Na-montmorillonite (Na-Mt) and $BDTA_4$-Mt organoclay are shown in figure 1. The X-ray diffraction patterns of the Na-Mt displays a broad (*00l*) reflection located around 9° (2θ) leading to an interlayer spacing close to 10 Å which underlines the dehydration of the clay mineral according to previous observations and numerical simulations on natural montmorillonites [13, 14]. Na-Mt is also characterized by several diffraction peaks at 20° (2θ) corresponding to a *(02.11)* reflection, and despite the purification process, contains quartz impurity, identified with Bragg peaks located at 21° (2θ). The X-ray diffraction patterns of $BDTA_4$-Mt shows several diffraction peaks at the 3° and 6° (2θ), corresponding to *(001)* and *(002)* reflections which reflect the existence of a well ordering in the silicate layers. This demonstrates that the intercalation of BDTA is stable and forms an ordered structure within the silicate layers. The insertion of BDTA expands the distance between the layers, giving a d_{001} spacing of ~35 Å. This value implies that inserted surfactant may adopt a perpendicular position between the clay platelets [7], which was confirmed through FTIR analysis.

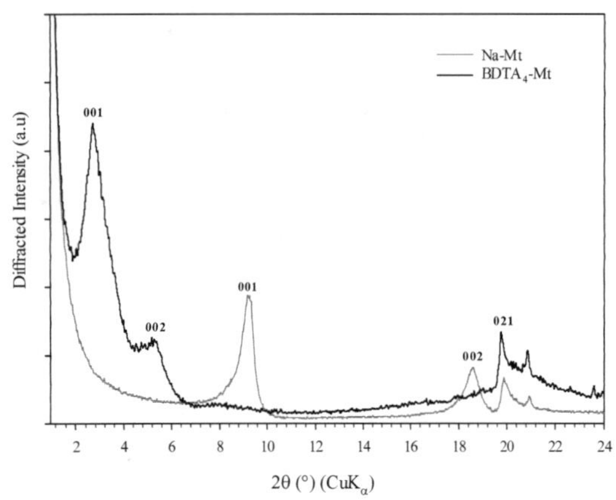

Figure 1. X-ray diffractograms of Na-Mt and $BDTA_4$-Mt organoclay.
The intercalation of BDTA is confirmed by the shift to low angle value of the *00l* reflections, expanding the d_{001} spacing from 1 (dehydrated Na-Mt) to 3.5 nm ($BDTA_4$-Mt).

Infrared spectroscopy gives a qualitative idea on the adsorption process. Figure 2 shows the infrared spectra between 650 and 4000 cm^{-1} of the dehydrated Na-montmorillonite, $BDTA_4$-Mt organoclays before and after being in contact with diclofenac which show similar feature: an intense absorption band corresponding to the Si-O stretching mode at 1030 cm^{-1} and the shoulder at 1117 cm^{-1}, several bands at 840, 890, and 918 cm^{-1} assigned to the AlMgOH, AlFeOH and AlAlOH deformations respectively, OH angular deformation observed at 1635 cm-1, whereas the OH stretching of water spreads out on a wide range of wavenumbers between 3200 and 3600 cm^{-1}. The last broad band at 3626 cm^{-1} is assigned to the OH stretching of structural hydroxyl groups. Moreover, to the assignment of the bands related to the clay mineral, the spectra reveal bands characterizing another crystalline or amorphous phases. Indeed, the weak band at 800 cm^{-1} is assigned to the Si-O stretching of quartz and/or amorpheous silica.

The spectrum of $BDTA_4$-Mt reveals expected bands of the Na-Mt with two additional bands related to the adsorption of BDTA onto the clay mineral surface. Indeed, the broad infrared absorption band at 1470 cm^{-1} is relative to the CH_2 scissoring mode whereas the intense absorption bands at 2920 and 2850 cm^{-1} correspond to the antisymmetric and symmetric CH_2 stretching mode of the hydrocarbon chains, respectively. These two CH_2 stretching bands are recognized to strongly depend on the density of confined surfactant within the interlayer space [12]. In addition to give an idea of the density of confined BDTA, the wavenumbers of CH_2 stretching bands stress out the conformational order of surfactant. The wavenumbers of both symmetric and antisymmetric absorption bands for the confined surfactant indicate that BDTA adopt an all-trans conformation within the interlayer space of the clay mineral.

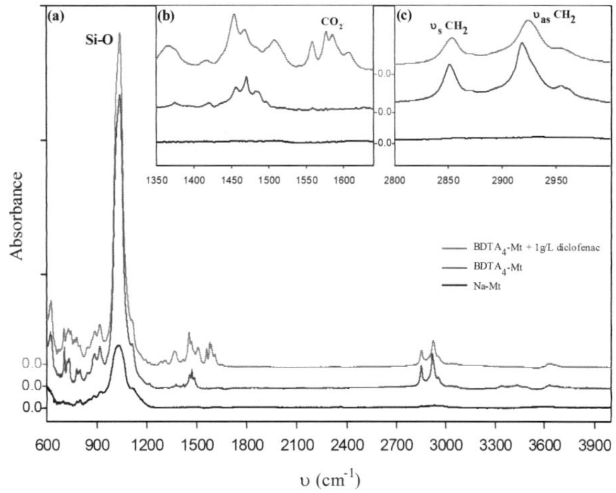

Figure 2. FTIR spectra of Na-Mt (black line), BDTA$_4$-Mt (blue line) and BDTA$_4$-Mt after interaction with a solution of diclofenac at 1 g L^{-1} (red line) **(a)** in the range 600-3900 cm^{-1}. **(b)** in the wavenumbers window of CO_2^- stretching bands of diclofenac in the range 1450-1650 cm^{-1}. **(c)** in the wavenumbers window of CH_2 symmetric (s) and asymmetric (as) stretching bands of BDTA surfactant in the range 2800-3000 cm^{-1}.

3.2. Environmental properties of BDTA$_4$-Mt: Diclofenac adsorption

Analytical GC/MS chromatogram results of solutions after interaction between the organoclay and the diclofenac (figure 3) shows the relative abundance of the different compounds in the supernatant. The picks in the range 15-24 minutes show the MTBSTFA compounds used for the sialylation of the sample before GC/MS analysis which allows a better thermal stability and a higher volatility for molecules. The chlorinated compound pick, visible at a retention time of 45 minutes, may correspond to a residue of diclofenac degradation during the process. The specific diclofenac pick is situated at a retention time of 54 minutes. The quantification of resulting diclofenac molecules after interaction in the supernatant requires the use of a standard. The 5-alpha-cholestan standard, at a concentration of 10.4 mg L^{-1}, has a retention time of 62 minutes. Area ratio between diclofenac and 5-alpha-cholestan picks provides the quantification of the diclofenac molecule present on the resulting solution. The analysis confirms a diminution of diclofenac amount, due to the adsorption onto organoclay.

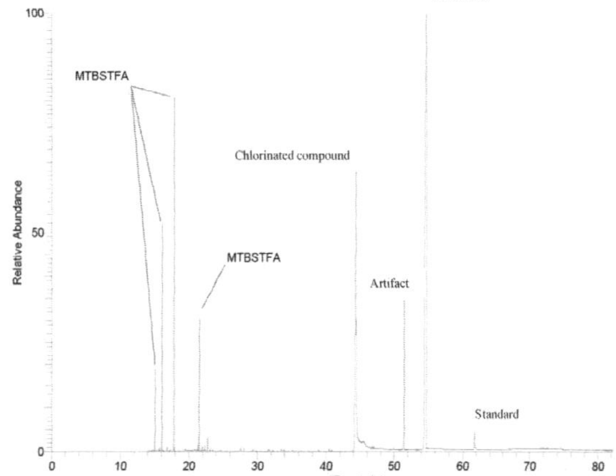

Figure 3. Total ion chromatogram obtained by GC/MS analysis after interaction between BDTA$_4$-Mt and 1 g L^{-1} solution of diclofenac at 25°C. The artefact pick, visible at a retention time of 52 minutes, is related to the GC/MS detector.

The spectrum of BDTA4-Mt after interaction with diclofenac at 1 g L^{-1} (insert in figure 2) reveals, new bands corresponding to the CO_2^- stretching bands of diclofenac in the range 1500-1600 cm^{-1} (in addition to organoclay absorption bands). The presence of CO_2^- stretching bands specific to the diclofenac, onto the resulting organoclay powder confirms the proper adsorption of diclofenac molecule within the organoclay determined through GC/MS analysis.

4. Conclusions

The use of BDTA cationic surfactant with a Na^+ exchanged clay for the preparation of organoclay contributes to enhance the adsorption of diclofenac defined as a micro-pollutant. The intercalation of surfactant at 4 times the CEC leads to an expansion of the interlayer space allowing a better access to diclofenac. Regarding the efficiency of these composite materials, the prepared organoclays in this study displayed some particular interests for the sequestration of diclofenac in wastewater or natural effluents.

Acknowledgments
The authors would like to thank the Region Centre Val de Loire (Project MONITOPOL2017 - 00117247) for their financial support.

References

[1] Thiebault T, Guégan R, Boussafir M 2015 Adsorption mechanisms of emerging micro-pollutants with a clay mineral: Case of tramadol and doxepine pharmaceutical products, *J. Colloid Interface Sci.* **453** 1-8.
[2] Ferrari B, Paxéus N, Giudice R L, Pollio A, Garric J 2003 Ecotoxicological impact of pharmaceuticals found in treated wastewaters: study of carbamazepine, clofibric acid, and diclofenac, *Ecotoxicol. Environ. Saf.* **55(3)** 359-70.
[3] Miege C, Choubert J M, Ribeiro L, Eusèbe M, Coquery M 2009 Fate of pharmaceuticals and personal care products in wastewater treatment plants–conception of a database and first results, *Environ. Pollu.* **57(5)** 1724-6.
[4] Verlicchi P, Zambello E 2014 How efficient are constructed wetlands in removing pharmaceuticals from untreated and treated urban wastewaters? A review, *Sci. Total Enviro.* **470** 1281-306.
[5] Guegan R 2011 Confinement effects on water structure in membrane lyotropic phases, *J. Colloid Interface Sci.* **358(2)** 485-90.
[6] Guegan R 2019 Organoclay applications and limits in the environment, *Comptes Rendus Chimie* **22(2-3)** 132-41.
[7] De Oliveira T, Guegan R 2016 Coupled organoclay/micelle action for the adsorption of diclofenac, *Environ. Sci. Technol.* **50(18)** 10209-15.
[8] Guegan R, Giovanela M, Warmont F, Motelica-Heino M 2015 Nonionic organoclay: A ‘Swiss Army knife’for the adsorption of organic micro-pollutants?, *J. Colloid Interface Sci.* **437** 71-9.
[9] Guegan R 2013 Self-assembly of a non-ionic surfactant onto a clay mineral for the preparation of hybrid layered materials, *Soft Matter* **9(45)** 10913-20.
[10] Guegan R, Sueyoshi K, Anraku S, Yamamoto Y, Miyamoto N 2016 Sandwich organization of non-ionic surfactant liquid crystalline phases as induced by large inorganic K 4 Nb 6 O 17 nanosheets, *Chem. Commun.* **52(8)** 1594-7.
[11] De Oliveira T, Fernandez E, Fougère L, Destandau E, Boussafir M, Sohmiya M, Sugahara Y, Guegan R, 2018 Competitive Association of Antibiotics with a Clay Mineral and Organoclay Derivatives as a Control of Their Lifetimes in the Environment, *ACS Omega* **3(11)** 15332-42.
[12] Guegan R, Véron E, Le Forestier L, Ogawa M, Cadars S 2017 Structure and dynamics of nonionic surfactant aggregates in layered materials, *Langmuir* **33(38)** 9759-71.
[13] Karaborni S, et al. 1996 The swelling of clays: molecular simulations of the hydration of montmorillonite, *Sci.* **271(5252)** 1102-4.
[14] Carli L N, Daitx T S, Guegan R, Giovanela M, Crespo J S, Mauler S 2015 Biopolymer nanocomposites based on poly (hydroxybutyrate-co-hydroxyvalerate) reinforced by a non-ionic

organoclay, *Polym. Int.* **64(2)** 235-41.

Optical Properties of Cd-Free Quantum Dots-Based Fluorescent Film

M Y Lin[1], W X Lin[1], S R Chung[1]

[1]Department of Materials Science and Engineering, National Formosa University, Yunlin 63201, Taiwan

srchung@nfu.edu.tw

Abstract. $CuInS_2$/ZnS Cd-free quantum dots (QDs) have many characteristics, such as high quantum yield (QY), wide emission peak and adjustable light color. They are suitable for optical conversion materials and applies in solid-state lighting (SSL). However, they are not easily dispersed during encapsulation, resulting in the deviation of light color out of the white light area after encapsulation. In this study, we use thermal injection synthesis method to prepare $CuInS_2$/ZnS and $CuInS_2$/ZnS:Al QDs. In order to improve the dispersion of QDs, solution type of QD/PS-PE-BR-PS copolymer (SEBS) mixture was prepared by mixing QD and SEBS with different ratios (10, 20, and 30 wt%) to form a fluorescent film. The experimental results show that the emission wavelengths and QY of $CuInS_2$/ZnS and $CuInS_2$/ZnS:Al QDs are 533 nm, 84 % and 536 nm, 97 %, respectively. The emission wavelength of $CuInS_2$/ZnS-based fluorescent film is 574 nm, while $CuInS_2$/ZnS:Al-based fluorescent film is 582 nm. The CRI, luminous efficacy, chromaticity coordinates and correlated color temperature of 20 wt% $CuInS_2$/ZnS-based fluorescent film excited by blue chip is 64, 42 lm/W, (0.33, 0.32), and 5443 K, respectively. On the other hand, we find that the CRI of $CuInS_2$/ZnS:Al-based fluorescent film can be improved from 64 to 73, and the luminous efficacy is 51 lm/W.

1. Introduction

The white light-emitting diode (WLEDs) has attracted extensive attention in the solid-state lighting (SSL) and backlight source technology because of its advantages of environmentally friendly, energy saving, and so on [1-3]. In SSL, there are at least combine two wavelengths of light to form white light, such as the white light is formed by combination a blue InGaN with the yellow YAG: Ce^{3+} phosphor [4]. However, due to the color temperature is higher than 4500 K and the lacks of red emission spectrum, the color rendering index (CRI) of YAG-based WLED is less than 80, which limits the application of indoor lighting [5]. Compared with phosphor, most of quantum dots (QDs) especially for Cd-based QDs exhibit tunable photoluminescence spectrum and high quantum yield (QY) [6, 7]. Narrow full width at half maximum (FWHM) and easy adjustment of three basic colors (red, green and blue) of QDs make them quite suitable to be used as backlight source [8]. However, high CRI is the main demand for SSL. So, the more important thing is to extend the emission spectrum of luminescent materials to covers entire visible range. In fact, the YAG-based WLED which is more popular device that often improve CRI by blending Cd-based red-emissive QDs to enhance the red band intensity in EL spectrum. Because of the well-known toxicity, Cd-based QD will be limited in the future. As time goes on, composition tunable of InP [9] and I-III-VI alloyed QDs ($CuInS_2$ [10], $CuInSe_2$ [11] and $AgInS_2$ [12]) which is Cd-free materials has been developed, but the FWHM of InP

QD is not wider enough to application in WLED [13]. Today, the emission wavelength of I-III-VI alloyed QDs can be adjusted in the range from visible to near infrared by doping process. Among these I-III-VI QDs, $CuInS_2$ (CIS) is a very suitable one for SSL applications because of its excellent absorption coefficient at 450 nm and wide emission spectrum in yellow green range [14]. In particular, the QY of CIS QDs can be increased from 2 to more than 70 % by coating ZnS shell, and the FWHM is larger than 80 nm [15]. Although the CRI of CIS/ZnS QDs modified YAG-based WLED is ultrahigh (>90), the luminous efficacy (<40 ml/W) is still lower than that of conventional WLEDs [16-18] and the emission wavelength red-shifts very significantly [19, 11], due to the agglomeration between QDs and lower QY compared to YAG. Moreover, the compatibility between silicone and CIS/ZnS is also poor, resulting in QDs tends to agglomerate. Pinhua et al. reported that the luminous efficacy of device enhances by adding Al [20]. On the other hand, the package form and encapsulates are also the key factors for devices performance. Remote-typed of QD-based WLED was produces which meaning that a fluorescent film is made and without contact with the blue InGaN chip [16-18]. Here, we report a facial chemical route to synthesize CIS/ZnS:Al QDs with high QY (97 %) by oxide layer passivation. The compatibility of $CuInS_2$/ZnS:Al QDs-based fluorescent films can be improved and luminous efficacy is 51 lm/W.

2. Experimental

2.1. Chemicals

Copper iodide (I) (CuI, 99.99 %), 1-dodecanethiol (DDT, 98 %), octadecene (ODE, 90 %), aluminum isopropoxide ($Al(IPA)_3$, 99.9 %) and zinc stearate ($Zn(SA)_2$, 10-12 % Zn basis) are provided by Aldrich. Indium acetate ($In(AC)_3$, 99.99 %) was purchased from Alfa Aesar. Styrene block copolymer (SEBS, model 01082) is provided by Phon Tech Industrial Company. Toluene (99.5 %), hexane (99.7 %), acetone (99.9 %), methanol (99.7 %), and isopropanol (99.5 %) were used directly.

2.2. Synthesis

2.2.1. Synthesis of $CuInS_2$ quantum dots. The CIS core QDs are synthesized by the hot injection method. 0.0625 mmol CuI, 0.5 mmol $in(AC)_3$ and 5 ml DDT were added to the 100 ml three-necked flask and heated at 100 °C for 30 minutes in vacuum first and then purged with Ar to form a yellow-green solution. Subsequently, the mixture was heated to 230 °C under the heating rate of 15 °C/min, and the CIS QDs can be obtained after reacts for 5 minutes.

2.2.2. Synthesis of CIS/ZnS quantum dots. The core/shell structure of CIS/ZnS QD is synthesized as follows. ZnS stock solution was prepared by mixing 4 ml ODE, 1 ml DDT, and 8.3 mmol $Zn(SA)_2$ in the three-necked flask and heated to 190 °C, then reacted for 50 minutes under Ar atmosphere. After that, the ZnS stock solution dropped into the CIS core QD solution at about 1 ml/min rate. Then the mixture was heated to 240 °C and reacted for 120 minutes. The CIS/ZnS core/shell QD was purified by excess methanol, and the supernatant was removed and the precipitate was collected. The precipitates were then dispersed in toluene and precipitated with excess methanol, and repeated precipitation/dispersion cycles three times to remove the byproducts and unreacted precursors. Finally, the precipitate was purified by toluene/acetone/methanol (1:1:2) cosolvent and the purified QD was dried at room temperature.

2.2.3. Synthesis of CIS/ZnS:Al quantum dots. The synthesis process of CIS/ZnS:Al was the same with CIS/ZnS. 4.15 mmol $Al(IPA)_3$ was dissolved in DDT. The $Al(IPA)_3$ and ZnS solutions were injected into CIS core solution simultaneously at 240 °C and reacted for 120 minutes. The precipitate was purified with hexane/isopropanol (1:5) cosolvent, and the purified QD was dry at room temperature.

2.3. Preparation of fluorescent film and white LED

The QDs, toluene, and PS-PE-BR-PS copolymer (SEBS) were mixed together and the mixture poured into the mold to form fluorescent film with 0.6 mm thickness, as showed in figure 1 (a). The

IOP Conf. Series: Materials Science and Engineering **739** (2020) 012025 doi:10.1088/1757-899X/739/1/012025

fluorescent film putted on the top of blue LED to form white LED, as showed in figure 1 (b), and the performance of device is measured by integral sphere (Isuzu Optics, ISM-360) under 20 mA injection current.

2.4. Characterizations
The optical properties of the samples were measured by Fluorescence Spectrometer (Hitachi F-7000) and ultraviolet visible spectrometer (UV-vis, Hitachi UH 5300). The relative QY of the samples was determined by comparing the emission area of samples to the organic dye (Rhodamine 6G in methanol). The structure of samples was measurement by X-ray diffractometer (XRD, Bruker D8A25) with Cu K_{α} (λ=1.5403 Å). Transmission electron microscope (TEM, JEM-2100F) was used to analyze the morphologies and size distributions of the samples.

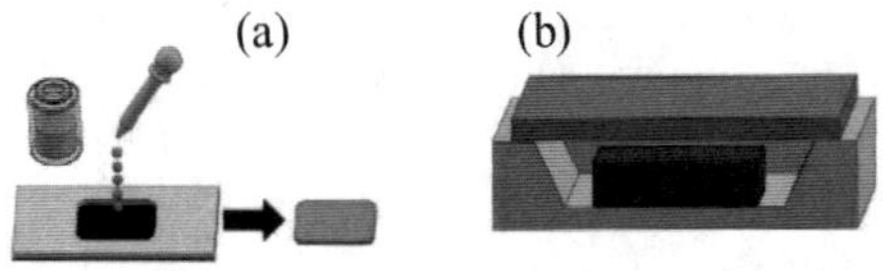

Figure 1. The schematic diagram of (a) fluorescent film and (b) fluorescent film-based WLED.

3. Results and Discussion
Figure 2 shows the excitation, emission and UV absorption spectra of CIS/ZnS and CIS/ZnS:Al QDs. In figure 2 (a), the absorption peak of CIS/ZnS QD is not obvious due to the multiple defect energy levels in QD. Moreover, the absorption peak of CIS/ZnS:Al QD, in figure 2 (b), is more prominent than that of CIS/ZnS QD. The FWHM of CIS/ZnS and CIS/ZnS:Al are 74 and 83 nm, respectively. This may be meaning that Al is doped into the CIS core, and produces a defect levels within the band gap, resulting in extends the FWHM. This is beneficial to increase the color rendering index (CRI) value of WLED. Both samples show the tail in UV absorption spectra meaning that the intrinsic character of defect emission mechanism [21]. Moreover, we also observe that a large Stokes shift and the wide FL emission range for both samples. The emission peaks of CIS/ZnS and CIS/ZnS:Al is 533 and 536 nm, respectively. The emission wavelengths of both samples are quite close, indicating that the doping of Al does not affect the band gap. We also can find that the QY can be enhanced from 84 to 97 % after Al doping, which might be due to the formation of Al_2O_3 on the surface, resulting in the passivation of the QDs surface more effective [20]. In figure 3 (a-c), the XRD pattern shows that the CIS QD has a chalcopyrite structure. With the coating of the ZnS shell, the main three diffraction peaks of CIS/ZnS move toward the ZnS zinc blende structure slightly. This displacement indicates the formation of $ZnCuInS_2$ alloy [21, 22], while the CIS/ZnS:Al diffraction peak is the same as CIS/ZnS, showing that a few amount of Al has little effect on the crystal structure. Based on figure 3 (d) and (e), we find that the particle size of CIS/ZnS:Al and CIS/ZnS are similar and about 3 nm. Moreover, the morphologies of samples are tetrahedral [20].

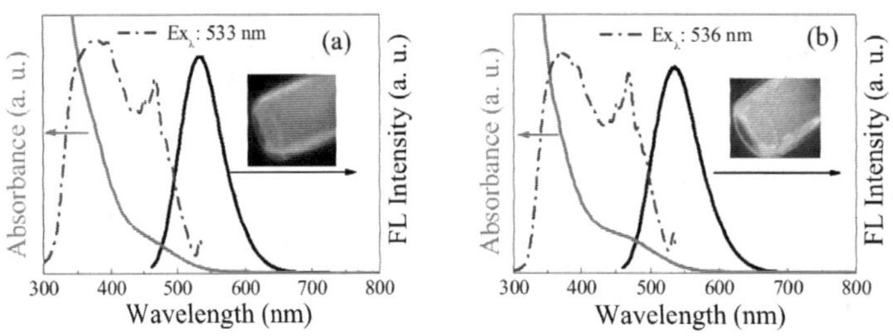

Figure 2. The absorption, excitation, and emission spectra of (a) CIS/ZnS and (b) CIS/ZnS: Al.

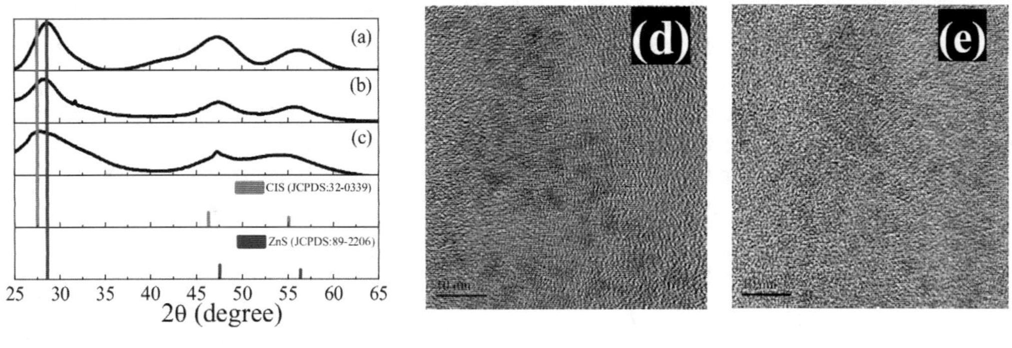

Figure 3. XRD patterns. (a) CIS/ZnS:Al, (b)CIS/ZnS, and (c) CIS. TEM images. (d) CIS/ZnS, (e) CIS/ZnS: Al.

A large amount of sulfenyl remains on the surface of the QDs, which makes the silicone gel unable to curing. Therefore, the SEBS copolymer, which is similar to silicone gel and it can be miscible with toluene and CIS QDs as well as the compatibility can be improved. Because the heat resistance of SEBS is about 140 °C, we use remote-typed to encapsulate it. In the package, we encapsulate the film with QD content of 10, 20, and 30 wt %, and make white light with two films assembly.

The EL spectra of CIS/ZnS- and CIS/ZnS:Al-based WLED are shown in figures 4 (a) and (b). We can find that the emission peak of CIS/ZnS:Al-based WLED is 584 nm, and the red-shift is about 50 nm. On the other hand, the emission wavelength of CIS/ZnS-based WLED is 573 nm, the red-shift is about 40 nm. The larger shift of CIS/ZnS:Al WLED may be due to the different residual precursors of the QDs. In addition, we can observe that the emission peak of CIS/ZnS:Al-based WLED does not change much by altering the content of QDs, and the difference between maximum and minimum emission peak is 4 nm, while the difference for CIS/ZnS-based WLED is 10 nm, illustrates that the dispersibility can be improved by Al doping, which change the surface properties of CIS/ZnS QDs.

Figures 4 (c) and (d) show the CIE coordinates of devices. The CIE of CIS/ZnS:Al-based WLED not change with QD concentrations and the correlated color temperature (CCT) is below 4500 K, and the highest luminous efficacy is 51 lm/W, while the CIE of CIS/ZnS-based WLED toward to warm light side, and the highest luminous efficacy is 42 lm/W. The results are summarized in table 1.

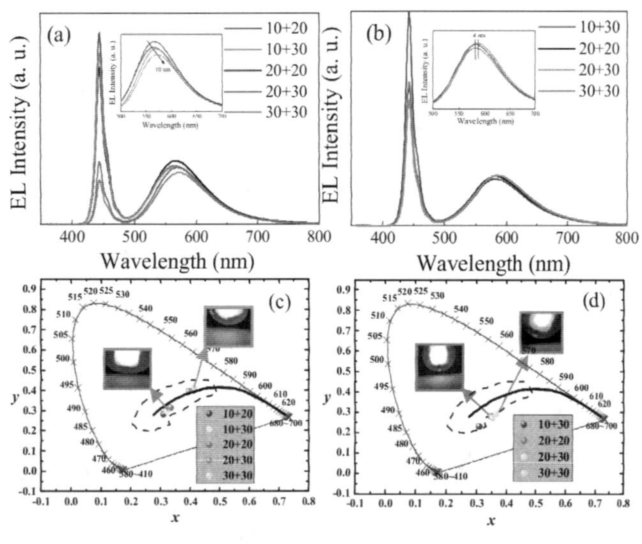

Figure 4. CIE coordinates of (a) CIS/ZnS and (b) CIS/ZnS:Al; Electroluminescence spectra of (c) CIS/ZnS and (d) CIS/ZnS:Al.

4. Conclusion

In this paper, CIS/ZnS and CIS/ZnS:Al core/shell QDs were synthesized by hot injection method. The emission wavelength, QY, and FWHM of CIS/ZnS QD is 533 nm, 84 %, and 74 nm, while that is 536 nm, 97 %, and 83 nm for CIS/ ZnS:Al QDs. The luminous efficacy and CRI of CIS/ZnS:Al WLED is 51 lm/W and 74, which is better than that of CIS/ZnS one. The Al doping is more effective to improve the QY, FWHM, and luminous efficacy. Therefore, the results show the evident to prove that the CIS/ZnS:Al QDs are quite suitable for SSL application.

Table 1. The performance of fluorescent film-based WLEDs.

Devices	QD content (wt%)	CRI	Efficacy (lm/W)	CCT (K)	CIE (x, y)
CIS/ZnS	10+10	0	40	0	(0.24, 0.18)
	10+20	67	39	7310	(0.31, 0.28)
	10+30	66	38	6380	(0.32, 0.30)
	20+20	64	42	5440	(0.33, 0.32)
	20+30	56	35	3770	(0.40, 0.40)
	30+30	55	32	3520	(0.41, 0.41)
CIS/ZnS:Al	10+10	0	44	0	(0.22, 0.12)
	10+20	0	50	0	(0.27, 0.19)
	10+30	71	49	10790	(0.31, 0.23)
	20+20	73	51	4460	(0.35, 0.28)
	20+30	70	50	3730	(0.36, 0.29)
	30+30	74	50	4020	(0.35, 0.27)

References

[1] Kong D S et al "Fine tuning of emission property of white light-emitting diodes by quantum-dot-coating on YAG: Ce nanophosphors" 2016 *Appl. Surf. Sci.* **379** 467-473
[2] Dai Q et al "Semiconductor Nanocrystals Based White Light-Emitting Diodes" 2010 *Small* **7** 1577-1588
[3] Li X et al "New Yellow $Ba_{0.93}Eu_{0.07}Al_2O_4$ Phosphor for Warm-White Light-Emitting Diodes through Single-Emitting-Center Conversion" 2013 *Light: Sci. Appl.* **2** e50
[4] Yin L et al "The thermal stability performances of the color rendering index of white light emitting diodes with the red quantum dots encapsulation" 2013 *Optical Mater.* **42** 187-192
[5] Cosgun A et al "Flexible Quantum Dot-PVA Composites for White LEDs" 2015 *J. Mater. Chem. C* **3** 257-264
[6] Dohnalová K et al "Surface Brightens up Si Quantum Dots: Direct Bandgap-like Size-Tunable Emission" 2013 *Light: Sci. Appl.* **2** e47
[7] Wonkeun C et al "YAG and CdSe/ZnSe nanoparticles hybrid phosphor for white LED with high color rendering index" 2011 *Mater. Chem. Phys.* **126** 162-166
[8] Zhang J et al "A New Strategy for Achieving White-Light Emission of Lanthanide Complexes: Effective Control of Energy Transfer from Blue-Emissive Fluorophore to Eu (III) Centres" 2015 *J. Mater. Chem. C* **3** 1799-1806
[9] Tamang S et al "Chemistry of InP Nanocrystal Syntheses" 2016 *Chem. Mater* **28** 2491-2506
[10] Da E N et al "Noninjection, one-pot synthesis of Cu-deficient CuInS2/ZnS core/shell quantum dots and their fluorescent properties" 2011 *J. Colloid and Interface Science* **361** 491-496
[11] Zhong H Z et al "Colloidal $CuInSe_2$ Nanocrystals in the Quantum Confinement Regime: Synthesis, Optical Properties, and Electroluminescence" 2011 *J. Phys. Chem. C* **115** 12396-12402
[12] Hamanaka Y et al "Photoluminescence Properties and Its Origin of $AgInS_2$ Quantum Dots with Chalcopyrite Structure" 2011 *J. Phys. Chem. C* **115** 1786-1792
[13] Bingkun C et al "From Large-Scale Synthesis to Lighting Device Applications of Ternary I-III-VI

Semiconductor Nanocrystals: Inspiring Greener Material Emitters" 2018 *J. Phys. Chem. Lett.* **9** 435-445

[14] Chuang P et al "Emission-Tunable $CuInS_2/ZnS$ Quantum Dots: Structure, Optical Properties, and Application in White Light-Emitting Diodes with High Color Rendering Index" 2014 *ACS Appl. Mater. Interfaces* **6** 15379-15387

[15] Sang H P et al "Highly Bright Yellow-Green-Emitting CuInS2 Colloidal Quantum Dots with Core/Shell/Shell Architecture for White Light-Emitting Diodes" 2015 *ACS Appl. Mater. Interfaces* **7** 6764–6771

[16] Abdelhay A et al "Ce-Doped YAG Nanophosphor and Red Emitting $CuInS_2/ZnS$ Core/Shell Quantum Dots for Warm White Light-Emitting Diode with High Color Rendering Index" 2014 *ACS Appl. Mater. Interfaces* **6** 252-258

[17] In S S et al "Stacked Quantum Dot Embedded Silica Film on a Phosphor Plate for Superior Performance of White Light-Emitting Diodes" 2014 *ACS Appl. Mater. Interfaces* **6** 5744-5748

[18] Namhun K et al "$CuInS_2/ZnS$ Quantum Dot-Embedded Polymer Nanofibers for Color Conversion Films" 2016 *J. Mater. Chem. C* **4** 2457-2462

[19] Hyunki K et al "Degradation Characteristics of Red Light-Emitting CuInS2/ZnS Quantum Dots as a Wavelength Converter for LEDs" 2011 *Electrochem. Solid-State Lett.* **14** K55-K57

[20] Pinhua R et al "Highly stable $CuInS_2$@ZnS:Al core@shell quantum dots: role of aluminum self-passivation" 2015 *Commun.* **51** 8757-8760

[21] Chen, B. K. et al "Highly emissive and color-tunable $CuInS_2$-based colloidal semiconductor nanocrystals: off-stoichiometry effects and improved electroluminescence performance" 2012 *Adv. Funct. Mater.* **22** 2081-2088

[22] Kim J H et al "Enhanced fluorescent stability of copper indium sulfide quantum dots through incorporating aluminum into ZnS shell" 2016 *J. Alloys and Compounds* **662** 173-178

Improving Stability of CIS/ZnS-Based White Light-Emitting Diodes by Silica Coating

Kawinthida Moolsarn[1], Meng-Yen Lin[1], Shu-Ru Chung[1]

[1]Department of Material Science and Engineering, National Formosa University, Yunlin, Taiwan

srchung@nfu.edu.tw

Abstract. Cadmium-free quantum dots (QDs) have been investigated for replacement cadmium-based QDs because of highly toxicity for humans and environmental such as $CuInS_2$/ZnS (CIS/ZnS) core/shell QDs that exhibit tunable emission wavelength and very broad emission spectrum. It often used in solid state lighting and white light-emitting diodes (WLEDs). This work reports on the synthesis of CIS/ZnS core/shell QDs using dodecanethiol (DDT) and zinc stearate as a precursor to form ZnS shell. Moreover, for improving the device stability, we coated SiO_2 on CIS/ZnS QDs surface. The crystal structure was analysis by transmission electron microscopy (TEM) and powder X-ray diffractometer (PXRD). The result shows that the emission wavelength of CIS/ZnS is 534 nm and with a high quantum yield (QY) of 107 %. The color rendering index (CRI), correlated color temperature (CCT) and luminous efficacy of SiO_2-coated WLED are 73, 4764 K, and 95 lmW^{-1}, respectively. Moreover, high stability of SiO_2-coated WLED can be obtained. The result indicates that CIS/ZnS@SiO_2 QDs have great potential to development of further WLED application.

1. Introduction

Semiconductor quantum dots (QDs) have widely attention in the field of solid-state lighting (SSL) [1], light emitting diodes (LEDs) [2-5], bioimaging [6] and solar cell [7] due to their appealing properties including excellent wavelength tunability, wide absorption and high quantum efficiency [8]. QDs are often made from II-VI systems, such as CdSe or ZnCdSe. However, Cd-based QDs are toxic for environment and humans, which limited for further industrialization [9, 10]. Non-toxic I-III-VI compounds have been replacement to Cd-based QDs such as ternary $CuInS_2$/ZnS (CIS/ZnS) core/shell QDs are investigated. For white light-emitting diodes (WLED), CIS/ZnS core/shell QDs have been intensively studied in several researches to improve quantum yields (QYs) and the luminescent properties, but the stability still low when fabricated on LED device. Recently, QDs-LED coated with SiO_2 have been developed to improving device stability and suppress energy transfer because of SiO_2 shell is highly transparent across wide range from infrared to visible zone [11, 12]. Here, we synthesized the CIS/ZnS QDs via a hot injection method with high QY 107% prepared SiO_2 shell coated on CIS/ZnS QDs surface by using 3-Aminopropyl-trimethoxysilane (APTES) and 3-Mercaptopropy-trimethoxysilane (MPS) to improve the device stability.

2. Experimental

2.1. Materials

Copper (I) iodide (CuI, 99.99 %), 1-dodecanethiol (DDT, 98 %), 1-octadecene (ODE, 90 %), and zinc stearate ($Zn(SA)_2$, 10-12 % Zn basis) were purchased from Aldrich. Indium acetate ($In(AC)_3$, 99.99 %), (3-Mercaptopropy) trimethoxysilane (MPS, 95 %), (3-Aminopropyl) triethoxysilane (APTES, 98 %) were purchased from Alfa Aesar. Toluene (99.5 %), hexane (99.7 %), acetone (99.9%), and methanol (99.7 %) were used as original without further purification.

2.2. Synthesis of $CuInS_2$/ZnS core/shell QDs
$CuInS_2$/ZnS (CIS/ZnS) QDs were synthesized by the hot injection method. Firstly, synthesized CIS core QDs. 0.0625 mmol of CuI, 0.5 mmol of $in(AC)_3$ were added into 100 mL of three-necked flask and vacuum for 30 minutes, then added 5 mL of DDT with heated at 100 °C for 30 minutes, after that purged with argon and cool down at room temperature to form yellow solution. Consequently, heated above solution to 230 °C under heating rate of 15 °C/min. CIS QDs can be obtained after reacts for 5 minutes. Secondly, synthesized CIS/ZnS QDs. ZnS stock solution was prepared by mixing 8.3 mmol of $Zn(SA)_2$, 1 mL of DDT, and 4 mL of ODE in three-necked flask and heated to 190 °C under argon atmosphere then reacted for 50 minutes. Finally, dropped ZnS solution into the CIS core solution at rate of 1mL/min. After that, heated to 240 °C and reacted for 120 minutes. The CIS/ZnS QDs was purified by excess methanol and collected precipitate, then purified by toluene/acetone/methanol (1:1:2) co-solvents and repeated this cycle for three times. The precipitates were dried at room temperature (RT).

2.3. Preparation of CIS/ZnS@SiO_2 QDs
The CIS/ZnS QDs was coated with silica by mixing 0.07 g of QDs, 0.05 mL of MPS, 0.16 mL of APTES, and 40 mL of hexane. Those chemical were added in 250 mL of three-necked flask then heating to 65 °C with 8 °C/min of heating rate for 25 minutes. After that, cool down the solution to RT and purified with ethanol two times.

2.4. Fabrication of CIS/ZnS@SiO_2 QDs-based WLEDs
The devices were fabricated in convert type by using 25 wt.% of QDs mixed with epoxy-based A: B gel, then dropped the mixtures into blue LED chip and cured its.

2.5. Characterization
The absorption spectrum of QDs were observed by using UV-Vis spectrophotometer (UV-vis, Hitachi UH5300). The emission wavelengths were observed by using fluorescence spectrophotometer (FL, Hitachi F-7000). The fluorescence QYs of QDs were calculated by integrating the FL area to compare with rhodamine 6G in ethanol solution. The crystal structure of QDs were measured by X-ray powder diffractometer (Bruker D8A25) with Cu Kα radiation (λ=1.5418 Å). The morphologies of QDs were observed by transmission electron microscope (TEM, JEOL JEM-2010).

3. Results and Discussion

3.1. Optical properties
The UV absorption and FL spectra of CIS/ZnS QDs are shows in figure 1 (a). We found that the emission wavelength of CIS/ZnS QDs is 534 nm and QY is 107 % with a full width at half maximum (FWHM) of 74 nm. Figure 1 (b) shows the FL spectra of CIS/ZnS@SiO_2 QDs. We can find that the emission wavelength of CIS/ZnS shifts from 534 to 548 nm after SiO_2 coating.

IOP Conf. Series: Materials Science and Engineering **739** (2020) 012026 doi:10.1088/1757-899X/739/1/012026

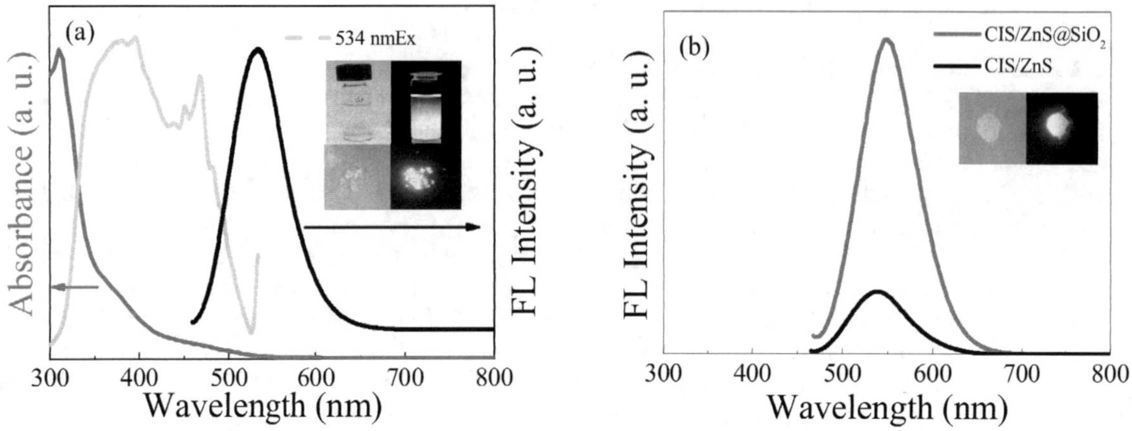

Figure 1. (a) UV and FL spectra of CIS/ZnS QDs.
(b) FL spectra of CIS/ZnS@SiO_2 QDs. compared with CIS/ZnS QDs.

3.2. Crystal structure and morphologies

The X-ray diffraction patterns of CIS, CIS/ZnS and CIS/ZnS@SiO_2 are presented in figure 2. By comparing the diffraction pattern of samples with JCPDS standard pattern of CIS and ZnS, we found that the diffraction peaks of CIS/ZnS QDs are 28.0°, 46.5° and 54.9°, respectively, which are corresponding to (112), (220), and (312) planes. It belongs to chalcopyrite phase. When coating with ZnS shell, the XRD peaks moved towards large angle, and close to zinc-blende phase, meaning that alloyed is formed. Figure 3 shows the TEM images of synthesized CIS/ZnS and CIS/ZnS@SiO_2 QDs. We found that the morphology of CIS/ZnS QDs are pyramid with average particle size 3.2 nm, the average size of QDs are increased to 3.5 nm after coating silica.

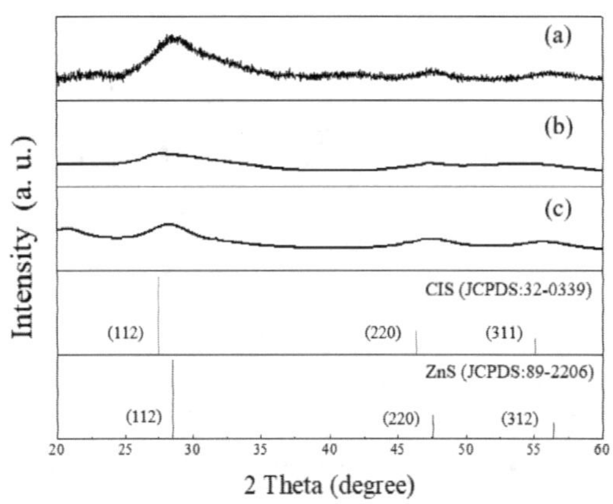

Figure 2. The XRD patterns of QDs. (a) CIS/ZnS@SiO_2, (b) CIS/ZnS, and (c) CIS core.

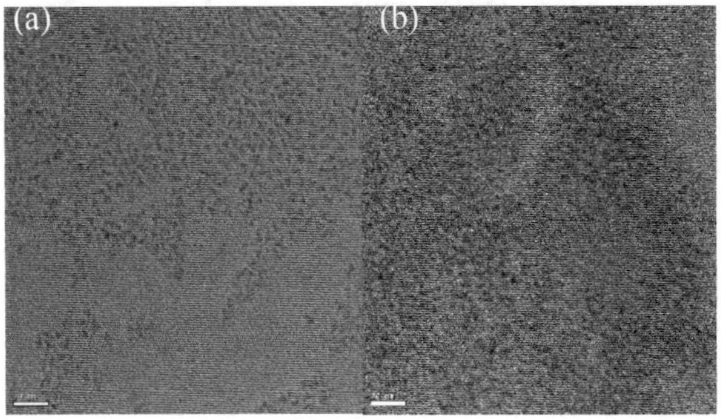

Figure 3. The TEM images of QDs. (a) CIS/ZnS and (b) CIS/ZnS@SiO_2.

3.3. Performance of WLEDs

For CIS/ZnS-based WLED, we found that it has CRI of 75, CCT of 6412 K, and luminous efficacy of 111 lmW^{-1}. The other hand, we can find that CRI of 73, CCT of 4764 K, and luminous efficacy decreasing to 95 lmW^{-1} after coated with silica. The results are summarized in table 1. Compared the device stability between before and after coating with silica refer in figure 4, we found that the luminous efficacy of CIS/ZnS-based WLED decreased rapidly, while the luminous efficacy of CIS/ZnS@SiO_2-based WLED keep unchanged after working 8 h, meaning that the device has high stability.

Table 1. The CIE color coordinates, CCT, CRI and luminous efficacy of WLED with forward current 20 mA.

Devices	**CIE (x, y)**	**CRI**	**CCT (K)**	**Efficacy (lm/W)**
CIS/ZnS	(0.32, 0.26)	75	6412	111
CIS/ZnS@SiO_2	(0.31, 0.26)	73	4764	95

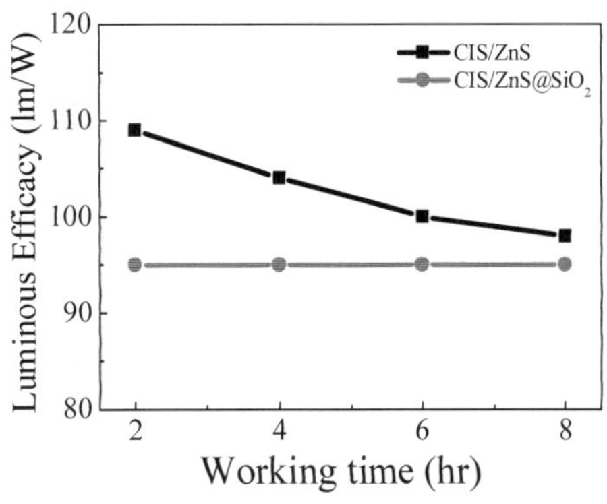

Figure 4. The luminous efficacy of QDs-WLED with forward current of 20 mA after working for 8 h.

4. Conclusion

In this study, the stability of CIS/ZnS core/shell QDs has been improved by coated with silica, and the CIS/ZnS@SiO_2 QD emitting yellow light at 548 nm, the crystal structure belongs to chalcopyrite phase and average particles size are 3.5 nm. The performance of CIS/ZnS@SiO_2-based WLED device have good improvement with high luminous efficacy value 95 lmW^{-1}, moderate CRI value of 73 and

CCT value of 4764 K with excellent device stability.

Acknowledgment
This work was supported by the Ministry of Science and Technology of Taiwan under Contract nos. 107-2221-E-150-034 and 108-2221-E-150-016.

References

[1] Chen N, Bai Z, Wang Z, Ji H, Liu R, Cao C, Wang H, Jiang F, Zhong H 2018 Low Cost Perovskite Quantum Dots Film Based Wide Color Gamut Backlight Unit for LCD TVs *Soc. Inf. Display* **49** 1657-1659

[2] Zhang J, Xie R, Yang W 2011 A simple route for highly luminescent quaternary Cu-Zn-In-S nanocrystal emitters *Chem. Mater.* **23** 3357-61

[3] Wang Y, Zhu G, Xin S, Wang Q, Li Y, Wu Q, et al. 2015 Recent development in rare earth doped phosphors for white light emitting diodes *L Rare Earth* **33**: 1-12

[4] Li F, You L, Nie C, Zhang Q, Jin X, Li H, et al. 2017 Quantum dot white light emitting diodes with high scotopic/photopic ratios *Opt Express* **25**: 21901-13

[5] Huang B, Dai Q, Zhuo N, Jiang Q, Shi F, Wang H, et al. 2014 Bicolor Mn-doped $CuInS_2$/ZnS core/shell nanocrystal for white light-emitting diodes with high color rendering index *J Appl Phys* **116**: 094303

[6] Foda M F, Huang L, Shao F, Han H Y 2014 Biocompatible and highly luminescent near-infrared $CuInS_2$/ZnS quantum dots embedded silica beads for cancer cell imaging *ACS Appl. Mater. Interfaces* **6** 2011-7

[7] Jiao S, Shen Q, Mora-Sero I, Wang J, Pan Z, Zhao K, Kuga Y, Zhong X, Bisquert J 2015 Band engineering in core/shell ZnTe/CdSe for photovoltage and efficiency enhancement in exciplex quantum dot sensitized solar cells *ACS nano* **9** 908-15

[8] Li X, Wu Y, Zhang S, Cai B, Gu Y, Song J, Zeng H 2016 $CsPbX_3$ quantum dots for lighting and displays: Room-temperature synthesis, photoluminescence superiorities, underlying origins and white light-emitting diodes *Adv Funct Mater* **26** 2435-45

[9] Zhang Y, Xie C, Su H, Liu J, Pickering S, Wang Y, et al. 2011 Employing heavy metal-free colloidal quantum dots in solution-processed white light-emitting diodes *Nano Lett* **11**:329

[10] Liu W, Zhang Y, Zhao J, Feng Y, Wang D, Zhang T, et al. 2015 Photoluminescence of indium-rich copper indium sulfide quantum dots *J Lumin* **162**: 191-96

[11] Acebron M, Galisteo-Lopez J F, Granados D, Lopez-Ogalla J, Gallego J M, Otero R, Lopez C, Juarez B H 2015 Protective ligand shells for luminescent SiO_2 coated alloyed semiconductor nanocrystals *ACS Appl. Mater. Interfaces* **7** 6935-45

[12] Li F, Guo C, Pan R, Zhu Y, You L, Wang J, et al. 2018 Integration of green $CuInS_2$/ZnS quantum dots for high-efficiency light-emitting diodes and high-responsivity photodetectors *Opt Mater Express* **8**: 314-23

IOP Conf. Series: Materials Science and Engineering **739** (2020) 012027 doi:10.1088/1757-899X/739/1/012027

Steam Integrated Flattening Machine for Bamboo Culms

A H Atienza[1], H Gutlay[1], P V Rodrigo[1], P Tamayo[1]

[1]Mapua Institute of Technology at Laguna, Malayan Colleges Laguna, Cabuyao City, 4025

ahsatienza@mcl.edu.ph

Abstract. Bamboo is a fast growing grass plant abundant in tropical and subtropical countries. Bamboo culms can be used in a variety of applications and thus needs to undergo different process one such process is the flattening process. Traditional flattening process uses the bolos, machetes and hand splitters to make slats of the bamboo. This paper aimed to present a flattening machine with a presser mechanism which utilizes heat and steam to help in the flattening process. The output of the machine was flattened semi-circular bamboo culms from varying radius a bamboo culm with minimal to no cracks. The parameters that were observed were the thickness of the culm, flattening time, heating temperature and length of the culm. From the testing, it showed that flattening time varies directly with the thickness of the bamboo culm.

1. Introduction

Bamboo as a fast growing plant is being used as the main material in handicrafts and in producing export-quality goods in the Philippines. The bamboo culm is the most utilized part of the bamboo in the industry. Flattening process for bamboo culms have been under study for years in order to yield a good quality flattened bamboo culm where waste and cracks are reduced. The objective of the study is to introduce new strategy and way of flattening semi-circular bamboo culms through the design and fabrication steam-integrated flattening machine for semi circular bamboo culms. The machine is focused to address the cracking of bamboo culms when flattened since mostly of the fabricated flattening machines results to cracked flattened bamboo.

Recent design o bamboo culm flattener only flattens the half of a bamboo culm through a series of rollers. The output of the flattener is cracked strip of bamboo [1]. It only made the traditional way of flattening easy. In the flattener designed, it aimed to minimize the cracks produced in flattening the bamboo culm through direct press, gradual press and continuous press [2]. Before it undergoes the process of pressing, bamboo culms are immersed in hot water to soften the bamboo culm. Another flattening machine where it flattens the bamboo culm by passing through a series rollers and heater beneath the rollers [3]. The bamboo culms are softened while it is being rolled flat. Other process requires the modification of the bamboo such as cutting the bamboo into square sticks then adhering them together [4].

2. Materials and methods

The flattening machine was fabricated on the residence of one of the researchers using available materials selected from qualitative material selection [5]. Testing of the machine after the fabrication was done and made some modifications and corrections before conducting an official test for experimentation.

The specie of the bamboo that was used in the testing and experimentation was Buho *(Schizostachyum lumampao),* which was cut 1/3 to ½ of a diameter of the bamboo culm. The length and thickness of the culm was also measured. In order to determine the degree of flatness of the bamboo culm, the method done by [6] was used as seen in figure 1. H is the height of the semi-circular bamboo culm which is equal to the radius of the bamboo culm at the initial condition. At the final condition after load is applied, the height should be equal to the culm wall thickness. Degree of flatness should result to a value of 1 which will indicate the successful flattening of the semi-circular bamboo culm.

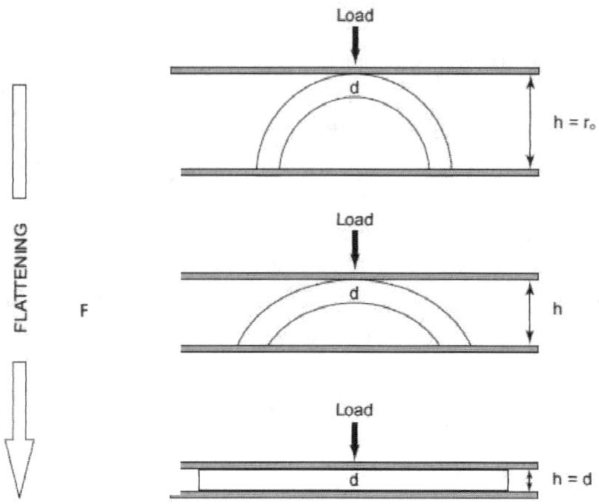

Figure 1. Schematic Representation of determination of the degree of flatness [5].

3. Design

The fabricated flattening machine is shown in figure 2. It consists of three heaters: Submersible heater for steam generation, and two flat plate heaters for bamboo. The presser plates were made of mild steel plate. The presser heater was designed so as the heaters could move accordingly to stay in contact with the surface of the bamboo while the bamboo culms are being flattened, aiming a constant and equally distributed heat on the outer surface of the bamboo. The bamboo table was made of mild steel plate with 9.50mm steam holes. The water tank was made from a 1-mm thick G.I sheet with insulation boards. The whole dimension of the flattening machine was 836.70 mm x 455.70 mm x 1240 mm.

Figure 2. Steam-Integrated Flattening Machine for Bamboo Culms.

4. Experimentation

The experiment was composed of mutiple trials. The lengths and radii of bamboo culm samples were measured. The operating heating temperature of steam used during flattening is at least 110ºC. The

IOP Conf. Series: Materials Science and Engineering **739** (2020) 012027 doi:10.1088/1757-899X/739/1/012027

final height and elapsed time of flattening were recorded. In order to test if the number of steam holes would affect the output of the machine, one section of the bamboo table has some of its holes on the center covered as shown in figure 3 while the other section has all holes opened as seen in figure 4. The holes were covered by a cork made from a piece of wood. A stopwatch was used to determine the time it takes to flatten the culms as well as the specific time where the bamboo culms started to show visible cracks. The test was made to have three trials by varying the number of holes on the center of the bamboo table to be covered. The bamboo culms that were used were ensured to all have the same dimensions. Three arrangements for the test were made all holes on the center were covered, alternating pattern of holes on the center were covered, and no holes were covered.

Figure 3. All Holes Covered on the Center.

Figure 4. Alternating Pattern of Holes Covered on the Center.

For the first trial shown in figure 5, the first bamboo culm started to show visible cracks after 30 minutes have elapsed on its flattening time and had fully cracked 10 minutes after. The bamboo culm was still allowed to be fully flattened. The flattening time for the first trial was 58.38 minutes. For the second trial as seen in figure 6, the second bamboo culm showed visible cracks after 36 minutes, however, the bamboo culm did not fully crack. The cracks were not directly formed on the center of the bamboo culm and were only visible on the inner surface. The flattening time for the second trial was 47.65 minutes. The third trial depicted in figure 7, the third bamboo culm did not show any visible cracks and was fully flattened after 40.26 minutes.

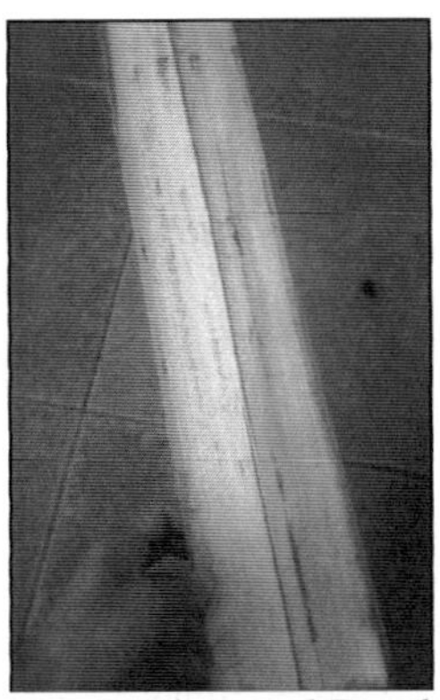

Figure 5. Flattened Bamboo Culm from the First Trial.

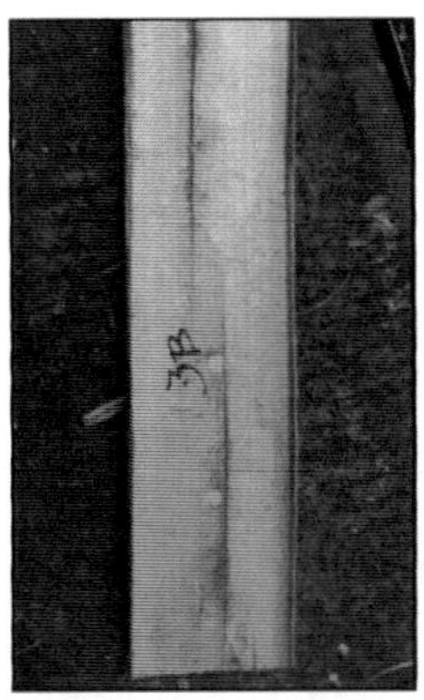

Figure 6. Flattened Bamboo Culm from the Second Trial.

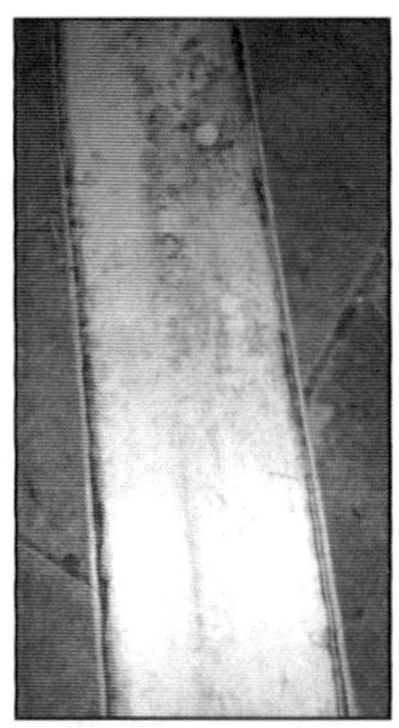

Figure 7. Flattened Bamboo Culm from the Third Trial.

5. Results and Discussion

Thickness affects the flattening of the bamboo culms, there is a direct relation with the thickness and the flattening time, as well as, the temperature needed to flatten the bamboo culms as seen in table 1 and figure 8. The thicker the bamboo, the more time it would require for the bamboo to be heated thoroughly and evenly, thicker bamboos also tend to be harder and stiffer. The length of the said bamboo culms also affected the flattening time and the output of the bamboo culms. As the length increased, there was an increase of time flattening; also, an increase in length entails the presence of nodes and these increased the formation of cracks in the flattened bamboo. This occurrence may be because the nodes are thicker and stiffer than that of the internodes, this creates an inconsistency in the culm. It is observed that a thicker bamboo requires a higher temperature and more time to flatten than a thinner one. The radius may also play into the flattening process as more area is needed to be heated thus increasing the flattening time, cutting the bamboo into smaller sections may be adapted to lessen the time.

Table 1. Testing and Experimentation Results.

Trial	Length (m)	Radius (mm)	Thickness (mm)	Final Height (mm)	Temperature (°C)	Time Elapsed (sec)	Time Elapsed (mins)
1	0.20	41.5	6.00	6.50	120	450	7.5
2	0.20	41.5	6.50	6.50	110	450	7.5
3	0.230	41.5	6.50	6.50	115	504	8.4
4	0.443	41.5	6.50	23.0	115	738	12.3
5	0.157	41.5	6.50	6.50	115	432	7.2
6	0.254	41.5	6.50	6.50	132.25	682	11.38
7	0.25.4	41.5	6.50	6.50	132.25	862	14.38
8	0.453	42.0	9.00	10.0	130	2280	38
9	0.452	50.0	10.5	10.5	134.5	2406	40.1
10	0.381	50.0	11.0	11.0	137.75	2413	40.23
11	0.254	50.0	11.0	12.0	137.75	2625	43.75
AVE	0.2983	43.9	7.80	9.60	125.41	1258.36	20.97

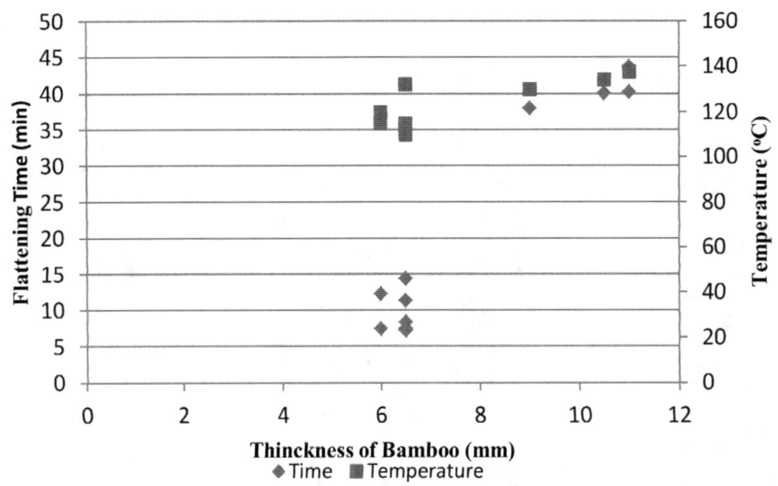

Figure 8. Relation of Bamboo Thickness to Flattening Time and Temperature.

This test proves that the number and location of steam holes designed for the bamboo table were the optimum. Based from the gathered results from table 2, the number of steam holes opened would affect the time it would take for the bamboo culms to be flattened. Moreover, based from the output of each trial, reducing the number of opened steam holes would also reduce the quality of the flattened bamboo.

Table 2. Effect of Number of Opened Steam Holes Results.

Trial	Length (m)	Radius (mm)	Thickness (mm)	Temperature (°C)	Time Elapsed (sec)	Time Elapsed (mins)
1	0.381	50	11.0	125	3480	58.38
2	0.381	50	11.0	120	2859	47.65
3	0.381	50	11.0	120	2415.6	40.26

6. Conclusion

Flattening time and temperature varies directly with the thickness of the bamboo culm as shown in the data. The designed number of steam holes is optimal to produce minimal to no cracks of flattened bamboo culms. Reducing the number of steam holes opened would increase the time to flatten a bamboo culm and also reduce the output quality. The fabricated flattening machine successfully flattened bamboo culms generating an efficiency of 95%. During the conduct of the experiment for the bamboo flattening, the researchers suggested the use of fresher bamboo culms with its nodes removed. Also, cutting the bamboo culms using a fine-toothed saw in order to prevent splints is recommended. In order to further boost the productivity of the flattening machine, it is more convenient to add automation system.

Acknowledgement

The authors would like to thank Malayan Colleges Laguna, A Mapua School for funding the presentation and publication of this research. Likewise, to the Mapua Institute of Technology at Laguna for all the support. Also their family and friends on unparalleled support and motivation.

References

[1] Bamboo Veneer Lathe and Bamboo Flattening Machine. (2018, August 3). DOST-FPRDI. Retrieved from http://fprdi.dost.gov.ph/index.php/technologies/bamboo-veneer-lathe-and-bamboo-flattening-machine

[2] Zhang Q 1988 Studies of bamboo Plywood-I. the softening and flattening of bamboo, *J. Nanjing Fores. Univ.* **4** 13-20.

[3] Park, C N 2015 Flattening of Bamboo by Thermo-mechanical Treatment. 10th World Bamboo Congress, Korea. Retrieved from https://worldbamboo.net/wbcx/Sessions/Theme%20Product%20Design%20Technology/Park,%20Choong%20Nyeon.pdf

[4] Madhavi M, Clouston P L, Arwade, S R 2011 Development of Laminated Bamboo Lumber: Review of Processing, Performance, and Economical Considerations, *Journal of Materials in Civil Engineering,* **23(7)**, 1036–1042.

[5] Farag M M 2015 Quantitative Methods of Material Selection, *Mechanical Engineers's Handbook,* 1-22.

[6] Parkkeeree T, Matan N, Kyokong B 2014 Flattening of half tubular bamboo culms and fixation of bamboo boards, *Journal of Tropical Forest Science* **26(1)**: 101–114.

IOP Conf. Series: Materials Science and Engineering **739** (2020) 012028 doi:10.1088/1757-899X/739/1/012028

Multi-Directional Forced Convection Kiln Oven Drying System for Bamboo Culms

A H Atienza[1], R Dela Peña[1], R Eroles[1], E Fandiñola[1]

[1]Mapua Institute of Technology at Laguna, Malayan Colleges Laguna, Cabuyao City, 4025

ahsatienza@mcl.edu.ph

Abstract. The bamboo industry in the Philippines is at the commencement of its development in terms of co-existing with the current technology. The study aims to develop and improve the processing of bamboo, specifically its drying process. The traditional drying process in the Philippines is sun drying which is a primitive way of drying the bamboo. The drying process presented in this study is the forced convection kiln oven drying. It offers fast drying time and allows the bamboo culm to be enclosed on a system to prevent dirt or other contaminants to accumulate on the culm. The prototype is composed of a blower that would exhibit the forced convection; a base considered to be the system for the bamboo to dry and a heating element that would induce the heat necessary for the drying. The heating source is an electrical component that requires electricity from an AC power source. In order to test the prototype, a fresh bamboo is to be dried inside the system and its moisture content would be monitored with respect to time. The result of the test would provide for the conclusion of the efficiency of the dryer to reduce the moisture content of the bamboo.

1. Introduction

Bamboo is a plant which can be utilized for many applications. Its shoots can be eaten while its culms can be used in the field of construction and furniture crafting. Bamboo culms, the stem part of the plant, are strong and durable which makes it ideal to be used as a construction material. Bamboo, being a grass, grows much faster than trees. Bamboo culms can be harvested within 3-5 years while an oak tree, also a common source for construction materials, can be harvested within 10-12 years. However, much like wood, bamboo is hygroscopic meaning it has the natural tendency to absorb moisture from the surrounding air. Bamboo culms can have a moisture content of more than 150% [1]. The presence of water in the bamboo promotes microbial activity. This microbial activity causes the decomposition of the fibers in the bamboo which effectively reduces its strength as a material. As a result, drying of bamboo culms becomes essential for the preserving and achieving the appropriate structural properties for its further utilization [2].

In the Philippines, the common drying method is open-sun drying which involves stacking the bamboo culms and exposing them to direct sunlight. However, this method is highly dependent on the climactic conditions. The Philippines, being a tropical country and an archipelago, has a naturally humid environment. The drying time decreases due to the fact the bamboo absorbs the moisture from the surrounding air. Other drying methods include forced convection drying, microwave drying, kiln drying, and vacuum-freeze drying. The best aspects of these drying methods have been incorporated into this project in order to determine the most efficient way to dry bamboo. Utilizing a kiln, which is a controlled environment for the bamboo to dry inside, is best suited [3]. A reliable and controllable heat source is needed. For this reason, an electric heater was used. Forced convection was utilized to decrease drying time [4]. Multi-directional air flow was used in order to decrease drying defects [2].

In addition to, moisture content was correlated bamboo color. Traditionally, bamboo dryness is determined from the weight of the culms. However, not everyone has access to weighing devices. For this reason, a visual method of determining moisture content was created. This is to allow small bamboo manufacturers, specifically those without access to weighing devices, to accurately determine the dryness of their bamboo.

2. Materials and Methods

The drying system used in this study is seen in figure 1 while the drying system process flow is shown in figure 2. The system takes in atmospheric air through the use of a fan. The fan then directs the air towards the heater and then towards the drying chamber. The drying chamber has a capacity of six 2-feet long bamboo culms. Two trays were used with three culms placed on each tray. The air flows through holes in the wall between the heater and drying chamber. The holes scatter the air flow and distribute the heat within the drying chamber. The door of the oven has a window for viewing the interior and bamboo color. Table 1 shows the materials used in the construction of the oven and were determined using the quantitative material selection [5].

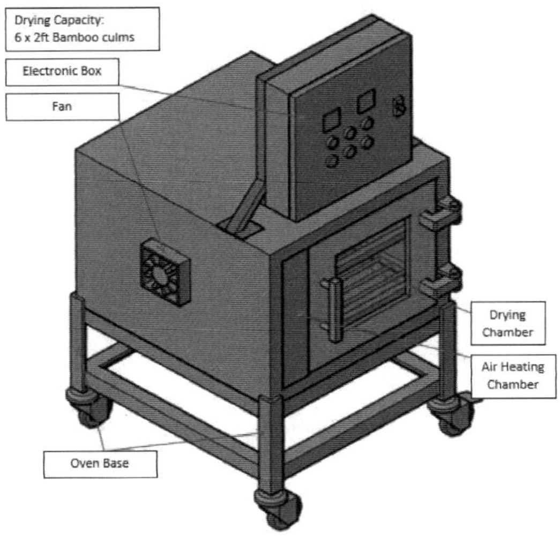

Figure 1. Drying System.

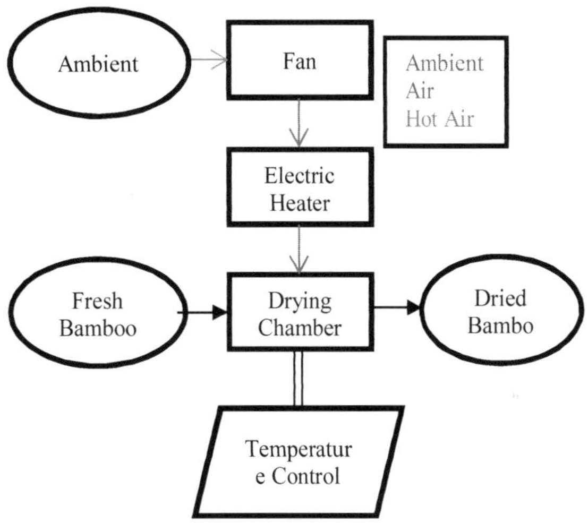

Figure 2. Drying System Process Flow.

Table 1. Oven Material Selection.

Oven Body	Mild Steel Plate
Outer Cover	B.I. Sheet
Drying Tray	Stainless Steel
Glass	Tempered Glass
Insulation	Rockwool
Oven Base	Angle Bar
Wheels	Swivel Caster Wheel

The drying temperature is normally set to an ideal value of 70°C [2]. The drying time is normally set to 2.5 hours. Both drying temperature and drying time can be set to other values. An interface displays the drying temperature and drying time of the oven. Temperature is regulated and controlled by an electronic system. The electronic system controls the electric heater. A thermocouple is used as the temperature sensor. Heater capacity was calculated from the moisture content of bamboo and specific heat capacity of bamboo [6]. Fan speed was calculated using heat transfer equations obtained from [7]. Theoretical weight loss was calculated using wood moisture calculations by [8] and psychrometric equations. The oven specifications were shown on table 2.

Table 2. Oven Specifications.

Heating Element	Electric Heater
Heater Capacity	1000 W
Fan Speed	1400 rpm; 1.34 m/s
Fan Diameter	115 mm
Temperature Sensor	Thermocouple
Drying Temperature	70°C (adjustable)
Drying Time	2.5 hours (adjustable)
Theoretical Moisture Removed	361

3. Results and Discussion

Upon completion of the fabrication, the prototype was tested. Sample weights were measured using an oven-dry weight basis. This involves measuring the initial weight of the culms and the weight at certain time intervals. Weights of the samples were measured every 30 minutes. This was done by removing them from the oven, weighing them and then putting them back into the oven. A weight sensor was not placed within the oven because of the limitations in the circuitry due to the high temperature. Variation of oven temperature with respect to drying time was also observed. This was done to observe the effect of opening the oven to measure the samples. The set drying temperature of 70°C was reached within 5 minutes. An average weight loss of 5.80 grams per hour was observed and shown in figure 3 and table 3.

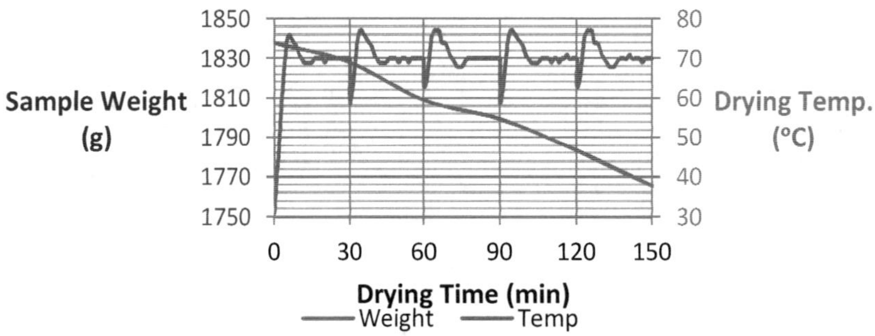

Figure 3. Variation of Sample Weight and Drying Temperature.

Table 3. Weight Loss in Bamboo Samples after 2.5 hours.

Sample No.	Initial Weight (g)	Final Weight (g)	Weight Loss (g)
1	717	638	79
2	606	577	29
3	1183	1149	34
4	853	819	34
5	1168.9	1129	39.9
6	648	620	28
			Average = 40.65
			Total = 243.9

Another test was conducted to compare the drying time of the oven to drying time of traditional open-sun drying. Bamboo samples were left to dry undisturbed within the oven which means the oven was not opened to measure sample weights. Only initial and final weights within the 2.5-hour drying period were measured. The average weight loss was 40.65 grams and the total weight loss was 243.9 grams. Bamboo samples of the same amount were laid out on the ground and exposed to sunlight.

Initial weight was measured. The culms were left to sun-dry for 24 hours. Final weight was measured. On average, the sun-dried culms lost 100 grams seen in table 4.

Table 4. Comparison of Sun-drying to Oven-drying.

Drying Method	Average Weight Loss (g)	Drying Time (hrs)	Drying Speed (g/hr)
Sun-drying	100	24	4.17
Oven-drying	40.65	2.5	16.26

It was found that color values, R-G-B, were somewhat linearly related to moisture ratio [9] referred to figure 4.

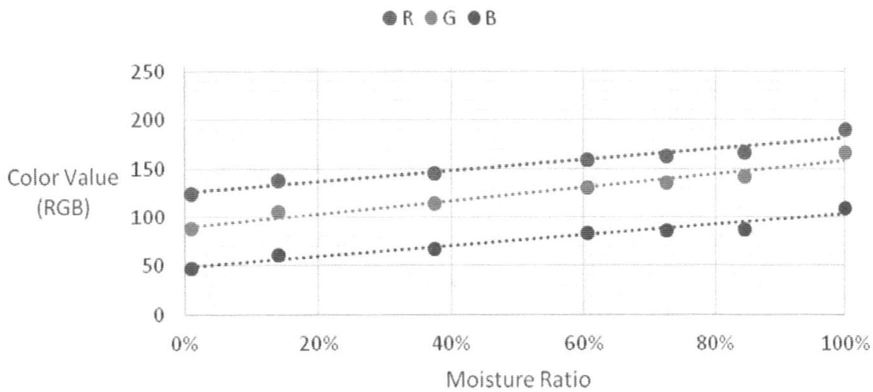

Figure 4. Changes in RGB Values as a Function of Moisture Ratio.

RGB values of various points in the culm were determined using an online image to RGB converter. RGB values were determined for different sample weights. It was found that bamboo changes color from green (fresh) to yellow brown (dried) described in table 5.

Table 5. RGB Values at Various Sample Weights.

Sample Weight (kg)	R	G	B	Color
1.849	43	80	39	
0.853	139	149	73	
0.606	153	158	95	
0.577	190	183	166	

By using a linear relationship between sample weight and color value, equations were created for the color values as a function of sample weight. By plotting straight lines in MATLAB using the established equations as the line color, a color scale was created in figure 5.

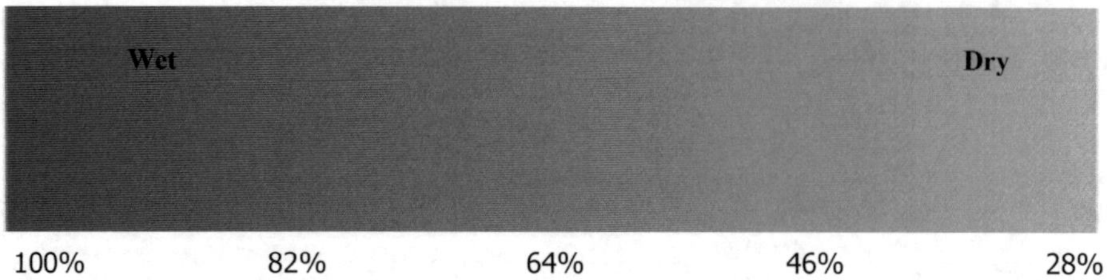

Figure 5. Bamboo Color Scale.

4. Conclusion

As shown in the results of the tests, there is a reduction in the weight of the bamboo signifying a reduction in the moisture content of the bamboo hence becoming dried. The drying speed of the bamboo culms was dependent on their position such that the culms on the top tray dried much faster than the culms at the bottom tray. The bamboo culms closer to the heating chamber also exhibited a faster drying time compared to the other culms. The weight of the culms on average was reduced by 5 percent. This corresponds to a 12 percent reduction in moisture within 2.50 hours. Using the forced convection method in the transferring the heat necessary for the drying process, no visible drying defects were observed on the bamboo culms tested. The forced convection was applied by the installation of a fan that transports air from the surrounding to the oven and by making use of holes that disperses the air flow in multiple directions, ensuring that the heat is spread evenly inside the drying chamber. The results of the tests show that the machine was able to dry the bamboo significantly faster than the sun-drying. The machine having an average rate of 40.65 grams per hour as weight reduction is 4 times faster than the traditional sun drying. The efficiency of the oven was also determined to be 67.56 percent, with an expected total moisture loss of 361 grams and an actual total moisture loss of 243.9 grams. In terms of energy consumption, the oven consumes 3 kWh for every 2.50 hours of drying.

Acknowledgement

The authors would like to thank Malayan Colleges Laguna, A Mapua School for funding the presentation and publication of this research. Likewise, to the Mapua Institute of Technology at Laguna for all the support. Also their family and friends on unparalleled support and motivation.

References

[1] Villanueva M M, Dolom P, Razal, R, Camacho S, Palacpac A 2010 State-of-the Art: Processing of Engineered-Bamboo Products in the Philippines. Forestry Development Center, College of Forestry and Natural Resources, University of the Philippines Los Baños, Laguna, Philippines

[2] Tang T, Welling J, Liese W 2013 Kiln drying for bamboo culm parts of the species Bambusa stenostachya, Dendrocalamus asper and Thyrostachys siamensis, *Journal of the Indian Academy of Wood Science,* **10(1)** 26-31. doi:10.1007/s13196-013-0089-4

[3] Cuaresma C (2018, January 25) Drying Technologies Innovation. DOST-FPRDI. Retrieved from http://www.fprdi.dost.gov.ph/index.php/priority-programs-projects/r-d/drying-technologies-innovation

[4] Ehl P, Banout J 2010 Using a double-pass solar drier for drying of bamboo shoots, *Journal of Agriculture and Rural Development in the Tropics and Subtropics* **111(2):**119-127

[5] Farag M M 2015 Quantitative Methods of Material Selection, *Mechanical Engineers's Handbook,* 1-22.

[6] Roxas C, Tangan F, Virtucio F 2000 *Ornamental Bamboos for Urban Parks* Ecosystems Research and Development Bureau, Department of Environment and Natural Resources.Retrieved from http://erdb.denr.gov.ph/wp-content/uploads/2015/06/denr_v9.pdf

[7] Holman J P 2010 *Heat Transfer (10th ed)* McGraw-Hill.

[8] Eckleman C A (1997, February 25) Wood Moisture Calculations. Forestry & Natural Resources. Retrieved from https://www.extension.purdue.edu/extmedia/FNR/FNR-156.pdf

[9] Bal L, Kar A, Satya S, Naik S 2011 Kinetic of colour change of bamboo shoots slices during

microwave drying, *Int. J. Food Sci. Technol.* **46(4)** 827-33.

IOP Conf. Series: Materials Science and Engineering **739** (2020) 012029 doi:10.1088/1757-899X/739/1/012029

Inorganic InP/ZnS Quantum Dots: Optical Properties and Fabrication in Display Application

Chaikit Chumchujan[1], Shu-Ru Chung[1]

[1]Department of Materials Science and Engineering, National Formosa University, Yunlin 63201, Taiwan

srchung@nfu.edu.tw

Abstract. Inorganic InP/ZnS quantum dots (QDs) is one kind of optoelectronic materials owing to the high quantum yield (QY) and non-toxicity. In this study, green light InP/ZnS QDs with high QY were synthesized by one-pot injection method and mixed with red light $CsPbBrI_2$ QDs as converting material. Those converting material blend with epoxy glue, and pumping by InGaN blue chip to form white light-emitting diode (WLED). The results show that the CIE of WLED provides at (0.31, 0.33) with NTSC of 90 %.

1. Introduction

In the past, fluorescent backlight is used and play an important role in display application. However, there are huge drawbacks [1, 2]. Firstly, it is high energy consumption due to high turn on voltage. Secondary, the color gamut of fluorescent lamp backlight is low, as 70 % of NTSC [3, 4]. Thus, it is replaced with WLED-based backlight. WLED-based backlight in displays are normally combined with blue LED and phosphors [5]. Choosing Cd-based quantum dots (QDs) as the phosphors due to QDs have a narrow full width at half maximum (FWHM), resulting in increasing the color gamut [6]. However, Cd-based has some disadvantages. It is high toxicity. Moreover, it has low PL quantum yield (QY) because of self-quenching [7]. Thus, InP nanoparticles are good choices because of they are non-toxicity, while the luminescence efficiency in near IR spectral range with very low QY due to easy to oxide [8]. Therefore, coating with a high band gap material such as ZnS and ZnSe on the InP surface to form InP/ZnS core/shell structure is necessary. It leads to increasing the QY and energy band gap of QDs. From the literature review, it provides the result that the QY of InP/ZnS QDs is between 30-60 % [9]. It depends on the molar ratios between myristic acid and indium, reaction temperature, as well as reaction time. The best condition in InP/ZnS synthesis is that the molar ratio of myristic acid: indium is 3.5:1 and the reaction time is 3-5 hours. Furthermore, increasing of reaction temperature also make increase QY [10].

2. Experimental methods

2.1. Chemicals

Zinc (II) iodide (ZnI_2, > 98 %) and indium (III) chloride ($InCl_3$, 99.995 %) were purchased from Acros Organics. Zinc (II) bromide ($ZnBr_2$, 98 %), Myristic acid (MA, 98 %) and Hexamethylphosphorous triamide (HMPT, 97 %) were purchased from Alfa Aesar. 1-dodecanethiol (DDT, > 98 %) 1-octadecene (ODE, 90 %) and oleylamine (OLA, 70 %) were provided by Aldrich.

2.2. Synthesis of green InP/ZnS QDs

In synthesis green InP/ZnS QDs, 0.72 mmol of $InCl_3$, 0.72 mmol of MA, and 2.16 mmol of Zn precursors (ZnI_2: $ZnBr_2$ = 1:1) were purged by Ar gas for 10 minutes, and then they were mixed with 10 mL OLA and ODE (the ratio of OLA: ODE was 6:4) in three-necked round-bottom flask. The mixture was stirred and degassed at 110 °C for 45 minutes followed by rapidly heated to 220 °C under Ar atmosphere. Next, 0.2 mL of HMPT was quickly injected above mixture and controlled temperature at 220 °C for 15 minutes, after that quickly quenched to 200 °C, then 3 mL of DDT was slowly injected and maintained the temperature at 200 °C for 6 hours. InP/ZnS QDs were precipitated with an excess of ethanol and purified with ethanol by centrifuge. Finally, purified InP/ZnS QDs were re-dispersed in hexane for characterization.

2.3. Fabrication of White Light-Emitting Devices

InP/ZnS QDs were blended with $CsPbBrI_2$ QDs and re-dispersed with hexane, then hexane was repelled in vacuum desiccator. The mixture of QDs were mixed in epoxy-based AB glue. The ratio of A: B glue was 1:2. Next, the mixture was stirred to be homogenous. Finally, it was put on 3020 InGaN blue chip and the device performance was measured by integrated sphere (Isuzu Optical, ISM-360) at a driving current of 20 mA.

2.4. Materials characterization

Purified InP/ZnS QDs were dispersed in hexane for optical properties. They were measured by fluorescence spectrophotometer (FL, Hitachi F-7000) and ultraviolet-visible spectrometer (UV-Vis, Hitachi UH5300). PLQY of the QDs was calculated by comparing the FL emission area of QDs, which were dispersed in hexane, with Rhodamine 6G in methanol as a reference sample at wavelength 450 nm. The morphology and size distribution of QDs were observed by transmission electron microscopy (TEM).

3. Results and discussion

3.1. Optical properties of InP/ZnS QDs

Figure. 1 shows the absorption, excitation and emission spectra of the synthesized InP/ZnS QDs. It displays that the emission wavelength of sample is 500 nm with nonobvious absorption edge. The PLQY of green QDs were measured to be 113 % with FWHM of 70 nm.

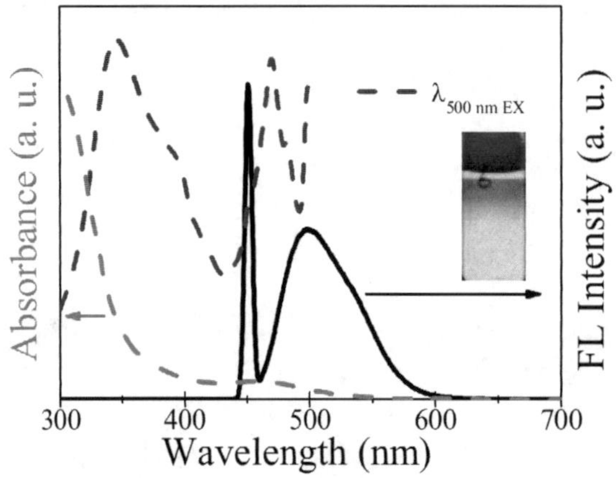

Figure 1. The fluorescence and absorption spectra of InP/ZnS under 450 nm. insert photograph shows the InP/ZnS QD under UV light.

3.2. Morphologies of InP/ZnS QDs

The TEM images of sample are presented in figure 2. It can be found that it has good dispersion and the morphologies are spherical shape. The particle size ranges from 2.6 to 3.0 nm.

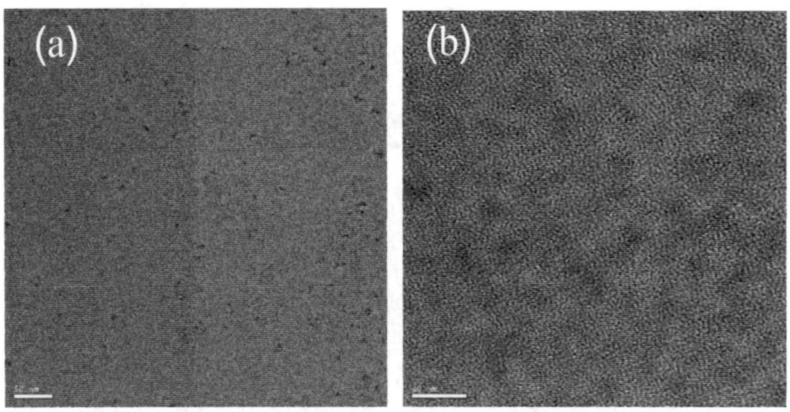

Figure 2. TEM images of InP/ZnS QDs in different magnifications.

3.3. InP/ZnS QDs-based WLED

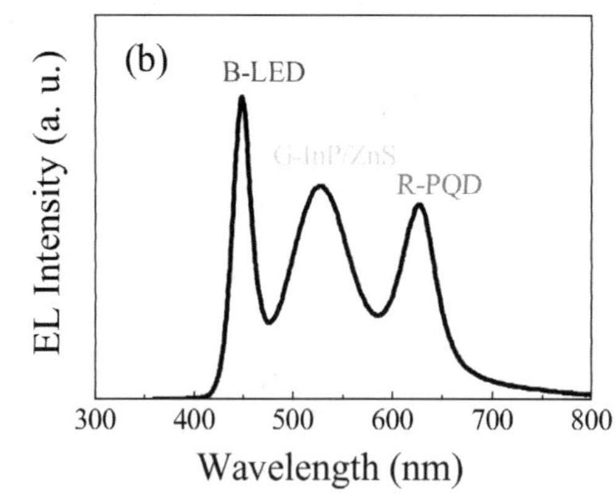

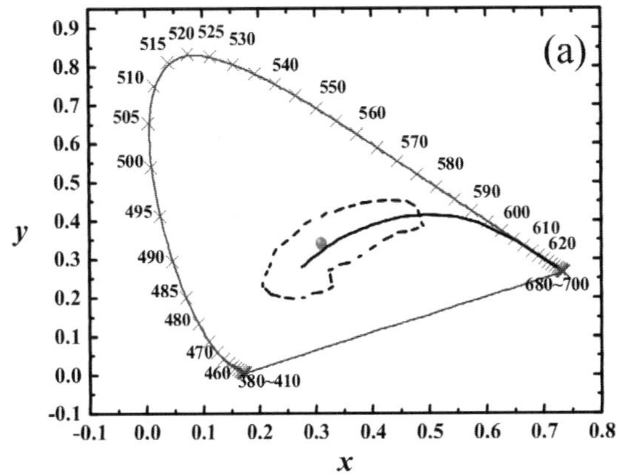

Figure 3. (a) CIE color coordinates and (b) EL spectra of WLED.

InP/ZnS QDs were mixed with $CsPbBrI_2$ and then this mixture was mixed with AB glue. The ratio of green light InP/ZnS QDs: red light $CsPbBrI_2$: AB glue is 0.503 g:0.006 g:1.000 g. Above mixture is putted on the blue LED and can produce white light under 20 mA current. The Commission International d'Eclairage (CIE) chromaticity coordinates shows in figure 3 (a) and EL spectra of device shows in figure 3 (b). It can be seen that the CIE chromaticity coordinates locates at (0.31, 0.34). Moreover, the luminous efficacy, CCT, and CRI of WLED are 11 lm/W, 6523 K, and 88, respectively. After filters the white light by color filter and calculate, the color gamut of WLED is 90 % of NTSC, as shown in figure 4.

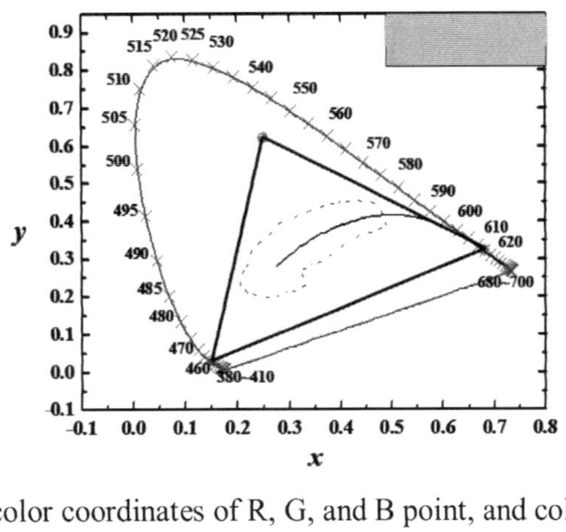

Figure 4. The CIE color coordinates of R, G, and B point, and color gamut of WLED.

4. Conclusion

In this paper, the green InP/ZnS QDs were successfully prepared with one pot injection method. The QY and particle size of InP/ZnS QD are 113 % and 2.6-3.0 nm. For the WLED performance, we can get the CIE color coordinates (0.31, 0.33) and 90 % of NTSC after blending with red light $CsPbBrI_2$ QD.

Acknowledgment
This work was supported by the Ministry of Science and Technology of Taiwan under Contract nos. 107-2221-E-150-034 and 108-2221-E-150-016.

References

[1] Yin L., Bai Y., Zhou J., Cao J., Sun X. W., Zhang J. H. 2015 The thermal stability performances of the color rendering index of white light emitting diodes with the red quantum dots encapsulation *Opt Mater* **42** 187-92.

[2] Yoshinori S., Kensho S., Yasunobu N. 1999 Light emitting device and display *US Patent* **5** 998, 925.

[3] Yang H. Lee D. K., Kim Y. S., 2009 Spectral vaiations of nano-sized $Y_3Al_5O_{12}$:Ce phosphors via codoping/substitution and their white LEDs characteristics *Mater. Chem. Phys.* **114** 665-9.

[4] Ito Y., Hori T., Kusunoki T., Nomura H., Kondo H. 2014 A phosphor sheet and backlight system providing wider color gamut for LEDs *J. Soc. Inf. Disp.* **22** 419-28.

[5] Wu Y., Wanlu Z., Guilin Z., Jiatao Z., Guoxing H., Ruiqian G. 2018 Super-high color temperature tunable white LEDs based on high quality InP/ZnS quantum dots via myristic acid passivation and Ag doping *Opt Com* **418** 46-50.

[6] Wu Y., Guoxing H., Shiliang M., Jiatao Z., Wanlu Z., Qiuhang C., Guilin Z., Ruiqian G. 2017 Controllable synthesis of dual emission Ag:InP/ZnS quantum dots with high fluorescence quantum yield *App Sur Sci* **423** 686-94.

[7] Han Z., Ning H., Zaiping Z., Qinli L., Fengjuan Z., Aiwei T., Yu J., Linsong L., Huaibin S., Feng T., Zuiliang D.. 2019 High-Efficiency Green InP Quantum Dots-Based Electroluminescent Device Comprising Thick-Shell Quantum Dots *Adv. Optical Mater.* **7** 1801602.

[8] F. Angel-Huerta, M.P. Gonzalez-Araoz, J.F. Sanchez-Ramirez, J. Diaz-Reyes, J.L. Herrera-Perez, J.S. Arias-Ceron. J.G. Mendoza-ALvarez 2018 Synthesis temperature-dependent optical properties of ZnS-shell formation on InP nanopaticles *Jou of Lum* **197** 277-84.

[9] Feng Z., Jicun R. 2012 Gas-liquid phase synthesis of highly luminescent InP/ZnS core/shell quantum dots using zinc phosphide as a new phosphorus source *J. Mater Chem.* **22** 1794-9.

[10] Liang L., Peter R. 2008 One-pot synthesis of Highly Luminescent InP/ZnS Nanocrystals without Precursor Injection *JACS Com.* **130** 11588-9.

IOP Conf. Series: Materials Science and Engineering **739** (2020) 012030 doi:10.1088/1757-899X/739/1/012030

Performance Seismic Design of the Retrofit of a University Library using Non-Conventional Methods

B Chavez[1], C Matias[1], G Huaco[2]

[1]Bachelor, Peruvian University of Applied Science UPC
[2]Professor, Peruvian University of Applied Science UPC

pccighua@upc.edu.pe

Abstract. This research proposes a structural retrofit of a reinforced concrete building older than 50 years. The structural system is dual using frames and walls, having an area of 1980 m^2 and 4 levels with a total height of 15.50 m. There are three unconventional methods to retrofit this infrastructure. The first is to install steel jackets with bolt anchors. The second methodology is through the use of jackets and anchors of Carbon Fiber Reinforced Polymers (CFRP). The third alternative is through reinforcement of walls with 2 diagonal struts made with CFRP sheets as struts, also installing CFRP anchors at each end of both struts to ensure that these sheets work up to their high levels of tension. These anchors together with the diagonal plates contribute to give ductility to the wall and in turn resist the effects of sliding at the base, a fault that is very common in rigid structures such as walls. Each unconventional reinforcement methodology used in this research is validated using laboratory tests results of reinforced columns and walls retrofitted with every of the three innovative methods. Nonlinear static analysis - Pushover of each alternative was performed and compared with the demand for the 1974 earthquake in Lima, scaling 3 seismic records to a PGA of 0.45g, which is the acceleration of design in Lima. It is shown that these proposals are effective in providing levels of shear base and displacement capacity in the inelastic design. For the 0.40mx0.80m columns, the results show that both reinforcements increase ductility by more than 10%; on the other hand, the proposal of reinforcement in the walls of 40cm thick, produced even better results increasing ductility by 100% and shear base by 100%.

1. Introduction

In structural engineering, the limits of inelastic deformation of individual members allow the structures to withstand a strong movement of the ground while these dissipate significant levels of seismic energy [1]. The formation of plastic hinges in columns can occur at the base of the frame where the damage incurred acts to dampen seismic loads. The ductile behavior of these is essential to avoid the complete collapse of the structure.

The collapse of structures in the face of earthquakes in recent years has highlighted the vulnerability of buildings to a seismic movement, these structures come to behave in their inelastic state. The structures of greatest vulnerability are presented in buildings built before 1970, which have design criteria below current earthquake-resistance design standards [2], that is, they do not have the ductility necessary to dissipate seismic energy during an earthquake of strong magnitude. The lack of rigidity and ductility in the structure causes the collapse of the building. Reinforcement research studies for buildings that are more than 50 years old have shown that the use of CFRP and Steel jackets improves seismic performance against an earthquake [3]. In addition, there are few studies where the main reinforcement is the use of CFRP anchors [4] used in CFRP jackets, a better response is obtained than reinforcement only using CFRP sheets.

The results of the analysis in columns and walls, and the anchoring system are also discussed to make recommendations for the use and improvements of this system in the future. it presents the case of structure analysis without beams because it is more conservative and in turn due to the limited information of the internal reinforcement of the columns and beams.

2. Background: infrastructure to reinforce

The building is a University Library in the city of Lima Peru built in 1969 and supported by soft ground. This is the National Agricultural Library (BAN). This facility is considered essential since it is required to remain operational after an important seismic event. Its structural system for lateral forces is made of porches and reinforced concrete walls in both orthogonal directions. The plan views are shown in figure 1a and elevation in figure 2.

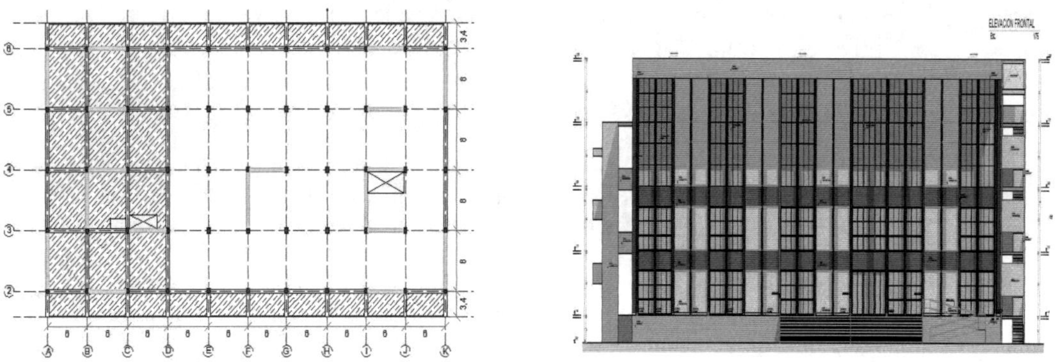

(a) (b)

Figure 1. (a) Detail in Plan of the National Agricultural Library (BAN).
(b) Elevation detail of the National Agricultural Library (BAN).

Data collection was obtained, obtaining essential information such as: material, structural system, height and year of construction. Likewise, the studies of representative trials such as diamantine extraction and sclerometer were performed at key points of the structure of the National Agricultural Library (BAN). These data show us the effort to compress the concrete "f'c" of the following structural elements; columns, beams and walls, finding fc values of less than 180kg/cm2 for columns, and less than 170kg/cm2 for beams. The walls have an average of 175kg/cm2 of concrete capacity. Limited information on the existing reinforcement was obtained, so for this investigation the capacity of each structural element was calculated considering a minimum number of rods that are similar to several of the construction practices of the 60s, that is, a smaller amount than that provided in Current design codes in Peru. This study began with the need to reinforce existing structures, with age over 50 years, which were built under design criteria that do not meet the technical design specifications of current regulations.

3. Structure Capacity vs. Seismic Demand

3.1. Estimation of Plastic Hinges for the retrofit methods

The capacity of the structure in its longitudinal direction was calculated for 4 cases: AS-BUILT, reinforcing only columns with steel jackets and metal bolts, reinforced only columns with jacket and CFRP anchors and finally reinforcing walls with diagonal struts in two directions using CFRP sheets and anchors. Pushover modal was performed specifying the modes attributed to the longitudinal direction. The software used was SAP2000 v.20.

3.1.1. As-Built. In order to determine the Ball joints for As-Built structures, the ASCE - 41[5] Standard was followed, in which the labels will be determined according to the parameters and conditions of the

National Agricultural Library (BAN). Table 1 presents a fragment of table 10-13 of ASCE Standard 41 [5], which according to the case of the evaluation of the structure, in our case will be condition ii; which refers to structures with short overlaps and a minimum amount.

Table 1. Modeling parameters a numerical acceptance criterion for nonlinear procedures – reinforced concrete columns [5].

Condition ii [a]										
$\frac{P}{A_g f'_c}$ [a]	$\rho = \frac{A_v}{b_w s}$	$\frac{V}{b_w d \sqrt{f'_c}}$ [e]								
≤ 0.1	≥ 0.006	≤ 3 (0.25)	0.032	0.060	0.2	0.005			0.045	0.060
≤ 0.1	≥ 0.006	≥ 6 (0.5)	0.025	0.060	0.2	0.005			0.045	0.060
≥ 0.6	≥ 0.006	≤ 3 (0.25)	0.010	0.010	0.0	0.003			0.009	0.010
≥ 0.6	≥ 0.006	≥ 6 (0.5)	0.008	0.008	0.0	0.003			0.007	0.008
≤ 0.1	≤ 0.0005	≤ 3 (0.25)	0.012	0.012	0.2	0.005			0.010	0.012
≤ 0.1	≤ 0.0005	≥ 6 (0.5)	0.006	0.006	0.2	0.004			0.005	0.006
≥ 0.6	≤ 0.0005	≤ 3 (0.25)	0.004	0.004	0.0	0.002			0.003	0.004
≥ 0.6	≤ 0.0005	≥ 6 (0.5)	0.0	0.0	0.0	0.0			0.0	0.0

3.1.2. Steel Jacket with Anchors – Columns. According to Aboutaha [6] he proposes a reinforcement with steel jackets and anchors (bolts) in columns with inadequate splicing overlap as shown in figure 2a, including lateral load versus drift ratio response, this is defined as tip displacement divided by column height. Figure 2b shows the profile of the column to be reinforced and the section of the column and the reinforcement with the Steel Jacket and Anchors.

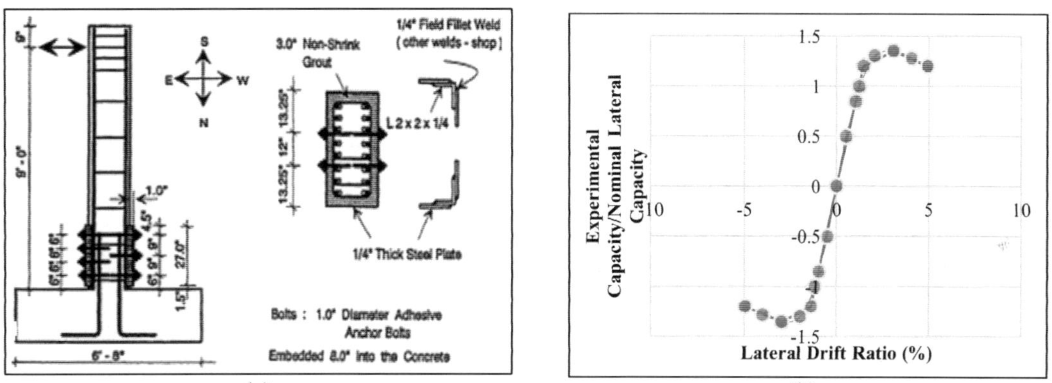

(a) (b)

Figure 2. (a) Details of the Column and section with Steel Jackets and Anchors (Bolts [6] (b) Standardized force-strain curves for columns reinforced with Steel Jacket with Anchors [6].

3.1.3. CFRP with Anchors – Columns. Huaco [7, 8] mentions in his research that he follows the guidelines available ASCE 41-13. It is presented in figure 3a the pattern of installation of the CFRP Jackets and CFRP Anchors for severe damaged columns, and figure 3b shows the bi-linear and tri-linear curve for the design of the capacity curve of the structure with reinforcement with carbon fiber jackets and anchors in columns.

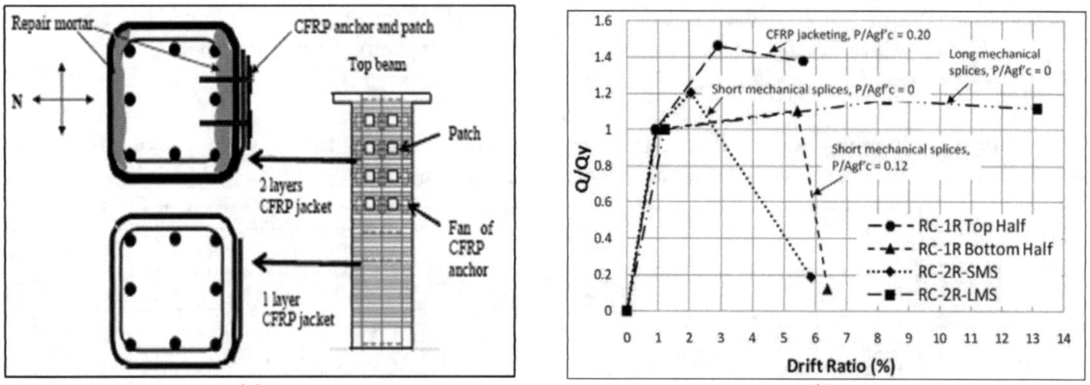

(a) (b)

Figure 3. (a) Section of the Column Reinforced with CFRP Jackets with Anchors, units in inches [7] (b) Standardized force-strain curves for columns with CFRP reinforcement with Anchors [8].

3.1.4. CFRP with Anchors – Walls. According to Huaco [8, 9] in his research, two types of behavior models were proposed for the reconstructed masonry walls: The Elastic - Plastic and Strength - Displacement models. The first idealized curves were developed from the main curves, and then normalized as ASCE 41 [5] provisions indicate. Figure 4a shows the details of the reinforcement in walls with 2 diagonal struts made of CFRP sheets, also installing CFRP anchors at each end of both struts to ensure that these sheets work to their high levels of tension. Figure 4b shows the bi-linear and tri-linear curve for the design of the capacity curve of the structure with reinforcement with struts of sheets and carbon fiber anchors in walls.

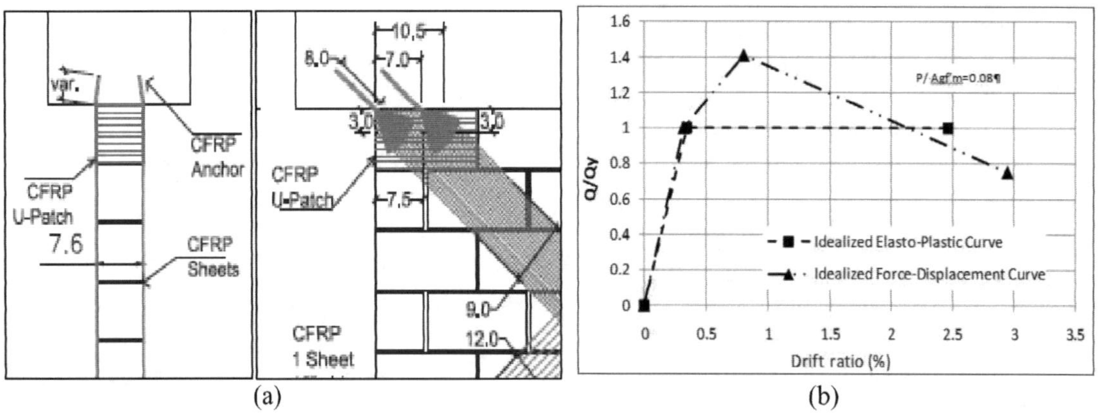

(a) (b)

Figure 4. (a) Detail of the sheet and anchors of CFRP in walls, units in inches [9] (b) Standardized force-strain curves for walls with CFRP reinforcement with Anchors [8].

3.2. Seismic Demand Curve Estimation

3.2.1. Seismic Catalog of Peru. It was used ground motions records from station near the location of the library, as well from earthquakes with PGA above 0.10g. No scaling was made to the records. According to the CISMID PERU database, the following seismic records were considered is shown in table 2. Spectral Acceleration vs. Period curves were calculated using a damping of 5% which correspond for concrete material. Figure 5 shows the variation of Sa vs. T. These curves represent the Seismic Demand Curves for the performance base design of the retrofit of the library.

Table 2. Ground Motions Records for Seismic Demand.

Event	Station	PGA (g)	Distance from Library
Lima 1966	Cercado L.	0.27	20km
Lima 1974	HUACO	0.21	6km
Lima 1974	UNALM	0.12	<1km

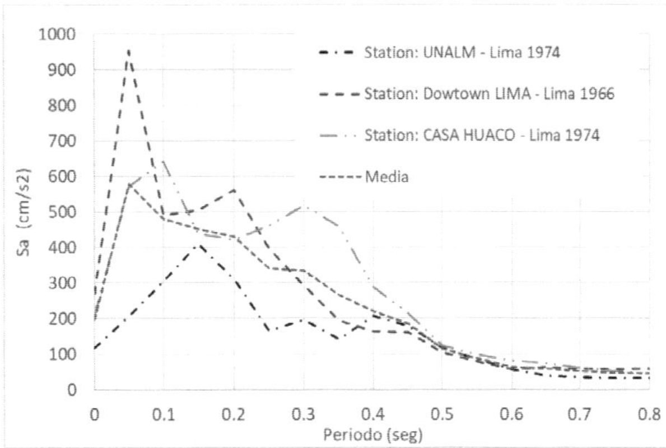

Figure 5. Pseudo Accelerations of the Seismic Catalog of Peru.

4. Results and discussions

In the collection and processing of data from the National Agricultural Library. The building capacity curves were obtained using the Non-linear Static Analysis - Pushover, in its longitudinal direction where the columns have lower inertia capacity and lower wall density. We proceeded with the As-built structure, and with each of the methodologies used. Likewise, several iterations were performed and finally obtaining the capacity curves for as-built and each retrofit procedure which are show in figure 6a. It can be seen the improvement in strength and ductility using CFRP at masonry walls compared to the installation of CFRP or steel jackets.

With the application of the ATC-40 [10] methodology by ADRS, the guidelines for the normalization of the demand of the Seismic Catalog of Peru are followed with the Shear Base (t) vs. Drift (%) axes. The Demand media curve was scaled to have a PGA 0.45g that correspond of design earthquake of Peruvian seismic code E030 [11] for Lima City located at Zone 4, severe earthquakes.

Having the structure capacity and seismic demand curves, the performance points are obtained as shown in figure 6b, which are the intersection points between the capacity curves and the demand curve according also to ATC-40 provisions.

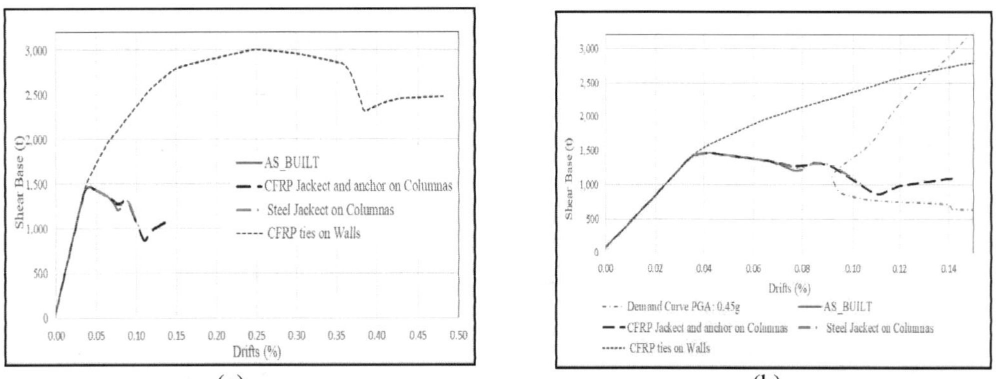

(a) (b)

Figure 6. (a) Structure Capacity Curves of the National Agricultural Library]
(b) Performance Points for each of the retrofit procedures applied to the building.

5. Conclusions

The use of CFRP materials in the reinforcement of structures over 50 years shows that it has a significant increase in capacity and a better response to an severe earthquake.

Retrofit of concrete walls with diagonal CFRP sheets and installations of CFRP anchors, is obtained better performance than reinforcement in columns with CFRP jackets. This shows that, in buildings over 50 years old with walls present, it is recommendable to retrofit the walls because it releases the greatest amount of seismic energy at the base of the building.

The seismic demands obtained from the Seismic Catalog of Peru were scaled up in the PGA of Zone 4 having a PGA 0.45g. This procedure is useful for evaluation criteria at the point of seismic performance.

The performance point of a structure in this case the National Agricultural Library BAN, defined as the intersection of capacity and demand curves, were determined from simple models of a degree of freedom. These models can reasonably represent the capacity and demand of the structure, either AS-BUILT and reinforced.

The reinforcement proposed in the National Agricultural Library in both columns and walls. It was found that the reinforcement of walls using CFRP sheets diagonally installed and CFRP anchors at each extreme of infill, it will increase the shear base capacity as well the ductility considerably in order to fulfill the objective of the continuity of the useful life of the structure due a severe earthquakes.

References

[1] Iacobucci R D, Sheikh S A, Bayrak O 2003 Retrofit of Square Concrete Columns with Carbon Fiber-Reinforced Polymer for Seosmic Resistance *Struc. J.* V. **100(6)** 785-94.

[2] Guerreiro J, et al. 2017 Bonding and Anchoring of a CFRP Reinforced Render for The External Strengthening of Old Masonry Buildings *Constr. Build. Mater.* **155** 56-64.

[3] Seible F, Priestley M. N 1997 Seismic Retrofit of RC Columns with Continuous Carbon Fiber Jackets *J. Compos. Constr.* **1(2)** 52-62.

[4] Kim Y, Ghannoum W M, Jirsa J. O 2015 Shear Behavior of full-scale Reinforced Concrete T Beams with CFRP Fibers and Anchors *Constr. Build. Mater.* **94** 1-9.

[5] ASCE/SEI Seismic Rehabilitation Standards Committee. 2007 Seismic rehabilitation of existing buildings (ASCE/SEI 41-06) *American Soci. Civ. Eng. Reston, VA.*

[6] Aboutaha R S 1996 Seismic Retrofit of non-ductile reinforced concrete columns using rectangular steel jackets 0970.

[7] Huaco G H, Jirsa J 2017 Innovative Procedures to Rehabilitate Severely Damaged Concrete Columns and Comparison with Alternative Methods *16th World Conf. Earthq. Eng. (WCEE) Santiago, Chile.*

[8] Huaco G D, Jirsa J 2015 Modeling Performance of Rehabilitated Extremely Damaged Concrete Columns and Masonry Wall for Analysis and Design. *Improv. Seism. Perform. Exi. Build. Other Struc*, 159-72.

[9] Huaco G H, Jirsa J 2017 Performance of Masonry Wall Retrofitted Using Cfrp Sheets and Anchors, 16th World Conference on Earthquake Engineering, *16th World Conf. Earthq. Eng. (WCEE) Santiago, Chile.*

[10] Council A T 1996 Seismic evaluation and retrofit of concrete buildings *Report No. SSC 96-01: ATC-40, 1.*

[11] Peru Construction Codes – Seismic Design Code E030, 2018. *National Institution Of Guidance To Construction Industry. SENCICO*, Peru.

IOP Conf. Series: Materials Science and Engineering **739** (2020) 012031 doi:10.1088/1757-899X/739/1/012031

Synthesis and Characterization of Monoclinic Gallium Oxide Nanomaterials for High-Concentration Ethanol Vapor Detection

H M B dela Torre[1], G N C Santos[1]

[1]iNano Research Facility, Physics Department, De La Salle University, 2401 Taft Avenue, Manila, Philippines

herbhelyne.delatorre@dlsu.edu.ph, gil.santos@dlsu.edu.ph

Abstract. This study was conducted to synthesize Gallium Oxide nanomaterials from its bulk form using a low-cost and repeatable method and investigate on its potential as sensing element for Ethanol (EtOH). Horizontal Vapor Phase Growth (HVPG) Technique was utilized to synthesize monoclinic Gallium Oxide (β-Ga_2O_3) nanowires. Images obtained from the Scanning Electron Microscope showed high density nanowires specifically in the area of highest temperature gradient. SEM also showed that the grown structures using HVPG Technique were at the nanoscale with diameters ranging from 51.60 to 908.38 nm. Higher surface-to-volume ratio was also noted in the area of highest temperature gradient which was subjected to an applied magnetics field. The Gallium to Oxygen ratio was verified via EDX to be approximately 2:3 which agrees with the atomic ratio of Ga_2O_3. The monoclinic structure of the grown nanomaterials was investigated using Raman Spectroscopy. Raman peaks of the samples were at 199 and 486 cm^{-1} which was accounted to the presence of two Raman-active modes of the b-polymorph of Ga_2O_3. The Raman spectrum of the grown nanowires confirmed that the material is monoclinic in structure and belongs to $C_{2/m}$ space group. The I-V Curve of the grown nanowires were also determined using two-point probe which illustrates a non-linear curve similar to that of a semiconductor material. Furthermore, additional fundamental properties such as resistivity and specific conductance of the materials were also determined via van der Pauw Technique. Results showed that the material has high specific conductance and low resistivity. The synthesized nanomaterial was responsive to Ethanol vapor. Exposure to the said compound increased its resistance. Graphs showed that there is a significant increase in the resistance after exposure to ethanol vapor.

1. Introduction

The synthesis of functional metal oxide nanomaterials has been widely studied due to its extensive potential especially in gas sensing. Semiconducting metal oxide nanostructures that are one-dimensional exhibits unique properties in chemical, optical, electrical and gas sensing. Due to high surface-volume ratio, nanomaterials have higher sensitivity and exhibit more surface reaction. The significant number of surface atoms is also accountable for greater sensitivity and selectivity which is a relevant factor in gas sensing applications. [1, 2] The use of nanotechnology for gas sensing devices results to lower power consumption and ease of use due to its compactness and portability. [3] Studies on Gallium Oxide, a wide-bandgap conductive oxide, are in the early stages, thus, very few reports are made about its crystal growth and material properties.

Successful synthesis of Gallium Oxide nanoparticles using various methods are reported, which include arc discharge, evaporation, and heating. [4, 5] The nanoscale Ga_2O_3 produced showed properties which are distinct from the bulk form. Five polytype of Gallium Oxide have been confirmed. These are denoted α, β, γ, δ and ε. The widely studied polytype is the β-polytype which is usually formed into a monoclinic crystal and stable material after undergoing annealing procedure. [6] The monoclinic Gallium Oxide has several useful characteristics, especially the wide bandgap of 4.8-4.9 eV and exhibits a good chemical and thermal stability which makes it an ideal material for gas sensors. [7] The gas sensing properties of the material synthesized in this study was described using Ethanol vapor obtained via static headspace gas sampling.

This paper reports the result of the synthesis of Gallium Oxide nanomaterials from its bulk form using a low-cost and repeatable method such as Horizontal Vapor Phase Growth (HVPG) Technique and investigate its potential as sensing element for Ethanol (EtOH) vapor.

2. Experimental Details

2.1. Synthesis of b- Ga_2O_3

Horizontal Vapor Phase Growth Technique, patented by Santos, et al, was used in the synthesis of the nanomaterial. The method employs a vapor-solid mechanism. Gallium Oxide bulk powder obtained from Sigma-Aldrich was contained in a fused silica quartz tube sealed under high-vacuum system while the pressure was maintained at approximately $1x10^{-6}$ Torr. Half of the sealed tube was placed in a horizontal tube furnace while the other half extending outside was subjected to a magnetic field (figure 2) to enhance the morphology of the crystal growth. [8] The temperature gradient created a path for the vapor deposition and the crystal growth was expected to be observed on the tube's inner walls at the region of relatively lower temperature. The growing time was set to 8 hours at growth temperature of 1200°C. The furnace was programmed at a ramp time (Δt) of 40 minutes as shown in figure 1. Samples from the two zones indicated in figure 2 were collected for characterization.

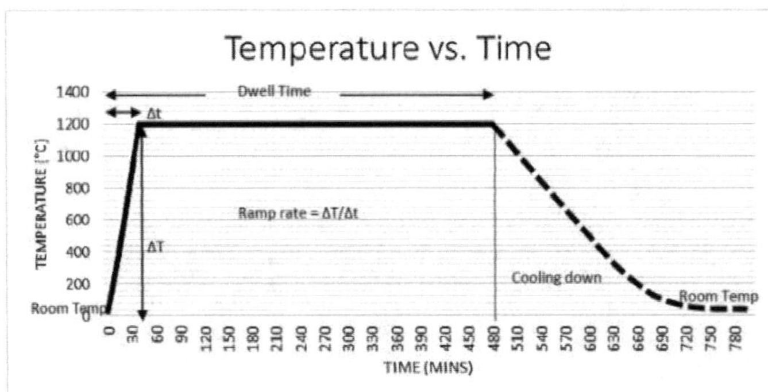

Figure 1. Temperature profile of the horizontal tube furnace with the ramp rate (Δt) set at 40 minutes and maximum temperature of 1200°C at 8 hours' dwell time.

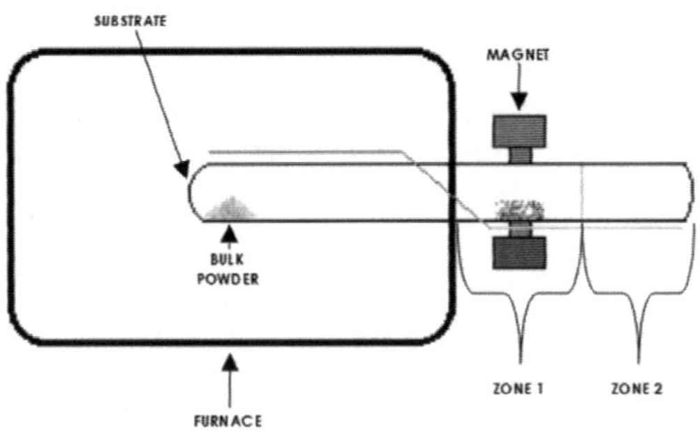

Figure 2. Diagram of the setup in synthesizing the nanostructures.

IOP Conf. Series: Materials Science and Engineering **739** (2020) 012031 doi:10.1088/1757-899X/739/1/012031

2.2. Characterization of the Grown Nanomaterials

Phenom Pro Desktop Scanning Electron Microscope was used to capture the SEM micrographs of the samples from zones 1 and 2. Data obtained from the surface morphology analysis were processed in Image J to determine the size and distribution of the as-grown nanomaterials on the inner surface of the substrate.

The samples were analyzed under EDX for elemental composition analysis. The obtained results were used to identify the elements found on the substrate and determine the Gallium to Oxygen atomic ratio.

Van der Pauw (vdP) technique, a general four-point probe technique, was employed to calculate the Resistivity and Specific Conductance of the arbitrarily-shaped samples. Using Keithley Source Measure Unit 2450 four-point probe, the voltage was measured across the substrate. From the readings obtained, Resistivity and Specific Conductance were calculated. Current-Voltage (I-V) curve of the same samples were obtained using Keithley Source Measure Unit 2450 two-point probe. Voltage sweep from -5 volts to +5 volts was graphed.

Raman spectrum of the grown Ga_2O_3 nanomaterials was determined to verify its crystal structure. The Raman peaks identified were cross-referenced with the gathered literatures about fabrication of β-Ga_2O_3.

The gas sensing property, specifically its sensitivity to high percent-by-volume concentration of Ethanol was investigated by measuring the change in resistance of the material when exposed to ethanol vapor. The gas sample was prepared via static headspace gas sampling and the resistance change was recorded using a Keithley Source Measure Unit 2450. Attributes investigated includes response time and sensitivity to the target gas.

3. Discussion

Gallium Oxide nanowires were grown using Horizontal Vapor Phase Growth Technique. SEM micrographs of the samples obtained from the two zones (figure 4) showed significant change in the morphology of the bulk (figure 3) samples after undergoing HVPG Technique. High-density nanowires were found in Zone 1. It was noted that this area has the highest temperature gradient and was also under the effect of an applied magnetic field. These two factors noted were noted to be accountable for the crystal growth zone preference. [8] This factor also contributed to the greater amount of yield in terms of the density of nanostructures found.

Figure 3. SEM micrographs of Gallium Oxide bulk powder.

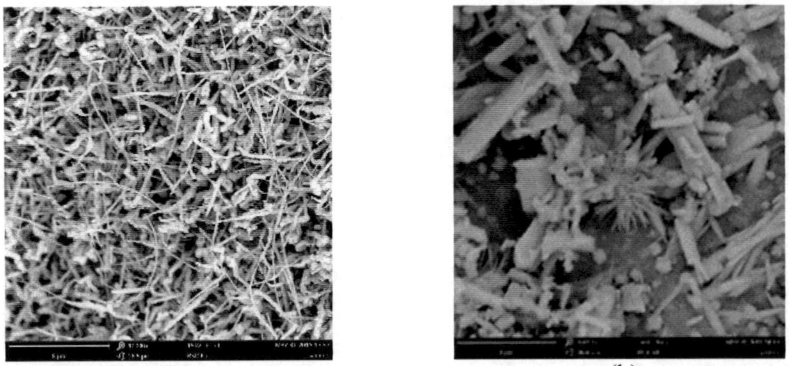

(a) (b)

Figure 4. SEM Micrograph of as-grown nanomaterials (a) Zone 1 (b) Zone 2.

Structures deposited in Zone 1 have diameters ranging from 51.60 – 908.38 nm with an average diameter of 237.50 nm. Most of the wires have diameter of 206.50 nm extending up to 435,019.36 nm. On the other hand, structures grown at Zone 2 have measurements between 2000 – 86,800 nm and mean diameter of 19,058.58 nm. A large number of particles have diameter of 16,400 nm in the said zone with maximum span length of 200,802.26 nm.

Higher surface-to-volume ratio was also noted in this area as compared to Zone 2. Note that this zone was subjected to applied magnetic field and the area of highest temperature gradient. The parameters employed to enhance the morphology of the nanowires included the use of 0.25 T variable gap neodymium magnet applied at the region closest to the middle of the tube. In the literature cited, the magnet was positioned in the region near the highest temperature gradient. The presence of the applied magnetic field was noted to assist the growth mechanism from the bulk material which enhanced its morphology. [8]

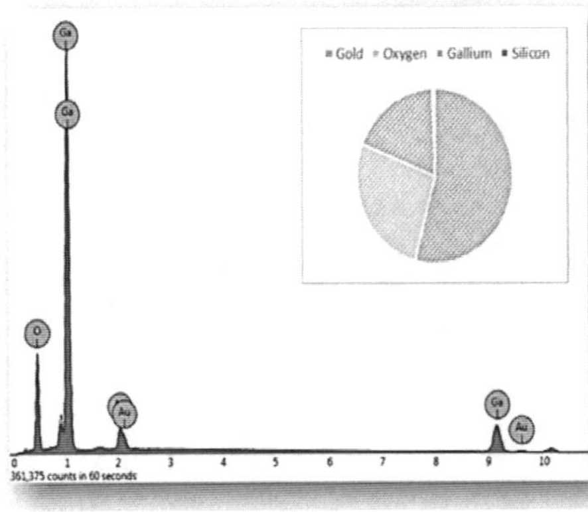

Figure 5. EDX Spectrum of Zone 1 showing peaks of Ga, O, and Au and corresponding pie chart.

Phenom XL Energy Dispersive X-ray Spectroscopy was used to investigate the materials elemental composition. Results confirmed the presence of Gallium, Oxygen and traces of Gold and Silicon found in Zone 1 which is shown in figure 5. Specifically, Zone 1 showed an atomic concentration of 26.65% Gallium, 38.17% of Oxygen and 35.19% of Gold. The traces of Gold atoms were attributed to the Gold sputtering during the sample preparation prior to the analysis, while the traces of Silicon were attributed to the substrate for the nanomaterials synthesis. The Gallium to Oxygen ratio was verified to be approximately 2:3 which agrees with the atomic ratio of Ga_2O_3.

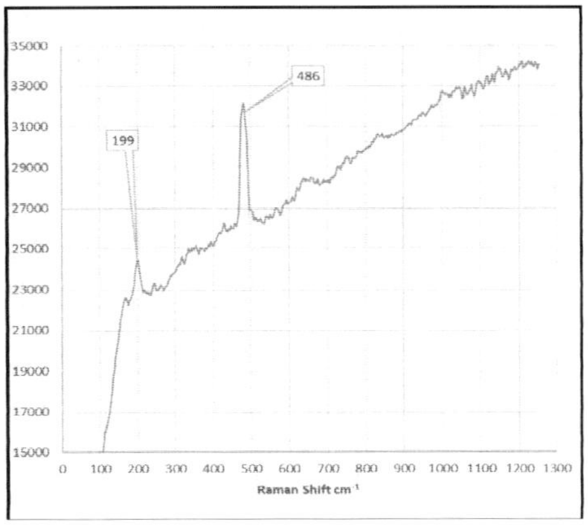

Figure 6. Raman Spectrum of Ga_2O_3 nanowires obtained from Zone 1 synthesized via Horizontal Vapor Phase Technique.

Raman scattering measurements were done using a laser with a wavelength of 565 nm to further investigate the structure of the synthesized nanomaterials. Fifteen Raman modes are expected in the vibrational spectrum of a β-polytype Ga_2O_3 which belongs to the space group C2/m [9]. The Raman-active modes of this polytype are clustered into three groups of frequencies: the low frequency liberation and translation of tetrahedron-octahedron chains with peaks below 200 cm-1, the mid frequency deformation of Ga2O6 octahedron with peaks 310 to 480 cm-1 and high frequency stretching and bending of GaO4 tetrahedron whose peaks are found at 500 to 770 cm-1. [10]. Figure 6 shows the Raman spectrum of the grown nanowires synthesized via HVPG Technique deposited on the Silica quartz tube. The position of the sharp peaks at 199 and 486 agrees with the reported Raman peaks in literature for β-Ga_2O_3. The peak at 199 means that the nanomaterial exhibited the liberation and translation of tetrahedron-octahedron chains. Raman scans also detected the presence of deformation of Ga2O6 octahedron chains. The peaks detected in the synthesized nanomaterial also imply that the crystal structures belong to the C2h group symmetry, hence the nanomaterial grown is confirmed to be a β-polytype.

I-V characterization was conducted at room temperature via a two-point probe using Keithley Source Measure Unit. The voltage sweep was set to 200 points and varying the source voltage from -5 to 5 V. The corresponding change in the current was automatically recorded and plotted in a graph using the SMU. Graph is non-linear and resembles the curve of a semiconductor diode. As shown in figure 7 the material has forward and reverse characteristics.

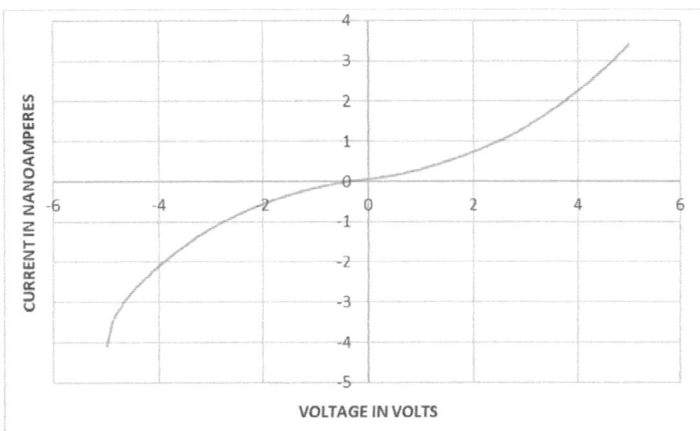

Figure 7. I-V Curve of Ga_2O_3 nanowires obtained from Zone 1.

Four-point probe measurements carried out on the samples also provided additional evidence of the fundamental electrical property of the nanomaterials. Since the sample is an arbitrarily-shaped material, voltage measurements was measured in eight permutations. Volume resistivity specific conductance was expressed in SI unit ohm-meter and Siemens per meter respectively. Results of the vdP Technique shows that the resistivity of the Ga_2O_3 nanomaterials grown in the Silica quartz tube substrate has a low resistivity and very high specific conductance. The materials tested have an average of 4.61×10^{-6} Ωm volume resistivity and an average of 2.17×10^{5} specific conductance.

Investigation on the synthesized nanomaterial's potential use for gas sensing application was conducted at room temperature. Ethanol vapor sensitivity of the synthesized nanomaterial was tested against the headspace gas obtained from 99.99% Ethanol. The exposure time was set to 90 seconds. Using Keithley 2450 SMU, a Resistance vs Time graph was recorded as shown in figure 8.

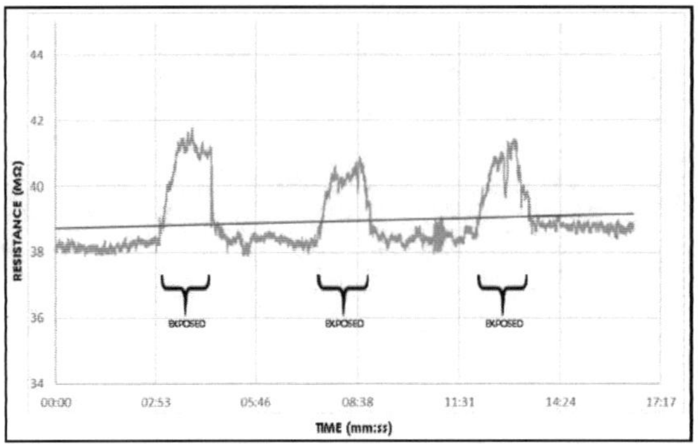

Figure 8. Resistance vs Time graph for the gas reference test at 99.99% concentration conducted at room temperature.

The average change in resistance upon exposure to EtOH vapor was 2.68 MΩ while the average response time recorded was 64.67 seconds. The Sensor Response S_g was determined by the formula

$$S_g = \frac{\Delta R}{R_g} \tag{1}$$

where ΔR = Ra − Rg, Ra is the electrical resistance in air ambient, and Rg is the resistance in Ethanol vapor. [11] The calculated Sensor Response was 0.069175.

4. Conclusion

The results showed that monoclinic Ga_2O_3 nanowires with diameters ranging from 51.60 to 908.38 nm were successfully synthesized via optimized HVPG Technique. The temperature gradient and the applied magnetic field was noted to be the primary contributor in the crystal structure and growth preference of the nanomaterial. The Gallium to Oxygen ratio was verified to be approximately 2:3 which agrees with the atomic ratio of Ga_2O_3. The Raman spectrum of the grown nanowires confirmed that the material is monoclinic in structure and belongs to $C_{2/m}$ space group. This confirms that the grown nanomaterial is a β-polytype. Electrical characterization of the as-grown nanomaterial suggests that it has high specific conductance and low volume resistivity.

The result of gas reference test demonstrates that the nanomaterial is responsive to the presence of ethanol. The data indicates that the resistance increased due to the presence of Ethanol vapor. Overall, the material showed potential for use in gas sensing and gas detection.

Acknowledgement
This study was supported by the Department of Science and Technology, Science Education Institute and USAID.

References

[1] Kolmakov A, Moskovits M 2004 Chemical sensing and catalysis by one-dimensional metal-oxide nanostructures *Annu. Rev. Mater. Res.* **34** 151-80.

[2] Duraia E M, Mansorov Z A, Tokmolden S 2009 Synthesis, characterization and photoluminescence of tin oxide nanoribbons and nanowires *Physica* B **404(21)** 3952-56.

[3] Cadena G J, Riu J, Rius F X 2007 Gas sensors based on nanostructured materials *Analyst* **132(11)** 1083-99.

[4] Kim H, Jin C, An S, Lee C 2012 Fabrication and CO gas-sensing properties of Pt-functionalized Ga2O3 nanowires *Ceramics Int.* **38(5)** 3563-67.

[5] Kim N H, Kim H W 2005 Gallium oxide nanomaterials produced on SiO2 substrates via thermal evaporation *App. Surface Sci.* **242(1-2)** 29-34.

[6] Hatanaka Y, Nakanishi Y, Ogita M, Saika N 1999 Ga2O3 thin films for high-temperature gas sensors *App. Surface Sci.* **142(1-4)** 188-91.

[7] Chun H J, Choi Y S, Bae S Y, Seo H W, Hong S J, Park J, Yang H 2003 Controlled structure of gallium oxide nanowires *J. Phys. Chem.* B **107(34)** 9042-46.

[8] Briones J C, Castillon G, Delmo M P, Santos G N C 2017 Magnetic-field-enhanced morphology of tin oxide nanomaterials for gas sensing applications *J. Nanomater.* 1-11.

[9] Filippo E, Siciliano M, Genga A, Micocci G, Tepore A, Siciliano T 2013 Single crystalline β-Ga2O3 nanowires synthesized by thermal oxidation of GaSe layer *Mater. Res. Bulletin* **48(5)** 1741-44.

[10] Wu X C, Song W H, Huang W D, Pu M H, Zhao B, Sun Y P, Du J J 2000 Crystalline gallium oxide nanowires: intensive blue light emitters *Chem. Phys. Lett.* **328(1-2)** 5-9.

[11] Ying Z, Wan Q, Song Z T, Feng S L 2004 SnO2 nanowhiskers and their ethanol sensing characteristics *Nanotechnol.* **15(11)** 1682-84.

IOP Conf. Series: Materials Science and Engineering **739** (2020) 012032 doi:10.1088/1757-899X/739/1/012032

Effect of Electric Arc Furnace Dust Treatment on the Properties of Fresh and Hardened Mortar

R Syahyadi[1], A Fauzi[1,2*], E Majuar[1], F Rizal[1,2], M Reza[1,2], dan Fajri[1,2]

[1]Politeknik Negeri Lhokseumawe, (PNL), Acheh, Indonesia
[2]Geopolymer and Green Technology Research Center, PNL, Acheh, Indonesia

amirfauzi@pnl.ac.id

Abstract. Electric arc-furnace dust (EAFD) is generated in considerable amounts by the electric arc steelmaking process. Use of EAFD in concrete improved many properties of concrete. Yet, the prolonged setting time even at low dosages of EAFD was a serious limitation. This phenomenon was attributed to the presence of transition metal in the material of EAFD. Therefore, it was necessary to stabilize the transition metal in an insoluble crystalline matrix. The calcination with high-temperature was done to fix the transition metal in a complex matrix with the cementitious properties. This material was called Engineered Dust (ED). The use of ED with different replacement levels of 0, 5, 10, 15 and 20% at w/b ratio of 0.5 was carried out. The results showed that the use of ED solved the prolonged setting time which the replacement up to 20% had a normal setting time of 5 to 6 hours. Furthermore, the material of ED was used up to 15% replacement level with the replacement of 20% ED that gave the lowest electric charge values in the RCPT.

1. Introduction

Electric arc furnace dust (EAFD) is a by-product material from the electric steelmaking industry. It was known that every ton production of steelmaking generated 2% EAFD in the environment. Furthermore, the material of EAFD was designated by the European Union (EU) and the EPA (The United States Environmental Protection Agency) as a hazardous waste due to was an abundant in the heavy metals such as zinc (Zn), lead (Pb), Chromium (Cr), and Cadmium (Cd). It meant that the material of EAFD may not be disposed of at landfills without the treatment [1]. Consequently, it was an environmental issue in the world.

The composition of EAFD depended on the scrap charge, the used additive, and the type of steel manufactured. For example, the use of galvanized steel scrap increased the zinc content in the EAFD significantly. Content of Zinc in EAFD varied in the range of 7% to 40%, and the content of lead content in the range of 4 to 9% [2]. The use of EAFD in the concrete has attracted the researcher to explore more information regarding the material of EAFD. It was due to the cementitious matrix in the concrete was able to encapsulate the heavy metal and reduced the leachability [3]. Some studies [14-15] reported that the enhanced concrete properties were at 3% replacement of EAFD. However, higher replacement levels of EAFD affected fresh and early hardened concrete properties that retarded the setting time of concrete and lower compressive strength at an early age [4]. De Vargas et al [5] investigated the effect of 5%, 15% and 25% EAFD on the setting time of cement paste. They showed that the increased of EAFD replacement brought to prolong the setting time of concrete up to 70 hours. De Souza et al [6] found that the use of 15% EAFD replacement showed the reduction of chloride penetration in concrete. Al-Mutlaq [7] investigated the use of local EAFD in concrete and reported the excessive setting time even in the presence of a chemical accelerator. It was due to the transition metal (ferric, zinc and manganese) was too high percentage so that they were entrapped in the link of a chemical structure during the initial hydration.

Based on the prolong setting time, the use of EAFD at a high replacement was not recommended as a material in the application. However, the transformation of EAFD to solve the impact on setting time of concrete was able to increase the consumption of EAFD in the environment. This study is investigated to analyze the effect of EAFD transformation in the fresh and hardening mortar.

2. Materials and Experimental work

The material of EAFD was obtained from the Sabic Steel Industry (Hadeed) plant in Jubail. Ordinary Portland cement (OPC) was produced by Al-yamamah cement factory in Riyadh which referred to ASTM C150. The transformation of EAFD was investigated due to the transition metal was entrapped in the link of chemical structure so that prevent their release to a cementitious matrix during initial hydration. The engineered transformation of EAFD was by the calcination of EAFD in the furnace with the temperature of 1000^0C within 5 hours which it was called the material of engineered dust (ED). The chemical composition of OPC, EAFD and ED material was analyzed by X-ray fluorescence (XRF). The fresh and hardening mortar was conducted by the sand-binder ratio of 2 and the water-binder ratio of 0.5 which the workability, the setting time, compressive strength and rapid chloride ion penetration test (RCPT) were used as a parameter.

3. Results and Discussions

3.1. Chemical composition

The chemical compositions of OPC, EAFD and ED material are given in Table 1. The result shows that the material of OPC is high in calcium oxide and silica oxide content. The material of EAFD is high in ferric oxide, zinc oxide, and manganese content, whereas the material of ED is high in silica oxide, ferric oxide, and zinc oxide. It is seen that the calcination of EAFD material with the temperature of 1000^0C reduced the content of transition metal. Even, it eliminates the content of manganese (Mn) from the chemical composition of EAFD. It was attributed to the unstable condition of transition metal in the material of EAFD which the calcination with high temperature transformed the material of EAFD to be more stable material. The calcination with high temperature improved significantly the content of Si in the material of ED, even more than the content of Si in the material of OPC.

Table 1. Chemical Composition of OPC, EAFD and ED.

Oxide Composition	**Analysis, % by weight**		
	OPC	**EAFD**	**ED**
SiO_2	20.2	1.91	36.54
Al_2O_3	5.49	0.26	3.9
Fe_2O_3	4.12	48.52	25.6
CaO	65.43	7.04	6.72
MgO	0.71	1.75	3.8
Na_2Oeq	0.06	3.22	0.96
SO_3	2.61	0.84	0.4
ZnO	-	27.73	18.2
MnO	-	3.04	-
LOI	1.38	5.68	3.88

3.2. Workbility

The flow diameter of fresh mortar mixtures was conducted by flow table apparatus with using the water-binder ratio (w/b) of 0.5 as summarized in Table 2. It is observed that the material of EAFD in this study was only used 3% replacement. It was suitable with the recommendation of the previous study mentioning that the use of EAFD in high replacement brought to prolong setting time. The table shows that the workability in the fresh mortar of EAFD particle is higher than the fresh mortar of OPC. It was attributed to the smooth surface of EAFD that did not absorb the water content from the fresh mortar mixture. The same trend of workability occurred in the fresh mortar mixture of ED. It is seen that the workability of 5% ED is almost similar to the workability of 3% EAFD. It means that the calcination with high temperature did not change the particle surface of EAFD. Furthermore, the increase of EAFD replacement improved the workability. It is due to the increase of EAFD replacement causing the reduction of OPC in the fresh mortar mixture so that it reduced the absorption of water content.

Table 2. Flow diameter of OPC, EAFD and ED in the fresh mortar mixture.

Mix ID	**Replacement Level (%)**	**Flow diameter (cm)**
OPC	-	23.1
EAFD3	3	23.7
ED5	5	23.6
ED10	10	23.7
ED15	15	23.9
ED20	20	24.0

3.3. Setting time

The setting time of OPC, EAFD, and ED in the fresh mortar mixture was shown in Figure 1. The setting time of OPC and EAFD is known about 6.4 hours and 26.4 hours, respectively. It is seen that there was contrast the setting time of OPC and EAFD. The prolong setting time was attributed to the presence of transition metal in the high percentage so that prevent their release to the cementitious matrix during the initial hydration. However, the calcination with the high temperature was succeeded to transform the material of EAFD so that was able to solve prolong the setting time in fresh mortar mixture. It can be proven in Figure 1 which the setting time of ED with the replacement of 5% to 20% is in the range of 5.6 to 5.9 hours. It means that the calcination with high temperature was able to reduce the percentage of metal content optimally which these metal in a high percentage had been entrapped in the link of the chemical structure. Furthermore, the increase of ED enhanced the setting time in the fresh mortar. It is linear with the workability. It means that the increase setting time was not due to the chemical content of ED but the drying time of initial water hydration. The result clearly demonstrated the potential of using ED material in term the restoring of setting time to its normal range.

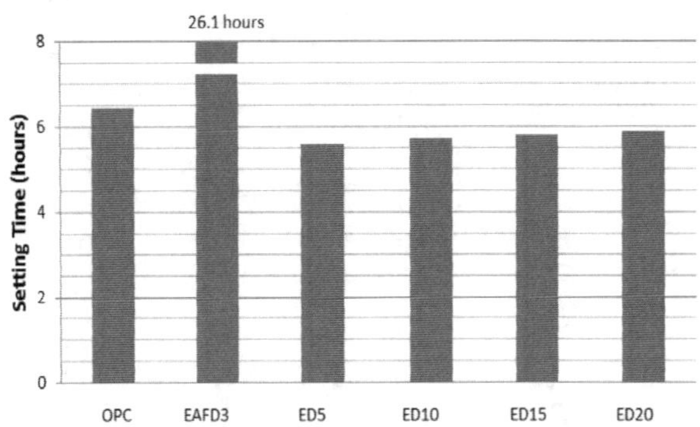

Figure 1. Setting time of OPC, EAFD and ED in the fresh mortar mixture.

3.4. Compressive strength

As mentioned in the previous study, the 1-day compressive strength of EAFD is not reported in Figure 2 since the fresh mortar mixture of EAFD was still unhardened condition. Furthermore, the compressive strength of OPC and 3% EAFD mixture at a later age are 60.4Mpa and 63.1Mpa. It means that the compressive strength of 3% EAFD mixture at a later age was higher than that given by the OPC mortar. This result gave the understanding that the material of EAFD was categorized as the cementitious material although it had not to impact at the early compressive strength. It was attributed to the high percentage of transition metal in the material of EAFD that caused the prevent of their release to the cementitious matrix during initial hydration so that retarded the pozzolanic reaction.

Furthermore, the compressive strength of ED with 5% to 20% replacement provides the early and later compressive strength. The early compressive strength of 5% to 20% ED is in the range of 16Mpa to 19Mpa, whereas the later compressive strength is in the range of 57Mpa to 65Mpa. It means that the pozzolanic reaction occurred in the short term which it was contrasted with the material of EAFD as described in the section of setting time. Furthermore, the material of ED improved the compressive strength up to 15% replacement. It was attributed to the increase of Si content during the calcination process which it reacted with $Ca(OH)_2$ to form the gel of C-S-H. However, the compressive strength of 20% ED is lower than that given by the OPC mortar. It means that the replacement of ED more than 15% was as the filler in the mortar and interfered the pozzolanic reaction. It was due to the optimum pozzolanic reaction that occurred in the replacement of 15 EAFD, whereas the rest of 5% replacement filled the pore. This result concludes that the use of 15% ED was considered as the optimum replacement level in the mortar mixture.

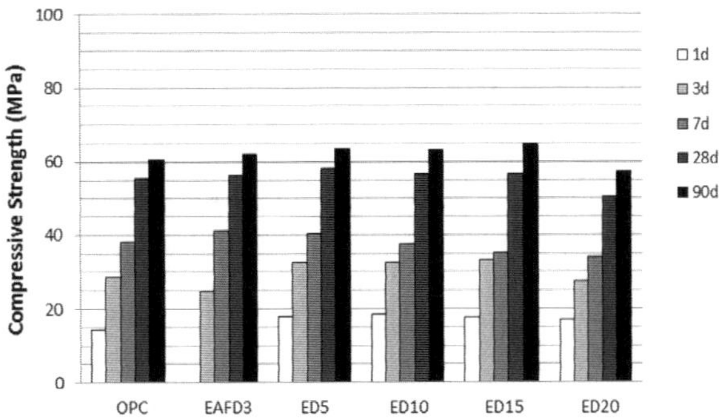

Figure 2. Compressive strength of OPC, EAFD and ED in the hardened mortar.

3.5. Rapid chloride-ion penetration test

The Rapid chloride-ion penetration test (RCPT) relies on the amount of electric charge passing through the mortar mixtures. The higher of electric charge passing into a mixture indicates the more permeable of mortar. Figure 3 shows that the material of EAFD with 3% replacement improves the chloride ion penetration resistance by 25% referred to the OPC mortar. It means that the material of EAFD was as the filler in the mortar. Furthermore, the material of ED with the replacement up to 10% ED has higher electric charge values (coulombs) compared to the OPC mortar whereas the material of ED with the replacement more than 10% ED has a lower coulomb compared to the OPC mortar. It confirms that the increase in compressive strength reduced the reading of RCPT. However, replacement of 20% ED did not provide the highest compressive strength. It is due to the replacement of 20% ED only improve the filler in the mortar but interfered the pozzolanic reaction.

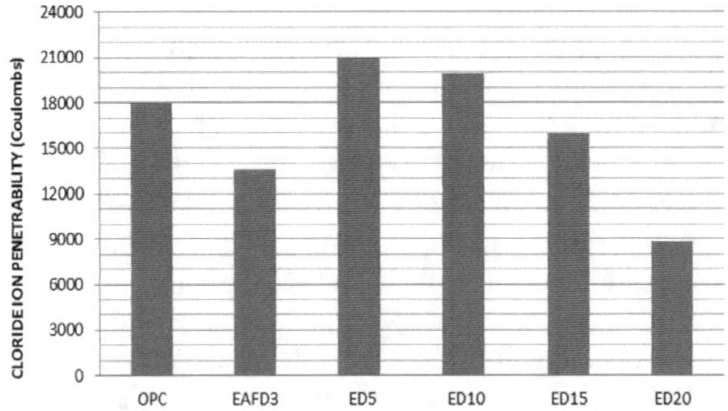

Figure 3. RCPT of OPC, EAFD and ED in the hardened mortar.

4. Conclusions

Based on the result of ED properties obtained from this study, the following conclusions can be drawn:

- The material of ED contained lower concentrations of transition metal than the material of EAFD;
- The production of ED was proven to successfully overcome the problem of prolonged setting time associated with the incorporation of 3% EAFD. This was confirmed by the normal range of setting times attained at all ED replacement levels up to 20%;
- The fresh mortar mixture of ED had slightly improved the flowability with the setting time in the range of 4 to 6 hours while the setting time of 3% EAFD was 26.1 hours;
- The use of 15% ED in the mortar mixture was the optimum replacement level which the RCPT of 15% ED was still higher than the EAFD of 3% replacement.

References

[1] Rizal F, Fauzi A, Yunus Y 2019 Feasibility using electric arc furnace dust as cementing material in high performance concrete, *IOP conf. series: Mater. Sci. Eng.* **536** 012107.

[2] Oustadakis P, tsakiridis P E, Katsiapi A, Leonardo S A 2010 hydro metallurgical process for zinc recovery from electric arc furnace dust: Part I: Characterization and leaching by diluted sulphuric acid, *J. hazardous mater.* **179(1-3)** 1-7.

[3] Harahsheh M, Nu'airat J, Otoom A, Hammouri I, Jabali H, Zoubi M, Al'asal S A 2019 Treatment of electric arc furnace dust and halogenated plastic wastes: A review, *J. environ. chem. Eng.* **7(1)** 102856.

[4] Fares G, Zaid R Z, Fauzi A, Hozaimy A M, Negheimish A I, Khan M I 2016 Performance of optimized electric arc furnace dust-based cementitious matrix compared to conventional supplementary cementitious materials, *Construct. build. mater.* **112** 210-21.

[5] Vargas A S, Masuero A B, Vilela A C F 2006 Investigations on the use of electric-arc furnace dust (EAFD) in Pozzolan-modified Portland cement I (MP) pastes, *Cement Concrete Res.* 2006. **36** 1833-41.

[6] Souza C D C, Machado A T, Lima L R P A, Cardoso R J C 2010 Stabilization of Electric-Arc Furnace Dust in Concrete, *Mater. Res.* **13(4)** 513-9.

[7] Almutlaq F M 20011 Aspects of the Use of Electric Arc Furnace Dust in Concrete, *Civil Eng.* p. 301.

IOP Conf. Series: Materials Science and Engineering **739** (2020) 012033 doi:10.1088/1757-899X/739/1/012033

Mechanical Splices for Seismic Retrofitting of Concrete Structures

G Huaco[1], J Jirsa[2]

[1]Professor, Peruvian University of Applied Science UPC
[2]Emeritus Professor, The University of Texas at Austin

pccighua@upc.edu.pe

Abstract. As an alternative to lap splicing, mechanical splices can be used for retrofit purposes. They are generally most economical than traditional lap splices when available spacing or length makes laps difficult to utilize. Mechanical splices are frequently used in new construction. However, their use is limited and not practical for use in retrofitted structures. However, if the bars to be joined do not need to be threaded in order to be connected with a special mechanical splice, such mechanical splices can be useful. It is presented a proposal of using two types of mechanical splices for retrofit purposes. Cycle Tension and cycle tension-compression tests are presented and discussed. It was found that mechanical splices are suitable and have acceptable response under seismic loads.

1. Introduction

Lap splices depend on quality of concrete and/or confinement by transverse reinforcement. Mechanical splicing provides load path continuity in the reinforcement, independent of the condition of the concrete. The one shown in figure 1 is composed of a hollow threaded steel tube to couple two reinforcing bars with threaded ends. This type of splice is not suitable for retrofit purposes since bars need to be threaded and threading bars embedded in concrete is nearly impossible.

Figure 1. Type of mechanical splices using threaded bars for new construction.

ACI 318 -14 [1] indicates that tension lap splices are not allowed for #14 or #18 bars. Lap splices add to congestion of the concrete section in elements near beam-column joints. An alternative solution is the use of mechanical splices.

Poorly confined lap splices cannot develop large ductility. The failure is brittle once slip between the spliced bars begins. Figure 3.3 represents the differences of deformation responses under tension axial load of a poorly confined lap splice and mechanical splice. Elwood, et.al, on the update of ASCE/SEI 41 Concrete Provisions [2] consider also the deficiencies related to the use of lap splices, especially for cases where the splice length is shorter than required by the code.

ACI 318-14 [1] code requires that mechanical splices develop at least 1.25fy (fy: nominal yield strength). For seismic applications, Type 1 splices must develop 1.25fy, however Type 2 splices must develop the specified tensile strength of the bar.

2. Mechanical splice for retrofitted uses - Replacing damaged existing bars

After an event that causes damage to a structure, the damaged bars may have to be removed if they are buckled and bent. Figure 2 shows damaged bent bars which can be repaired by installing new bars that are mechanically spliced (coupled) to the existing bars. For this study, splices which do not need end-bar preparation were considered. Once the bar is removed the splice sleeve can be positioned over the existing bars and a new bar is introduced where the bent bar was removed. The splice sleeve is then moved to center on the location where the existing and new bars meet. The bars are held tightly in the sleeve with bolts that are torqued to a prescribed level. The system is designed to develop 100% or 125% of nominal yield strength of the bars as ACI 318-14 [1] requires. There are different types of couplers that vary according to the process of the installation.

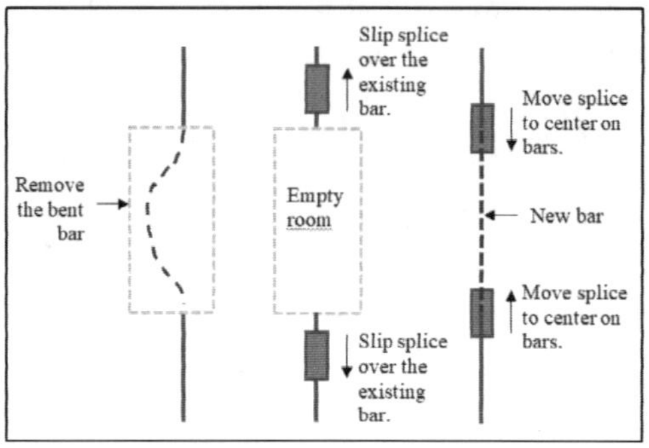

Figure 2. Advantage of mechanical splice to replace bent bars.

Two different mechanical splice configurations were evaluated. The first is the Short Mechanical Splice Lenton SMS and Long Mechanical Splice Lenton LMS produced by Erico. There are different number of bolts depending on size of bar. Short Mechanical Splice - SMS consists of a 6.8 in. steel cylinder with 6 bolts. These bolts are located into one longitudinal line which go through along the splice. Seating of the pointed bolts reduced the cross sectional area of the bar about 5%. The material used for this type of splice is also Grade 60. Long Mechanical Splices – LML consists of a 10 in. length steel cylinder with 8 bolts, two outer bolts with rounded ends and 6 interior bolts with pointed ends. Similar to the short mechanical splices, the rounded and pointed end bolts are located along a plane one through the splice. Rounded point bolt at the end of the sleeve in intended to prevent the bars from fracturing in the splice so that the requirement for specified ultimate strength is achieved. Figure 3 shows SMS and LMS, and figure 4 shows the interface between the rebar and the mechanical splice.

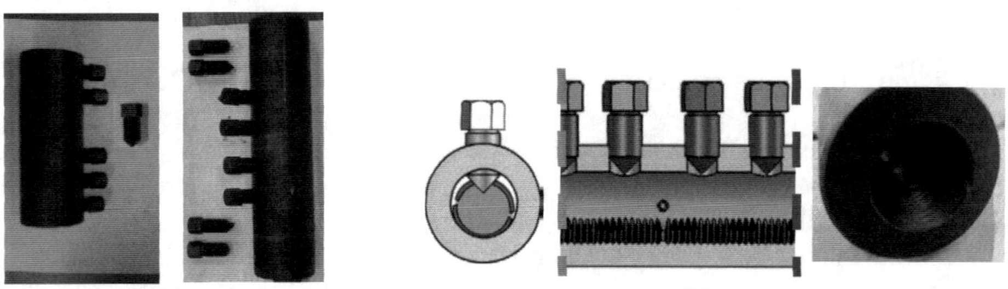

(a) (b)

Figure 3. Short (left) and Long (right) Mechanical Splices for retrofitting purposes (a), Bolts tighten the bar with the interior sleeve of the mechanical splice (b).

3. Mechanical splices in special moment frames in high seismic zones

The manufacture's data for the mechanical splices indicate that the long mechanical splice meets the requirements Type 2 and Type 1 splice (large deformation and high strength). The bars to be coupled should be A-615 and A-706. For short mechanical splice, it meets the requirement for Type 1 splice when the bar is A-615 (high strength), however this short mechanical splice could be used as Type 2 when the bar is A-706 (large deformation). Table 1 summarizes the requirements for the selection to the type of mechanical splice.

Table 1. Provisions for the selection of the mechanical splices [1].

Type of Splice	ASTM Specification for bars	ACI-318 Seismic Provisions Type (location)	Requirement Strength of bars (ACI 318-11)	Requirement Strength of bars (AASTHO)
Short Mechanical Splice	A-706	Type 2	1.25 fy & 1.0fu	1.35fy
Short Mechanical Splice	A-615 & A-706	Type 1	1.25 fy	1.35fy
Long Mechanical Splice	A-615 & A-706	Type 1 and Type 2	1.25 fy & 1.0fu	1.35fy

For rehabilitation purposes, it should be observed that these bars can come from structures that may be quite old. Properties of the bar are very important in order to choose the appropriate mechanical splice. Table 2 shows comparisons of different old types of reinforcing bars. Long mechanical splices are capable to work with old bars A-615 of building build from 1968, and short mechanical splices are allowed to be used with bars used in structures build from 1974. However, if it is used a bar with the same characteristics of A-615 (min yield and tensile strength), the long mechanical splice can be applied. An example is the bar type is A432 Grade 60 which has same value for minimum yield and tensile strength, then it can be repaired a structure built after 1959 using the long mechanical splices.

Short mechanical splices could be used to retrofit columns for structures build after 1974, considering the splices as Type2. However, considering the short mechanical splices as a Type 1, they can be used in the rehabilitation of structures built after 1959, similar to the long mechanical splice case.

Table 2. Reinforcing bars 1911 to present, ASTM specification, minimum Yield and Tensile Strengths in psi [3].

ASTM Spec	Years		Steel Type	Grade 33 (Structural)		Grade 40 (Intermediate)		Grade 50 (Hard)		Grade 60		Grade 75	
	Start	End		Min. Yield	Min. Tensile	Min. Yield	Min. Tensile	Min. Yield	Min. Tensile	Min. Yield	Min. Tensile	Min. Yield	Min. Tensile
A15	1911	1966	Billet	33,000	55,000	40,000	70,000	50,000	80,000				
A408	1957	1966	Billet	33,000	55,000	40,000	70,000	50,000	80,000				
A432	1959	1966	Billet							60,000	90,000		
A431	1959	1966	Billet									75,000	100,000
A615	1968	1972	Billet			40,000	70,000			60,000	90,000	75,000	100,000
A615	1974	1986	Billet			40,000	70,000			60,000	90,000		
A615	1987	Present	Billet			40,000	70,000			60,000	90,000	75,000	100,000
A16	1913	1966	Rail					50,000	80,000				
A61	1963	1966	Rail							60,000	90,000		
A616	1968	1999	Rail					50,000	80,000	60,000	90,000		
A160	1936	1964	Axle	33,000	55,000	40,000	70,000	50,000	80,000				
A160	1965	1966	Axle	33,000	55,000	40,000	70,000	50,000	80,000	60,000	90,000		
A617	1968	1999	Axle			40,000	70,000			60,000	90,000		
A996	2000	Present	Rail, Axle			40,000	70,000	50,000	80,000	60,000	90,000		
A706	1974	Present	Low-Alloy							60,000	80,000		
A955M	1996	Present	Stainless			40,000	70,000			60,000	90,000	75,000	100,000

IOP Conf. Series: Materials Science and Engineering **739** (2020) 012033 doi:10.1088/1757-899X/739/1/012033

4. Test Program

4.1. First Set of Mechanical Splices - Cyclic Tension Axial Load Test

A set of 4 spliced bars was tested under cyclic tension axial load. Two Long Mechanical Splice and two Short Mechanical Splices were tested [4]. The nomenclature used for each specimen is MS-L1 and MS-L2 for the two long mechanical splices and MS-S1 and MS-S2 for the short mechanical splice. Table 3 summarizes the specimens tested. ASTM specification of each bar is listed. Length of the specimen and extensometer used is also listed too. The type of steel bar was A-706 Grade 60 for all the new bars used in the specimens. It was studied the Type 2 mechanical splice for Short Mechanical Splices and Long Mechanical Splices. Figure 4a shows the test setup.

Table 3. Specimen details for the cyclic tension test.

	Condition		ASTM Specification	
Specimen	**Top Bar**	**Bottom Bar**	**Top Bar**	**Bottom Bar**
MS-L1	New	New	A-706	A-706
MS-L2	New	New	A-706	A-706
MS-S1	New	New	A-706	A-706
MS-S2	New	New	A-706	A-706

It was found that the SMS specimens failed at the contact of the point end bolt because the reduction of bar's area due the pointed bolt, however the rupture of LMS specimens was at the rebar 4in outside the sleeve since the rounded bolts do not reduce the bar's area as pointed ones do. Figure 5b shows failure pattern of both types of mechanical splices.

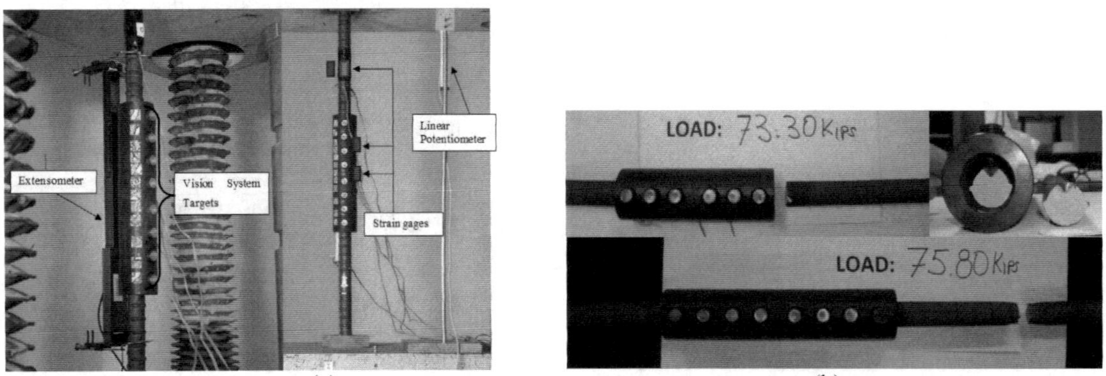

(a) (b)

Figure 4. Instrumentation for MS-L1 Test (a), Failure pattern for SMS and LMS (b) [5].

Table 4 presents the different values in loads response of each case of specimen. It can be seen that the value of yielding stress is similar for long and short mechanical splice. The fractures stresses are very similar among each other because the failure of each specimen was under the rupture of the steel bar. It is also noticed that both short and long splices meet the requirements of ACI318-11 and AASTHO for Type 1 and Type 2 splices.

Table 4. Results of first set of mechanical splice and acceptance criteria [4].

Area
Bar ϕ 1" 0.79 in2

Specimen	ASTM bar	Yield Stress of bar (fy) (ksi)	Fracture Stress of bar (fu) (ksi)	Type of Failure
MS-L1	A-706	60.92	96.75	Rupture of bar 4in above the splice
MS-L2	A-706	61.36	96.51	Rupture of bar 4in above the splice
MS-S1	A-706	62.22	87.83	Rupture of bar in edge of splice
MS-S2	A-706	62.54	93.33	Rupture of bar in edge of splice

Every mechanical splice meets the requirements for Type 2 and Type 1 splice

4.2. Second Set of Mechanical Splices - Compression-Tension Cyclic Axial Load Test

This set of tests includes 6 mechanical splices, 2 long mechanical splices and 4 short mechanical splices. In the two first specimens, new bars were used, and in the last 4 specimens with short mechanical splices, a new bar and a bar previously yielded were used together. This is reproducing the condition of a member which was repaired after suffered buckling of the longitudinal rebar. The nomenclature used for each specimen is MS-L3 and MS-L4 for the two long mechanical splices and MS-S3, MS-S4, MS-S5 and MS-S6 for the short mechanical splices. A-615 Grade 60 bars were used. Table 5 contains how previously deformed the rebar was for each specimen of SML and LMS. The tension-compression machine and the installed specimen is shown in Figure 5a.

Table 5. Results of first set of mechanical splice and acceptance criteria [4].

	Condition		ASTM Specification	
Specimen	Top Bar	Bottom Bar	Top Bar	Bottom Bar
MS-L3	New	New	A-615	A-615
MS-L4	New	New	A-615	A-615
MS-S3	New	4Yield	A-615	A-615
MS-S4	1.5Yield	New	A-615	A-615
MS-S5	4Yiel	New	A-615	A-615
MS-S6	New	1.5Yield	A-615	A-615

During the tests, there was no fracture of any part of the mechanical splices. The bolts behaved as one piece together with the mechanical splice. However, buckling of the bars appeared when the deformation reached -3□y at compression loads, Figure 5b shows the specimen MS-S3 after buckling of the bar that had been previously yielded. The mechanical properties of this bar were changed under the previous compression-tension axial load test. It was reached tension deflections more of +8□y and under compression deflections more than -3□y. The specimen failed by the fracture of the previously yielded bar in the contact zone with the last bolt of the mechanical splice as Figure 5c shows.

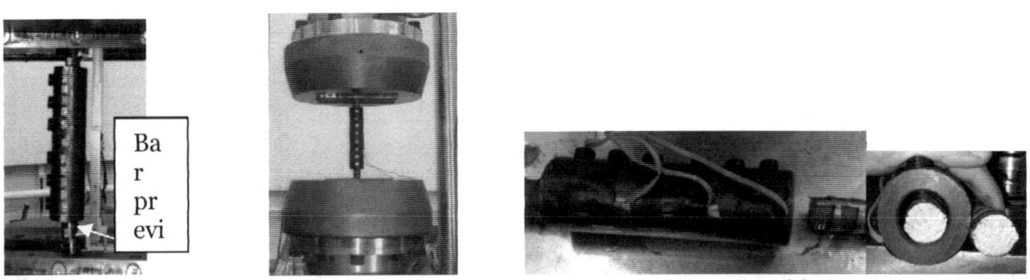

(a) (b) (c)

Figure 5. Test Setup for the compression-tension cycle test
and details of the mechanical splices tested [4] (a), Buckling failure under compression load [4] (b),
Failure under tension load: previously bar yielded was fractured [4](c).

It can be seen at Figure 6 the hysteretic response of the LMS and SMS specimens, where a drop of load while the direction of load is presented. It may be due la accommodation of both bars and the sleeves together to the bolts of the splice, especially the previously yielded bars. Table 6 summarizes the values of axial load and strain for tension and compression yielding loads and strain values

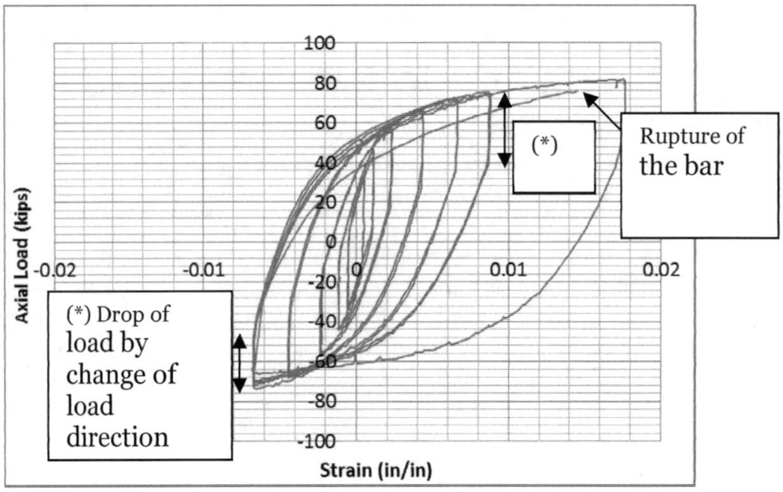

Figure 6. Strain Deformation of the system measured by the Extensometer [5].

Table 6. Results of first set of mechanical splice and acceptance criteria [4].

Specimen	Fracture Stress of bar (fu) (ksi)	Meet requirements for Type 1 ?	Meet requirements for Type 2 ?	Type of Failure
MS-L3	109.24	Yes	Yes	Rupture of bar above the splice
MS-L4	109.54	Yes	Yes	Rupture of bar above the splice
MS-S3	96.77	Yes	Yes	Rupture of 4 Yield bar on the edge of splice
MS-S4	85.75	Yes	No	Rupture of 1.5 Yield bar on the edge of splice
MS-S5	72.09	No	No	Shear failure of the bolts on the new bar zone
MS-S6	82.76	Yes	No	Rupture of 1.5 Yield bar on the edge of splice

#8 bar Grade60 A-615

5. Conclusions

For selection of mechanical splice for the rehabilitation of columns, short mechanical splice behavior proved to be suitable for hinge areas with large deformations and high forces especially where there may be difficulty installing the long couplers. These splices can satisfy Type 2 and Type 1 splice requirements if A-706 is used. However, the long mechanical splices should be used wherever there is sufficient room for installation and are suitable for both A-615 and A-706 bars.

More test data is recommended to optimize mechanical splices. The behavior of short splices might be improved if the pointed end of the last bolt is replaced with a rounded end. The reduction of the bar area will not be as great and the splice may meet the requirement for Type 2 splices using A615 bars.

It is also recommended that long mechanical splices using previously yielded bars and new bars be tested. Despite the fact that long mechanical splices performed adequately under high loads and large deformations, previously yielded bars were not tested in the long splices.

References

[1] ACI Committee, International Organization for Standardization 2008 Building Code Requirements for Structural Concrete (ACI 318-08) and commentary American Concrete Institute.

[2] American Society of Civil Engineers, *Structural Engineering Institute ASCE41-16/SEI Seismic*

Rehabilitation of Existing Buildings, Reston, Virginia. EEUU

[3] Concrete Reinforcing Steel Institute 2001 Evaluation of Reinforcing Bars in Old Reinforced Concrete Structures, *Eng. Data Rep.* 48.

[4] Huaco G, Jirsa J 2013 Procedures to Rehabilitate Extremely Damaged Concrete Members using Innovative Materials and Devices *Diss. PhD Dissertation. The University of Texas at Austin.*

[5] Huaco, G (2013): *Procedures to Rehabilitate Extremely Damaged Concrete Members using Innovative Materials and Devices,* Ph.D. Dissertation, The University of Texas at Austin. USA, 649pp

IOP Conf. Series: Materials Science and Engineering **739** (2020) 012034 doi:10.1088/1757-899X/739/1/012034

Gravel Sampling for Testing

Ayad Al-Rumaithi[1], Mohannad Al-Sherrawi[1]

[1]Department of Civil Engineering, University of Baghdad, Baghdad 17001, Iraq

Dr.Mohannad.Al-Sherrawi@coeng.uobaghdad.edu.iq

Abstract. Sampling is the selection of a representative portion of a material, and it's as important as testing. The minimum weight of gravel field or lab sample depends on the nominal maximum particle size. The weight of the sample will always be greater than that portion required for testing. The approximate precision desired for the testing will control the weight of the gravel sample. In this study, gravel sample has been simulated by using multilinear approximated function for Fuller's curve on the logarithmic scale. Gravel particles are divided into classes according to their medium diameter and each class was simulated separately. A stochastic analysis, by using 100 realizations in sampling, has been done and the root mean square error for the errors between sampled and target curve has been discussed for two selected samples of coarse aggregate.

1. Introduction

Gravel (coarse aggregate) is the natural or artificial inorganic granular material used with sand (fine aggregate) and a cementing medium to form concrete.

Gravel is a combination of stone particles between a maximum and a minimum sieve size. These sizes are varying in different specifications.

One of the main contributing factors to the quality of concrete is the quality of aggregate used therein. It is, therefore, imperative that due consideration is given to sampling procedures, which will assist in the accurate and objective evaluation of the quality of the sampled aggregate. Sampling is perhaps the most important step in assuring that good quality aggregates are being used.

A sample is a representative small portion from a larger amount of a material. It is essential that a truly representative sample of the material be obtained for testing purposes. Gravel sampling is of equal importance to its testing and a man drawing samples shall use every precaution to obtain a sample that will show the true nature and condition of the gravel, which it represents.

A field sample is supposed to be sufficiently a large amount of gravel for full representation of the whole supply. Specifications require only a portion of the field sample for the test sample.

The field sample should be reduced (split) to a sample size (test sample) that can be quickly tested. Two different methods are used to reduce a field gravel sample to the proper test size; mechanical splitter and quartering.

The test sample size must fall within the minimum and maximum weight limits provided by a specific specification [1].

In a gravel gradation, a sample of weighed dry gravel is separated through a series of sieves with progressively smaller openings. Once separated, the weight of retained particles on each sieve is valued and compared to the weight of gravel sample. Particle size distribution is then expressed as a percent retained by weight on each sieve size. Results are usually expressed in tabular or graphical format. Graphical displays use traditionally semi- logarithmic or the standard power gradation graph.

There is different gravel grading specified by specifications in the world. The adopted gravel grading is based on the uses and the availability. Gradation is measure of size range of gravel in specified range.

Several common terms are used to classify gravel gradation; fine gradation, coarse gradation, uniformly graded, open graded, gap graded and dense or well-graded (preferable for maximum density).

2. Aggregate simulation

In order to get a realistic representation of aggregate sample and confirm a minimum mass of test portion for sieve analysis, it should be simulated according to a grading curve. Grading curves can be either empirical from design codes or explicit functions such as Fuller's curve. In this study Fuller's curve was used with the equation shown below:

$$Passing\,\% = \left(\frac{d}{dmax}\right)^{\alpha} \quad (1)$$

Where Passing % is the passing percentage for (d) sieve size, dmax is the maximum aggregate diameter and α is Fuller's curve exponent which can be between 0.45 and 0.5 [2] or may be between 0.33 and 0.5 [3].

In the early 1960s, the American Federal Highway Administration used α equals 0.45 in the standard gradation graph [4].

Aggregate particles are assumed to be ellipsoids with three principal radii (r_1, r_2, and r_3) with $r_1 \geq r_2 \geq r_3$. The medium diameter $2 * r_2$ is the main diameter which is determinant for passing the aperture of the sieve. r_1 and r_3 are assumed to be related to r_2 by the following equations [5]:

$$r_1 = \left(1 + u_1 * \frac{m-1}{m+1}\right) * r_2 \quad , \quad 0 < u_1 < 1 \quad (2)$$

$$r_3 = \left(1 - u_3 * \frac{m-1}{m+1}\right) * r_2 \quad , \quad 0 < u_3 < 1 \quad (3)$$

Where u_1 and u_3 are realizations of independent uniform random variables. m is variable that determines the flatness of aggregate particles. If m=1 the particles will be spheres and higher values will increase flatness.

The mass of individual aggregate particles is: $m=4\rho\pi r_1 r_2 r_3$, where ρ is the density. r_2 should be chosen so the cumulative distribution for the particle masses follow the grading curve. In order to simplify the problem, the curve is approximated as multilinear function on the logarithmic scale with $2\times r_2$. In this case, particles are divided into classes according to their medium diameter and each class is simulated separately. Equation (4) is used in order to get linear distribution in logarithmic scale within each diameters class.

$$d_{eqv} = 2 * r_2 = \frac{d_i\, d_{i+1}}{\sqrt{u_2 d_i^3 + (1-u_2) d_{i+1}^3}} \;, 0 < u_2 < 1 \quad (4)$$

Where u_2 is the realization of uniform random variable ranging from 0 to 1. d_i and $d_{(i+1)}$ are the boundaries of aggregate diameters class (i).

The mass of each aggregate diameter class should be:

$$m_i = \frac{\left(\frac{d_i+1}{dmax}\right)^{\alpha} - \left(\frac{d_i}{dmax}\right)^{\alpha}}{1-\left(\frac{dmin}{dmax}\right)^{\alpha}} * m_{total} \quad (5)$$

where dmin is the minimum diameter of aggregate involved and m_{total} is the total mass of aggregate.

The process can be summarized as: The mass proportion is determined for highest diameters class using equation (5). Then individual aggregates are simulated and their mass is computed using equations (1-4). The procedure is repeated until the required mass is exceeded. The difference between the assigned mass and the simulated mass is subtracted from the mass of the next aggregate class to ensure total mass will not exceed. Then the process is repeated for the next diameter class.

3. Simulation results and sampling study

3.1. Sample Simulation

Two samples of coarse aggregate were simulated with the parameters in table 1, by using the method that described earlier. The grading of the selected samples is according to ISS 45 [6].

Table 1. Coarse Aggregate Simulation Parameters.

	5-20 Aggregate	**5-40 Aggregate**
$dmin\ (mm)$	5	5
$dmax\ (mm)$	20	40
α	0.5	0.5
m	3	3
$\rho\ (g/mm^3)$	2.6	2.6
$m_{total}\ (kg)$	100	100

Figure 1 shows the distribution curves for the multilinear approximation of Fuller's curve (Target Curve) and the simulated curves. It shows high level of matching.

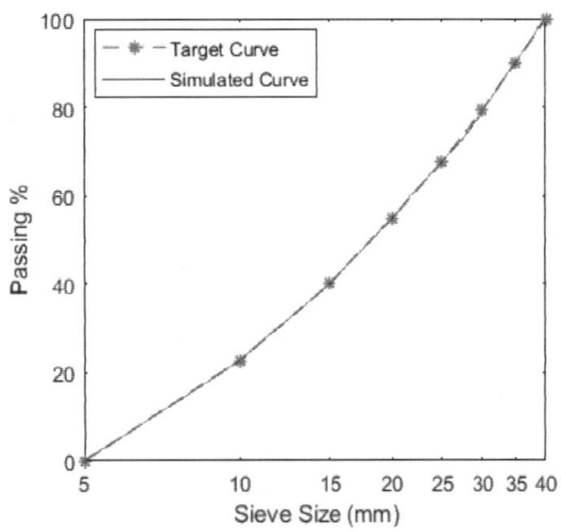

a) 5-20 Aggregate

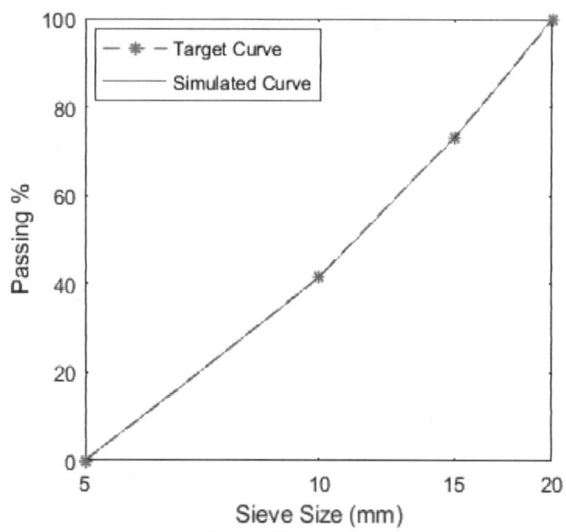

b) 5-40 Aggregate

Figure 1. Target vs. Simulated Curve.

3.2. Sampling Study

In order to study the effect of suggested sampling on the distribution of coarse aggregate, the two samples were sampled by taking random gravel particles from the sample till reaching a certain sampling weight. The sampling weights were 0.5 kg, 1 kg, ...10 kg for 5-20 aggregate and 1 kg, 2 kg, 20 kg for 5-40 aggregate. Figure 2. a) and b) shows the distribution of sampled gravel for some different weights.

Figure 2. c) and d) shows the difference between the sampled and target curve (error). It is clear from the figures that as the sample weight increases the error decreases. Also the error is greater in higher sieve sizes than small sieve size due low number of large gravel particles.

Time will not allow testing the total gravel. The key to sampling is to ensure that the weight of sample taken representative of the material in the stockpile with acceptable error. The approximate precision accepted by the lab for the testing will control determination of the weight of the gravel sample.

3.3. Stochastic Analysis (100 Realizations)

Since the stock pile simulation and sampling study described earlier are random processes, it is important to repeat them many times to get the statistics. So, 100 realizations have been repeated with each realization having curves different than curves in figure 1 and 2 figure 3 shows how distribution of sampled weight can be different for a three realizations for the two samples.

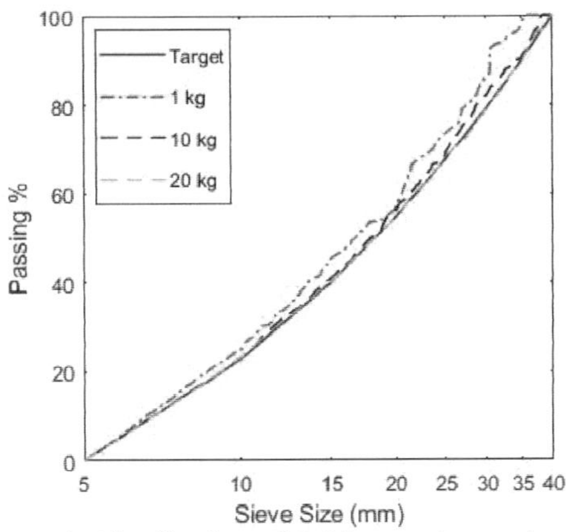

a) Distribution of 5-20 gravel samples

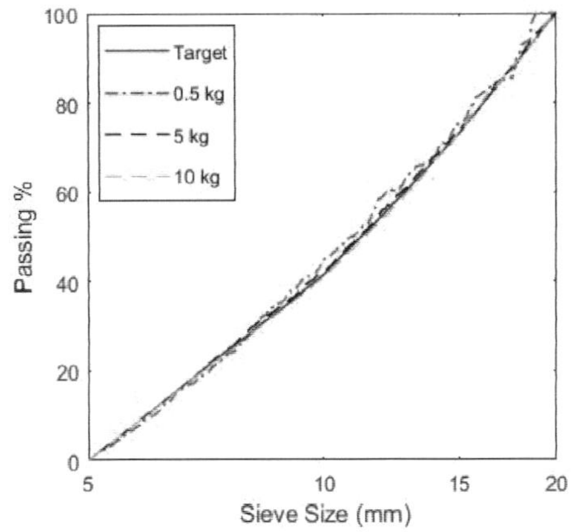

b) Distribution of 5-40 gravel samples

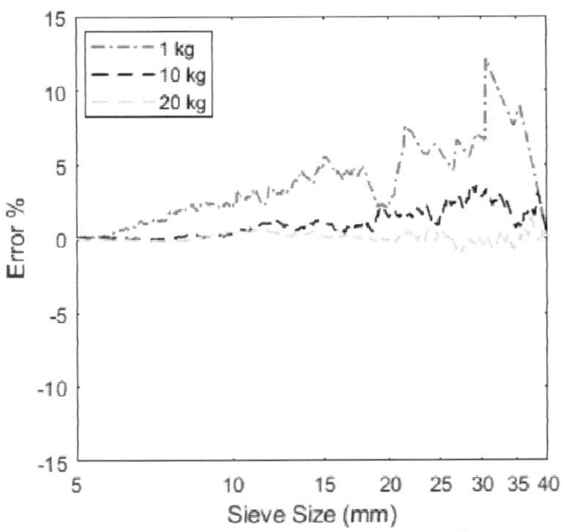

c) Error of 5-20 gravel samples

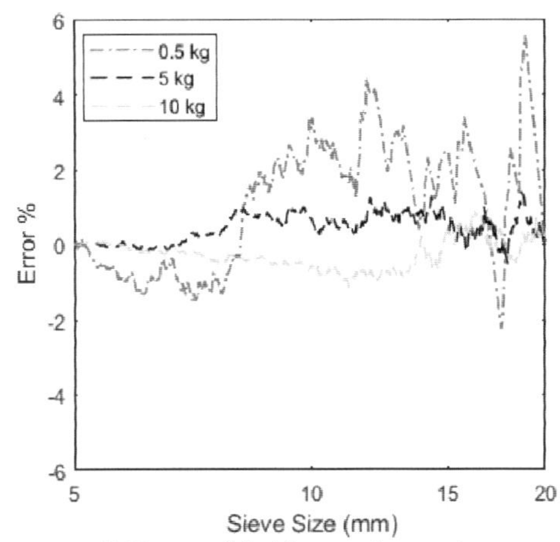

d) Error of 5-40 gravel samples

Figure 2. Distribution and error of the sampled gravel.

It is irrational to show all the 100 realizations and their errors, so it is more convenient to find root mean square error (RMSE) for the errors between sampled and target curve. It is clear that RMSE decreases with the increase of sampling weight and it is greater for higher diameters. Figure 4. shows how RMSE changes with diameter can be different for three realizations.

IOP Conf. Series: Materials Science and Engineering **739** (2020) 012034 doi:10.1088/1757-899X/739/1/012034

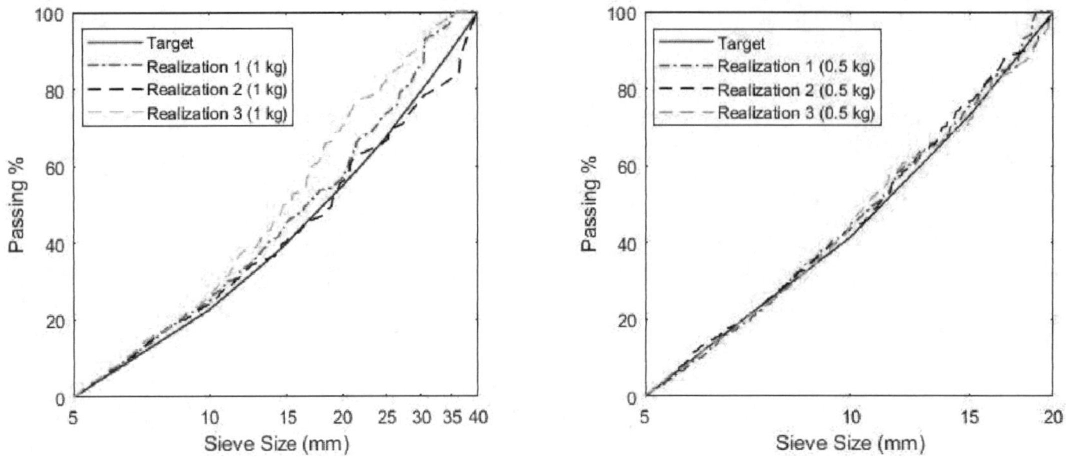

a) 5-20 Aggregate b) 5-40 Aggregate

Figure 3. Distribution of sampled weight for three different realizations.

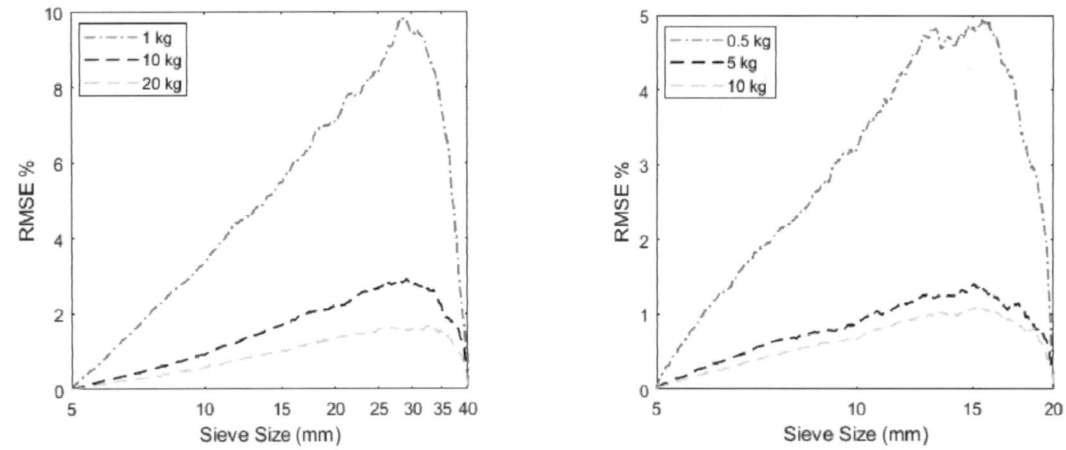

a) 5-20 Aggregate b) 5-40 Aggregate

Figure 4. RMSE change with diameter.

To see the effect of sampling weight on RMSE, the sampling weight is plotted against maximum values in RMSE curves as shown in figure 5.

The plotted results show that the proposed multilinear approximated function for Fuller's curve produces a good estimate of gravel sample weights. It achieves a decreasing RMSE for the two samples.

Theoretically, there exists a particular aggregate gradation that, for a given nominal maximum particle size, will yield the maximum material density. This gradation would involve a particle arrangement where successively small particles fill the spaces between the larger particles.

The approximate precision desired for the estimate should be prescribed by the sampler. That is, it must be decided what maximum deviation can be tolerated between the estimate to be made from the sample and the result that would be obtained by measuring every unit in the lot or process.

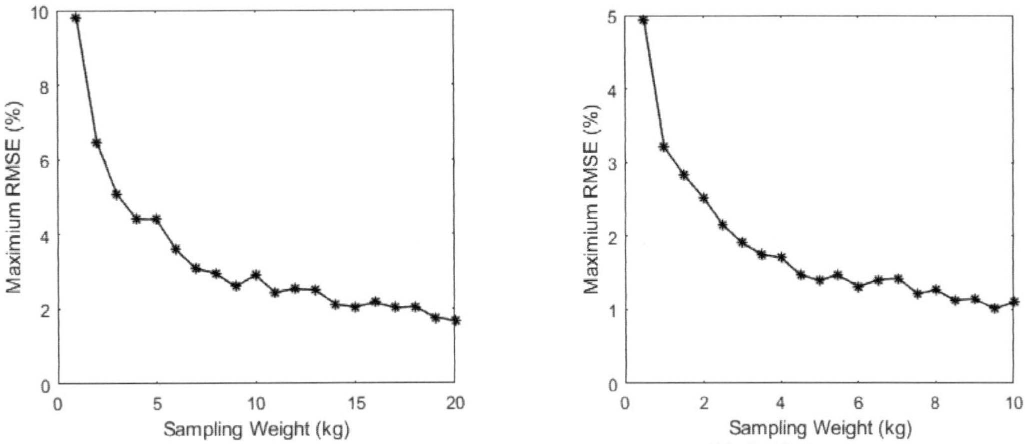

a) 5-20 Aggregate b) 5-40 Aggregate

Figure 5. Maximum RMSE change with sampling weight.

4. Conclusions

A multilinear approximated function for Fuller and Thompson's equation on the logarithmic scale has been developed in random processes to simulate gravel test sample for sieve analysis (grading). Gravel particles are assumed to be ellipsoids and the medium diameter of the gravel particles has been employed in the sampling procedure.

The desired precision for the gravel sieve analysis results and the nominal maximum particle size will control the weight of the gravel sample.

References

[1] AASHTO R 90, Standard Practice for Sampling Aggregate Products.

[2] Fuller W B, Thompson S E 1907 The laws of proportioning concrete, Transactions of the ASCE, 159.

[3] Andreasen A H M and Andersen J 1930 Über die Beziehung zwischen Kornabstufung und Zwischenraum in Produkten aus losen Körnern (mit einigen Experimenten). Colloid & Polymer Science, Vol. 50 (3), pp. 217-228.

[4] Roberts F L, Kandhal P S, Brown E R, Lee D Y, and Kennedy T W 1996 Hot Mix Asphalt Materials, Mixture Design, and Construction. National Asphalt Paving Association Education Foundation. Lanham, MD.

[5] Unger Jörg F, Stefan Eckardt 2011 Multiscale modeling of concrete, Arch Computat Methods Eng. 18:341.

[6] Iraqi Standard Specification No 45, Natural Sources for Gravel that is Used in Concrete and Construction, Baghdad-Iraq, 1984.

IOP Conf. Series: Materials Science and Engineering **739** (2020) 012035 doi:10.1088/1757-899X/739/1/012035

Bending Strength of Polyester Composites Reinforced with Stitched Random Orientation and Plain Weave Abaca Fiber

M Iqbal[1], Y Aminanda[2], T Firsa[1], M Ali[1]

[1]Department of Mechanical and Industrial Engineering, Faculty of Engineering, Universitas Syiah Kuala, Banda Aceh, Indonesia
[2]Mechanical Engineering department, Universiti Teknologi Brunei

mohd.iqbal@unsyiah.ac.id

Abstract. This paper discusses the bending strength of Fiber Reinforced Polymer (FRP) composite made from polyester resin matrix and abaca fiber. The abaca fiber was prepared in the form of stitched random orientation and plain weave woven fabric. The abaca FRP composite was produced by using hand-lay-up method with the variation of weight percentage of the fiber. The weight density of the FRP composite was measured to determine the effect of fiber percentage to the weight density of the composite panel. The three points bending test was performed in order to examine the maximum bending force and flexural strength of the abaca FRP composite. The result of the experiment shows that the fiber weight percentage gave significant effect to the weight density and flexural strength of the abaca composite. Furthermore, the comparison between the two types of fiber form shows that the abaca plain weave woven fiber produces higher flexural strength compare to stitched random orientation woven fiber at all levels of fiber contents.

1. Introduction

The use of natural composite in industry has been growth in the present years due to their advantages such as higher specific mechanical properties, lower weight, better insulating and thermal resistance, lower cost and biodegradability compare to synthetic fiber [1-3]. The amount of natural composite used in America and Europe has increased more than 20% in 2010 to 2015 compared to the earlier five years range. The applications of natural fiber reinforcements have grown faster as well due to the improvement of manufacturing technology and process treatments. As first being produced to building interior panels in 1950s, lately the natural fiber has been used to produce car door panels (from flax-sisal fiber mat reinforced epoxy), car rear shelf trim panels (from flax fiber reinforced polypropylene composite), ceiling and partition boards in structural applications [4, 5].

The composition properties of natural fiber and polymer synthetic matrix material significantly contributes to the performance of the natural fiber reinforced composite. The fiber mostly contributes to the strength and stiffness of the structural load whereas the matrix contributes to the shape, surface and environment resistance. Many researchers reported that adding the fiber content increasing the strength of the FRP composite. However, the fiber is bounded by the polymer matrix, therefore increasing the fiber content decreases the fiber-matrix bonding [6].

Abaca fiber is a bast fibre and is obtained from the pseudo-stem of banana plant (Musa textiles) that belongs to the Musacea family of plants [7]. The fibre is extracted from the trunk of the plant and went several manual processes, i.e. decorticating, washing, drying and cleaning. The banana plant growths naturally 2 m to 4 m tall in tropical forests (see figure 1. a), in Filipina, Indonesia, Thailand and Vietnam. Traditionally, it has been used to make twins and ropes for hundreds years but nowadays it is gaining importance as a good material for tea bag, bank note, food pack and reinforcement for composite material. Abaca fiber is also known as Manila hemp since Filipina supplies almost 60% of the world market demand of abaca fiber. Indonesia, Thailand, Vietnam and other countries share the rest of 40% market demand.

The development of fiber structure become major method to improve the performance of fiber reinforced composites. Stitching, knitting, braiding and weaving has become alternative technologies in increasing through thickness mechanical properties of laminate composites, improving the impact resistance of fiber reinforced composites and reducing fabrication cost [6, 8, 9].

The study by Ghazilan et al [10] shows that the tensile properties of stitched empty fruit bunch oil palm reinforced epoxy composite has better performance in both tensile strength and tensile modulus compared to unstitched one. Reis et al [11] reported that fiber orientation affected the growth of delamination and the extension shear coupling effects. The delamination growth divides the composite into sublaminate, resulting in a higher occurrence of delamination buckling failures. The extension shear coupling effects induces the damage accumulation in the matrix, which result in the non-linear compressive response of the composite subjected to off-axis loading conditions. Dahale et al [12] studied the effect of weave parameters to the mechanical properties of woven glass composites. They reported that higher weft density specimens showed a more compact structure with less crimp that resulted in higher tensile and compressive properties of the weave fiber composite.

(a) (b)

Figure 1. (a) The banana tree in Aceh, Indonesia (b) the abaca fiber.

(a) (b)

Figure 2. The abaca fiber (a) stitched random orientation, (b) plain weave woven.

(a) (b)

Figure 3. The fabrication of abaca FRP composite (a) the mould (b) the process.

This study aimes to investigate the bending strength of abaca Fiber Reinforced Polymer composite. The fiber was prepared in two different types, i.e. the stitched random orientation and plain weave woven type in order to compare the bending strength of the composite configuration from two different fiber types. The fiber content varies as 17, 19 and 21 of weight percentage. The weight density and the bending strength of each weight percentage level was investigated to find out the fiber weight percentage that gives the highest bending strength.

2. Methods

The abaca fiber was supplied by CV Royal Nanggrou Diamond, Aceh, Indonesia. The fiber was originally from the banana field that naturally growth in Lhok Nibong, North Aceh District, Aceh Province, Indonesia, as shown in figure 1 (a) and 1 (b). The fiber was manually produced by local farmers using decorticator machines, washed in well or river water and dried naturally under the open sun light. No chemical treatment was applied to the fiber in both the farm processes and laboratory experiment. According to the regular users and traders, the said fiber is one of the best quality abaca fiber that produced in Aceh.

Two forms of abaca fiber were prepared for the filler of the abaca FRP composite, i. e. stitched random orientation and plain weave woven form, as shown in figure 2(a) and 2(b) respectively. The stitched random orientation fabric was made by laying the raw abaca fiber in random orientation. The fiber then sewed using cotton sewing thread to keep the fiber uniform in similar size with the composite mould. Sewing lines provided only to one direction with 50 mm interval in order to make sure that the sewing thread does not affect the fracture force of the abaca composite. The plain weave woven fabric was made by manual weaving process using 300 mm x 300 mm wood frame. The yarn consists of one layer of abaca fibers with 5 mm width. The weaving process was conducted line by line to produce intermittent pattern with 90^0 orientation.

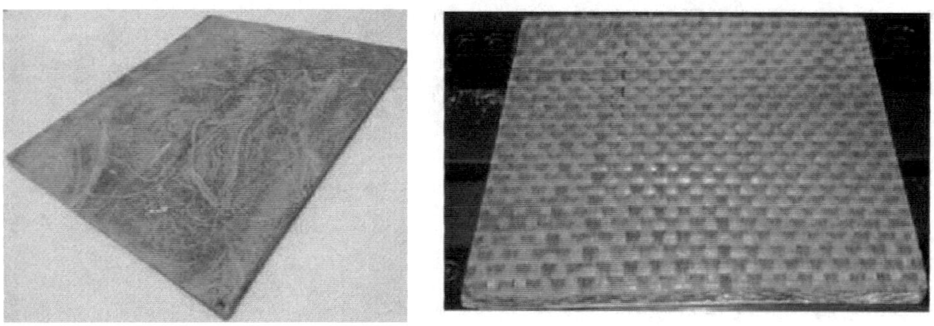

(a) (b)

Figure 4. Abaca composite panels (a) with stitched random orientation fiber, (b) plain weave woven fiber.

(a) (b)

Figure 5. Samples of abaca composite specimen (a) with stitched random orientation fiber, (b) plain weave woven fiber.

The FRP composite was made using hand-lay-up method in an open mould. The mould was made from acrylic with the open pocket dimensions of 200 mm x 170 mm x 10 mm depth, as shown in figure 3(a). The process started by applying a thin layer of wax to the surfaces of the mould pocket. Polyester resin (with 4% of hardener) was poured into the mould until 1 mm thickness. The fiber then was put into the mould followed by resin until the mould pocket was full, as shown in figure 3(b). The fiber and resin was chopped smoothly using a soft painting brush to make sure the fiber and resin were mixed uniformly.

The cover acrylic plate was put on the top of the mould pocket. Five kg of load was put on the cover plate to press it down to the target thickness (5 mm). Some of the resin licked out from the mould pocket through the small gap between the cover plate and the side plate of the mould pocket. The bubbles inside the resin were removed by sliding a yarn under the cover plate. After natural curing process for 8 hours, the abaca composite could be easily removed from the mould pocket. The abaca composite panel with random fiber and woven fiber were shown in figure 4(a) and 4(b) respectively.

Three different levels of fiber weight has been prepared for the composite panel, i.e. 30 gr, 40 gr and 50 gr, resulting the composite panels with the fiber weight percentage of 17.11%, 18.9% and 20% respectively. Each panel was cut into 9 specimens with 165 mm x 19 mm x 5 mm as per ASTM D790 using Bosch GDC 120 semi auto cutting machine. Samples of abaca composite specimen from random fiber and woven fiber are shown in figure 5 (a) and 5 (b) respectively. Three specimens were selected randomly from each group for weight density measure and bending test. The average of the three measures was taken as the result.

The weight density of the composite specimen was calculated using the following equation:

$$\rho = m / v \quad (1)$$

Where ρ is density (gr/ m3), m is the mass of the specimen that measured using weighing scale, and v is the volume of the specimen that measured using standard scale glass. Three points bending test was carried out at Hung Ta HT-8503 universal testing machines with constant speed of 22 mm/min, as shown in figure 6. The applied flexural stress was automatically recorded during the bending test and the stress vs strain of the test could be plot for further analysis.

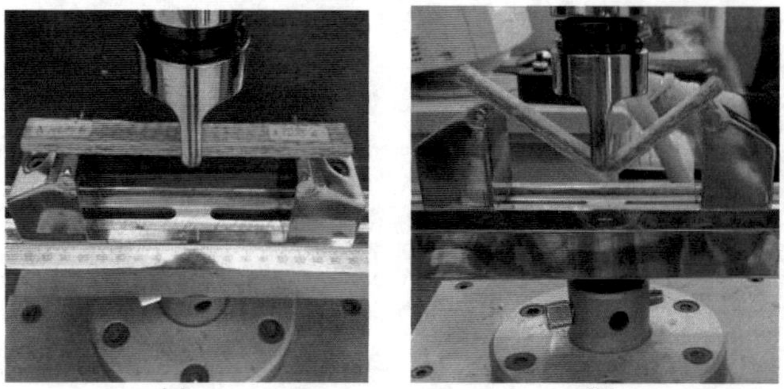

(a) (b)

Figure 6. Three points bending test (a) the starting position (b) the ending position.

Table 1. Result of the experiment.

No	Fiber Content (%)	Weight Density (gr/m^3)	Flexural Strength (MPa)	
			Stitched random orientation fiber	Plain weave woven fiber
1	17	1,260	50,11	54,01
2	19	1,150	34,11	41,48
3	21	1,090	27,95	31,38

The flexural strength σ_f (maximum bend stress) was calculated for test specimen using the following equation:

$$\sigma_f = (3\ F_m\ L) / (2\ h\ d^2) \quad (2)$$

Where Fm is the applied flexural force, L is the distance between the support points (110 mm for all the testes), h is the specimen width (19 mm for all the specimen) and d is the specimen thickness.

3. Result and Discussions

Based on the result of the experiment, the weight density and the flexural strength were calculated, as presented in table 1. The density of the abaca composite panels was greater than 1 for all of the fiber content levels, which is higher than the density of abaca fiber. However, they varied from one level to the others. In order to show the effect of the fiber content, the density was plot into a graph as shows in figure 7(a). It was clearly seen that the increasing of the fiber content reduced the density of the abaca composite. It is because the density of the abaca fiber is lower than the density of the polyester resin and increasing the fiber content will decrease the resin content in the composite panel.

Figure 7(b) shows the flexural strength of the abaca FRP composite with stitched random orientation fiber (blue line), and abaca FRP composite with plain weave woven fiber (red line) for the three fiber content levels. It is clearly seen that the fiber content gave significant effect to the flexural strength of the abaca composite panels for both fiber forms. The increasing of the fiber content decreases the flexural strength of the abaca composite panel. The possible reason was because the properties of the abaca fiber is lower than the properties of the polyester matrix, therefore higher content of fiber decreases the flexural strength. Contrary, when the properties of the fiber is higher than the properties of the matrix, higher content of fiber increases the flexural strength. As reported in references [5, 13, 14], the authors used glass or carbon fiber where its properties are higher compared to the polyester or epoxy matrix, therefore higher content of fiber increases the flexural strength of the composite. However, more experiments are need to be conducted in the future, in order to find the optimum fiber content for the better mechanical properties of the abaca composite panel.

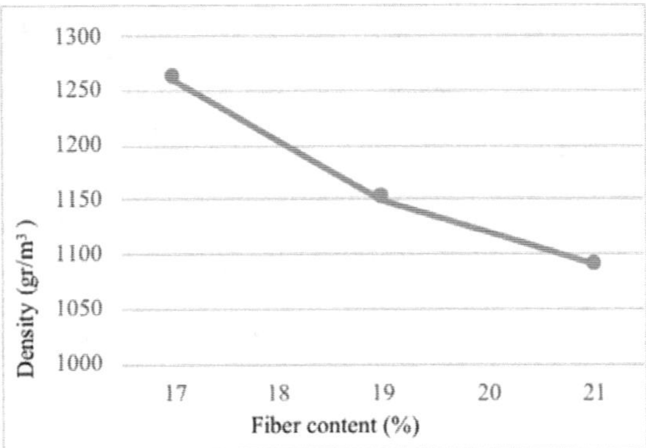

(a)

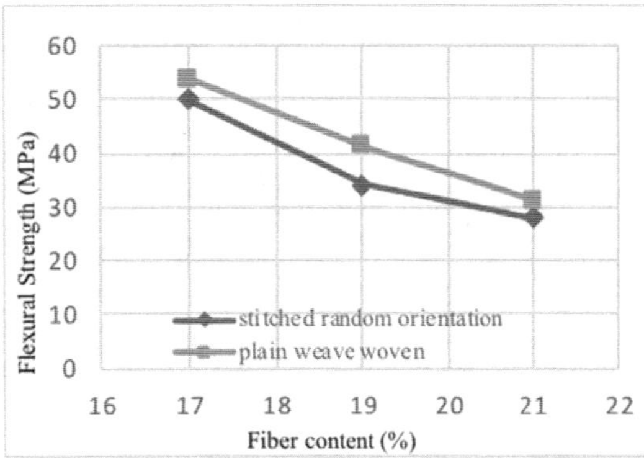

(b)

Figure 7. Effect of fiber content to (a) the density (b) the flexural strength of the abaca composite.

Figure 7(b) also shows the comparison between the flexural strength of abaca FRP composite with stitched random orientation fiber and abaca FRP composite with plain weave woven fiber. It could be seen that the flexural strength of plain weave woven fiber is higher than the flexural strength of stitched random orientation one at all fiber content levels. It is because the weaving pattern in woven form produce a higher bounding force among the fiber compare to the one in the random form [6, 12]. However, plain weave woven fiber requires more production process, longer production time, and extra production cost. In the case where the flexural strength is sufficient for product requirements, the stitched random orientation fiber would be economically recommended due to a small difference of flexural strength with woven one.

4. Conclusion

The study of the bending strength of abaca FRP composite has been reported. The result of the experiment show that the bending strength decrease continuously when the fiber content increase from 17% to 21%. The comparison between different fiber woven type shows that the plain weave woven fiber has higher bending strength compare to the stitched random orientation type. The authors wish to thank Ministry of Research, Technology and Higher Education Republic of Indonesia and Syiah Kuala University, Banda Aceh, Indonesia for financial support for this research project.

References

[1] Dittenber D B, Rao H V 2012 Critical review of recent publications on use of natural composites in infrastructure, *Comp. Part A: App. Sci. Manuf.* **43(8)** 1419-29.

[2] Iqbal M, Bahri S, Akram A 2019 Effect of cutting parameter on tool wear of HSS tool in drilling of Kevlar composite panel, *IOP Conf. Ser. Mater. Sci. Eng.* **523(1)** 012078.

[3] Nazaruddin N, Akram A, Hasanuddin I, Iqbal M, Kurniawan R, Putra R 2019 Mechanical properties of glass fiber reinforced polyester resin for use as the wall of the Acehnese boat 'Thep-Thep', *IOP Conf. Ser. Mater. Sci. Eng.* **523(1)** 012080.

[4] Bourmauda A, et al. 2019 Main criteria of sustainable natural fibre for efficient unidirectional biocomposites, *Comp. Part A: App Sci. Manuf.* **124** 105504.

[5] Pickering K L, Efendy M G A, Le T M 2016 A review of recent developments in natural fibre composites and their mechanical performance, *Comp. Part A: App. Sci. Manuf.* **83** 98-112.

[6] Uhlig K, Bittrich L, Spickenheuer A, Humberto J, Almeida Jr S 2019 Waviness and fiber volume content analysis in continuous carbon fiber reinforced plastics made by tailored fiber placement, *Comp. Struct.* **222** 110910.

[7] Summerscalesa J, et al. 2010 A review of bast fibres and their composites. Part 1–Fibres as reinforcements, *Comp. Part A: App. Sci. Manuf.* **41(10)** 1329-35.

[8] Zhang S, Caprani S, Heidarpour A 2018 Influence of fibre orientation on pultruded GFRP material properties, *Comp. Struc.* **204** 368-77.

[9] Wang H W, et al. 2014 Analysis of effect of fiber orientation on Young's modulus for unidirectional fiber reinforced composites, *Comp. Part B: Eng.* **56** 733-9.

[10] Ghazilan A L A, Mokhtar H, Dawood M S I S, Aminanda Y, Ali J S M 2017 Tensile Mechanical Property of Oil Palm Empty Fruit Bunch Fiber Reinforced Epoxy Composites, *IOP Conf. Ser. Mater. Sci. Eng.* **184(1)** 012046.

[11] Reisa V L, Opeltc C V, Cândidoa G M, Rezendec M C, Donadona M V 2018 Effect of fiber orientation on the compressive response of plain weave carbon fiber/epoxy composites submitted to high strain rates, *Comp. Struc.* **203** 952-9.

[12] Dahalea M, Nealea G, Lupicinib R, Casconeb L, McGarriglea C, Kellya J, Archera E, Harkin-Jonesa E, McIlhagger A 2019 Effect of weave parameters on the mechanical properties of 3D woven glass composites, *Comp. Struc.* **223** 110947.

[13] Glóriaa G O, et al. 2017 Bending test in epoxy composites reinforced with continuous and aligned PALF fibers, *J. Mater. Res. Technol.* **6(4)** 411-6.

[14] Ramnath B V, et al. 2013 Evaluation of mechanical properties of abaca–jute–glass fibre reinforced epoxy composite, *Mater. Des.* **51** 357-66.

Numerical Simulation and Experiment of a MEMS Ultrasonic Cavitation Model

Zihong Liu[1, 2, 3], Hui You[1, 3*], Liangchao Li [2], Ronghui Lin[1], Ping Zhang[1]

[1]Department of Precision Instrumentation and Machinery, University of Science and Technology of China, Hefei 230026, Anhui, China
[2]Key Laboratory of Testing Technology for Manufacturing Process of Ministry of Education, Southwest University of Science and Technology, Mianyang 621010, Sichuan, China
[3]Institute of Intelligent Machines, Chinese Academy of Science, Hefei 230031, Anhui, China

usmlhy@iim.ac.cn

Abstract. This paper aims at analyzing the microbubble ultrasonic cavitation from the designed Bowl-shaped piezoelectric transducer in a micro channel. The cavitation conditions were derived and three methods of analysis were employed: (i) numerical analysis of piezoelectric parameters for the bubble cavitation using 2D piezo-acoustic coupling modelling. (ii) numerical analysis of microbubble dynamics and activities under ultrasound irradiation using 2D CFD simulation. (iii) experimental analysis of ultrasonic cavitation using designed micro-fluid chip by acoustic chemistry method. With the cavitation condition, the cavitation phenomenon of a lone micro bubble was analysed by the classic mathematical modelling, by which the piezoelectric transducer was optimized for ultrasonic cavitation in the micro channel. From the CFD results, the velocity of acoustic streaming increased to 58.6m/s, and the movement of the bubble wall was affected by the ultrasonic characters, which was in good agreement with the literature. The ultrasonic cavitation was proved to exist in the micro channel by the acoustic chemistry experiment.

1. Introduction

Cavitation [1, 2] is used to increase transiently the cell membrane permeability [3, 4] during exposure to ultrasound. The efficiency of the sonoporation is enhanced by the microbubble cavitation [5-7], which is a potential method used for gene delivery and drug delivery. The microbubbles are useful for improving the efficiency of the vivo sonoporation [8]. And the mechanisms of sonoporation caused by microbubbles has been revealed [9-12], the cell perforation and repair have been studied [13, 14]. However, most of the previous researches were carried out either in macro volumes with conventional ultrasonic equipment or in micro experiment model; there have been few studies on a Micro-Electro-[Mechanical-System(MEMS) device promising lab-on-chip approach to ultrasonic plasmid or DNA delivery, and the numerical analysis with the bubbles cavitation in the micro channel of the lab-on-chip device has not been elucidated.

To elucidate the mechanisms and the effect of the device parameters on the ultrasonic cavitation in the micro channel, we studied a MEMS device. In the micro channel, the core element of the MEMS device is a MEMS ultrasound transducer, which is designed to be a Bowl-shaped ZnO thin film structure to obtain a strong focusing effect. To analytically proof the feasibility of ultrasonic cavitation in the micro channel, the necessary and sufficient condition of the ultrasonic cavitation was deduced by using Blake threshold pressure and the Keller-Miksis equation [15]. For the necessary and sufficient condition of ultrasonic cavitation, the optimized geometric and physical parameters of the ultrasonic transducer was determined by simulated analysis using multiphysics soft COMSOL 5.3.

In the numerical simulation of the ultrasonic cavitation, it is assumed that there is a full-compressible gas dynamic in the bubble, the motion of the bubble wall was described by the Navier-Stokes equation and continuity equation in the micro-fluid channel. The boundary conditions near the moving bubble wall in the micro channel was set according to Rayleigh-Plesset equation and fluid-structure interaction in CFD simulation.

According to the optimized parameters and the analysis of the numerical simulation, the MEMS device was manufactured, on which the ultrasonic cavitation was proved to exist in the micro channel by acoustic chemistry experiment.

2. Methodology

2.1. Experimental model

Figure 1 shows the MEMS cavitation model, which utilizes ultrasonic cavitation to form reversible porosity in cell membrane for plasmid or DNA drug transfer. The core element of the system is a concave sphericity(Bowl-shaped) piezoelectric ultrasound transducer situated on bottom of the micro channel. The cross-sectional dimensions of the micro channel are 400um*400um.

In the cavitation experiment, the given solution is delivered into the micro channel by a peristaltic pump, and is collected by a collector. The micro bubble in the solution grows and compresses following the ultrasound in the micro channel. And the cavitation comes out as the micro bubble stored enough energy when compressed small enough.

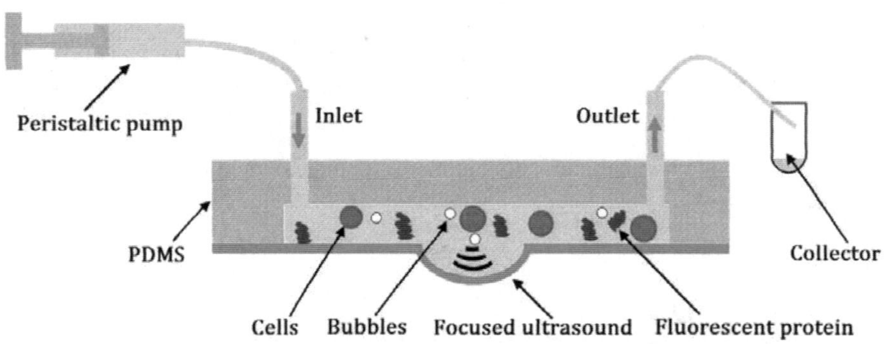

Figure 1. The cavitation experimental model of a micro-fluid system.

2.2. Mechanistical modeling base

According to the van der Waals equation [16], the pressure in or out the micro bubble is equal when the bubble is in equilibrium.

$$P_V + P_g\left(\frac{R_0}{R}\right)^3 = P_h + \frac{2\sigma}{R} \tag{1}$$

IOP Conf. Series: Materials Science and Engineering **739** (2020) 012036 doi:10.1088/1757-899X/739/1/012036

Where, P_V is the vapour pressure in the bubble, P_g is the gas pressure in the bubble, P_h is the fluid pressure near the bubble, and σ is the coefficient of surface tension. The critical radius of the bubble in equilibrium is obtained by taking the derivative of the R in equation (1). The critical pressure P_k of the fluid near the bubble is given by bringing the critical radius into equation (1):

$$P_k = P_V - \frac{2}{3}\left[\frac{\left(\frac{2\sigma}{R_0}\right)^3}{3P_g}\right]^{\frac{1}{2}} \tag{2}$$

By taking $R = R_0$ and $P_V = 0$, P_k is calculated through equation (1) and equation (2):

$$P_k = -\frac{2}{3}\left[\frac{\left(\frac{2\sigma}{R_0}\right)^3}{3(P_h + \frac{2\sigma}{R_0})}\right]^{\frac{1}{2}} \tag{3}$$

P_k is negative in equation (3), which means that the negative pressure acting on the bubble is necessary for generating the bubble with initial radius R_0. Make $P_k = P_h - P_B$, the Blake threshold pressure P_B is defined by equation (3):

$$P_B = P_h + \frac{2}{3}\left[\frac{\left(\frac{2\sigma}{R_0}\right)^3}{3(P_h + \frac{2\sigma}{R_0})}\right]^{\frac{1}{2}} \tag{4}$$

In the study, the radial dynamics of the micro bubble was described by the Keller-Miksis equation. In this model, liquid compressibility and radiation dumping were considered through the time derivative of gas pressure [16]:

$$\frac{P_V + P_g\left(\frac{R_0}{R}\right)^3 - P_h^{'}}{\rho} = R\frac{d^2R}{dt^2} + \frac{3}{2}\left(\frac{dR}{dt}\right)^2 + \frac{4\mu}{R}\frac{dR}{dt} + \frac{2\sigma}{\rho R} \tag{5}$$

In this equation, $P_h^{'}$, ρ and μ denote fluid pressure near the bubble wall, fluid density and shear viscosity of the fluid, respectively. Considering the effect of the ultrasound in the fluid, $P_h^{'}$ $(= P_h - P_a)$ is depended on the static pressure (P_h) and the ultrasonic pressure (P_a). where, P_a $(= -P_A \sin(\omega t))$ is determined by the acoustic pressure amplitude P_A and the acoustic pressure frequency $f(=\frac{\omega}{2\pi})$.

When ignoring the shear viscosity of the fluid, equation (5) is denoted:

$$R\frac{d^2R}{dt^2}+\frac{3}{2}\left(\frac{dR}{dt}\right)^2=\frac{1}{\rho}\left[\left(P_h+\frac{2\sigma}{R_0}\right)\left(\frac{R_0}{R}\right)^3-\frac{2\sigma}{R_0}-P_h+P_A\sin\omega_a t\right] \tag{6}$$

The equation (6) is expended to the $\frac{1}{R_0}$ power when the radial of the micro bubble is R $(=R_0+r,\ r<R)$:

$$r^2+\omega_r^2 r=\frac{P_A}{\rho R_0}\sin\omega_a t \tag{7}$$

Where, ω_r $(=2\pi f_r)$ is the resonance frequency. The normal resonance frequency (f_r) was given by Minneart according to equation 7[17]:

$$f_r=\frac{1}{2\pi R_0^2}\left[\frac{3}{\rho}\left(P_h+\frac{2\sigma}{R_0}\right)\right]^{\frac{1}{2}} \tag{8}$$

The cavitation is obvious when the ultrasonic frequency (f_a) is equal to f_r. The change in radial could be obtained by equation 7:

$$r=\frac{P_A}{\rho R_0\left(\omega_r^2-\omega_a^2\right)}\left(\sin\omega_a t-\frac{\omega_a}{\omega_r}\sin\omega_r t\right) \tag{9}$$

According to equation 9 and equation 3, for smaller bubble, the bubble could collapse as $f_a<f_r$ and $P_A>P_B$.

2.3. Computational Fluid Dynamic

2.3.1. Ultrasonic in the micro fluid. Figure 2 shows the 2D geometric model of the axially symmetrical micro channel, in which the cross section of the channel is extended to 1000um*400um for numerical analysis of geometry parameters. The piezoelectric domain is made of the crystal ZnO, which is a common material in MEMS piezoelectric transducers. An AC electric potential of 100V is applied to the inside of the of the Bowl-shaped, while the outside is grounded. At the interface between the water and piezoelectric domain, the boundary condition for the acoustics interface is that the pressure is equal to the normal acceleration of the piezoelectric domain:

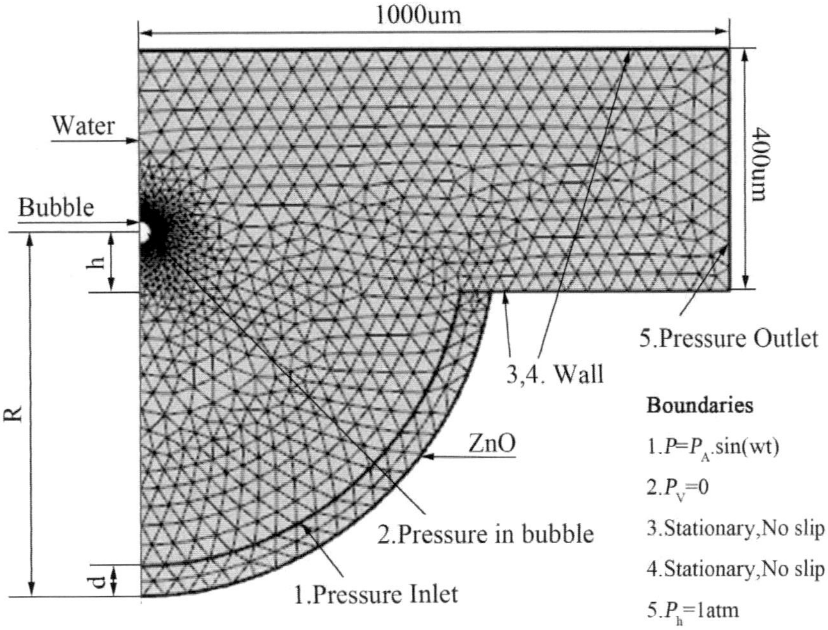

Figure 2. Geometric model used for ultrasound and bubble in the micro fluid.

$$\mathbf{n}\cdot\left(\frac{1}{\rho}(\nabla p)\right)=\mathbf{a}_{\mathrm{n}} \tag{10}$$

Because there are no sources present, the wave equation in the water domain simplifies to:

$$\nabla\cdot\left(-\frac{1}{\rho}(\nabla p)\right)-\frac{\omega^2 p}{\rho c^2}=0 \tag{11}$$

Where c , $\mathbf{n}$ and $\mathbf{a}_{\mathrm{n}}$ denote the velocity of sound in water, the normal vector, and the normal acceleration of the piezoelectric transducer. The ultrasonic pressure in the water could be obtained by Eq. equation (10) and equation (11), which changes in sine.

2.3.2. Micro-bubble dynamic. Since the flow rate in the micro channels is very low, it is assumed that the intensity of the sound field is constant, and the liquid in the micro channels is a kind of incompressible continuous media under normal temperature.

In the microfluidic channel, the motion of the bubble wall could be solved by the continuity equation and Navier-Stokes equation with the above governing equations.

Continuity equation:

$$\frac{\partial\rho}{\partial t}+\nabla(\rho\cdot\mathbf{u})=0 \tag{12}$$

Navier-Stokes equation：

$$\rho\frac{\partial \mathbf{u}}{\partial t}+\rho(\mathbf{u}\cdot\nabla)\mathbf{u}=\nabla\cdot[-p+\mu\left(\nabla\mathbf{u}+(\nabla\mathbf{u})^T\right)-\frac{2}{3}\mu(\nabla\cdot\mathbf{u})]+F \tag{13}$$

In this equation, F is the interfacial momentum. In this model (figure 2), the interface pressure between the piezoelectric transducer and water is:

$$p=p_h-P_A\sin(2\pi ft) \tag{14}$$

Where, P_h is the fluid pressure(=1 atm), and the ultrasonic pressure($=P_A\sin(2\pi ft)$) is obtained by equation (10). P_A and f refer to the ultrasonic amplitude and frequency. By taking the equation (14) into equation (13), the velocity of the bubble wall could acquire by equation (13) and equation (14).

2.3.3. Geometry, boundary conditions and numerical solution of equations. Equation (10) - (14) were numerically solved by COMSOL Multiphysics (Version 5.3). Figure.2 shows the geometric model and All details of the boundary, which is symmetrical. The residuals for all predicted parameters were considered to be less than 10-3. Furthermore, 189 boundary elements and 2416 region elements with minimum element quality of 0.0412um were considered. The Acoustic-Piezoelectric Interaction(acpz) and the Fluid-Structure Interaction(FSI), which are based on the Finite Difference Method(FDM), were used to solve the governing equations and the boundary [18].

3. Results and discussion

3.1. Geometry parameters of the acpz model

According to equation (4), Blake threshold pressure was P_B ($=1.1026\times10^5 P_a$) with the initial bubble radius R_0 (=4um) when the constant P_h ($=1.013\times10^5 P_a$) and σ ($=0.076 Nm^{-1}$) were determinate, which was greater than actual value. P_B was far below the gasification pressure of the liquid, which was conducive to the formation of bubble nuclei and could cause instantaneous cavitation. The bubble location h(=0.1mm) in figure 2 was set for numerical comparison.

Figure 3 shows ultrasound pressure distribution on the axis of Bowl piezoelectric model and plate piezoelectric model. The ultrasonic pressure produced from Bowl is about 1.5 times bigger than the pressure from Plane with same conditions, which are both greater than P_B in the micro channel when the ultrasonic frequency is equal to the resonance frequency of the bubble. According to the cavitation condition mentioned above, bigger ultrasonic pressure is necessary for cavitation, and the Bowl piezoelectric ultrasonic transducer is designed. In the numerical analysis of Acoustic-Piezoelectric Interaction(acpz) model, the width of micro channel in the axis is designed to 4mm without distortion for two ultrasonic periods.

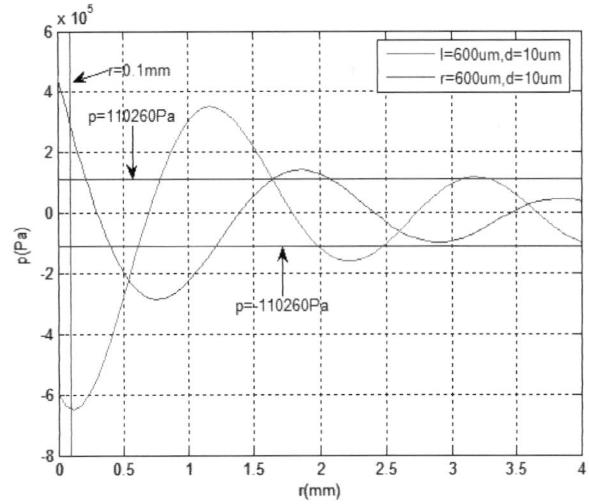

Green line(Bowl): h=0.1mm, r=0.6mm, d=0.01mm.
Blue line(Plane): h=0.1mm, l=0.6mm (equal to chord length of Bowl), d=0.01mm.

Figure 3. Axial acoustic pressure distribution from Bowl or Plate piezoelectric model.

Figure 3 shows that the amplitude of the axial ultrasonic pressure is bigger than Blake threshold P_B in the micro channel. In particular, the amplitude is about 6 times bigger than P_B in the location(h=0.1mm) of the bubble as blue line shown in figure 3.

Figure.4 shows ultrasound pressure distribution on the axis with different piezoelectric thickness d, arc chord distance h and radius r of the Bowl piezoelectric ultrasonic transducer. d is inversely proportional to the axial ultrasonic pressure (figure 4 a). h and r are proportional to the axial ultrasonic pressure (figure.4 b, figure.4 c).

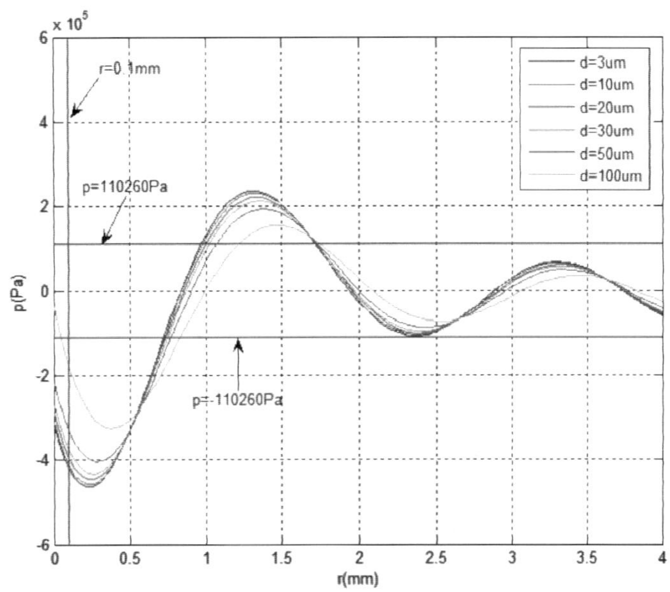

(a)

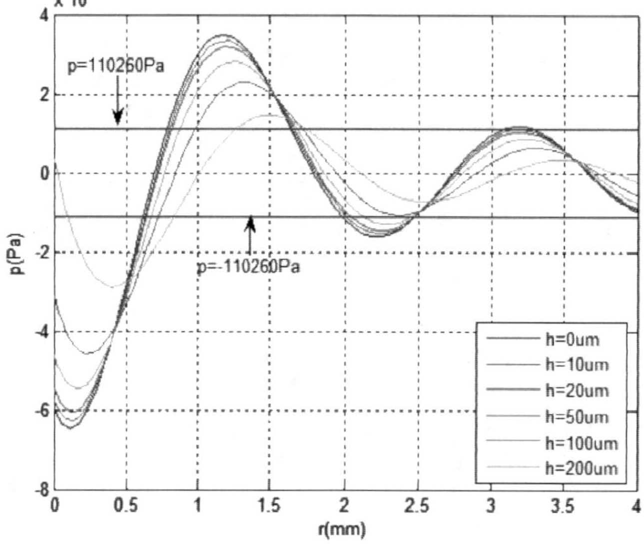

(b)

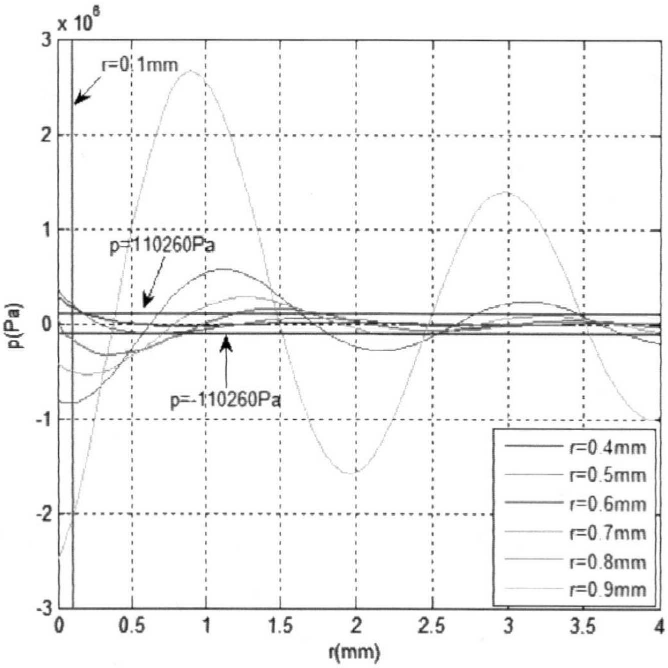

(c)

(a) h=0.1mm, r=0.6mm. (b) r=0.6mm, d=0.01mm. (c)h=0.1mm, d=0.01mm.

Figure 4. The axial ultrasonic pressure distribution with different transducer geometry parameters.

3.2. Micro bubble dynamics and activities under ultrasound irradiation

As acpz model mentioned above, the bubble (R=4um) was put to the sphere center of Bowl transducer in the FSI model. In order to better observe the bubble dynamics, the size of micro channel was reduced to 0.1 times without changing the ultrasonic and fluid characteristics around the bubble.

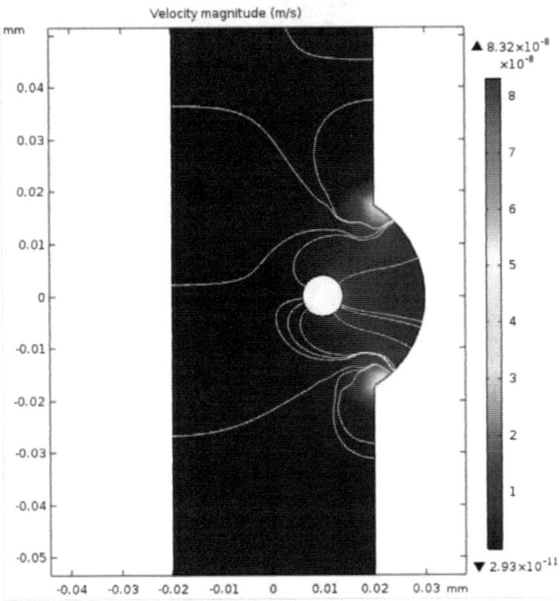

(a)

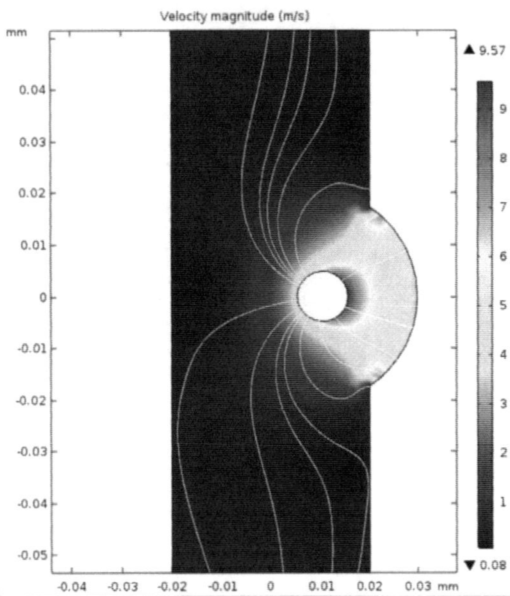

(b)

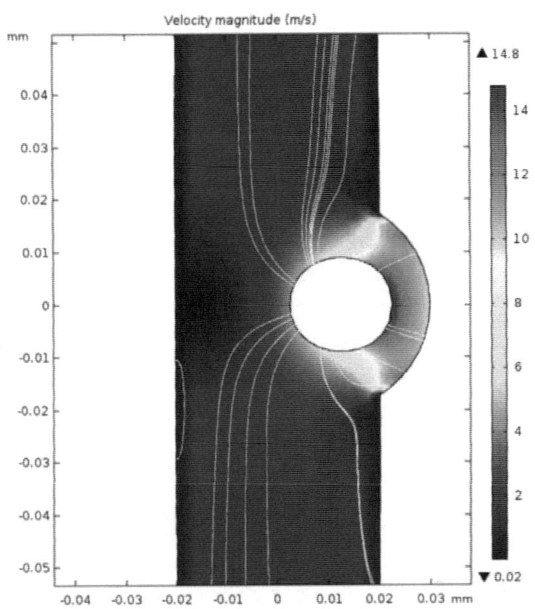

(c)

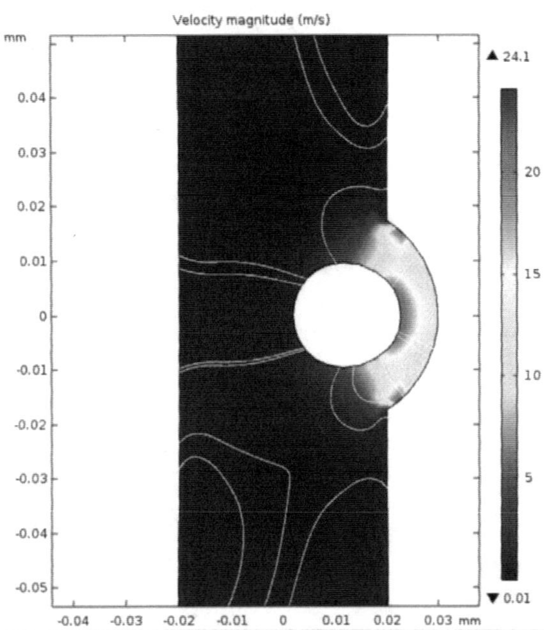

(d)

IOP Conf. Series: Materials Science and Engineering **739** (2020) 012036 doi:10.1088/1757-899X/739/1/012036

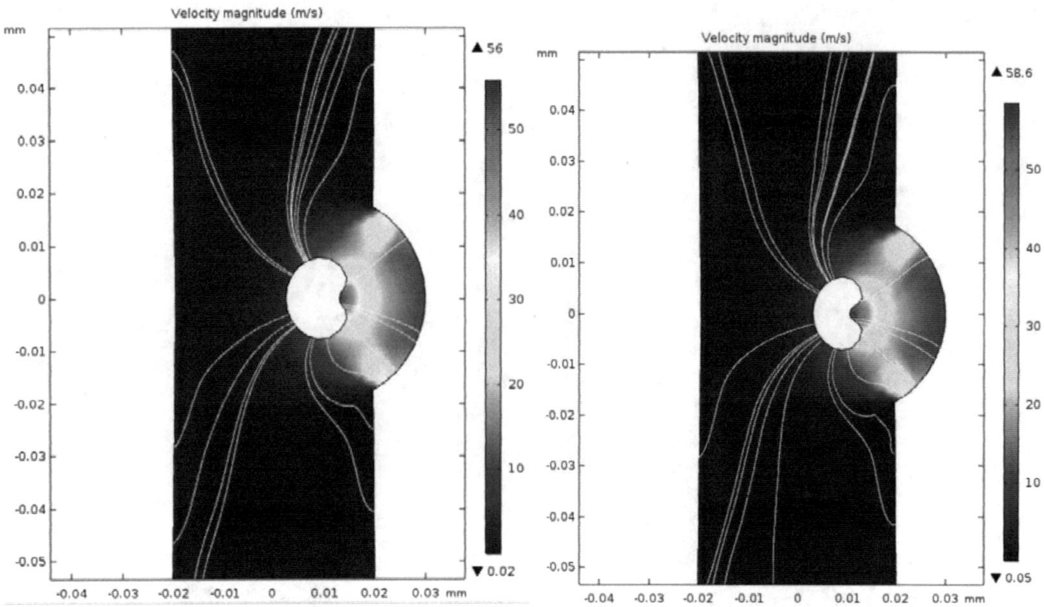

(e) (f)

(a) t=0. (b) t=0.25T. (c) t=0.5T. (d) t=0.75T. (e) t=0.875T. (f) t=0.89T

Figure 5. Micro bubble dynamics and activities under ultrasound irradiation.

Figure 5 shows the micro bubble dynamics and activities under ultrasound irradiation. It was observed that the bubble grew to the largest at 0.5T (T is ultrasonic period), and the invagination appeared near the side of piezoelectric transducer at 0.75T. The mesh of bubble wall collapsed at moment after 0.89T, and cavitation generated. The dynamics and activities of the micro bubble were consistent with the literature [19, 20]. According to equation (6), the radial dynamics of the micro bubble is highly consistent with figure.5, which confirmed the formation of cavitation in the micro channel.

3.3. The MEMS system and experiment

Figure 6 shows the experiment system [21]. In the micro-fluid experiment, the MEMS system is constituted of ultrasound transducer and micro channel. The substrate of the ultrasound transducer is the concave polyimide, which is made by a precision steel ball using embossing method. The piezoelectric thin film on the substrate is made by ZnO magnetron sputtering, on which two layers of Al are respectively electrodeposited as the positive electrode and negative electrode of the transducer. And the micro channels are made through soft imprinting method using polydimethylsiloxane(PDMS).

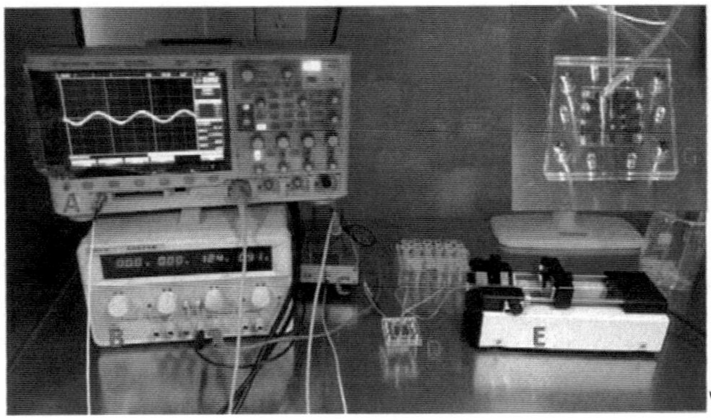

A. oscilloscope. B. stabilized voltage supply. C. power amplifier circuit.
D. microfluidic chip. E. peristaltic pump. F. collector. G. enlarge figure of microfuidic chip.

Figure 6. The experiment system of MEMS ultrasonic fluid system.

In the cavitation experiment, the hydroxyl terephthalic acid ions (HTA) has the characteristics of high fluorescence and stable chemical properties, by which the ultrasonic cavitation efficiency could be marked [22]. Figure 7 shows the relationship between fluorescence intensity of HTA (the solution is 8.0×10^{-3}mol/L) and the ultrasonic cavitations efficiency, which is tested by repeated experiment with the MEMS system (figure 6). Through a number of experiments, it is shown that the cavitation is generated in the liquid, and the ultrasonic cavitation is proportional to the fluorescence intensity.

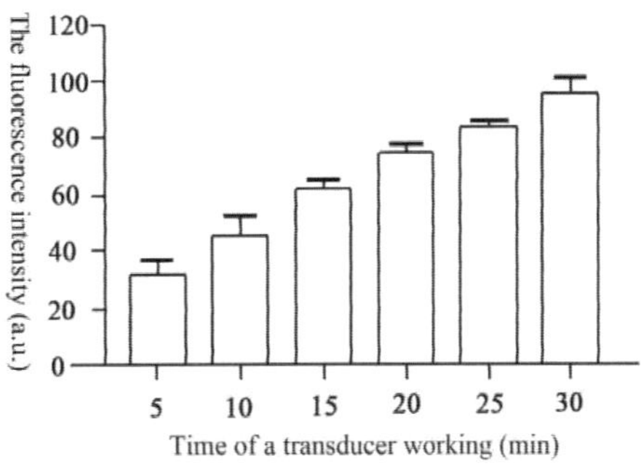

Figure 7. Ultrasonic cavitation experiment of MEMS transducer in the micro channels [23].

4. Conclusion

In the present study, the geometric parameters of the Bowl-shaped piezoelectric transducer are optimized according to the derived cavitation conditions, micro bubble dynamics and activities under ultrasonic irradiation was analyzed by CFD simulation, and ultrasonic cavitation is proved to exist in the MEMS chip by a number of acoustic chemistry experiments.

The axial ultrasonic pressure is inversely proportional to thickness d, proportional to arc chord distance h and radius r of the Bowl-shaped piezoelectric ultrasonic transducer, and the amplitude of the ultrasonic pressure in the circle center of the transducer is about 6 times bigger than Blake threshold pressure with designed geometric parameters (h=0.1mm, r=0.6mm, d=0.01mm). The concave of the bubble wall appeared near the side of piezoelectric transducer, and the velocity of acoustic streaming increased to 58.6 m/s in the concave area according to the CFD simulation result. Dynamics and activities of the bubble were in great agreement with the literature [19, 20].

Acknowledgements
The authors are grateful for the support provided by National Natural Science Foundation of China (Grant No.61176105), and Sichuan Education Department Program, China (Grant No.14ZB0112).

References

[1] Young F. R 1999 Cavitation *World Scientific.*

[2] Leighton T 2012 The Acoustic Bubble *Academic Press.*

[3] Newman C. M, Lawrie A, Brisken A. F and Cumberland D. C 2001 Ultrasound gene therapy: on the road from concept to reality *Echocardiog.* **18(4)** 339-47.

[4] Bao S, Thrall B. D and Miller D. L 1997 Transfection of a reporter plasmid into cultured cells by sonoporation in vitro *Ultrasound Med. Biol.* **23(6)** 953-9.

[5] Nobuki K, Okada K and Yamamoto K 2009 Sonoporation by Single-Shot Pulsed Ultrasound with Microbubbles Adjacent to Cells *Biophysical J.* **96(12)** 4866-76.

[6] Greenleaf W. J, Bolander M. E, Sarkar G, Goldring M. B and Greenleaf J. F 1998 Artificial cavitation nuclei significantly enhance acoustically induced cell transfection *Ultrasound Med. Biol.* **24(4)** 587-95.

[7] Lawrie A, Brisken A. F, Francis S. E, Cumberland D. C, Crossman D. C, Newman C. M 2000 Microbubble-enhanced ultrasound for vascular gene delivery *Gene Therapy.* **7(23)** 2023-27.

[8] Taniyama Y, Tachibana K, Hiraoka K, Namba T, Yamasaki K, et al. 2002 Local delivery of plasmid DNA into rat carotid artery using ultrasound *Circulation* **105(10)** 1233-9.

[9] Ohl C. D and Ikink R 2003 Shock-wave-induced jetting of micron-size bubbles *Phys. Rev. Lett.* **90(21)** 204502.

[10] Postema M, Van Wamel A, Lancee C. T and De Jong N 2004 Ultrasound-induced encapsulated microbubble phenomena *Ultrasound Med. Biol.* **30(6)** 827-40.

[11] Okaka K, Kudo N, Niwa K and Yamamoto K 2005 A basic study on sonoporation with microbubbles exposed to pulsed ultrasound *J. Med. Ultrasonics.* **32(1)** 3-11.

[12] Prentice P, Cuschieri A, Dholakia K, Prausnitz, M and Campbell P 2005 Membrane disruption by optically controlled microbubble cavitation *Nat. Phys.* **1(2)** 107-10.

[13] Kudo N, Okada K and Yamamoto K 2009 Sonoporation by Single-Shot Pulsed Ultrasound with Microbubbles Adjacent to Cells *Biophys. J.* **96(12)** 4866-76.

[14] Deng C. X, Sieling F, Pan, H and Cui J 2004 Ultrasound-induced cell membrane porosity *Ultrasound Med. Biol.* **30(4)** 519-26.

[15] Leighton T. G 2007 Derivation of the Rayleigh-Plesset Equation in Terms of Volume *University of Southampton.*

[16] Lofstedt R, Weninger K, Puterman S, Barber B. P 1995 Sonoluminescing bubbles and mass diffusion *Phys. Rev. E* **51(5)** 4400.

[17] M.Minneart, Philos.Mag.16(1933) 235.

[18] Sajjadi B, Raman A. A. A, Ibrahim S and Shah R. S. S. R. E 2012 Review on gas liquid mixing analysis in multiscale stirred vessel using CFD *Rev. Chem. Eng.* **28** 171-89.

[19] Kudo N, Okada K, Yamamoto K 2009 Sonoporation by single-shot pulsed ultrasound with microbubbles adjacent to cells *Biophys. J.* **96(12)** 4866-76.

[20] Blake J. R, Taib B. B, Doherty G 1986 Transient cavities near boundaries. Part 1. Rigid boundary *J. Fluid Meechan.* **170** 479-97.

[21] Lin R 2016 Self-focused Piezoelectric Micromachined Ultrasonic Transducer and Its Application in Cell Delivery *Uni. Sci. Technol, China.*

[22] Mason T. J, Lorimer J. P, Bates D. M, Zhao Y 1994 Dosimetry in sonochemistry: the use of aqueous terephthalate ion as a fluorescence monitor *Ultrason. Sonochem.* **1(2)** S91-5.

[23] Zhang P 2016 Research on Gene Transfection by Ultrasound Method under the Microfluidic Environment *Uni. Sci. Technol, China.*

IOP Conf. Series: Materials Science and Engineering **739** (2020) 012037 doi:10.1088/1757-899X/739/1/012037

Thermal Simulation and Sonoporation Experiment Based on a Focused Ultrasonic System

Zihong Liu[1, 2, 3], Hui You[1,3*], Ping Zhang[1], Ronghui Lin[1], Yongjia Chang[3]

[1]Department of Precision Instrumentation and Machinery, University of Science and Technology of China, Hefei 230026, Anhui, China
[2]Key Laboratory of Testing Technology for Manufacturing Process of Ministry of Education, Southwest University of Science and Technology, Mianyang 621010, Sichuan, China
[3]Institute of Intelligent Machines, Chinese Academy of Science, Hefei230031, Anhui, China

usmlhy@iim.ac.cn

Abstract. This paper aims at analyzing ultrasonic thermal effect and sonoporation based on a focused ultrasonic experiment system. The numerical analysis, about cavitaion conditions and ultrasonic thermal effect using the coupling model of pressure acoustic and bioheat transfer, were given, in which the temperature rise trend is identical to the conclusion in the literature [2]. According to the simulation analysis of ultrasonic thermal effect, temperature rise is approximately 18.5K near the focal spot with Ultrasonic radiation for 60s, which had little effect on 293T cell viability. In the acoustic chemical experiment, 293T cell viability is about 85% at 60s of the ultrasound irradiation, which was observed by 200× fluorescence microscope. The cell states of survival and abnormal was observed and analyzed in the sonoporation experiment at 60s of ultrasonic time.

1. Introduction

When the ultrasound passes through a volume of cell solution, some of the primary energy is absorbed locally by the solution and turned into heat, which results in a temperature increase [1-5]. In terms of the focused ultrasound and ultrasonic thermal effect, focused ultrasound has been applied in ultrasound hyperthermia [6, 7] and focused ultrasound surgery [8, 9], and theory study on temperature rise has been given [1, 2]. However, the numerical analysis of ultrasonic thermal effect and cavitation in cell solution has not been elucidated, and there have been few studies on sonoporation experiment involving ultrasonic thermal effect.

To elucidate ultrasonic thermal effect and sonoporation in cell solution under high frequency ultrasonic transducer, a focused ultrasonic experiment system was set up [10]. Based on the ultrasonic cavitation conditions deduced by Blake threshold pressure and the Keller-Miksis equation [11], the ultrasonic parameters in coupling simulation model was determined, and the numerical analysis of ultrasonic thermal effect was given based on homogeneous Helmholtz equation [12] and Pennes' Bioheat Transfer equation [13].

In the experiment, sonoporation experiment with ultrasonic thermal effect was carried out by acoustic chemistry testing [14, 15]. Cell perforation and cell viability were tested respectively using PI and FDA, which was observed by fluorescence microscope.

2. Experimental setups and methods

2.1. Experimental model

Figure 1 shows sonoporation experimental system with ultrasonic induced heating based on a focused ultrasonic transducer. In this experimental system, the cell solution in the beaker was put in the focused sound field, and the geometric center of the cell solution was almost at the focus of the transducer. In this study, when the focused ultrasound from the ultrasonic transducer passed though the cell solution, some of the energy was absorbed locally by the solution and turned into the heat or facilitated cavitation which was used to transitorily increase sonoporation in cell membrane permeability [16, 17]. In order to reduce the heating effect of the transducer components, the convection channel was designed as shown in figure 1.

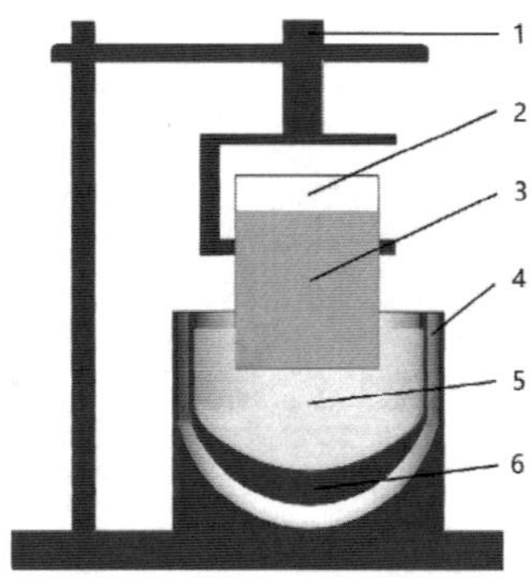

1.Screw micrometer. 2.Beaker. 3.The cell solution.
4.Convection channel. 5.Water. 6.Focused ultrasonic transducer.

Figure 1. Experimental setup used for sonoporation with ultrasonic induced heating.

2.2. Mechanistical modeling base

According to the van der Waals equation [18], the pressure in or out the micro bubble is equal when the bubble is in equilibrium.

$$P_g\left(\frac{R_0}{R}\right)^3 = P_h - P_V + \frac{2\sigma}{R} \tag{1}$$

Where, P_V is the vapour pressure in the bubble, $P_g(= P_h + 2\sigma / R_0)$ is the gas pressure in the bubble, P_h is the fluid pressure, and σ is the coefficient of surface tension.

To obtained minimum pressure out the bubble when bubble radius changes, the critical radius could be solved by taking the derivative of R in equation (1). And Blake threshold pressure P_B could be gained from the critical radius and equation (1):

$$P_B = P_h + \frac{2}{3}\left[\frac{\left(\frac{2\sigma}{R_0}\right)^3}{3(P_h + \frac{2\sigma}{R_0})}\right]^{\frac{1}{2}} \tag{2}$$

In the study, the radial dynamics of the micro bubble was described by the Keller-Miksis equation [11]:

$$\rho\left[R\frac{d^2R}{dt^2} + \frac{3}{2}\left(\frac{dR}{dt}\right)^2\right] = \left[P_g - P_h - P(t)\right] - \frac{4\mu}{\rho R}\frac{dR}{dt} - \frac{2\sigma}{R} \tag{3}$$

In this equation, ρ and μ denote the fluid density and shear viscosity of the fluid. Here, driving pressure $P(t)$ applied on bubbles can be approximated by acoustic pressure P_a ($= -P_A \sin(\omega t)$). P_A refers to the acoustic pressure amplitude and $f(=\frac{\omega}{2\pi})$ refers to the acoustic pressure frequency.

By ignoring the shear viscosity of the fluid, the equation (3) is expended to the $1/R_0$ power when bubble radial is R ($= R_0 + r$, $r < R$):

$$r^2 + \omega_r^{\ 2} r = \frac{P_A}{\rho R_0}\sin \omega_a t \tag{4}$$

According to equation (4), the resonance frequency $f_r\left(=\frac{\omega_r}{2\pi}\right)$ of the bubble was given by Minneart. And bubble vibration is obvious when the ultrasonic frequency(f_a) is equal to f_r

$$f_r = \frac{1}{2\pi R_0}\left[\frac{3}{\rho}\left(P_h + \frac{2\sigma}{R_0}\right)\right]^{\frac{1}{2}} \tag{5}$$

According to equation (2) and equation (4), microbubble cavitation could occur when $P_A > P_B$ and $f_a < f_r$.

2.3. Computational model and boundary conditions

2.3.1. Acoustics and thermodynamics coupling based on the system. It is assumed that the acoustic wave propagation is linear. Nonlinear effects and shear waves are neglected. In 2D axisymmetric cylindrical coordinates, the wave equation of the ultrasonic pressure is solved by the homogeneous Helmholtz equation:

$$\frac{\partial}{\partial r}\left[\frac{r}{\rho_c}\left(\frac{\partial p}{\partial r}\right)\right] + r\frac{\partial}{\partial z}\left[\frac{1}{\rho_c}\left(\frac{\partial p}{\partial z}\right)\right] - \left[\left(\frac{\omega}{c_c}\right)^2\right]\frac{rp}{\rho_c} = 0 \tag{6}$$

In this equation, the density ρ_c and the speed of sound c_c are complex-valued to account for the material's damping properties. In the acoustic field obtained by equation (6), the heat source Q $(= 2\alpha I)$ for thermal simulation could be by gained by the following equation:

$$Q = 2\alpha \left| \mathrm{Re}\left(\frac{1}{2} p \upsilon \right) \right| \tag{7}$$

In this equation, the ultrasonic parameters α , I , p and υ denote the ultrasonic absorption coefficient, ultrasonic intensity magnitude, ultrasonic pressure and the ultrasonic particle velocity vector, respectively. By taking ultrasonic heat source Q into the Pennes' Bioheat Transfer equation, the temperature of the cell solution could be obtained.

$$\rho c_p \frac{\partial T}{\partial t} = \nabla \cdot (k \nabla T) - \rho c_p \omega_p (T - T_0) + Q + Q_m \tag{8}$$

Where, ρ , c_p and k respectively denote the density, the specific heat and the thermal conductivity, . And ω_p, Q and Q_m respectively refer to the solution perfusion rate (here, $\omega_p = 0$), the heat source obtained by equation (7) and the biologic heat source.

2.3.2. Simulation model and Boundary conditions. Figure 2 shows the geometry of the ultrasonic transducer and cell solution phantom in the simulation model. Both the cell solution in the utensil and the bowl-shaped transducer are immersed in water. Four cylindrical perfectly matched layers(PMLs) (1-3 and 7 in figure 2) are used to absorb the outgoing ultrasonic waves. In the simulation model, the normal displacement of the transducer was set to 3.8nm according to the transducer characteristics (WHQ2018).

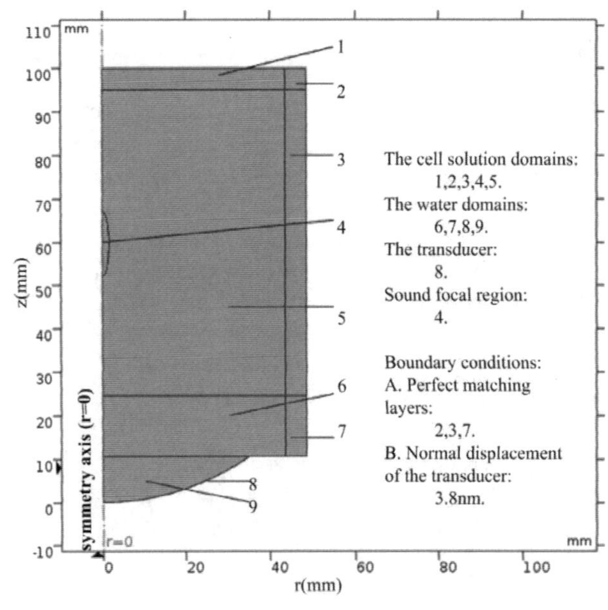

Figure 2. Geometric and physical model for numerical simulation.

The software COMSOL Multiphysics (Version 5.3) was used to carried out the grid division and numerical simulation. Based on the finite difference method(FDM) [19], pressure acoustics (acpr) and bioheat transfer (ht) were employed, which the pressure acoustics simulation is performed in all domains and the heat transfer model is only applied in the cell solution domain. Through the grid division, 703221 DOF was obtained in acoustic pressure model and 8126 DOF was obtained in bioheat transfer model.

2.4. Test methods of sonoporation experiment

Due to thermal effect of the ultrasound, the temperature rises with ultrasonic action time, which determines the characteristics of cell perforation and cell activity. With suitable cell solution temperature determined by numerical analysis, cell membrane perforation and cell activity could be separately test using propidium iodide (PI) and fluorescein diacetate(FDA). According to the sonoporation, the permeability of cell membranes is improved under ultrasound, which permits the macromolecules and DNA fragments near the cell to penetrate the cell membrane into the cell [20, 21]. However, the cell membrane would be deadly damaged when exposed to ultrasound, which is beyond repair. To conform the reates of cell membrane perforation and cell activity, four tested solutions were employed with 200× fluorescence microscope (Lecia DMI4000B, Germany):

(i). The cell solution (2.0×106/ml), cultured using human embryonic kidney cells(293T), is used to quantify the ultrasonic thermal effect in the experimental platform.

(ii). The 293T cell solution (2.0×106/ml) with PI is used to test cell membrane perforation. The testing mechanism is that the PI permeates only through damaged cell membranes and produces red fluorescence.

(iii). The 293T cell solution (2.0×106/ml) with FDA is used to test cell activity. The testing mechanism is that FDA permeates only through living cell membranes and produces green fluorescence when the cell is living.

(iv). The 293T cell solution (2.0×106/ml) with PI and FDA is used for comparative testing.

In our experiment, the rates of cell membrane damaged, cell viability and cell transfection were tested respectively by above solution with the same condition.

3. Results and discussion

3.1. Numerical simulation

According to equation (7) and equation (3), the resonance frequency of the bubble is $f_r\,(=1.34MHz)$ with the initial bubble radius R_0 (=3.6 μm) when P_h ($=1.013\times10^5 P_a$) and σ ($=0.076Nm^{-1}$) are determined, and Blake threshold pressure is $P_B\,(=1.159\times10^5 P_a)$ which is greater than actual value in the experiment system. The transducer, drive frequency 1.045MHz and ultrasonic power 3.78w, is employed in the experiment.

Figure 3 shows acoustic pressure distribution in the cell solution in numerical simulation. According to ultrasonic cavitation conditions deriveed above, the cavitation comes into being in some regions where the acoustic pressure P_A is greater than P_B with $f_a < f_r$ (as figure 3.(a) shown), especially in and near the sound focal area as figure 3.(b) and figure 3.(c) shown.

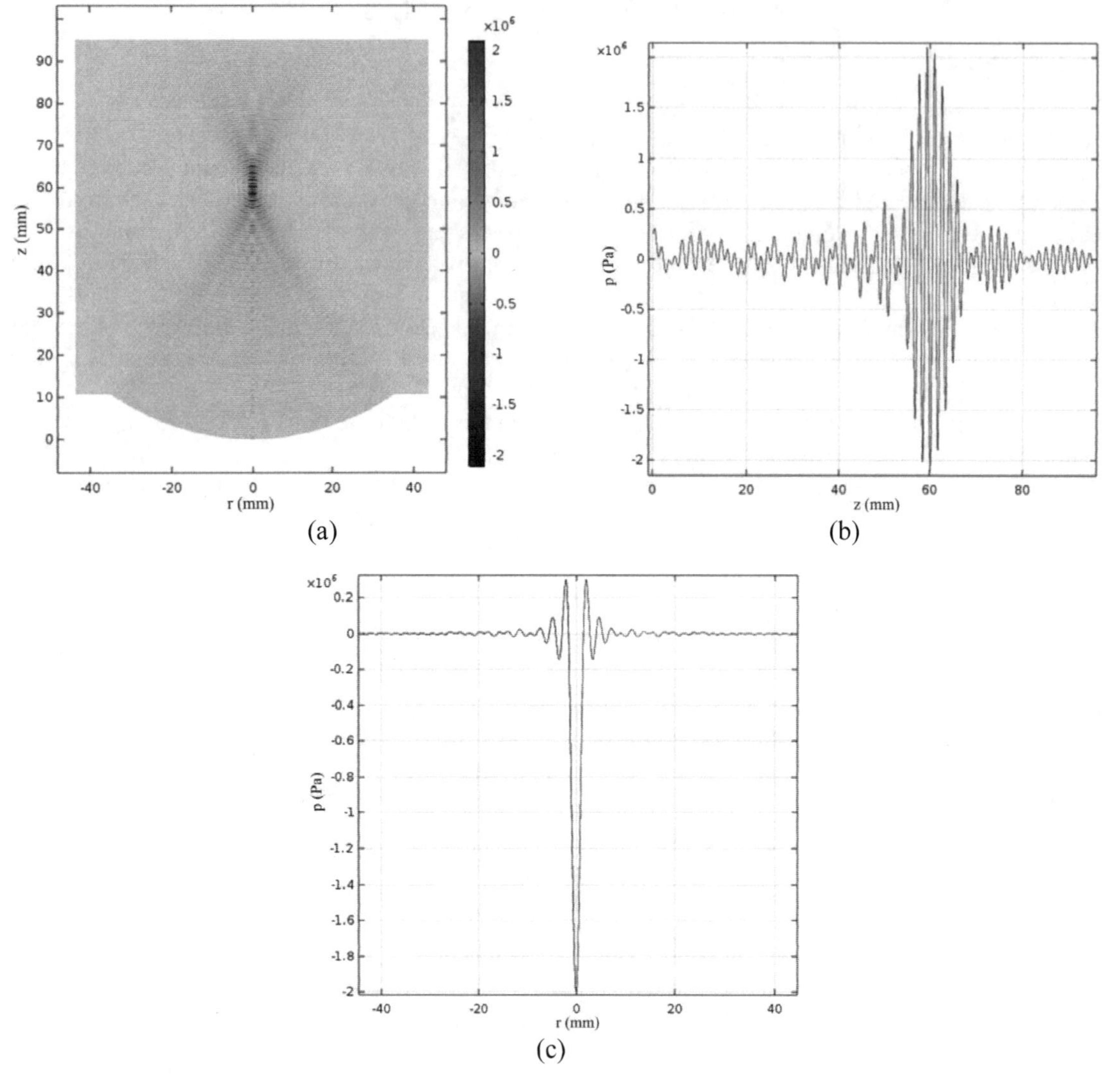

(a). Acoustic pressure in cell solution. (b). Acoustic pressure on z axial.
(c). Acoustic pressure on radius direction at sound focal spot

Figure 3. acoustic pressure distribution in cell solution (fa=1.045MHz).

In numerical simulation about ultrasonic thermal effect in cell solution, the values of the physical constants used are given in Table 1 [1, 2].

Table 1. Values of physical properties.

Name	Property	Value	Unit
ρ	Density	1044	Kg/m^3
c_p	Heat capacity at constant pressure	3370	$J/(kg \cdot K)$
k	Thermal conductivity	0.45	$W/(m \cdot K)$
T0	Room temperature	273	K
α	Ultrasonic absorption coefficient	8	1/m

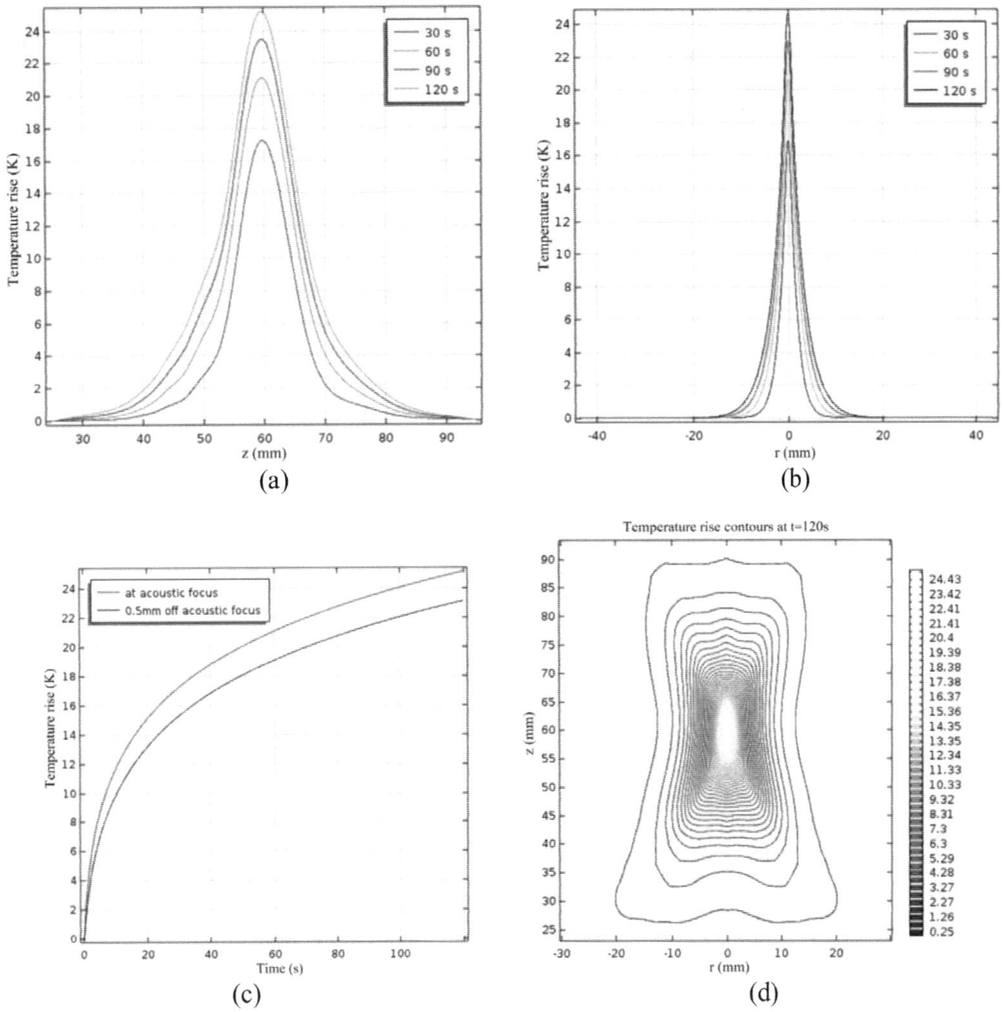

(a). Temperature rise on z axial. (b). Temperature rise on radius direction at sound focal spot. (c). Temperature rise at and 0.5 ㎜ off acoustic focus. (d). Temperature rise contours at 120 sec.

Figure 4. Temperature rise under ultrasonic irradiation in cell solution.

Figure 4 shows temperature rise under ultrasonic irradiation in cell solution. According to the simulation results shown in figure 4. (a) and figure 4. (b), ultrasonic focused spot is the high-temperature area and the fastest temperature rise area, which is an ellipse region with axial length of 15mm and 3mm. Figure.4. (c) shows temperature rise at acoustic focus and 0.5mm off acoustic focus, which indicates that temperature rise trend in the figure is identical to the conclusion in the literature (See figure 3 in the literature [2]). As the temperature difference increases in the cell solution, temperature rises slowly after 20s (As shown in figure 4. (c)). In the numerical simulation of ultrasonic thermal effect, the temperature is 24.43K hotter than room temperature T0(273K) when ultrasonic action time is 120s in and near the ultrasonic focal spot area. Figure 4. (d) shows temperature rise contours at 120s, in which the high-temperature area is at ultrasonic focused ellipse region.

3.2. Sonoporation experiment

Figure 5 shows 293T cell viability under ultrasonic in the acoustic chemistry experiment. Due to ultrasonic thermal effect and ultrasonic cavitation in the cell solution, cell viability declines rapidly after 60s of ultrasonic irradiation. In the acoustic chemistry experiment, cell viability is about 85% at 60s, which is meaningful for study on sonoporation.

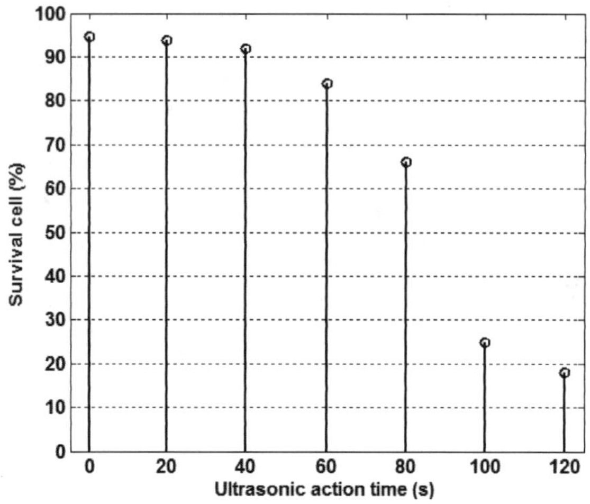

Figure 5. Cell viability with ultrasound action time.

The experimental temperature of 293T cell solution should be similar to human body temperature for the sonoporation experiment, that ultrasonic action time was set to no more than 60s in sonoporation experiments according to ultrasonic thermal simulation.

Figure 6 shows the images of cell marked with FDA fluorescence or PI fluorescence, which is obtained by experimental treatment at 60s of ultrasonic radiation [22]. In the acoustic chemistry experiment, the cavitation and sonoporation come into being, which could be testified by cell membrane damage shown in figure 6. 293T cells with green fluorescence indicate that the cells are living, which contain cells of membranes perforated and cells of repaired membranes. 293T cells with red fluorescence indicate that the cell membrane is damaged by bubble cavitation, which is dead or will repair the cell membranes itself.

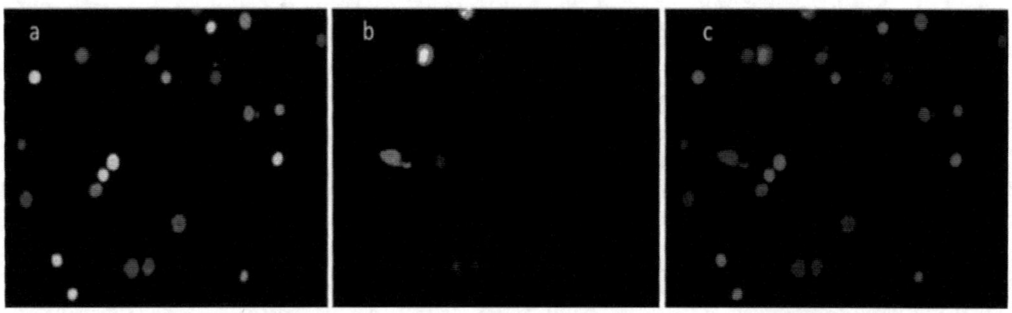

(a). Living cell with FDA fluorescence. (b). Damaged cell with PI fluorescence.
(c). Cell with fluorescence form FDA and PI.

Figure 6. Acoustic chemistry experiment on 293T (PI and FDA) (with a 200× lens).

4. Conclusion

This study shows that thermal effect of ultrasound in cell solution is obvious and cell membrane permeability is increased in the focused ultrasonic experimental system.

According to the result of numerical analysis, cavitation comes out under ultrasound ($f_a = 1.04MHz$) and temperature rise is efficient, which is highly consistent with ultrasonic thermal simulation. The temperature rise is about 18.5K when 293T cell solution is under ultrasound 60 seconds according to ultrasonic thermal simulation, which is benefit for cell to live and sonoporation experiment. In line with the acoustic chemistry experiments, the percentage of cell viability is approach to 85% with ultrasonic radiation for 60s.

Acknowledgements

The authors are grateful for the support provided by National Natural Science Foundation of China (Grant No.61176105), and Sichuan Education Department Program, China (Grant No.14ZB0112).

References

[1] Parker K J 1983 The thermal pulse decay technique for measuring ultrasonic absorption coefficients, *J. Acoustic. Soc. America* **74(5)** 1356-61.

[2] Clarke R L, Ter Haar G R 1997 Temperature rise recorded during lesion formation by high-intensity focused ultrasound, *Ultrasound Medic Bio.* **23(2)** 299-306.

[3] Ter Haar G 2007 Therapeutic applications of ultrasound, *Prog. Biophys. Molec. Bio.* **93(1-3)** 111-29.

[4] Webb H, Lubner M G, Hinshaw J L 2011 Thermal ablation, *Semin. Roentgenol.* **46(2)** 133-41.

[5] Ahmed M, Brace C L, Lee Jr F T, Goldberg S N 2011 Principles of and advances in percutaneous ablation, *Radiology* **258(2)** 351-69.

[6] Hynynen K, Edwards D K 1989 Temperature measurements during ultrasound hyperthermia, *Medi. Physi.* **16(4)** 618-26.

[7] Seip R, Vanbaren P, Ebbini E S 1994 Dynamic focusing in ultrasound hyperthermia treatments using implantable hydrophone arrays, *IEEE Transa. Ultrason. Ferroelect. Freq. Contr.* **41(5)** 706-13.

[8] Tempany C M, Stewart E A, Mcdannold N, Quade B J, Jolesz F A, Hynynen K 2003 MR imaging–guided focused ultrasound surgery of uterine leiomyomas: a feasibility study, *Radiology* **226(3)** 897-905.

[9] Funaki K, Fukunishi H, Takayama T 2012 M481 symptom improvement after magnetic resonance-guided focused ultrasound surgery for uterine myomas, *Int. J. Gynecol. Obstetr.* **119** S685.

[10] Wang X, Gao J, Zhang P, Lin R, Yang Y, You H, Wang X 2015 Ultra-high frequency sonochemistry and its dosimetry, *Int. Conf. Info. Sci, Mach. Mater. Energ.* 1851-5.

[11] Mason T J, Lorimer J P, Bates D M, Zhao Y 1994 Dosimetry in sonochemistry: the use of aqueous terephthlate ion as a fluorescence monitor, *Ultraso. Sonochem.* **1(2)** S91-5.

[12] Moiola A, Hiptmair R, Perugia I 2011 Plane wave approximation of homogeneous Helmholtz solutions, *Journal of Applied Methematics and Physics.* **62(5)** 809-37.

[13] Ezzat M A, Alsowayan N S, Al-Muhiameed Z I, Ezzat S M 2014 Fractional modelling of pennes' bioheat transfer equation, *Heat Mass Transfer* **50(7)** 907-14.

[14] Blake J R, Taib B B, Doherty G 1986 Transient cavities near boundaries part 1 rigid boundary, *J. Fluid Mech.* **170** 479-97.

[15] Kudo N, Okada K, Yamamoto K 2009 Sonoporation by Single-Shot Pulsed Ultrasound with Microbubbles Adjacent to Cells, *Biophysic. J.* **96(12)** 4866-76.

[16] Stevenson D J, Gunnmoore F J, Campbell P, et al. 2010 Single cell optical transfection, *J. Roya. Soc. Interf.* **7(47)** 863-71.

[17] Heiser W C 2004 Delivery of DNA to cells in culture using particle bombardment, *Gene Deliv.*

Mannal. Cells. Human Press **245** 175-84.
[18] Lofstedt R, Weninger K, Putterman S, Barber B P 1995 Sonoluminescing bubbles and mass diffusion. *Phys. Rev. E* **51(5)** 4400.
[19] Sajjadi B, Raman A A A, Ibrahim S, Shah R S S R E 2012 Review on gas-liquid mixing analysis in multiscale stirred vessel using CFD, *Rev. Chem. Eng*. **28(2-3)** 171-89.
[20] Miller D L, Pislaru S V, Greenleaf J F 2002 Sonoporation: mechanical DNA delivery by ultrasonic cavitation, *Somatic Cell Molec. Geneti*. **27(1-6)** 115-34.
[21] Deng C X, et al. 2004 Ultrasound-induced cell membrane porosity, *Ultras. Medic. Bio*. **30(4)** 519-26.
[22] Zhang P 2016 Research on Gene Transfection by Ultrasound Method under the Microfluidic Environment, *Uni. Sci. Technol., China.*

Numerical Analysis on Seismic Performance of a New Type of Fully Assembled Beam-Column Joints

W Zhai[1], W Jiang[1], J H Yang[1], J F Yang[2, 3], K Qu[2], L Chen[2]

[1]PowerChina Hubei Electric Engineering Corporation Limited, Wuhan, 430040
[2]School of Civil Engineering, Xi'an University of Architecture & Technology, Xi'an, 710055, China
[3]Key Lab of Structural Engineering and Earthquake Resistance, Ministry of Education (XAUAT), Xi'an, 710055, China

yjf9807@126.com

Abstract. To investigate the seismic performance of a new type of fully assembled beam-column joints, finite element models are established by ABAQUS with the nonlinear and geometric nonlinear effects considered to simulate 5 joints based on the quasi-static test. Through the comparison between finite element analysis and experimental results, the validity of finite element models is verified. Then the correlation parameters and their effects on the seismic performance of joints are analysed.

1. Introduction

Assembled steel structure building systems have rapidly developed under the promotion of national policies and developed more improved, modular and industrialized residential building systems based on the traditional steel structures [1]. Assembled steel structures refer to the multi-high-rise buildings constructed with the industry-produced steel columns and beams as framework and the lightweight, heat insulating and high-strength wall materials as enclosure system [2], in which the beam-column joints are very important to the overall mechanical behaviour of assembled steel structures [3]. At present, a lot of researches have been done on the assembled beam-column joints at home and abroad [4-9]. Literature [4] and [5] introduce a new type of beam-column joint connected with T-plate and high-strength bolt to improve the assembly efficiency of steel structures. In literature [6] and [7], the assembly of joint is further improved through the application of single side locking bolts. However, too many bolt holes on the columns of above-mentioned joints markedly reduce the bearing capacity. To solve this problem, a standard joint with well industrialization production and high seismic performance is proposed by the AISC and analysed in literature [8] and [9]. In this paper, finite element models of a new type of assembled beam-column joint based on the pseudo-static test are established, and the effects of correlation parameters on the seismic performance of joint are analysed and compared to provide a certain reference for the design joints.

2. Establishment of finite element model

2.1. Joint specimen design

As presented in figure 1, the new type of beam-column joint used for assembled steel structure is composed of rectangular tube column, H-beam, channel plate, angle iron, T-plate and high-strength bolt.

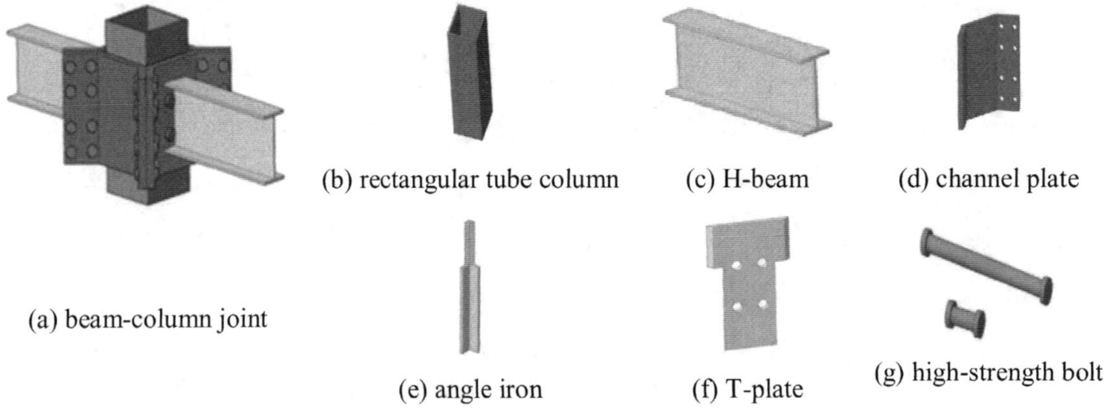

(a) beam-column joint (b) rectangular tube column (c) H-beam (d) channel plate (e) angle iron (f) T-plate (g) high-strength bolt

Figure 1. Beam-column joint and components.

Welded rectangular steel tube column and H-shaped steel beam are connected by 10.9-grade friction-type high-strength bolts, and all components are produced with Q345B steel. In literature [10], pseudo-static test with the same axial compression ratio of 0.2 was conducted on 5 scale models of 1:2 ratios. Specific dimensions of these specimens are shown in table 1.

Table 1. Specific dimensions of specimens.

Group	Dimension		Thickness			Bolt specification		RBS
	Column	Beam	T-plate	Angle iron	Channel plate	Shearing bolt	Bending bolt	
JD1	200×10	H200×100×5.5×8	12	12	28	M20×4	M24×8×2	NO
JD2	200×10	H200×100×5.5×8	12	12	28	M16×4	M24×8×2	NO
JD3	200×10	H200×100×5.5×8	12	12	20	M20×4	M24×8×2	NO
JD4	200×10	H200×100×5.5×8	12	12	28	M20×4	M20×8×2	NO
JD5	200×10	H200×100×5.5×8	12	12	28	M20×4	M24×8×2	YES

2.2. Finite element model

Finite element models corresponding to the 5 beam-column joints are established by ABAQUS and the finite element analysis results are compared with the testing results to prove the validity of the finite element model. The hysteretic and skeleton curves of 5 joints obtained from finite element models and tests are compared in figure 2 and 3, respectively.

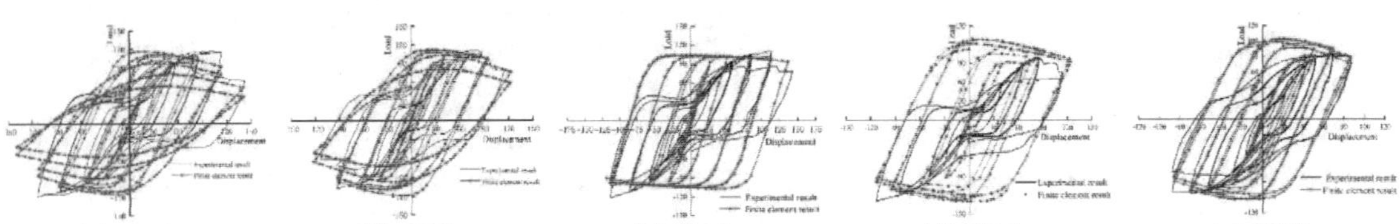

(a) JD1 (b) JD2 (c) JD3 (d) JD4 (e) JD5

Figure 2. Experimental and finite element hysteretic curves.

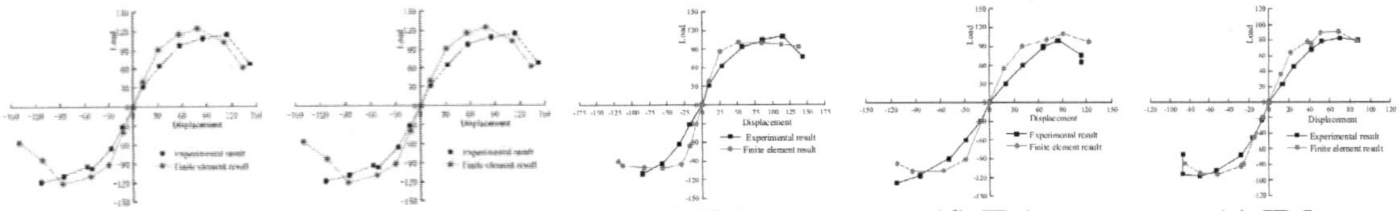

(a) JD1 (b) JD2 (c) JD3 (d) JD4 (e) JD5

Figure 3. Experimental and finite element skeleton curves.

The finite element results agree well with the experimental results because both skeleton curves are close to each other and present converted S shape, indicating that specimens has gone through 3 stages of elastic, elastoplastic and plastic failure under the action of low frequency cyclic loading. Besides, comparisons of the failure modes and the parameters between finite element analysis and experimental results are presented in figure 4 and table 2, respectively.

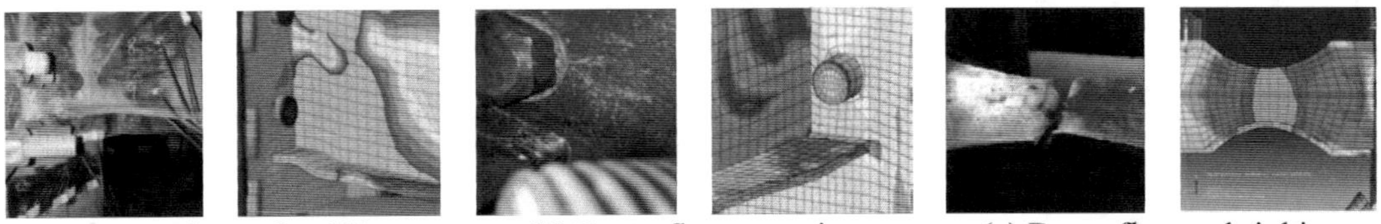

(a) Beam flange buckling (b) Beam flange tearing (c) Beam flange shrinking

Figure 4. Experimental and finite element failure modes.

Table 2. Comparison of the parameters between experimental results and finite element analysis.

Group	Yield					Ultimate				
	Experimental		Finite element		Ratio	Experimental		Finite element		Ratio
	P_{yE} /kN	Δ_{yE} /mm	P_{yF} /kN	Δ_{yF} /mm	P_{yF}/P_{yE}	P_{uE} /kN	Δ_{uE} /mm	P_{uF} /kN	Δ_{uF} /mm	P_{uF}/P_{uE}
JD1	98.4	58.8	106.4	46.0	1.08	118.6	114.0	123.9	79.2	1.04
JD2	89.2	48.4	94.1	39.6	1.05	110.2	76.0	103.3	57.8	0.94
JD3	90.5	57.3	89.0	30.0	0.98	121.1	114.0	101.2	83.6	0.84
JD4	87.9	56.9	92.8	47.4	1.06	128.6	114.0	111.4	90.6	0.87
JD5	69.5	37.5	79.0	37.9	1.14	95.2	70.0	90.7	68.9	0.95

In table 2, P_{yE} is the test yield load, Δ_{yE} is the test yield displacement, P_{yF} is the finite element yield load, Δ_{yF} is the finite element yield displacement, P_{uE} is the test ultimate load, Δ_{uE} is the test ultimate displacement, P_{uF} is the finite element ultimate load and Δ_{uF} is the finite element ultimate displacement. The difference less than 10% between finite element and experimental results can verify the accuracy of modelling, and thus the finite element simulation can be used to study the performance of joints.

3. Parameter analysis

3.1 Ratio of axial compression stress to strength

4 finite element models (Group JD-1) with different axial compression ratios (0.2, 0.4, 0.6 and 0.8) are designed to study the relation between the ratio of axial compression stress to strength and the seismic performance.

3.1.1. Stress analysis. In figure 5, JD-1 presents a typical failure mode that the channel plate is destroyed earlier than beam and column. That is, the joint fails before members. The ultimate failure

IOP Conf. Series: Materials Science and Engineering **739** (2020) 012038 doi:10.1088/1757-899X/739/1/012038

mode of joint is the destruction of channel plates and the local buckling of beam with a small axial compression ratio, as well as the failure of channel plates and column with large ratio.

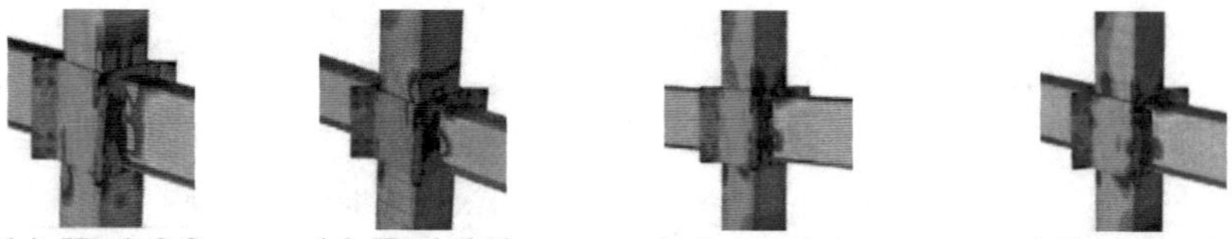

(a) JD-1-0.2 (a) JD-1-0.4 (c) JD-1-0.6 (d) JD-1-0.8

Figure 5. JD-1's Mises stress nephogram at the limited state of bearing capacity.

3.1.2. Skeleton and hysteretic curve. The skeleton and hysteretic curves of JD-1's column top loading point are drawn in figure 6 and 7, respectively. When the axial compression ratio increases from 0.2 to 0.8, the skeleton curves are in good agreement, while the hysteretic curves gradually shrink, which indicates the stepwise reduction of energy dissipation capacity.

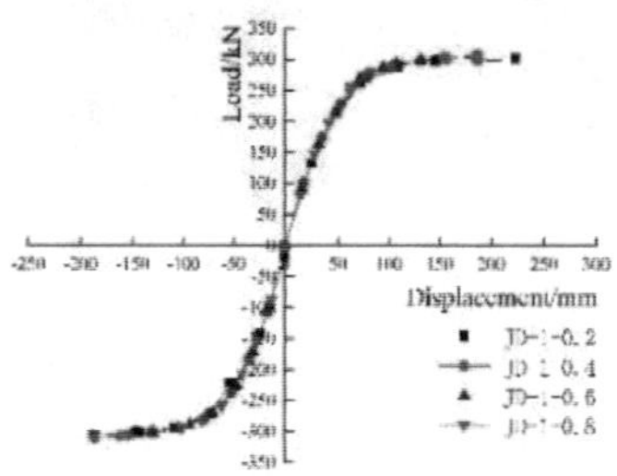

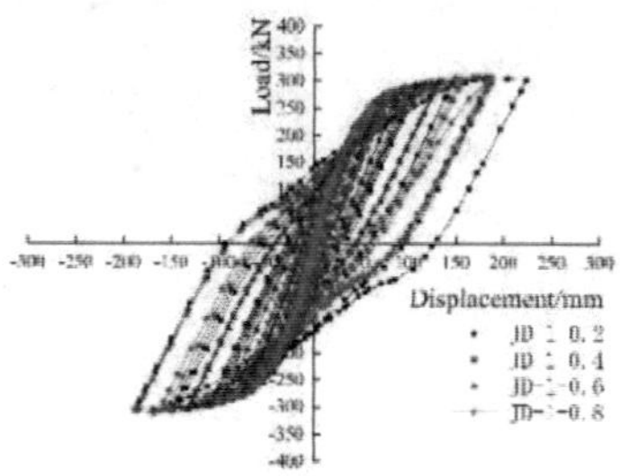

Figure 6. JD-1's skeleton curves. **Figure 7.** JD-1's hysteretic curves.

3.1.3. Ductility and energy dissipation capacity. As shown in table 3, when the ratio of axial compression stress to strength is less than 0.4, ductility coefficients are all greater than 3, indicating the good ductility of joints. Because the ductility coefficient decreases with the increase of axial compression ratio, high axial compression ratio can greatly weaken the ductility of joints. Besides, those energy dissipation coefficients below 2 correspond to passive energy dissipation modes that the channel plate bends before the beam yields fully.

Table 3. Ductility and energy dissipation capacity of JD-1.

No.	Axial compression ratio	Ductility coefficient μΔ	Energy dissipation coefficient Ce	Failure mode
JD-1-0.2	0.2	3.34	1.42	Channel plate and column core region buckling
JD-1-0.4	0.4	3.06	1.32	Channel plate and column core region buckling
JD-1-0.6	0.6	2.33	0.87	Channel plate buckling
JD-1-0.8	0.8	2.07	0.68	Channel plate buckling

3.2. Thickness of channel plate

3.2.1. Stress analysis. The Mises stress nephogram of JD-2 is shown in figure 8. As the thickness of channel plate increases, the stiffness grows accordingly, and increasingly radical plastic deformation appears in column and beam intensively. The failure mode of joint gradually changes from the buckling of channel plate to the formation of plastic hinge at the beam end, and the buckling of channel plate and column tends to aggravate.

(a) JD-2-20 (b) JD-2-25 (c) JD-2-30 (d) JD-2-35

Figure 8. JD-2's Mises stress nephogram at the limited state of bearing capacity.

3.2.2. Skeleton and hysteretic curve. The skeleton and hysteretic curves of JD-2's column top loading point are drawn in figure 9 and 10, respectively. With the thickness of channel plate increasing, the hysteresis loop expands gradually, indicating the reinforcement of energy dissipation capacity. Meanwhile, the deformation capacity of joint is enhanced, but joint begins to shrink and slip with the increase of loading displacement. This is mainly because bolts tend to loosen under the action of low frequency cyclic loading when used for fully bolted joint and there is a smooth stage when the load remains while the displacement decreases in the unloading process after the channel plate being pulled open.

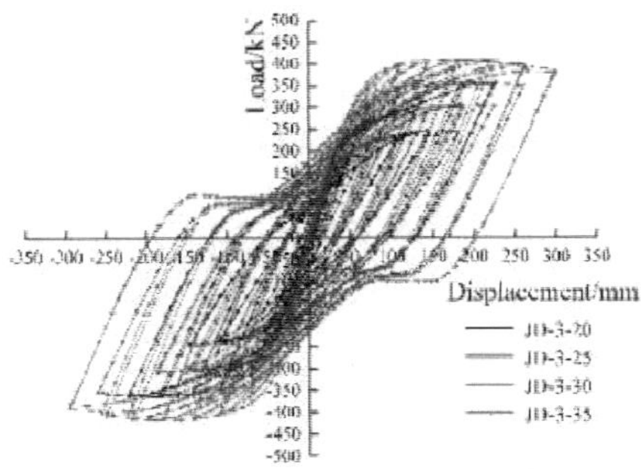

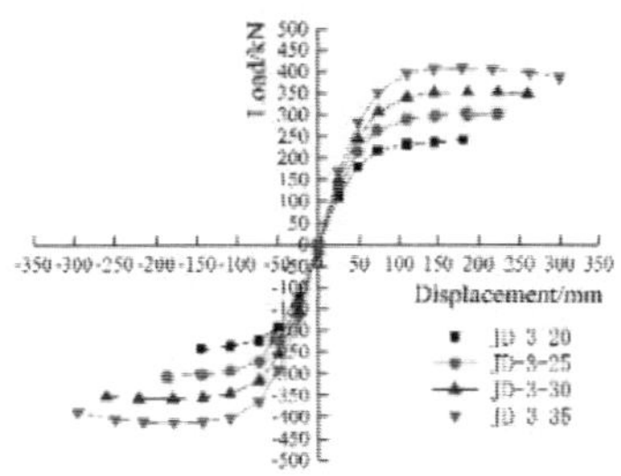

Figure 9. JD-2's hysteretic curves. **Figure 10.** JD-2's skeleton curves.

3.2.3. Ductility and energy dissipation capacity. As shown in table 4, when the thickness of channel plate grows from 20mm to 35mm, the ductility coefficient has a sharp rise, indicating that the increase of channel plate thickness can strengthen the ductility of joint efficiently.

Table 4. Ductility and energy dissipation capacity of JD-2.

No.	Axial compression ratio	Ductility coefficient μΔ	Energy dissipation coefficient Ce	Failure mode
JD-2-20	0.2	3.01	1.00	Channel plate and column core region buckling
JD-2-25	0.2	3.34	1.32	Channel plate and column core region buckling
JD-2-30	0.2	3.78	1.69	Beam buckling
JD-2-35	0.2	4.32	1.69	Beam buckling

4. Conclusions

Basing on the experiment of a new assembled joint between column and beam, finite element models are established to simulate this joint and analyse the effects of correlation parameters on its seismic performance. Following conclusions may be drawn:

(1) The joint studied has good ductility, energy dissipation capacity and bearing capacity when the ratio of axial compression stress to strength is not beyond 0.4.

(2) Increasing the thickness of channel will significantly raise the yield and ultimate load, initial stiffness, ductility and energy dissipation coefficient of joint.

References

[1] Hao J P, Sun X L, Xue Q, Fan C L 2017 Research and application of assembled steel structure building systems, *Eng. Mech.* **34** 1-13.

[2] Yan W, Cao Y H, Li G. R 2004 Development of assembly-type RC structure and building industrialization *J-Chongqingv Jianzhu Uni* **26** 131-36.

[3] Liu X. C, Pu S. H, Xu A. X, Ni Z, Zhang A L, Yang Z W 2015 Experimental study on static and seismic performance of bolted joint in modularized multi-layer and high-rise assembled steel structures, *J. Build. Struct.* **36(12)** 43-51.

[4] Wang X, Li H, Jiang C 2003 Hysteretic behavior of angles beam-column connections, *J. Huazhong Uni. Sci. Technol. (Natur. Sci. Edi.)* **31(8)** 13-5.

[5] Xin B U, Qian G U, Wang X W 2017 Cyclic Behavior of Spatial Beam-to-side Column Joint with T-stub, *DEStech Transc. Mater. Sci. Eng. (ICMSEA_MCE).*

[6] Li L M, Chen Y Y, Li N, Cai Y C 2010 Seismic performance of outer-shell connection of cold-formed square tubular column and H-shaped steel beam, *J. Jilin Uni (Eng. Technol. Edi.)* **40(1).**

[7] Li G Q, Duan L, Lu Y, Zhang L 2015 Experimental and theoretical study of bearing capacity for extended endplate connections between rectangular tubular columns and H-shaped beams with single direction bolts, *J. Build. Struct.* **36** 91-100.

[8] Yang C, Yang J F, Su M Z, Liu C Z 2016 Numerical study on seismic behaviours of ConXL biaxial moment connection, *J. Constr. Steel Res.* **121** 185-201.

[9] Rezaeian A, Omidi M J, Shahidi F 2014 Seismic behavior of ConXL rigid connection in box-columns not filled with concrete, *J. Constr. Steel Res.* **97** 79-104.

[10] Yang J F, Chen L, Cheng J P, Zhan Y H, Yan X F 2017 Experimental study on seismic behavior of a new type of fully assembled beam-column joints, *Eng. Mech.* **34** 75-86.

Stable Colloidal Silica Particles Doped with a Fluorescent Dye

Tomiris Mulikova[1], Laura Khamkhash[2], Kanat Dukenbayev[3], Anara Molkenova[1], Timur Sh. Atabaev[1]

[1]Department of Chemistry, Nazarbayev University, Astana 010000, Kazakhstan
[2]Nazarbayev University Core Facilities, Astana 010000, Kazakhstan
[3]Electronic and Computer Engineering Department, Nazarbayev University, Astana 010000, Kazakhstan

timur.atabaev@nu.edu.kz

Abstract. In this study, we described the quick synthesis strategy of stable colloidal silica particles encapsulated with a fluorescent dye. Atomic Force Microscopy AFM, Fourier-Transform Infrared Spectroscopy FTIR, and Fluorescence Spectroscopy FS were utilized to characterize the obtained silica particles. AFM analysis revealed that obtained silica particles have a spherical shape with a mean size of 210 ± 13 nm. FS analysis showed that fluorescent dye was successfully incorporated into a silica matrix. In particular, eye-visible green fluorescence emission with a peak maximum at 526 nm was detected. The emission stability of prepared particles was tested in a biologically-relevant pH range, under constant UV irradiation and at different storage time intervals.

1. Introduction

To date, various organic fluorescent dyes are widely utilized for bioimaging thanks to their good optical properties and nontoxicity [1]. On the other hand, organic dyes have several drawbacks such as chemical instability, temperature and pH sensitivity, and quick fluorescence quenching [2, 3]. To address these issues, a fluorescent dye can be incorporated into a protective media. For example, indocyanine green dye (ICG) can be successfully encapsulated within the polymeric and silica nanoparticles [4, 5]. However, encapsulation of organic dye in a polymer matrix is still a challenging task because of the low incorporation process. From this point of view, silica is an ideal material for introducing the fluorescent molecules thanks to the preparation simplicity, low-cost, chemical inertness, and biocompatibility [3]. Silica matrix can be an effective barrier protecting the dye from the surrounding environment. Furthermore, biocompatible silica particles with excellent optical properties can be of great interest for biomedical imaging, sensing and security applications [3, 6-8]. Therefore, one of the objectives of this study was the development of a quick and user-friendly synthesis of colloidal silica particles encapsulated with a fluorescent dye. In addition, we also studied the photostability of the prepared sample in terms of storage time and pH range.

2. Materials and Methods

For the synthesis of colloidal SiO_2 particles encapsulated with a fluorescent dye, 30 ml of ethanol (96% purity) was mixed with 10 ml of deionized (DI) water. Later on, 5 mg of 2, 7 – dichlorofluorescein dye (~ 90% purity) and 0.5 ml of tetraethyl orthosilicate (TEOS, 99.999% purity) were added. The obtained solution was heated to 30 ^{0}C and vigorously stirred (700 rpm) for 5 min. Next, 300 μl of

NaOH (0.5 M) was added and left to react for 2 hours. Formed silica particles were collected using centrifugation (4000 rpm), washed several times with DI water and dried.

Atomic Force Microscope (AFM, SmartSPM 1000) was used to analyze the morphology and size of prepared SiO_2 particles. IR transmission measurements were carried out using a Fourier-transform infrared spectrometer (FTIR, Nicolet iS5). The optical properties of nanoprobes were examined using a fluorescence spectrophotometer (Agilent Cary Eclipse). All measurements were performed at room temperature 23 ± 1 ^{0}C.

3. Results

AFM was utilized to analyze the size and shape of prepared silica particles. Figure 1 shows that synthesized particles have a spherical morphology with a mean size of 210 ± 13 nm. In some cases, small clusters consisting of several particles can be observed as well. The colloidal solution prepared from silica powder can be stable for 2-3 days. After that, the precipitation process can be observed. However, one can disperse the precipitated particles by simple sonication process for 5-10 seconds.

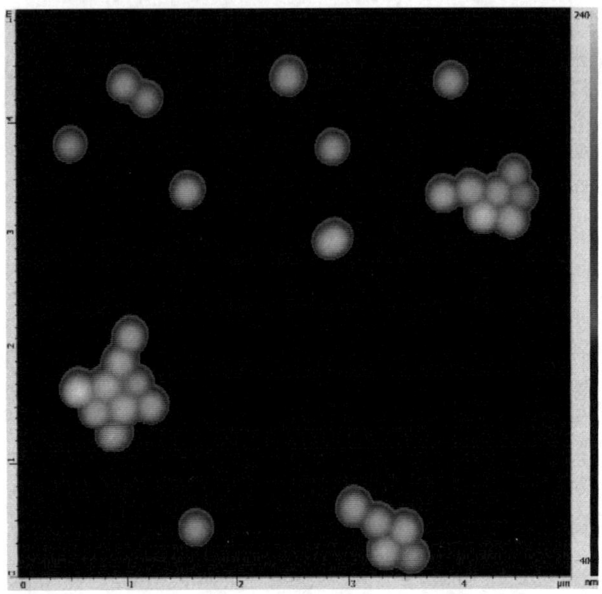

Figure 1. AFM image of prepared fluorescent silica particles.

Figure 2 shows the FTIR analysis of the prepared particles. Strong absorption bands at 800 and 1076 cm^{-1} confirmed the formation of SiO_2 structure. The band near 800 cm^{-1} arises from symmetric stretching vibration of Si-O-Si bond [9, 10]. The characteristic band at 1076 cm^{-1} was assigned to the asymmetric stretching vibrations of Si-O-Si bond in the SiO_4 tetrahedron [9, 10].

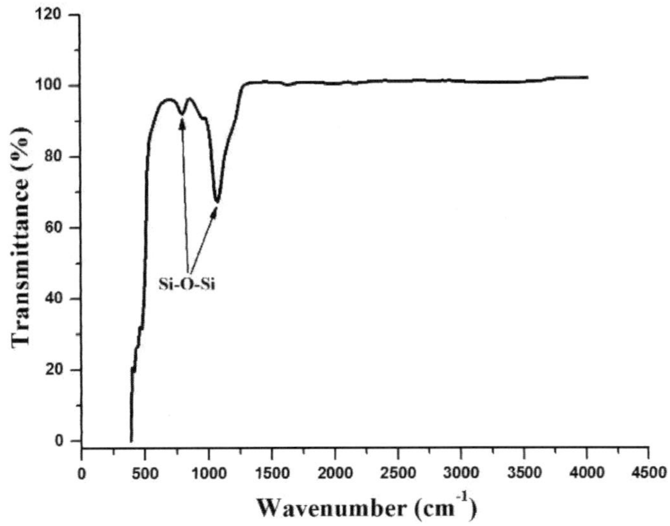

Figure 2. FTIR analysis of prepared fluorescent silica particles.

Figure 3 shows the fluorescence excitation and emission spectra of the silica particles. One can notice that the emission and excitation spectrum consist of two well-resolved peaks. The maximum of the excitation was detected at 336 nm, while the maximum of emission peak was detected at 526 nm. One can notice that the emission peak is nearly similar to those of the parent 2, 7 – dichlorofluorescein dye (529 nm in 0.1 M Tris pH 8.0).

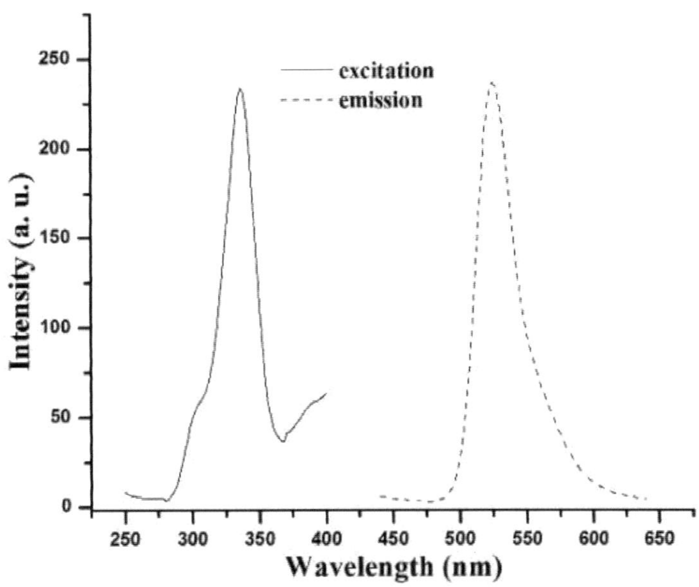

Figure 3. Excitation and emission curves of prepared fluorescent silica particles.

The fluorescence stability is an important parameter for numerous applications such as bioimaging, sensing, and security. The photostability of prepared silica particles was investigated in biologically-relevant pH range of 4-8 (figure 4). It was found that these silica particles can provide a stable emission in the pH range of 6-8, and decreased at lower pH range.

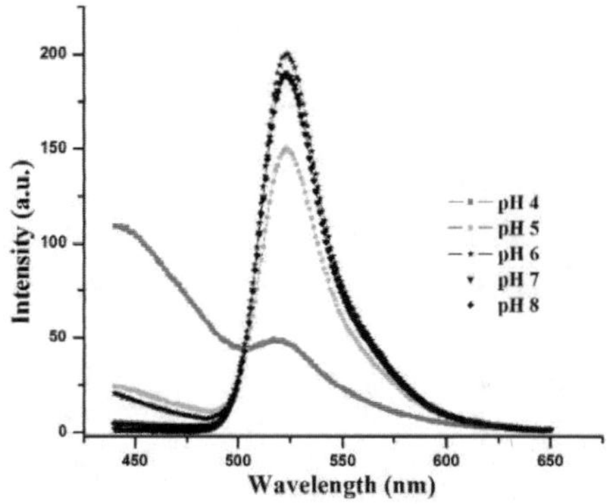

Figure 4. pH-dependent emission of prepared fluorescent silica particles.

4. Discussion

This study proposed a quick synthesis strategy for the preparation of stable fluorescent silica particles. We showed that prepared silica particles have a spherical morphology with a mean diameter of 210 ± 13 nm. FTIR analysis further confirmed the formation of pure silica structure. Interestingly, no traces of absorbed water or 2, 7 – dichlorofluorescein dye on the surface of silica particles were detected. The prepared dye-doped silica particles have yellowish-orange color, while undoped silica particles have a white color. Also, FS analysis reveals that the dye-doped silica particles are fluorescent and the emission pattern was similar to that of the bare dye. Therefore, it can be concluded that the dye molecules were successfully incorporated into silica particles. Incorporation of the dye inside the silica can be quite beneficial in terms of the preservation of the organic dye from the surrounding environment. It was also found that the silica particles are photostable in the range of 6-8 only. It can be explained by partial etching of the amorphous silica matrix in acidic conditions. Figure 5 shows the UV irradiation stability of silica particles measured at pH 7. UV lamp (λ=365 nm) with a power 6 W was used as an irradiation source. One can notice that the emission intensity slightly decreasing with the UV irradiation time. Deterioration of the emission signal is mainly related to the UV-induced degradation of 2, 7 – dichlorofluorescein dye in a silica matrix. Nevertheless, around 67.3 % of initial intensity was still preserved after 30 min of harsh UV irradiation.

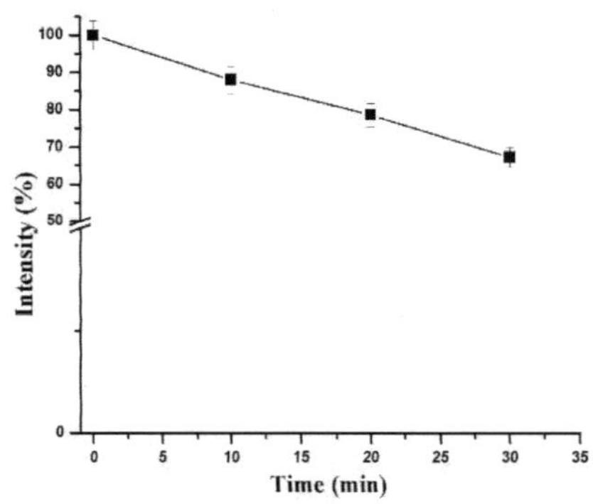

Figure 5. Emission stability of prepared fluorescent silica particles under UV-irradiation.

We also found that prepared silica particles can be stable in terms of storage time. Prepared dried particles were kept in the dark at ambient conditions (T = 22 ± 2 ^{0}C, humidity 30 – 40 %) for 1 month, and colloidal solutions of silica particles with the same concentrations were prepared every 10 days. The experiments were repeated three times for each trial. Figure 6 shows that no significant fluorescence quenching was observed in all cases. The fluorescence intensity deviation was negligible (± 3.7 %). Thus, it can be concluded that fluorescent silica particles can be stored safely for some time without significant fluorescence loss.

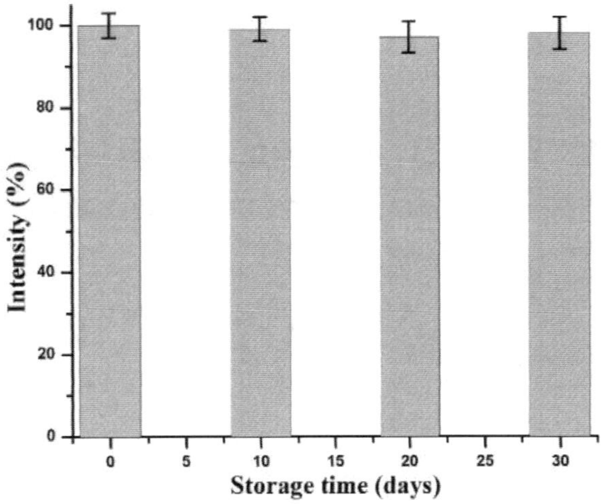

Figure 6. Storage time-dependent emission of prepared fluorescent silica particles.

5. Conclusions

In this study, we describe the one-pot synthesis method for the preparation of fluorescent silica particles with the mean diameter of 210 ± 13 nm. FS revealed that these particles have strong emission with a maximum at 526 nm. We found that prepared silica particles are photostable in the pH range of 6-8. In addition, we also found that the sample is photostable in terms of storage time. Thus, prepared fluorescent silica particles can be utilized for bioimaging, sensing and security applications.

Acknowledgements

Anara Molkenova would like to acknowledge the NU postdoc program.

References

[1] Ettinger A, Wittmann T 2014 Fluorescence live cell imaging, *Methods Cell. Biol.* **123** 77-94.

[2] Bae S W, Tan W, Hong J I 2012 Fluorescent dye-doped silica nanoparticles: new tool for bioapplications, *Chem. Commun.* **48(17)** 2270-82.

[3] Atabaev T S, Urmanova G, Ajmal M, Hong N H 2013 Fabrication of non-toxic dye-embedded silica particles for live cell imaging purposes, *BioNanoSci.* **3(2)** 132-6.

[4] Han Y H, Kankala R K, Wang S B, Chen A Z 2018 Leveraging Engineering of Indocyanine Green-Encapsulated Polymeric Nanocomposites for Biomedical Applications, *Nanomater.* **8(6)** 360.

[5] Quan B, Choi K, Kim Y H, Kang K W, Chung D S 2012 Near infrared dye indocyanine green doped silica nanoparticles for biological imaging, *Talanta* **99** 387-93.

[6] Xu J, Liang J, Li J, Yang W 2010 Multicolor dye-doped silica nanoparticles independent of FRET, *Langmuir* **26(20)** 15722-5.

[7] Ha C T, Lien N T H, Anh N D, Lam N L 2017 Development of Natural Anthocyanin Dye-Doped Silica Nanoparticles for pH and Borate-Sensing Applications, *J. Electron. Mater.* **46(12)** 6843-7.

[8] Korzeniowska B, Nooney R, Wencel D, McDonagh C 2013 Silica nanoparticles for cell imaging and intracellular, sensing *Nanotechnol.* **24(44)** 442002.

[9] Zhang Q, Chen C, Wang M, Cai J, Xu J, Xia C 2011 Facile preparation of highly-dispersed cobalt-silicon mixed oxide nanosphere and its catalytic application in cyclohexane selective oxidation, *Nanoscale Res. Lett.* **6(1)** 586.

[10] Atabaev T S, Lee J H, Han D W, Choo K S, Jeon U B, Hwang J Y, Yeom J A, Kang C H, Kim H K, Hwang Y H 2016 Multicolor nanoprobes based on silica-coated gadolinium oxide nanoparticles with highly reduced toxicity, *RSC Adv.* **6(24)** 19758-62.

Earthquake Vulnerability Assessment of High-Rise Buildings in Surabaya using RViSITS Android Application

Wahyu Riyanto[1, a], Djoko Irawan[1, b], Tri Joko Wahyu Adi[1, c], Data Iranata[1, d], Aniendhita Rizki Amalia[1, 2, e]

[1]Department of Civil Engineering, Faculty of Civil, Environmental and Geo Engineering, Institut Teknologi Sepuluh Nopember, Surabaya, Indonesia
[2]Graduate School of Science and Technology, Kumamoto University, Kumamoto, Japan

[a]wahyuriyanto303@gmail.com, [b]djoko_i@ce.its.ac.id, [c]tri_joko@ce.its.ac.id, [d]data@ce.its.ac.id, [e]aniendhita.ra@gmail.com.

Abstract. This article presents about vulnerability assessment of high-rise building in Surabaya, the second largest city in Indonesia. In 2007, the Earthquake National Research Center -Indonesia's Ministry of Public Works released the latest earthquake fault map. Include the discovery of Surabaya fault and Waru fault that across Surabaya. According to that new discovered fault, it is possible that there is a higher potential earthquake hazard for buildings structures in Surabaya. Therefore, the efforts to assess the vulnerability of buildings towards the earthquake need to be conducted, especially for the high-rise buildings. An android based RViSITS (Rapid Visual Survey by Institut Teknologi Sepuluh Nopember) is an application that has been developed based on FEMA 154. That application helps the vulnerability assessments of buildings can be conducted quickly and efficiently. Since, it does not require a lot of personnel, quite short research time and cheaper cost than having to use a manual form that is considered less effective for a large number of buildings. This research itself is conducted to assess the vulnerability of high-rise buildings (8-15 floors) throughout Surabaya city using the RViSITS android application. Based on the results of this study on 98 high rise buildings, it is found that 21 buildings (21.43%) are in the prone category since they had an index value < 2, while 77 buildings (78.57%) are in the safe category with an index ≥ 2. The result obtained in this research can be used as a database of local government institution for mapping the vulnerability of high rise buildings and initial input for mitigation plan toward the higher possibility of an earthquake hazard in Surabaya.

1. Introduction

Geologically, the Indonesian archipelago is at the confluence of two major earthquake pathways. It is Pacific Circum and Alpide Transasiatic earthquake path. Whereas geographically, the Indonesian archipelago is at 6° N-11° South and 95° East-141° East. In addition, Indonesia is located at the confluence of three crustal plates that is the Eurasian plate, Pacific plate and Australian Indian plate. Therefore, the Indonesian archipelago is in an area that has high tectonic and volcanic earthquake activities.

Indonesian government has stipulated several regulations relating to earthquakes such as SNI 1726-2012 concerning earthquake loads on building structures and SNI 2874-2013 regarding on planning of concrete building structures that are resistant to earthquake loads. However, in Indonesia there are many buildings that are still built without regard to these regulations. The more buildings are built without considering the existing regulations; the number of buildings that are vulnerable to damage will increase. It is added by the absence of data related to the vulnerability of each building that has been built, therefore the risk of building damage due to the earthquake is greater [1].

In fact, nowadays, earthquake is still become a threat of disasters for Indonesian people. The impact caused by earthquake is very diverse. It can cause harm to death for humans and the damage of buildings [2]. An earthquake is a disaster that cannot be predicted and prevented by humans, but the impact caused by an earthquake can be minimized by choosing appropriate steps and methods [3].

Earthquake National Research Center -Indonesia's Ministry of Public Works has released the latest earthquake fault map update in 2017. It causes the emergence of sources and potential new earthquake hazards in Indonesia. One of them is the emergence of Surabaya fault and Waru fault across Surabaya city [4]. Thus, some efforts should be made as early as possible for the assessment of vulnerability, especially for buildings which is still functioning to minimize seismic damage due to earthquakes [5].

Building vulnerability assessment is distinguished by a thorough check and quick check called rapid visual screening (RVS). On a thorough check requires a team of structural and earthquake experts, besides this method requires an expensive cost and takes a long time to conduct. Moreover, examination using the RVS method does not require a team of experts and many personnel. It can be completed in a short time and cheaper cost [1,6]. The application of RVS method is used to ensure the condition of building whether it is vulnerable or safe and as a basis for providing follow-up recommendations on the vulnerability of buildings for mitigation plans [7]. Therefore, the RVS method is quite effective to be applied to evaluate building vulnerability in large numbers [8].

In recent digitalization era, the RVS team of Institut Teknologi Sepuluh Nopember, Surabaya, Indonesia, has successfully developed an android-based application called RViSITS. This application is the development of manual form on FEMA 154. Building vulnerability assessment using manual forms is not effective for a large number of buildings [9]. Then, by the existence of RViSITS android application, building vulnerability assessments can be conducted simpler and faster. The surveyor only needs a smartphone in practice without using the manual form for large numbers of building being examined. This application is equipped with features that are connected to the camera, so that during the survey, it can document the object examined. This application is also connected to Global Positioning System (GPS) to determine the coordinates of the building location and the results can be accessed easily by various government sectors as well as the public sectors [1,9]. Therefore, the main purpose of this study is to map the vulnerability value of buildings (8-15 floors) in Surabaya due to the 2017 earthquake map update and the Surabaya fault and the Waru fault crossing Surabaya using the RViSITS android application. Furthermore, conclusions and recommendations based on the research can be accessed by all sectors, both local governments and the public.

2. Material and Method

2.1. Rapid Visual Screening/RVS

Rapid visual screening (RVS) method is a procedure to quickly assess the vulnerability of buildings due to earthquake. In the 1980s, the RVS method for buildings is first developed by Applied Technology Council (ATC) [10]. RVS is one of the recommended building vulnerability assessment procedures since it can be done without detailed structural calculations. In practice, RVS is easier than analytical calculations that require many detailed calculations with many scenarios [11]. This method can also be conducted with a simple step; by surrounding the building from the outside and ascertain the level of vulnerability of building whether it requires further check or nor based on the final scores [1].

The RVS method has been used by many researchers to evaluate seismic vulnerability of buildings [1,2,6–8,12–21]. In addition, this method can help the authorities in strengthening buildings that are very vulnerable in anticipation of reducing the risk of damage [11]. Although the RVS method can be conducted quickly with a low cost, but this method has several shortcomings. These shortcomings include the need for a long time during the survey process and the final score calculation for a large number of buildings. It takes a long time to compile survey data into digital data. The results of

IOP Conf. Series: Materials Science and Engineering **739** (2020) 012040 doi:10.1088/1757-899X/739/1/012040

manual calculations are prone to errors, while if there are errors in recording the survey results, it is necessary to change the new form. Thus, this method is considered ineffective and inefficient [1,6,9].

The assessment form in RVS method is divided into three kinds, namely low seismicity, medium seismicity, and high seismicity. On each of the form are having difference value parameters [10,11]. In the figure 1 above shows one of manual form example for high seismicity category commonly used to assess the vulnerability of buildings using the RSV method [10].

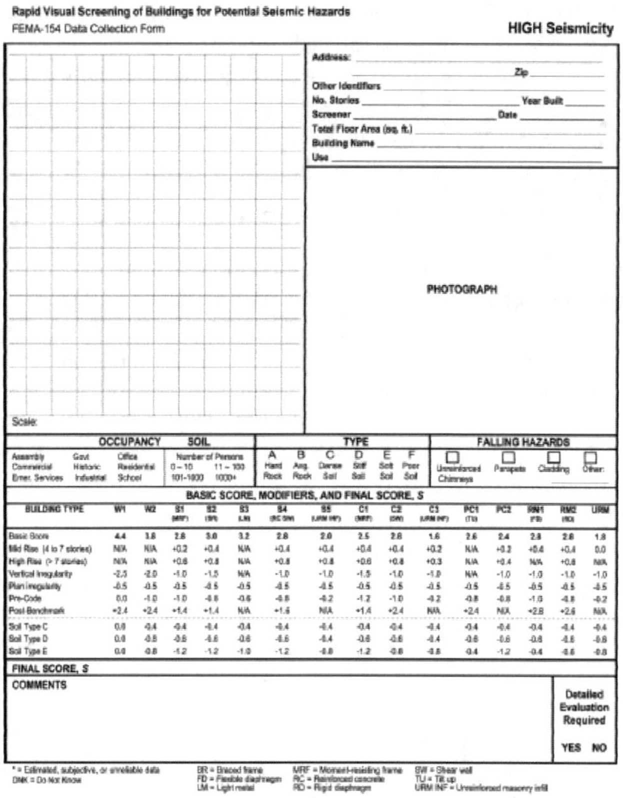

Rapid Visual Screening of Buildings for Potential Seismic Hazards
FEMA-154 Data Collection Form

HIGH Seismicity

Address: ________ Zip ________
Other Identifiers ________
No. Stories ________ Year Built ________
Screener ________ Date ________
Total Floor Area (sq. ft.) ________
Building Name ________
Use ________

Scale:

PHOTOGRAPH

OCCUPANCY			SOIL		TYPE						FALLING HAZARDS			
Assembly	Govt	Office	Number of Persons		A	B	C	D	E	F	Unreinforced Chimneys	Parapets	Cladding	Other:
Commercial	Historic	Residential	0 – 10	11 – 100	Hard Rock	Avg. Rock	Dense Soil	Stiff Soil	Soft Soil	Poor Soil				
Emer. Services	Industrial	School	101-1000	1000+										

BASIC SCORE, MODIFIERS, AND FINAL SCORE, S

BUILDING TYPE	W1	W2	S1 (MRF)	S2 (BR)	S3 (LM)	S4 (RC SW)	S5 (URM INF)	C1 (MRF)	C2 (SW)	C3 (URM INF)	PC1 (TU)	PC2	RM1 (FD)	RM2 (RD)	URM
Basic Score	4.4	3.8	2.8	3.0	3.2	2.8	2.0	2.5	2.8	1.6	2.6	2.4	2.8	2.8	1.8
Mid Rise (4 to 7 stories)	N/A	N/A	+0.2	+0.4	N/A	+0.4	+0.4	+0.4	+0.4	+0.2	N/A	+0.2	+0.4	+0.4	0.0
High Rise (> 7 stories)	N/A	N/A	+0.6	+0.8	N/A	+0.8	+0.8	+0.6	+0.8	+0.3	N/A	+0.4	N/A	+0.6	N/A
Vertical Irregularity	-2.5	-2.0	-1.0	-1.5	N/A	-1.0	-1.0	-1.5	-1.0	-1.0	N/A	-1.0	-1.0	-1.0	-1.0
Plan Irregularity	-0.5	-0.5	-0.5	-0.5	-0.5	-0.5	-0.5	-0.5	-0.5	-0.5	-0.5	-0.5	-0.5	-0.5	-0.5
Pre-Code	0.0	-1.0	-1.0	-0.8	-0.6	-0.8	-0.2	-1.2	-1.0	-0.2	-0.8	-0.8	-1.0	-0.8	-0.2
Post Benchmark	+2.4	+2.4	+1.4	+1.4	N/A	+1.6	N/A	+1.4	+2.4	N/A	+2.4	N/A	+2.8	+2.6	N/A
Soil Type C	0.0	-0.4	-0.4	-0.4	-0.4	-0.4	-0.4	-0.4	-0.4	-0.4	-0.4	-0.4	-0.4	-0.4	-0.4
Soil Type D	0.0	-0.8	-0.6	-0.6	-0.6	-0.6	-0.4	-0.6	-0.6	-0.4	-0.6	-0.6	-0.6	-0.6	-0.6
Soil Type E	0.0	-0.8	-1.2	-1.2	-1.0	-1.2	-0.8	-1.2	-0.8	-0.8	-0.4	-1.2	-0.4	-0.6	-0.8
FINAL SCORE, S															

COMMENTS

Detailed Evaluation Required

YES NO

* = Estimated, subjective, or unreliable data
DNK = Do Not Know
BR = Braced frame
FD = Flexible diaphragm
LM = Light metal
MRF = Moment-resisting frame
RC = Reinforced concrete
RD = Rigid diaphragm
SW = Shear wall
TU = Tilt up
URM INF = Unreinforced masonry infill

Figure 1. Manual Form of RVS Method in accordance with FEMA 145.

2.2. RViSITS Method

RViSITS android application is a form of manual form development based on FEMA 154. It has function to assess the vulnerability of buildings caused by earthquakes in smartphone application. This application is firstly created by RVS Team of Sepuluh Nopember Institute of Technology, Surabaya, Indonesia [1]. The development of this application has purpose to develop information system integrated for efficiency and provide solution from several obstacles face while using building vulnerability inspection system caused by earthquakes conventionally [9]. Afterwards, the development is made by wholly adapting all of the regulation in FEMA 154 with adjustment based on regulation that exists in Indonesia. The RViSITS application has provide simplicity and used to simplify and fasten the process of assessment sent via internet. Further, there will be server as collecting and processing the data. From the data, it can be obtained vulnerability mapping as well as the area that affected by disasters. Thus, it can be accessed by the public, government and private organization [1].

There are 8 steps in the assessment of building vulnerability using RViSITS namely [1,10]: 1) performing update data and information dealing with building that is reviewed based on existing data obtained; 2) walking around the building and take pictures to know the condition of buildings; 3) determining the seismicity level; 4) determining the capacity of peoples who can entering inside the buildings; 5) determining the soil types; 6) identifying the falling hazard; 7) determining the building function; 8) determining the building types on the basic score (BS); 9) identifying and click on the list of choices in M score (as like the amount of floors, building types, determining the pre–code or post benchmark and the soil type); 8) in the final score (S) it can be determined by adjusting the basic score (BS) with the modification score (MS) as mentioned in the step 7, and thus the result score will be emerged automatically in the RViSITS application. If necessary, the detail evaluation about the building can be inputted in the "insert comment."

3. Research Method

The main purpose in this research is to assess the vulnerability of high rise buildings which has eight floor up to fifteen floor in Surabaya which spread along North, South, Center, East and West Surabaya in accordance with the figure 2. The assessment of the building is carried out on 98 buildings which represented five areas in Surabaya City namely 33 buildings in East Surabaya, 7 buildings in West Surabaya, 27 buildings in Central Surabaya, 7 buildings in North Surabaya, and 24 buildings in Surabaya South. The building chosen has eight to fifteen floors. Chosen buildings are including commercial, residential, public service, office, industrial and collage buildings.

This research is divided into five steps such as 1) collecting the reference and regulation which supports the research; 2) conducting introduction survey to get the information about the buildings reviewed; 3) conducting assessment of building vulnerability using RViSITS application; 4) processing the results data; 5) serving the observation result in form of conclusion and recommendation.

Figure 2. Buildings Points Reviewed.

The research steps using RViSITS can be seen in the Figure 3 (a) to (n) [9,22].

Figure 3. The Steps in Using RViSITS Androrid Program (a) to (n).

The score in RVS is divided into three kinds. Basic score (BS) is determined based on the building structure reviewed. In the Modifiers Score (MS) is influenced by several factors including the building type (mid-rise building and high rise building), the shape of building (vertical and plan irregularity), pre-code, post-benchmarks, and the soil type (C, D, and E type). Meanwhile, in the last score (S) is obtained based on the addition of basic score (BS) and modifiers score (MS) chosen. The final score is amounted in range from 0 to 7. The higher the score shows that the performance structure is better. Based on the current design seismic criteria, the final score (S) is permitted not less than 2 score. The buildings with S score is less than 2 should be assess more by the professional structure expert in the earthquake planning field [1,10]. Here is the pattern to obtain the final score as like equation (1).

$$Final\ Score\ (S) = Basic\ Score\ (BS) + \ Modifiers\ Score\ (MS) \quad (1)$$

The potential of building damage is categorized into five grades in accordance with the final score (S) as mentioned in the table 1.

Table 1. Damage Potential based on Structural Score [11].

RVS Score	Damage Potensial
$S \leq 0.3$	High probability of grade 5 damage; very high probability of grade 4 damage
$0.3 < S \leq 0.7$	High probability of grade 4 damage; very high probability of grade 3 damage
$0.7 < S \leq 2.0$	High probability of grade 3 damage; very high probability of grade 2 damage
$2.0 < S \leq 2.5$	High probability of grade 2 damage; very high probability of grade 1 damage
$S > 2.5$	Probability of grade 1 damage

In the grade 1, there is little bit damage on the building, thus it can be ignored. In the grade 2, there is medium damage in which the damage looks like cracks in the wall. In the grade 3, the building is predicted will heavily damage with the cracks in the column and shell column. In the grade 4, the damage categorize as very heavy in which the several column or upper floor will collapse, meanwhile in the grade 5, the ground floors collapsed [11,14].

4. Result and Discussion

In the analysis process using RViSITS application, the basic score is influenced by several parameters, namely: 1) Seismicity level: based on earthquake map 2017, the acceleration of the period of 0.2 second of Surabaya City is 0.9 g, and the acceleration of the 1 second period is 0.3 g. It means that the Surabaya City in High Seismicity zone according on FEMA 154, so this paper used the RVS High Seismicity level.; 2) Buildings type: 97 buildings are C2 (concrete shear wall), and 1 building is S2 (steel bracing frame).

On modifiers score, all buildings reviewed are included in the high rise building category. Type of soil being soft soil (Type E). Other factors are influenced by shape of building (18 buildings including vertical irregularity, 7 buildings including plan irregularity, 20 buildings including both, and 53 other buildings including typical buildings vertically and horizontally). In terms of the year of construction, 21 buildings were built before 2002 where earthquake regulations have not been implemented in Indonesia. So the score modifier is selected post code, and 77 buildings are selected post benchmark.

Based on the result analysis using RViSITS application is described in the figure 4.

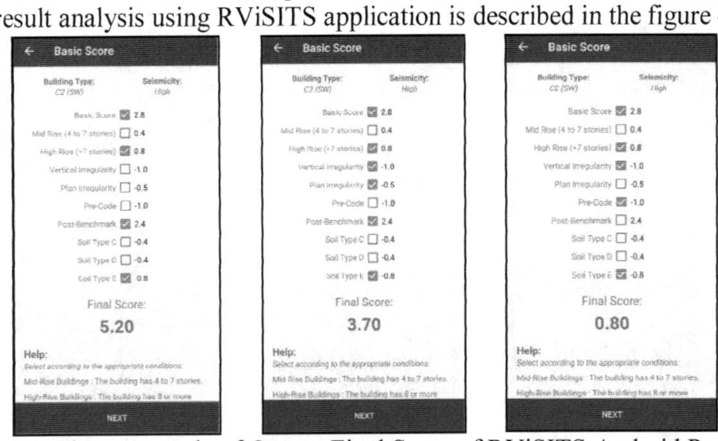

Figure 4. Three example of Output Final Score of RViSITS Android Program.

The final score is obtained automatically by the RViSITS android application based on the sum of the values of the basic score and modifiers score as in formula1. From the figure 4 the final scores for all the buildings being reviewed are presented in a diagram like in figure 5 (a) to (e). It shows the result analysis using RViSITS application found that there are 21 buildings which including into vulnerability category since the final score (S) is < 2. Otherwise, 77 other buildings included into save category with the final score (S) is ≥ 2.

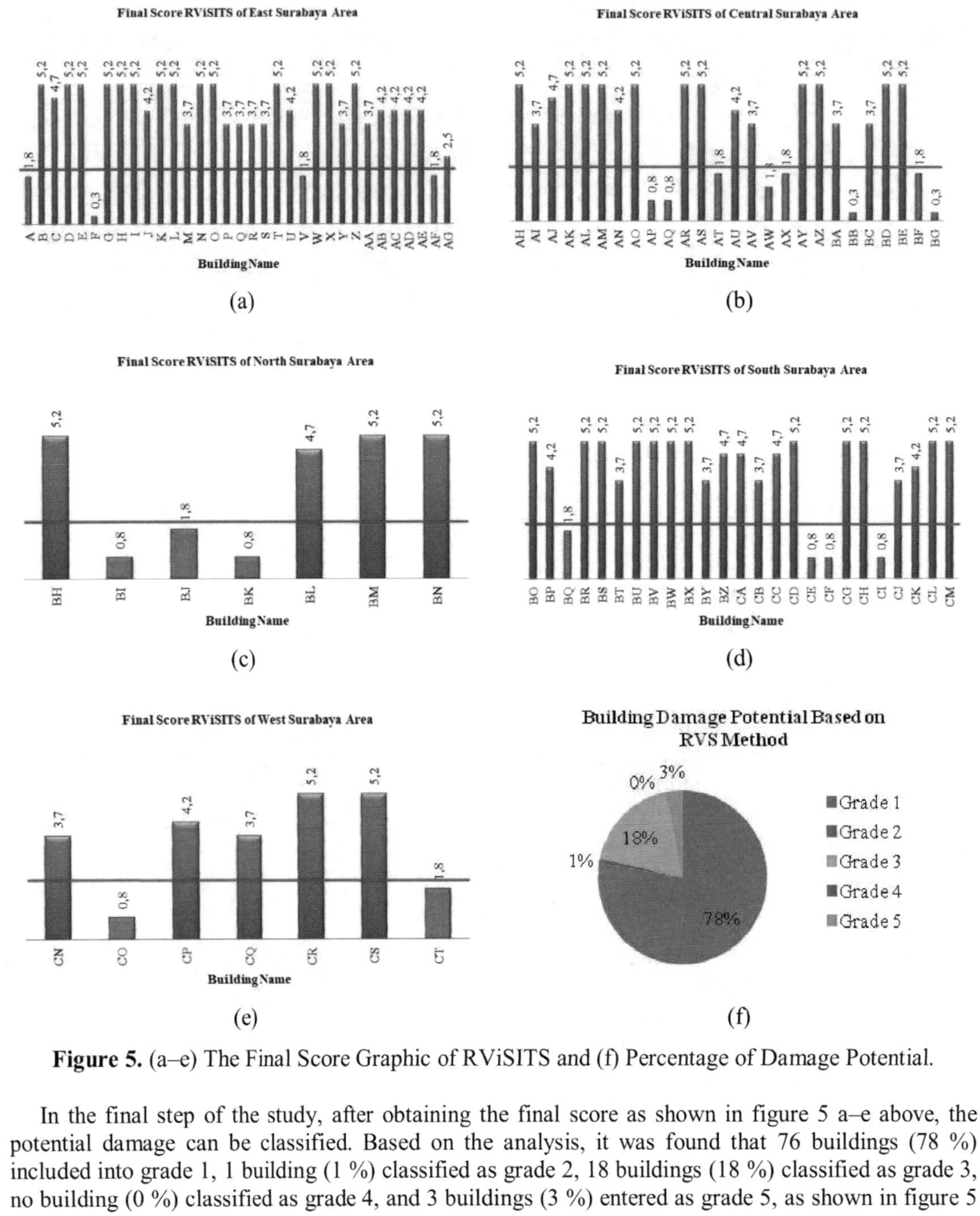

Figure 5. (a–e) The Final Score Graphic of RViSITS and (f) Percentage of Damage Potential.

In the final step of the study, after obtaining the final score as shown in figure 5 a–e above, the potential damage can be classified. Based on the analysis, it was found that 76 buildings (78 %) included into grade 1, 1 building (1 %) classified as grade 2, 18 buildings (18 %) classified as grade 3, no building (0 %) classified as grade 4, and 3 buildings (3 %) entered as grade 5, as shown in figure 5 (f).

5. Conclusions

The building vulnerability assessment was carried out using the RViSITS application on 98 high rise buildings (8-15 floors) in Surabaya. Based on the assessment it was found 21 buildings have fall into the vulnerable category are caused because all the buildings were built before 2002 in which the earthquake regulations have not been applied in Indonesia [1]. Moreover, the other causes are influenced by irregular shape of buildings both vertical and horizontal. While reviewing the possibility of the potential damage of 21 building (21%) included in grades 3 to 5, where it is possible that heavy damage will occur if an earthquake occurs on a large scale. Furthermore, the proposed recommendation is that there is a need for planning data collection, as well as further inspection of 21 buildings that are included in the vulnerable category.

Acknowledgement

This research has been funded by Penelitian Unggulan Institut Teknologi Sepuluh Nopember - 2019.

References

[1] Wahyuni E, Iranata D, Suswanto B, Nurcahyo C B and Sutrisno W 2018 Assessment of vulnerable buildings due to earthquake loading using Rapid Visual Screening smartphone application *Int. J. Adv. Sci. Eng. Inf. Technol.* **8** 567–72

[2] Perrone D, Aiello M A, Pecce M and Rossi F 2015 Rapid visual screening for seismic evaluation of RC hospital buildings *Structures* **3** 57–70

[3] Ningthoujam M C and Nanda R P 2018 Rapid visual screening procedure of existing building based on statistical analysis *Int. J. Disaster Risk Reduct.* **28** 720–30

[4] Indonesian Earthquake National Research Center Team 2017 *Peta Sumber dan Bahaya Gempa Indonesia Tahun 2017* ed S Masyur Irsyam, Sri Widiyantoro, Danny Hilman Natawidjaja, Irwan Meilano, Ariska Rudyanto, Sri Hidayati, Wahyu Triyoso, Nuraini Rahma Hanifa, Didiek Djarwadi, Lutfi Faizal (Jakarta, ID)

[5] Barbat A H, Carreño M L, Pujades L G, Lantada N, Cardona O D and Marulanda M C 2010 Seismic vulnerability and risk evaluation methods for urban areas. A review with application to a pilot area *Struct. Infrastruct. Eng.* **6** 17–38

[6] Yadollahi M, Adnan A and Zin R 2012 Seismic vulnerability functional method for rapid visual screening of existing buildings *Arch. Civ. Eng.* **58** 363–77

[7] Sadat M R, Huq M S and Ansary M A 2010 Seismic vulnerability assessment of buildings of Dhaka city *J. Civ. Eng.* **38** 159–72

[8] Shah F, Ahmed A, Kegyes-Brassai O, Alghamdi A, Kegyes-B O K, Al-Ghamadi A and Ray R P 2016 A Case Study Using Rapid Visual Screening Method to Determine the Vulnerability of Buildings in two Districts of Jeddah, Saudi Arabia *15th Int. Symp. New Technol. Urban Saf. Mega Cities Asia*

[9] Team R R 2016 *Training Module - Rapid visual screening of Building using Android Based Apps. RViSITS* ed R Team (Surabaya, ID)

[10] Rojahn C, Scawthorn C, Anagnos T, Wong K and Poland C D 2002 *Rapid visual screening of buildings for potential seismic hazards, Handbook, FEMA 154* (Washington, DC: Applied Technology Council)

[11] Mohamad I I, Mohd Yunus M Z and Harith N S H 2019 Assessment of building vulnerability by integrating rapid visual screening and geographic information system: A case study of Ranau township *IOP Conference Series: Materials Science and Engineering* **vol 527** (Kuala Lumpur, Malaysia: IOP Publishing) pp 1–8

[12] Formisano A and Marzo A 2017 Simplified and refined methods for seismic vulnerability assessment and retrofitting of an Italian cultural heritage masonry building *Comput. Struct.* **180** 13–26

[13] Karapetrou S T, Fotopoulou S D and Pitilakis K D 2017 Seismic Vulnerability of RC Buildings under the Effect of Aging *Procedia Environmental Sciences* **vol 38** (Thessaloniki, Greece: Elsevier B.V.) pp 461–8

[14] Shah M F, Kegyes-B. O K, Ray R P, Ahmed A and Al-ghamadi A 2018 Vulnerability

Assessment of Residential Buildings in Jeddah: A Methodological Proposal *Int. J. GEOMATE* **14** 134–41

[15] Ajay Kumar S, Rajaram C, Mishra S, Pradeep Kumar R and Karnath A 2017 Rapid visual screening of different housing typologies in Himachal Pradesh, India *Nat. Hazards* **85** 1851–75

[16] Rahman N, Ansary M A and Islam I 2015 GIS based mapping of vulnerability to earthquake and fire hazard in Dhaka city, Bangladesh *Int. J. Disaster Risk Reduct.* **13** 291–300

[17] Achs G and Adam C 2012 Rapid seismic evaluation of historic brick-masonry buildings in Vienna (Austria) based on visual screening *Bull. Earthq. Eng.* **10** 1833–56

[18] Achs G and Adam C 2012 A Rapid-Visual-Screening Methodology for the Seismic Vulnerability Assessment of Historic Brick-Masonry Buildings in Vienna *15th World Conference on Earthquake Engineering (15WCEE)* **vol 154** (Lisboa)

[19] Sarmah T and Das S 2018 Earthquake Vulnerability Assessment for RCC Buildings of Guwahati City using Rapid Visual Screening *Procedia Engineering* **vol 212** (Bangkok, Thailand: Elsevier B.V.) pp 214–21

[20] Hancilar U, Çaktö E, Erdik M, Franco G E and Deodatis G 2014 Earthquake vulnerability of school buildings: Probabilistic structural fragility analyses *Soil Dyn. Earthq. Eng.* **67** 169–78

[21] Srikanth T, Kumar R P, Singh A P, Rastogi B K and Kumar S 2010 Earthquake vulnerability assessment of existing buildings in Gandhidham and Adipur Cities Kachchh, Gujarat (India) *Eur. J. Sci. Res.* **41** 336–53

[22] Riyanto W, Irawan D, Adi T J W, Iranata D and Amalia A R 2019 Structural Performance Monitoring System Using Android RViSITS For High Rise Building *J. Phys. Conf. Ser.* **1381** 1–8

IOP Conf. Series: Materials Science and Engineering **739** (2020) 012041 doi:10.1088/1757-899X/739/1/012041

A Comparative Study of Hyperelastic Constitutive Models to Characterize the Behaviour of a Biopolymer Material for Diaphragm of Blood Pump Manufacturing

L. V. Belyaev[1], **A. V. Zhdanov**[1], **A. B. Ivanchenko**[1], **A. V. Stepenkin**[1], **D. O. Kochetov**[1]

[1]Department of the mechanical engineering, Vladimir State University, Gorky str. 87, Vladimir, 600000, Russia

Abstract. This article presents the results of the stress-strain state modelling of a blood pump diaphragm manufacturing from polyurethane 'Vitur' T-1413-85 for pulsatile type Ventricle Assist Device systems. Based on the analysis of the most distributed models of hyperelasticity, a model which most accurately describes the behaviour of the specified material has been determined ($R^2 = 0.9999$ between numerical simulation and experimental uniaxial tensile tests data). Experimental studies have been carried out to determine the displacement of the diaphragm with a thickness of 0.3 mm and 0.5mm under the influence of control pressure. A comparison of the modelling results of the blood pump diaphragm stress-strain state and the conducted experimental studies on the determination of displacements confirmed the adequacy of the chosen model ($R^2 = 0.96$ between numerical simulation and experiment) for both of the studied diaphragm thickness.

Introduction

According to statistical data, every year in Russia are born about 10000 children with congenital heart diseases. For every 1000 newborns, there are 10 children who need heart surgery [1]. The mortality rate for children on the waiting list under the age of 18 is several times higher than for other age groups, which is largely due to the lack of long-term mechanical circulatory support systems. The designing process of paediatric Ventricle Assist Device (VAD) is a more difficult task than directly reducing the existing systems which have a large volume of ejection [2]. A comparative analysis of statistical data about the use of paediatric pulsatile and non-pulsating type VAD systems showed that the negative effects began to appear precisely in pulsatile type VAD systems [3, 4] due to the fact that these systems were not designed for use in paediatrics, but older age groups VAD systems were adapted. Therefore, it is necessary to note the need for a separate development of the VAD system for each age group of patients.

One of the types of systems that are most often used for these purposes are pulsatile type paracorporal VAD systems with a pneumatic drive [5, 6], which are designed for long-term use and should provide an estimated service life of at least five years or about 200 million cycles of the blood pump works; and contain a blood compartment and an air chamber separated by an elastic diaphragm, creating a pulsatile flow.

Thus, the main element of the blood pump, critically affecting the life of its work, is the diaphragm, which must not only support the above number of work cycles, but also has to provide the required hemodynamic characteristics of its work [7].

Currently, to determine the number of cycles of the blood pump, an approach based on the operation numerical modelling of these systems using virtual experimental stands can be used [4] to predict the performance of the VAD systems at the design stage.

However, for the successful application of this approach, it is necessary to use adequate mathematical models that describe the behaviour of a particular material used to manufacture the blood pump diaphragm.

To considerate the working conditions of the diaphragm special requirements for structural materials are required for their manufacture. They should have a high wearing capacity, the required damping, thermophysical, electrical and elastic properties [8], and a resistance to biological fluids; and provide biocompatibility, a good shape restoration after deformation and not lose them after sterilization [9]. Materials that most fully meet the specified requirements are thermoplastic polyurethanes, which have the indicated properties, the appropriate certificates and are technologically advanced to manufacture [10].

This article presents the results for determining the model of hyperelasticity for the correct mathematical description of the 'Vitur' grade polyurethane behaviour for the further virtual modelling of the blood pump diaphragm behaviour in order to predict the required service life for its work.

Materials and methods

The diaphragm of blood pump is made of polyurethane 'Vitur' T-1413-85 (LLC SPF 'VITUR') [11]. Uniaxial tensile tests were performed, according to the standard test methods (standard ISO 37:2011), on dog-bone specimens (n = 10) excised from the diaphragm. Experiments were carried out using a Lloyd tensile test machine (AMETEK TCI, Berwyn, PA) with a load cell of 5 N and a strain rate of 20 mm/min. Air was modelled as an ideal gas with a density of 1.29×10^{-3} kg/m^3.

To obtain a constitutive model that can reproduce with the maximum possible accuracy the behaviour of the 'Vitur' T-1413-85 material, four classic hyperelastic models (and corresponding submodels) were analysed. It was Mooney–Rivlin (2 and 5 parameters) [12, 13], Neo–Hookean [14], Ogden (N = 1, N = 2) [15] and Yeoh (3 parameters) [16].

According to the well-known experimental loading curve, the constants in the strain potential energy were determined in the ANSYS program (ANSYS, Inc., USA) using the «Curve Fitting» procedure. To determine the quality of each of the models, the R^2 correlation coefficient was calculated. As a result of the models, a strain energy density curve and a stress–strain curve were obtained. This stress–strain curve will be compared with the actual material curve. The best hyperelastic model will be the one whose average values R^2 (strain energy density and stress) is closer to 1.

After determining the model of hyperelasticity, on the basis of the criterion discussed above, the stress-strain state of the blood pump diaphragm has been simulated on the basis of its 3D model. The diaphragm 3D-model is a cylindrical disk with an outer diameter of 80 mm, a working diameter of 50 mm and 0.5 mm and 0.3 mm of thickness. Due to the diaphragm geometric symmetry and the acting load, its ¼ part has been considered. The thickness of the diaphragm was represented by three layers of three-dimensional 8-node finite elements. The diaphragm finite element model contains 77 068 and 20 7884 nodes, 56 409 and 153 744 elements with a thickness equal to 0.5 mm and 0.3 mm, respectively. The boundary conditions corresponded to the membrane fixation conditions of during experimental studies and the selected calculation scheme. The load was set by a linearly increasing pressure for 3 seconds, from 0 to 1.1 atm (10 7873 Pa) spaced at intervals of 0.1 s.

To estimate the accuracy of the simulation results, a comparison of the calculated and experimental data has been carried out. The experimental data were obtained at the test bench, whose diagram is shown in figure 1. The diaphragm made from the 'Vitur' T-1413-85 material with a thickness of 0.5 mm and 0.3 mm is secured between the aluminium base and the clamp with six screws. On the one hand, a pressure gauge is connected to the aluminium base to control the pressure; on the other, the pressure regulator providing the air supply of the required pressure. To measure the displacements (h) of the membrane midpoint in the vertical plane, we used an electro-optical sensor with high sensitivity installed in the arm clamp, and the measurement results have been displayed on the monitor. Five experiments have been carried out for each aperture thickness.

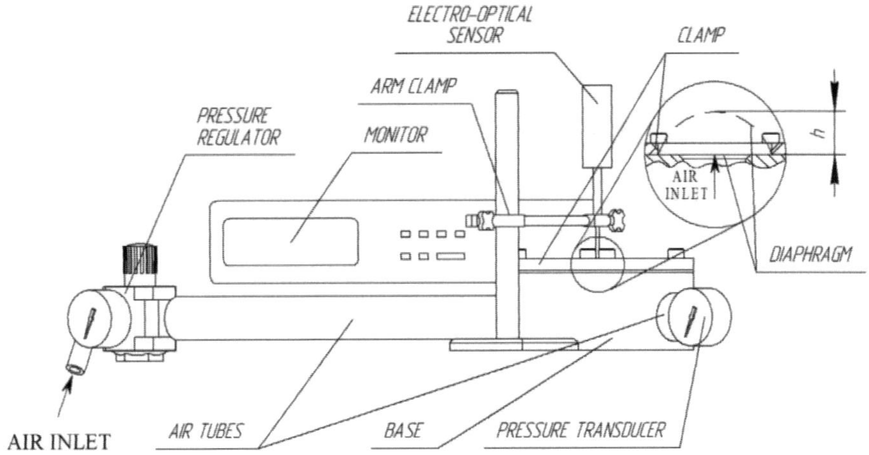

Figure 1. Schematic of experimental setup for diaphragm deformation assessment.

Results

In this section, the results obtained for each of the models (and submodels) are presented. The stress–strain curves and strain energy density function curves are shown in figure 2.

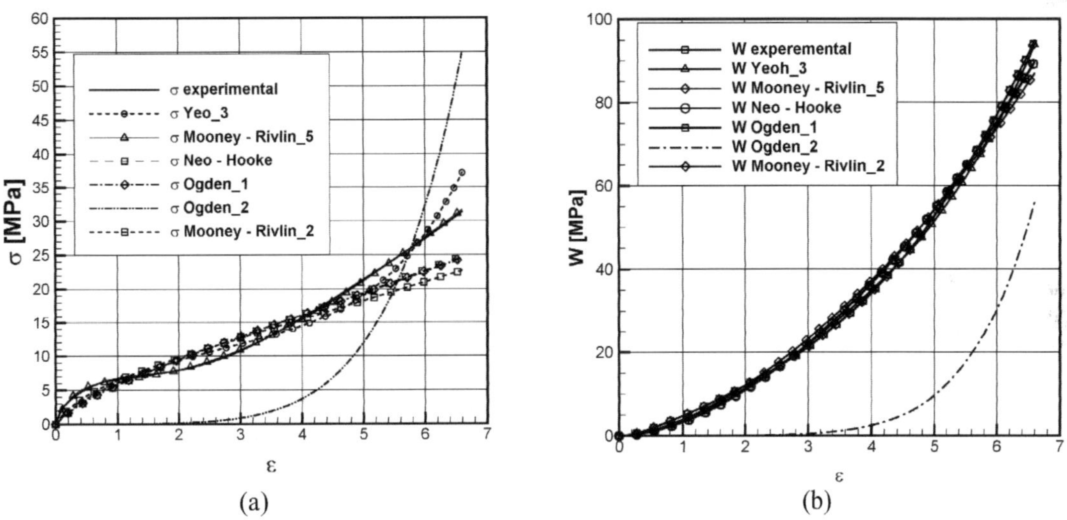

Figure 2. Comparison of experimental and computation data:
Stress-strain (a) and Strain energy density - strain (b) plots.

The values of the material constants and R^2 coefficient are shown for each of the studied models in table 1.

The analysis of the data presented in table 1 shows that the behaviour of the material under study can be described by all the considered models of hyperelasticity, except for the Ogden model (N = 2), which has a coefficient value of $R^2 < 0{,}9$. However, the most accurate description is given using the five parametric model Mooney–Rivlin (average value $R^2 = 0.9999$) and it was used to simulate the stress-strain state of the blood pump diaphragm.

As a result of numerical and experimental research, the displacements of the central point of the diaphragm have been obtained (figure 3). The analysis of the graphs shows that the experimental values of the displacements have larger values than those obtained in the modelling process. The large displacements values of the experimental (14.88 mm) and those obtained in the modelling process

(12.6 mm) are characteristic of the smaller aperture thickness: 0.3 mm. The corresponding movements for an aperture thickness of 0.5 mm are: 12.05 mm and 10.09 mm. respectively.

Table 1. The values of the material constants and R^2 coefficient.

Model	Material parameter	Value	R^2 (σ)	R^2(W)	Average [R^2(σ);R^2(W)]
Mooney–Rivlin 2 parameters	C_{10}	1 373 972.9308 Pa	0.9619	0.9967	0.9793
	C_{01}	926 605.6951 Pa			
Mooney–Rivlin 5 parameters	C_{01}	4 262 541.1172 Pa	0.9997	1.0000	0.9999
	C_{02}	–552 059.6224 Pa			
	C_{10}	–139 000.0739 Pa			
	C_{11}	277 624.5084 Pa			
	C_{20}	–17 835.0520 Pa			
Neo-Hookean	μ	3 260 238.7247 Pa	0.9683	0.9980	0.9832
Ogden N = 1	α_1	1.9837	0.9671	0.9979	0.9825
	μ_1	3 336 940.0532 Pa			
Ogden N = 2	α_1	7.4353	0.8910	0.8964	0.8937
	α_2	7.4372			
	μ_1	58.6609 Pa			
	μ_2	58.6610 Pa			
Yeoh 3 parameters	C_{10}	1 800 959.9809 Pa	0.9809	0.9994	0.9902
	C_{20}	–16 902.1001 Pa			
	C_{30}	275.9376 Pa			

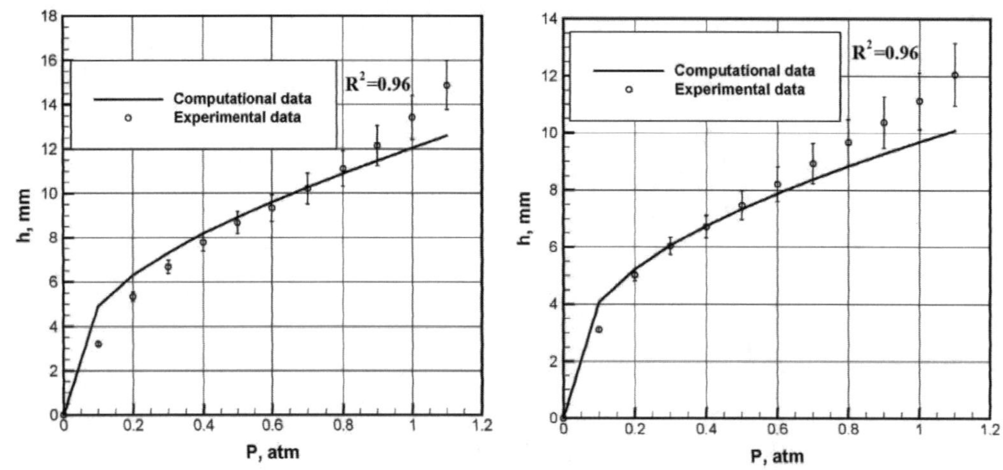

(a) (b)

Figure 3. Comparison of experimental and computation data of diaphragm inflation with different thickness: 0.3 mm (a) and 0.5 mm (b).

Conclusions

The results showed that a better accuracy provides a description of the material behaviour using the five parametric Mooney–Rivlin model. A comparison of the modelling results of the blood pump diaphragm stress-strain state and the conducted experimental studies on the determination of displacements confirmed the adequacy of the chosen model (R^2 = 0.96 between numerical simulation and experiment) for both of the studied diaphragm thickness, which allows us to talk about its potential use for a further behaviour virtual modelling of the blood pump diaphragm made of the polyurethane brand 'VITUR' in order to predict the required resource for its performance.

Acknowledgments
Financial support: This research was supported by the Presidential Program of Support for Young Candidates of Science (project no. MK-1246.2018.8).

References
[1] Ministry of Public Health of the Russian Federation, Statistical Account, Moscow (2012).
[2] Belyaev L V, Zhdanov A V, Kulikov N I 2013 Fundament *Issled* **1** 676-81.
[3] Blume, Elizabeth D, et al. 2016 Outcomes of Children Implanted with Ventricular Assist Devices in the United States: First Analysis of the Pediatric Interagency Registry for Mechanical Circulatory Support (PediMACS), *J. Heart Lung Transplant.* **35(5)** 578-84.
[4] Caimi A, Sturla F, Good B, et al. 2017 Toward the Virtual Benchmarking of Pneumatic Ventricular Assist Devices: Application of a Novel Fluid-Structure Interaction-Based Strategy to the Penn State 12 cc Device, *J. Biomech. Eng.* **139(8)** 081008.
[5] Potapov, Evgenij V, Stiller B, Hetzer R 2007 Ventricular Assist Devices in Children: Current Achievements and Future Perspectives, *Pediatr. Transplant.* **11(3)** 241-55.
[6] Weiss, William J, et al. 2012 Chronic in Vivo Testing of the Penn State Infant Ventricular Assist Device, *Asaio J.* **58(1)** 65.
[7] Donahue T H, Dehlin W, Gillespie J, Weiss W J, Rosenberg G 2009 Finite element analysis of stresses developed in the blood sac of a left ventricular assist device, *Med. Eng. Phys.* **31(4)** 454-60.
[8] Hayahi K 1994 Fatigue properties of segmented polyether polyurethanes for cardiovascular applications, *Biomat. Mechan. Prop. ASTM Int.* 9-19.
[9] Abraham G A, Frontini P M, Cuadrado T R 1997 Physical and mechanical behavior of sterilized biomedical segmented polyurethanes, *J. Appl. Polym. Sci.* **65(6)** 1193-203.
[10] Belyaev L V, Zhdanov A V, Morozov V V 2017 Materials and technologies for pulsative Russian artificial heart ventricle manufacturing, *2017 Int. Conf. Mech. Syst. Contr. Eng. (ICMSC)* 22-6.
[11] LLC SPF "VITUR" Available at: http://www.vitur33.ru/. (accessed: July 2019) (in Russian).
[12] Mooney M 1940 A theory of large elastic deformation, *J. App. Phys.* **11(9)** 582-92.
[13] Rivlin R S, Saunders D W 1951 Large elastic deformations of isotropic materials VII. Experiments on the deformation of rubber, *Philos. Trans. Roya. Soc. London. Ser. A, Math. Phys. Eng. Sci.* **243(865)** 251-88.
[14] Treloar L R G 1943 The elasticity of a network of long–chain molecules. I, *Trans. Farad. Soc.* **39** 36-41.
[15] Ogden R W 1972 Large deformation isotropic elasticity on the correlation of theory and experiment for incompressible rubberlike solids, *Proc. Roya. Soc. London. A. Math. Phys. Eng. Sci.* **326(1567)** 565-84.
[16] Yeoh O H 1990 Characterization of elastic properties of carbon black filled rubber vulcanizates, *Rubb. Chem. Technol.* **63(5)** 792-805.

IOP Conf. Series: Materials Science and Engineering **739** (2020) 012042 doi:10.1088/1757-899X/739/1/012042

Sn-Doped Hematite Nanoparticles for Potential Photocatalytic Dye Degradation

Svetlana Em[1], Mussa Yedigenov[1], Laura Khamkhash[2], Anara Molkenova[1], Timur Sh. Atabaev[1]

[1]Department of Chemistry, Nazarbayev University, Astana 010000, Kazakhstan
[2]Nazarbayev University Core Facilities, Astana 010000, Kazakhstan

timur.atabaev@nu.edu.kz

Abstract. Recently, low-cost hematite (α-Fe_2O_3) metal oxide semiconductor has attracted enormous attention for the photoelectrochemical water splitting and photocatalytic applications. The main purpose of this work is to study the effects of Sn-doping on the photocatalytic properties of hematite nanoparticles (NPs). Morphology and physicochemical properties of prepared samples were analyzed using X-ray diffraction analysis (XRD), transmission electron microscope (TEM), and UV-vis spectroscopy. Rhodamine B (RB) dye was used as a model organic pollutant to test the photocatalytic activity of prepared samples. It was found that the photocatalytic activity of Sn-doped hematite NPs was higher than that of the undoped hematite NPs. The recyclability tests suggested that 3 mol% Sn-doped hematite NPs can be used as a promising reusable photocatalyst for dyes degradation.

1. Introduction

Recently, pollution of water resources with organic dyes has grown significantly due to textile and dye manufacturing industries. The photocatalytic decomposition of organic pollutants on the surface of semiconductor nanostructures can be considered as one of the effective methods of wastewater treatment. In particular, electron-hole pairs can be generated in semiconductor nanostructure under the sunlight illumination followed by generation of the highly reactive OH radicals. These highly reactive OH radicals degrade the organic pollutants and attack the microorganisms, bacteria, fungi, etc. [1]. Thus, wastewater purification from organic pollutants and disinfection can be performed simultaneously. To date, various metal oxide micro-/nanostructures such as TiO_2 [2], ZnO [3], $SrTiO_3$ [4], WO_3 [5], and SnO_2 [6] have been widely used as photocatalysts for dye degradation. On the other hand, TiO_2, ZnO, $SrTiO_3$, WO_3, and SnO_2 have relatively large bandgaps, thus, mainly absorb only UV and part of the blue spectrum. Therefore, extending the absorption range of photocatalyst towards the visible spectrum will be highly beneficial in terms of construction of a low-cost photocatalytic reactor. In this reactor, the photocatalyst can utilize the solar photons only to purify and disinfect the wastewater. In this regard, the hematite (α-Fe_2O_3) can be a promising material because of its excellent chemical stability, low-cost, abundance and suitable bandgap range (1.9-2.2 eV). On the other hand, high recombination rate and poor conductivity are seriously limiting the widespread applications of hematite as a photocatalyst [7]. To resolve these issues, hematite can be doped with some elements such as Ti [7], Sn [8], Ni and Zn [9], Mo and Cr [10], etc. Previous study suggested that Sn-doping (5%) can significantly improve the conductivity and charge transfer in hematite film [8]. On the other hand, hematite for photocatalytic degradation of pollutants must be sufficiently small in size to achieve high surface area and shorten the pathway of the photogenerated charge carriers to the surface.

To the best of our knowledge, the synthesis of Sn-doped hematite nanoparticles by combustion method has been not reported so far. Thus, the synthesis of Sn-doped hematite using a facile combustion method was reported in this study. Furthermore, the effects of Sn-doping concentration on the photocatalytic properties of hematite were investigated in this study.

2. Materials and Methods

$Fe(NO_3)_3{\bullet}9H_2O$ (INN, ≥98%), Polyethylene glycol (PEG, Mn = 400), and $SnCl_2{\bullet}2H_2O$ (TCD, ≥98%) were purchased from Sigma-Aldrich and used as received. In a typical preparation procedure, 9.68 g of INN and 1 g of PEG were dissolved in 50 ml of deionized water. Next, various amounts of TCD (0, 1, 3, and 5 mol% of Sn in relation to Fe) were added to the solution under vigorous stirring. The prepared solution was dried at 80 °C for 24 h, and then inserted into a preheated furnace (550 °C) for 1 h. The obtained foamy mass was collected, crushed in the agate mortar and then used as a photocatalyst.

Rhodamine B (RB) dye was used as a model pollutant dye to evaluate the photocatalytic activity of prepared Sn-doped hematite NPs. In particular, 1 g of the photocatalyst was added to the 100 ml of RB solution (20 mg/L) in a glass vessel. First, the mixture was stirred for 30 min in the dark to reach the equilibrium adsorption-desorption state between the RB and photocatalyst surface. Next, H_2O_2 (1 ml, 30%) as an oxidant was also added to the mixture solution under continuous stirring. The solution was then exposed to the simulated light from a xenon lamp (300 W). Every 15 min, 2 ml of the solution was removed from the mixture solution for analysis. The RB dye degradation rate was monitored at the 553 nm using the UV-vis. All measurements were performed at room temperature.

3. Results and discussion

The morphology and size distribution of prepared samples were analyzed using a TEM. Figure 1 shows the typical images of (a) undoped and (b) 3 mol% Sn-doped hematite NPs. The majority of formed NPs have a scabrous quasi-spherical form with size distribution from ~ 30 to 180 nm, however, particles with bigger sizes were also observed. One can see that Sn-doped hematite NPs in average have slightly smaller size than undoped hematite NPs. The average size reduction was probably due to Sn-doping effects and observed also for other Sn-doped hematite NPs (not shown here).

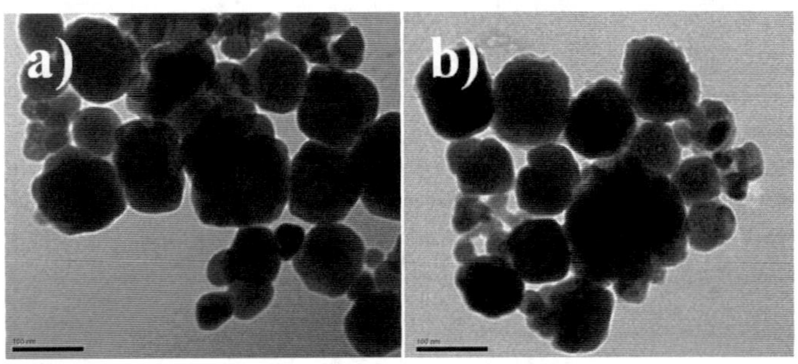

Figure 1. TEM images of (a) undoped and (b) 3 mol% Sn-doped hematite NPs.

XRD analysis was performed to confirm the formation of hematite structure. Figure 2 shows the typical XRD patterns of hematite NPs doped with different Sn concentrations. All detected peaks were easily indexed to the rhombohedral phase of α-Fe_2O_3 (JCPDS No. 33-0664) [7]. One can see that the prepared samples were highly crystalline since sharp and narrow peaks were observed. In addition, no impurity peaks were detected, suggesting that Sn was probably doped into a hematite matrix due to low concentration.

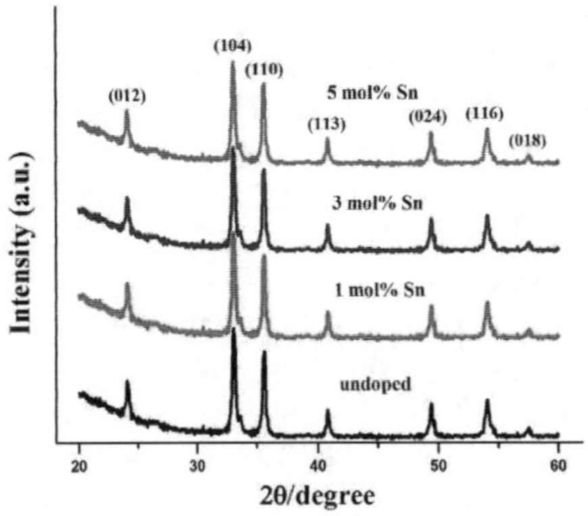

Figure 2. XRD patterns of hematite NPs doped with different Sn concentrations.

The mean crystallite sizes for all samples were calculated by Scherrer's equation. It was found that the mean crystallite sizes decrease as the Sn-concentration increasing (33.7; 31.2; 29.4; and 27.1 nm for 0, 1, 3 and 5 mol% doped samples). The size reduction can be associated with Sn atoms that can locate near/in hematite boundaries and decrease the diffusion rate. As a consequence, the mean crystallite size of the Sn-doped samples become smaller than that of the undoped one. EDX elemental mapping was employed to confirm the successful Sn-doping. Figure 3 shows the mapping results of 1 mol% Sn-doped hematite NPs taken over a relatively large area. One can see that all three elements, i.e. Fe, O, and Sn were detected, even though the sample with small doping concentration was tested (1 mol% Sn). According to the distribution uniformity, it can be concluded that Sn element was uniformly doped into a hematite NPs.

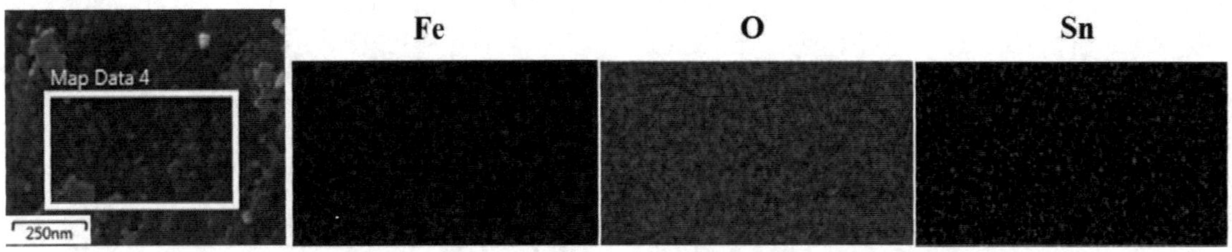

Figure 3. EDX elemental mapping of 1 mol% Sn-doped hematite NPs.

To investigate the photocatalytic activity of prepared Sn-doped hematite NPs, RB dye was used as a model pollutant for degradation. Figure 4 shows the degradation of RB without and with the presence of Sn-doped hematite NPs under constant simulated light illumination for 1 hour. It was found that without the photocatalyst addition RB concentration was decreased to about 89%. On the other hand, enhanced degradation rate was observed when photocatalyst was added to the solution. For example, the concentration of RB decreased to about 61 % in the presence of undoped hematite NPs. In case of the doped hematite NPs, the degradation rate was found to be Sn-concentration dependent. In hematite nanostructure, increased donor concentration can be formed at low Sn concentration when Fe^{3+} substituted with Sn^{4+}. The increased donor concentration suppresses the electron-hole recombination rate and improves the hematite conductivity [7-10]. On the other hand, high Sn concentration may decrease the conductivity and increase the recombination rate due to numerous defects in the hematite structure. In addition, Sn-doped hematite has enhanced absorption in visible region because of the reduced bandgap [8]. Therefore, the observed improved photocatalytic activity can be attributed to the numerous of factors such as: a) high surface area, b) improved conductivity, c) reduction of electron-hole charge recombination, d) bandgap alteration. It is obvious that 3 mol% Sn-doped hematite NPs have higher photocatalytic activity compared to other samples (RB concentration decreased to around 33%). Thus, 3 mol% Sn-doped hematite NPs was considered as an optimal doping concentration for Sn-doped hematite NPs prepared by this method. Hematite NPs with optimal doping concentration was further used for time-dependent degradation rate and recyclability experiments.

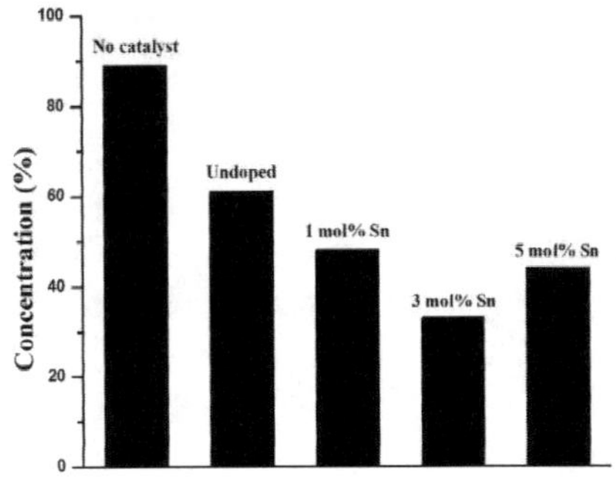

Figure 4. Degradation of RB dye without and with the photocatalyst under constant simulated light illumination for 1 hour.

The time-dependent degradation rate of colorant can be expressed using the following equation:

$$I=C/C_0 \times 100\%$$

where, I – signal intensity (a.u.), C_0 – initial concentration of RB dye, and C – concentration at any time moment. It is obvious that the peak intensity depends on the amount of RB dye in solution. Thus, with gradual degradation of RB dye the absorption intensity decrease too. The absorbance intensity of RB was monitored at maximum point (553 nm). Figure 5 shows the normalized time-dependent degradation rate of RB dye in dark conditions, without and with 3 mol% Sn-doped hematite NPs. It was found that no changes occur with RB dye in dark conditions. The RB was partially self-degraded (89% remain after 1 h of constant light illumination). On the other hand, the degradation rate increased significantly with addition of photocatalyst (33% remain after 1 h of constant light illumination). Thus, we concluded that dye pollutants degraded more efficiently in the presence of Sn-doped hematite NPs.

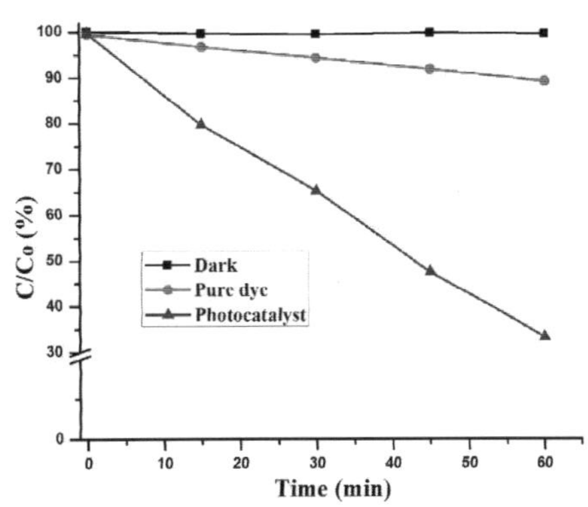

Figure 5. Time-dependant photocatalytic degradation of RB under constant light illumination.

The photocatalytic activity of the recycled sample (3 mol% Sn-doped hematite NPs) was tested under the same conditions more than three times. No significant changes (deviation $\leq$ 6%) were observed, indicating that prepared sample can be reusable.

4. Conclusions

In summary, a combustion method was used to fabricate Sn-doped hematite NPs. It was shown that the photocatalytic properties of hematite NPs are strongly depended on the Sn concentration. According to the experimental study, 3 mol% Sn-doped hematite has the highest photocatalytic activity compared to other samples. In addition, the reusability tests showed that 3 mol% Sn-doped hematite NPs can be successfully used several times under the direct solar light illumination.

Acknowledgements
Anara Molkenova would like to acknowledge the NU postdoc program.

References

[1] Linsebigler A L, Lu G, Yates Jr J T 1995 Photocatalysis on TiO_2 Surfaces: Principles, Mechanisms, and Selected Results, *Chem. Rev.* **95(3)** 735-58.

[2] Akpan U G, Hameed B H, 2009 Parameters Affecting the Photocatalytic Degradation of Dyes Using TiO_2-Based Photocatalysts: A Review, *J. Hazard. Mater.* **170(2-3)** 520-9.

[3] Daneshvar N, Salari D, Khataee A R 2004 Photocatalytic Degradation of Azo Dye Acid Red 14 in Water on ZnO as an Alternative Catalyst to TiO_2, *J. Photochem. Photobiol. A Chem.* **162(2-3)** 317-22.

[4] Da Silva L F, Lopes O F, De Mendonça V R, Carvalho K T, Longo E, Ribeiro C, Mastelaro V R 2016 An Understanding of the Photocatalytic Properties and Pollutant Degradation Mechanism of $SrTiO_3$ Nanoparticles, *Photochem. Photobiol.* **92(3)** 371-8.

[5] Han F, Li H, Fu L, Yang J, Liu Z 2016 Synthesis of S-Doped WO3 Nanowires with Enhanced Photocatalytic Performance towards Dye Degradation, *Chem. Phys. Lett.* **651** 183-7.

[6] Hu W, Yuan X, Liu X, Guan Y, Wu X 2017 Hierarchical SnO2 Nanostructures as High Efficient Photocatalysts for the Degradation of Organic Dyes, *J. Sol-Gel Sci. Technol.* **84(2)** 316-22.

[7] Atabaev T S, Ajmal M, Hong N H, Kim H K, Hwang Y H 2015 Ti-Doped Hematite Thin Films for Efficient Water Splitting, *Appl. Phys. A* **118(4)** 1539-42.

[8] Qin D D, Li Y. L, Wang T, Li Y, Lu X Q, Gu J, Zhao, Y X, Song, Y M, Tao C L 2015 Sn-Doped Hematite Films as Photoanodes for Efficient Photoelectrochemical Water Oxidation, *J. Mater. Chem. A* **3(13)** 6751-5.

[9] Barrero C A, Arpe J, Sileo E, Sánchez L C, Zysler R, Saragovi C 2004 Ni- and Zn-Doped Hematite Obtained by Combustion of Mixed Metal Oxinates, *Physica B: Conden. Matt.* **354(1-4)** 27-34.

[10] Kleiman-Shwarsctein A, Hu Y S, Forman A J, Stucky G D, McFarland E W 2008 Electrodeposition of α-Fe_2O_3 Doped with Mo or Cr as Photoanodes for Photocatalytic Water Splitting, *J. Phys. Chem. C* **112(40)** 15900-7.

IOP Conf. Series: Materials Science and Engineering **739** (2020) 012043 doi:10.1088/1757-899X/739/1/012043

On-line Monitoring by Acoustic Emission of Butt Joint Structure on the Composite Fuselage Frame

X F Qi[1], Y Yang[1], W P Kang[1], Q Wang[1], G Zhao[1]

[1]Aircraft Strength Research Institute of China, Xi'an 710065, China

741612110@qq.com

Abstract. Acoustic emission (AE) technique is an online and dynamic non-destructive testing technology by diagnosing sounds, applicable to damage identification of all sorts of materials and structures. In order to fulfill the requirement of on-line monitoring for butt joint structure on the composite fuselage frame in static loading, Acoustic Emission (AE) is employed in this article. The experiment research is focused on examining the damage evolution process of the specimen subjected to mechanical tests by constantly monitoring the AE and strain responses. Combined with strain results, the recorded acoustic signals are analyzed using AE characteristic parameters such as AE amplitude and AE accumulative energy to identity five different damage stages and conclude the corresponding damage style during the entire test. The outcomes of this article can be available for supplementation and improvement of the current on-line non-destructive testing research for composite structures.

1. Introduction

In recent decades, especially after the 1950s, acoustic emission (AE) technology has developed rapidly and been widely used in various engineering fields since it is a dynamic non-destructive testing method that provides a new means for technicians to conduct non-destructive testing[1]. It is particularly worth mentioning here that in contrast with other non-destructive testing methods, AE possesses the capability of achieving early damage recognition, allowing real-time monitoring of damage process, and identifying different kinds of damage mechanisms[2], which is attributed to its high sensitivity, passivity and dynamic nature. By the use of the AE method, on-line monitoring for the damage evolution of the specimen can be implemented with the assistance of other non-destructive testing methods. For example, the initiation point, expansion speed and expansion time of the damage can be comprehensively understood by means of AE. Hence, this article investigated the damage behavior of composite structure based on the AE technique.

At present, researchers have conducted a great quantity of investigation to explore the AE technique application in many kinds of engineering fields. For example, Yongwei Z et al. [3] investigated the relationship between AE characteristic parameters and rock failure process, and accordingly determined the crack type by the average frequency and the reciprocal of rise angle. In an effort to identify the damage mechanisms in metal plates, Kordatos E Z [4]recorded AE activity during fatigue experiments in metal CT specimens with a V-shape notch which were loaded in fatigue until final failure, and thus discovered that AE parameters exhibited a sharp increase approximately 1000 cycles before than final failure. In the fatigue life prediction for steel bridges, Yu J et al. [5]performed fatigue tests to detect AE signals from fatigue cracks and then investigated AE characteristic parameters and crack growth rate of the steel bridge. Accordingly, the study indicates that the correlation of AE absolute energy rate with crack growth behavior is of paramount importance and prognosis for in-service steel bridges.

In particular, in the case of composite materials, to attempt the relationship between AE response and the corresponding damage mechanism, many investigations based on the AE technique have been performed up to now. After examining AE response of loaded composite material, Sause MGR [6]considers the highest frequency AE signals to be related to fiber breakage, while the lower frequency AE signals are in general caused by mechanisms such as matrix fracture or fiber pull-outs. In the studies of Masmoudi S et al. [7], the AE signals were analyzed by using the classification k-means method to identify the different damage mechanisms and to follow the evolution of these mechanisms for composite materials. In addition, after employing broad band and high fidelity sensors to acquire abundant AE signals, Prosser W H et al. [8]adopted a waveform based AE response handling system to study the initiation of transverse matrix cracking in cross-ply graphite/epoxy composites.

In this context, this article proposed the use of AE technique to implement the on-line monitoring for butt joint structure on the composite fuselage frame in an attempt to attain the damage evolution process of the specimen. Combined with strain measurement technique, the research utilized AE characteristic parameters such as AE amplitude and AE accumulative energy to analyze the recorded AE signals and to identify five different damage stages and the corresponding damage style of the specimen during the entire test. In the light of the investigation, the five damage stages were summarized as elastic deformation stage, initial damage stage, stable damage development stage, rapid damage expansion stage and overall failure stage. The research results can provide real-time damage prediction for composite structure verification tests and valuable data sources for further improvement of AE technology. The outcomes of this article can be used for supplementation and improvement of the current on-line non-destructive testing research for composite structures.

2. Acoustic emission technique

Acoustic Emission is a common physical phenomenon, which is transient elastic wave generated by the rapid release of energy from localized sources within a material. Under the action of excitation (such as force, temperature, electricity, etc.), the material will behave local stress concentration, resulting in energy release and further generating acoustic emission waves coupled with damage activation mechanisms. After transforming AE waves into the corresponding electrical signals by a sensor, the information about the intensity and activity of the sound source can be obtained by analyzing the signals and, thus the evaluation of material integrity can be realized.

The commonly used procedure for AE monitoring is as follows[9]. Firstly, piezoelectric sensors are employed to convert surface vibrations caused by acoustic emission waves transmitted to the surface of the material into electric signals. Secondly, these electric signals are collected and analyzed by acoustic emission detection equipment. Finally, after investigating AE characteristic parameters, relevant information about the AE source can be obtained, and further the evaluation of the activity, intensity and severity of the sound source can be realized. The parameters commonly used for the analysis of AE signals include threshold, peak amplitude, energy, rise time, duration, counts, etc.

3. Experimental procedure

3.1. Material and geometry of the specimen

IMA_M21E unidirectional strip material was used for the main parts of the specimen, including the skin, the stringer, the frame, etc. The material meets the specification of CMS-CP-309, and the thickness of single layer is 0.184mm. The specimen is of C-shaped thin frame, and its geometry is displayed in Figure 1, where the radius of the skin on the curved frame is 2960 mm, the span angle is 28.45°, and the width is 620 mm. In this experiment, the area of concern is located on the butt joint section in the middle of the specimen, marked by a red rectangle in Figure 1. Furthermore, in order to allow more focused observation during the test, two embedded defects were prefabricated in the butt joint structure of the specimen in the process of manufacturing, as shown in Figure 2. Accordingly, the

AE sensor was mounted on the location indicated by the red dot in Figure 1 in order to carry out the on-line monitoring for the damage evolution of the butt joint structure in the specimen.

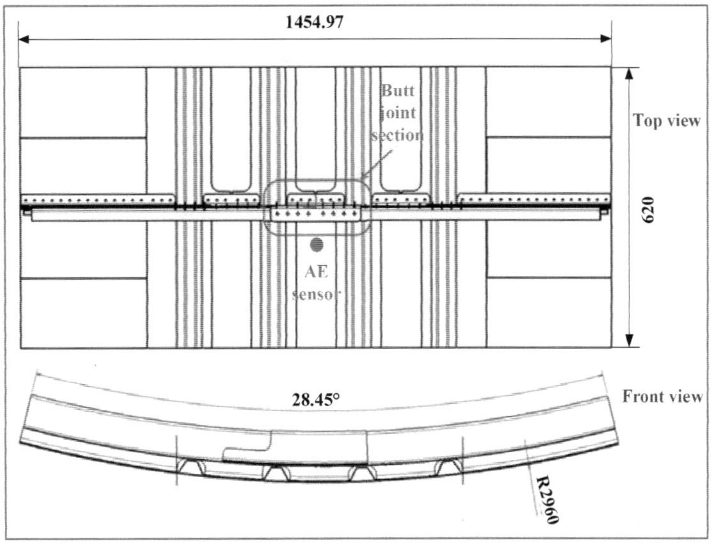

Figure 1. The specimen geometry.

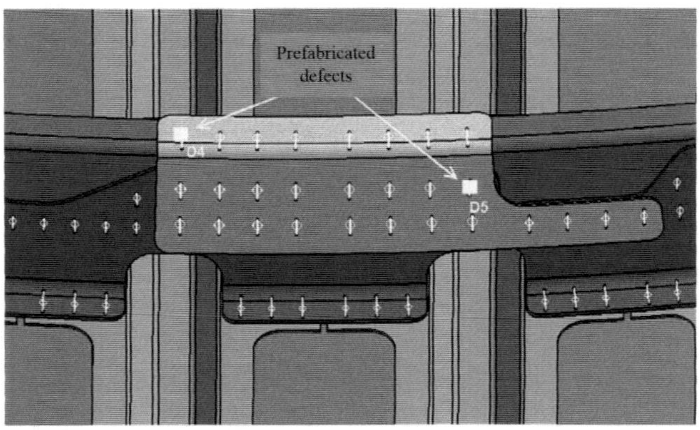

Figure 2. Schematic diagram of the locations of embedded defects.

3.2. Experiment procedure

At ambient temperature, the specimen was subjected to reverse four-point bending in static loading until failure. As depicted in Figure 3, two supporting points of the fixed platform on the lower part of the load machine support the specimen upwards while two loading points on the upper part of the load machine force the specimen downwards, so as to form the force coupling and achieve the reverse bending effect. Hence, the inner edge of the frame would be pulled while the skin compressed under the loading condition.

3.3. AE monitoring system

In the test, AE signals were sensed by AE sensor R15α from Physical Acoustics Corporation (PAC) with a frequency range of 50 kHz-200 kHz and a resonant frequency of 150 kHz, and then were amplified by the preamplifier provided by PAC with a 40dB uniform gain across all frequencies. Subsequently, the acquisition of AE signals was carried out by the master computer that utilizes SAMOS PCI-8 data acquisition board with 8 channels. In particular, All AE signals were bandpass filtered before recording between 100kHz and 400kHz which effectively removes the mechanical noise of low frequency and electronic noise of high frequency, and all signals were recorded with a Peak Definition Time (PDT), Hit Definition Time (HDT) and Hit Lockout Time (HLT) of 50, 200, 300μs, respectively at a sampling rate of 3MHz.

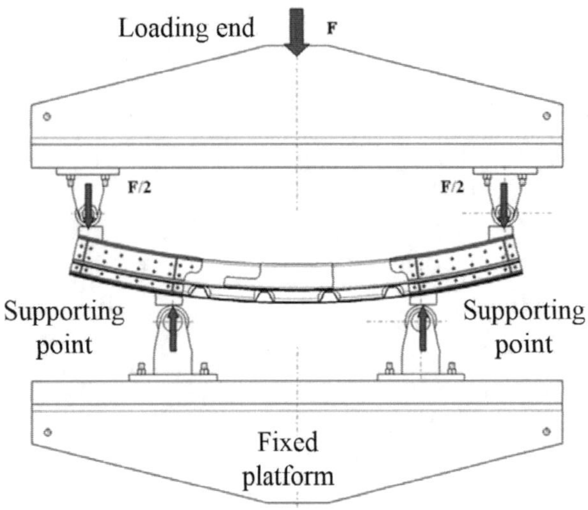

Figure 3. Schematic diagram of the specimen tested in reverse four-point bending

4. Results and analysis

In this article, we adopted the method based on the characteristic parameters to analyze the recorded AE signals during the entire test. Figure 4 shows the relationship between AE characteristic parameters and times, including AE amplitude vs. time and AE accumulative energy vs. time. In this figure, the black vertical solid lines represent some special time labels, and the corresponding descriptions can be seen in Table 1.

Table 1. Time labels information.

Time label position(s)	Load value	remark	Time label position(s)	Load value	remark
69	40%		260	38 kN	
82	50%		290	40 kN	
95	60%		340	42 kN	Persons could hear sounds.
140	30kN		480	45 kN	
165	32 kN		530	45.6 kN	Sounds became very loud.
200	34 kN		560	46.718 kN	Failure of the specimen.
220	36 kN				

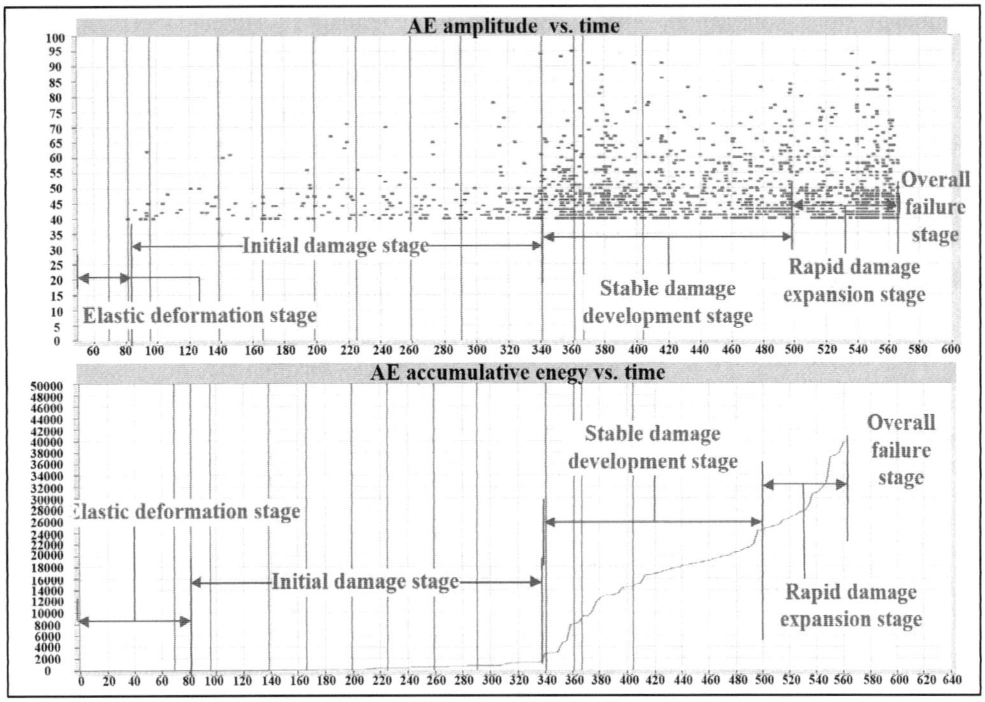

Figure 4. AE different characteristic parameters vs. time during the entire test.

Besides, the strain behavior of the strain gauge near the fracture location, identified by the red oval in Figure 5, is also illustrated in Figure 6. The combination of Figure 4, Table 1 and Figure 6 was utilized to identify the damage evolution process in the static test for the specimen, which is remarked in Figure 4 and Figure 6 and comprises five different stages.

The first stage (0-82s) is the elastic deformation stage. It can be seen from the diagram of AE amplitude vs. time in Figure 4 that there is no acoustic emission event, representing no new damage in the composite material. This conclusion can be confirmed by the curve in Figure 6 where the load level of the first obvious turning point is 12, and the corresponding load is 60% load (as shown in Table 1, the time position of the AE history diagram corresponding to 60% load is 95s). The strain change rate before turning point is larger, which implies the elastic deformation of the specimen During this period, corresponding to the elastic deformation stage of acoustic emission. The strain change rate after the turning point becomes smaller, which represents plastic deformation caused by the damage of composite materials. In addition, the above results also show that acoustic emission monitoring (83s) can detect the damage initiation earlier than strain monitoring (96s).

The second stage (83-335s) is the initial damage stage. From Figure 4, it can be seen that the AE amplitude in this stage is basically below 65dB, and the energy released by AE activities is also at a low level. A large number of low-amplitude and low-energy AE signals are caused by the damage of the composite matrix, and micro-cracks begin to appear in the composite matrix during this stage. The damage types in this stage mainly include delamination at weak joints caused by the embedded defects, extrusion damage and primary micro cracks of the matrix, etc.

In the strain-load level curve of Figure 6, the initial damage stage is between 13-328 load level, corresponding to 60% load-42kN load. According to Table 1, this stage corresponds to the range of 95-340s of AE monitoring time. Thus, it is found that AE monitoring (83-335s) is also ahead of strain monitoring (95-340s) in the initial stage of damage perception.

The third stage (336 ~ 500s) is the stable development stage of damage. As can be seen from Figure 4, the AE activity is more frequent in this stage, the amplitude of AE events increases, and the energy begins to increase, which means that the damage is becoming more and more serious. From Figure 4, it can be seen that the AE signal with amplitude ranging from 65 dB to 95 dB increases obviously, and the energy vs. time curve also increases rapidly. In this stage, the damage mainly comes from the further propagation of the primary crack in the matrix formed in the first stage, which results in the debonding of the matrix and the fibers at the front of the crack tip, the buckling of the interface and the instability of the specimen. A more careful observation of Figure 4 shows that in the latter half of this stage the amplitude of AE events falls back and the energy growth slows down. This is mainly because after the release of most energy the phenomenon of local stress concentration is greatly alleviated, thus causing the damage expansion of the specimen to be relatively flat.

This stage is between 329 and 453 load level in Figure 6, corresponding to 42kN to 45.4kN load. According to Table 1, this period is about 341-515s of AE monitoring time. It can be seen that the AE monitoring (336-500s) is also ahead of the strain monitoring (341-515s) to perceive this stage.

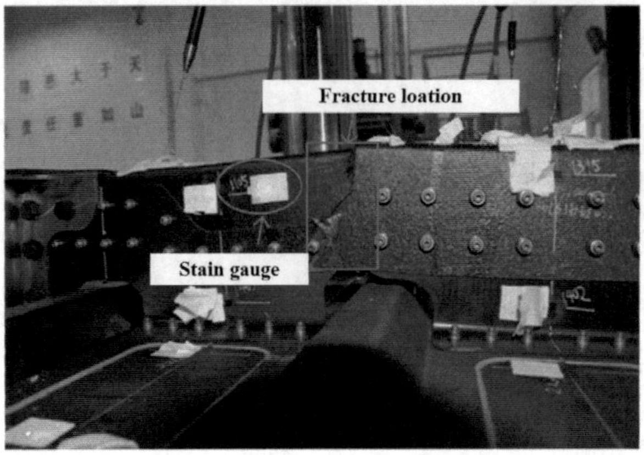

Figure 5. Picture of strain gauge location and fracture location.

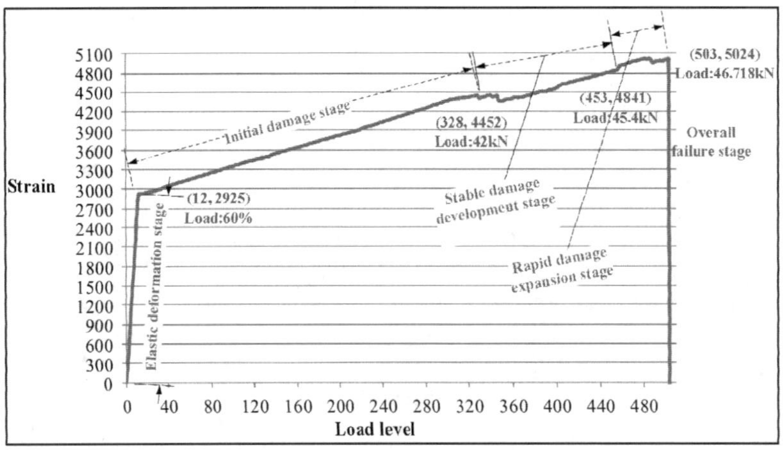

Figure 6. Strain-load level curve of strain gauge near the fracture location.

The fourth stage (501-560s) is the rapid expansion stage of damage. As can be seen from Figure 4, at this stage the AE energy increases sharply and AE activity becomes frequent again. This is because with the increase of load, damage continues to accumulate, resulting in a rapid decline in the bearing capacity of the structure and further rapid expansion of damage. At this stage, the damage begins to expand unstably, and the matrix damage is further aggravated, accompanied by delamination and fiber breakage. Since the matrix has been completely broken before the fiber breaks, the AE events in the mid-term of this stage decrease significantly. In the later stage, with the further increase of load, the fibers are pulled out from the matrix, even some fibers break, and a greater stress concentration is formed near the fracture surface.

In Figure 6 , this stage is between 453 and 503 load level, and the corresponding load is 45.4 kN to 46.718 kN, which is about 516 s to 560s of AE monitoring time. Consequently, AE monitoring (501-560s) is also ahead of strain monitoring (516s-560s) in sensing this stage.

The fifth stage (the moment of fracture) is the failure stage of the specimen. At this stage, a large number of reinforcing fibers in the specimen were pulled off instantaneously, the AE energy increased rapidly to the maximum value, and ultimately the beam of the specimen was destroyed at the embedded defect D5 (see Figure 2 and Figure 5). At this time, the specimen had been in the stage of overall failure. It can also be seen from Figure 6 that the strain decreases sharply at the moment of fracture, which is because the strain gauge is also destroyed at the same time.

5. Conclusions

In this article, AE technology was employed to monitor the docking test of composite fuselage frame, and it can be found that the process of crack initiation, propagation and destruction of composite materials can be identified by using the analysis of AE signal characteristic parameters, which verifies that AE technology can be used in the on-line monitoring of composite structures. According to the preceding investigation, the following conclusions can be drawn.

1) By combining the AE signal characteristic parameters and the strain-load level curve of the specimen, the damage evolution process of the specimen can be achieved and divided into five stages: elastic deformation stage, initial damage stage, stable damage development stage, rapid damage expansion stage and overall failure stage. The damage accumulates in the first stage, the crack originates in the second stage, buckling begins in the third stage, rapid propagation is in the fourth stage, and destruction is in the fifth stage.

2) The main types of damage in each stage are as follows: no new damage occurs in the first stage; the second stage mainly includes delamination damage at weak joints caused by embedded defects, extrusion damage and primary micro cracks of matrix; the third stage mainly includes matrix crack, matrix and fiber debonding; the fourth stage mainly includes matrix damage, delamination damage and a large amount of fiber breakage; in the fifth stage, a large number of reinforced fibers were pulled off instantaneously and the specimen was broken.

References

[1] Huguet S, Godin N, Gaertner R, Salmon L, and Villard D 2002 Use of acoustic emission to identify damage modes in glass fibre reinforced polyester *Composites Science & Technology*, **62** 1433-1444.

[2] Sarasini F and Santulli C 2014 Non-destructive testing (NDT) of natural fibre composites: acoustic emission technique *Natural Fibre Composites* **71** 273-302.

[3] Zhou Y, Zhang P, Zhao Y, Xiong S, Gang L and Tao Q 2017 Underlying law behind acoustic emission waveform parameters during rock fracture process *Journal of Heilongjiang University of Science & Technology*

[4] Kordatos E Z, Aggelis D G and Matikas T E 2011 Monitoring of fatigue damage in metal plates by acoustic emission and thermography *Proceedings of Spie - The International Society for Optical Engineering* **7982** 79820V.

[5] Yu J, Ziehl P, Zárate B and Caicedo J 2011 Prediction of fatigue crack growth in steel bridge components using acoustic emission *Journal of Constructional Steel Research* **67** 1254-60.

[6] SAUSE M G R, GRIBOV A, UNWIN A R and HORN S 2012 Pattern recognition approach to identify natural clusters of acoustic emission signals *Pattern Recognition Letters* **33** 17-23.

[7] Masmoudi S, Mahi A E and Turki S 2016 Fatigue behaviour and structural health monitoring by acoustic emission of E-glass/epoxy laminates with piezoelectric implant *Applied Acoustics* **108** 50-8.

[8] Prosser W H, Jackson K E, Kellas S, Smith B T and Friedman A 1995 Advanced, Waveform Based Acoustic Emission Detection of Matrix Cracking in Composites *Ndt & E International* **30** 1052-8.

[9] Rice J R 1980 Elastic wave emission from damage processes *Journal of Nondestructive Evaluation* **1** 215-24.

IOP Conf. Series: Materials Science and Engineering **739** (2020) 012044 doi:10.1088/1757-899X/739/1/012044

Effect of Method of Soil Drying On Atterberg Limits and Soil Classification

Ahmed Raad Al-Adhadh[1], Haider Kadhem Sakban[2], Zahraa Tawfiq Naeem[1]

[1]Department of Civil Engineering, Engineering College, Al-Muthanna University, Al-Muthanna, Iraq
[2]Department of Civil Engineering, Engineering College, University of Thi-Qar, Thi-Qar, Iraq

ahmad_al_iraqi2000@yahoo.com

Abstract. One of the most significant factors that effects the soil classification is Atterberg limits, liquid limit and plastic limit. Atterberg limits were developed by a Swedish scientist at the early 1900's called Atterberg. These limits could express the consistency of fine-grained soils due to variety of water content. These limits divide the soil into four major states, solid, semi-solid, plastic, and liquid state. According to American Standard for Testing and Materials (ASTM), to check liquid limit and plastic limit tests for a soil, the soil should be dried before the test for preparation purpose. ASTM specified two ways to dry the soil specimens, oven dry and air weather dry and both should give same results. Most of engineers will go with dry oven method to speed up the specimen preparation process assuming there is no any difference between these two methods of drying. In this research, the effect of the drying method has been studied. The results showed that the drying method has a significant effect on the liquid and plastic limits and then on the classification of soils. The soil specimens of this research were brought from all over Iraq cities to ensure studying different soils that could exhibits different behaviors.

1. Introduction

The classification of soil is considered major factor for designing geotechnical structures, no matter what the use of soil is, either supporting soil or as a constructional soil. Supporting soil could be existing under shallow foundations such as spread footings or mat foundations, around and below deep foundations such as piles and drill shafts, behind the retaining walls etc. on the other hand constructional soil could exist in all earth structures such as earth dams. All these geotechnical structures required deep study about the soil physical properties before getting started the structural design. The physical properties of soil and then the design of the geotechnical structure significantly depend on the plasticity of soil, which can be expressed by Atterberg limits [1]. Atterberg limits, liquid limit and plastic limit, were firstly developed by the Swedish scientist namely Atterberg at the early 1900's [2]. The consistency limits are greatly important to classify the soils, Serge Leroueil [3] has made a study on Illinois soil properties showed a relevant relationship between Atterberg limits and some engineering properties of soil. For example, there are more than a relation among the consistency limits and some physical and chemical properties such as the organic content matter, percentage of clay particles smaller than 0.002 mm, percent of Illite and Montmorillonite in the clay, percent of silt particle with size range between 0.05 and 0.002 mm. Haigh [4] has correlated the liquid limit with the clay strength and the plastic limit with the soil capillary suction. Therefore, the importance of Atterberg limits could be clearly visualized due to the previous studies. However, not many research studies are available to estimate the correlation between the effect of way of calculating consistency limits and soil classification. In this paper, a study has been made to show the effect of the drying method of soil on the values of liquid and plastic limits and then on the classification of soil.

1.1. Atterberg Limits

Fine-grained soils can be remolded in presence of water without crumble if clay minerals are existing. This phenomenon happened due to the cohesion exists between particles because of the water surrounding them [5]. There was a method to describe the consistency of fine-grained soil concerning the change of water contents. The Swedish scientist Atterberg developed this method. The consistency limits were proposed to distinguish among four states of soil. These four states of soil are solid, semi-solid, plastic, and liquid state, stated from low to high moisture content respectively [6]. The consistency limits among them are shrinkage limit to separate between solid and semi-solid state, plastic limit to separate between semi-solid and plastic state, liquid limit to separate between plastic and liquid state as shown in figure 1. In this study, it is focused on the effect of liquid and plastic limits due to their important on the classification of fine-grained soils.

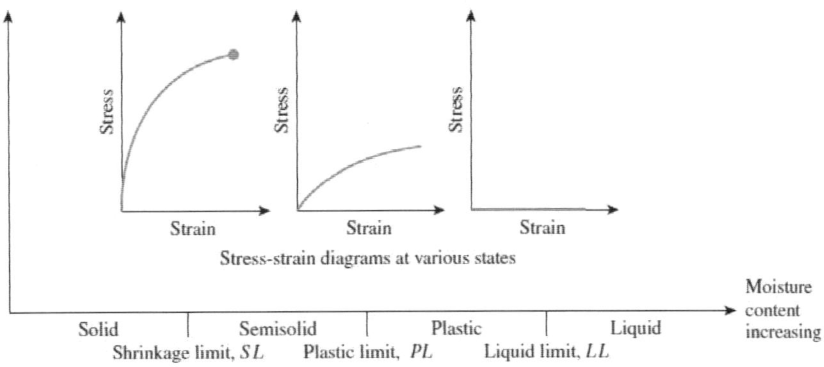

Figure 1. The consistency limits and soil states Das [5].

1.1.1. Liquid Limit. The liquid limit is can be described by the water content that transmits the soil from plastic to liquid state. In other words, the soil is transmitted to a state like a liquid if the water content increased passing the liquid limit, Sridharan [7]. Two methods are specified by ASTM D4318 [8] to make liquid limit test. These two methods are the one-point liquid limit and multipoint liquid limit. The first method is not quite precise as long it takes one test in consideration and the number of blows should be ranged between 20 and 30. Otherwise, a second method depends on setting up the moisture content to exhibit the required number of blows. Therefore, the second method is more precise than the one-point liquid as long as it depends more than one point for evaluating the liquid limit. Three liquid limit test must be made at least to ensure the required precise. The first test is made with a water content corresponding to a number of bows of 25 to 35 blows. The second test is made with a water content corresponding to a number of 20 to 30 blows and the last one with is done with a water content required to achieve 15 to 25 blows. Multipoint liquid limit method is the method that has been used in this study due to its accurate.

1.1.2. Plastic limit. The ASTM D4318 [8] procedure is strictly followed. Two method can be followed to calculate the plastic limit. First method is made by rolling the soil with hand at sufficient pressure by the palm of hand or the fingers, Haigh [2]. Second method is the rolling machine method. The first method, hand roll method, has been used in this study.

1.2. Soil preparation method

According to ASTM D4318 [8], there are two ways for preparing the specimens, dry preparation method and Wet preparation method. In this paper, dry preparation method will be dependent. Two different drying methods are mentioned in the ASTM D4318 [8], weather temperature drying method and oven drying method. The weather temperature drying method includes that the soil remains in the weather for days until its weight becomes constant with time. On the other hand, the oven drying method consists of placing the soil in the oven at temperature of about 60°C until the soil pulverized easily. After drying, the soil should be pulverized with a rubber hummer to avoid crushing the soil particles. After that, the soil is sieved on sieve No. 40 taking in considerations that pushing the soil particles to make them pass the sieve is forbidden. Two ways are allowed to remove the particle stuck in the sieve opening, brushing the sieve with a brush or washing the sieve, and the first one is dependent in this study. The sample could be divided into two parts, one of them dried in oven and the other exposure to the weather temperature to be dried naturally. The portion of the sample used for the liquid limit test is mixed with different amount of water to achieve the corresponding number of blows. After the liquid limit test is done, small amount of soil, about 20 g could be used for the plastic limit test [9]. The 20 g is used for the plastic limit test after reducing the amount of water in the soil to make it easy to be rolled without sticking in hands. Reducing the amount of water could be done by spreading the soil and remixing it on any surface.

1.3. Clasification of soil
The main two systems of soil classification considered in the civil engineering are Unified Soil Classification System (USCS) and Classification of Soils for Highway Construction Purposes [10]. Both of these systems will be discussed in this study.

1.3.1. Unified Soil Classification System (USCS). The unified soil classification system, ASTM D2487 [11], was used to classify the soil as this system describes the classification of mineral and organomineral soils for engineering uses [12]. This system is the modified one of the Airfield classification system which is developed by Casgrande at the 1940s. If the precision is in demand while classifying the soil, particle size distribution characteristics and plasticity characteristics, liquid limit and plasticity index, are required. The fine-grained soil, silt and clay, is defined in the ASTM 421 [13] as that portion of soil passing sieve No.200 size (0.075 mm). The difference between clay and silt is that the clay in presence of water shows a respectable strength due to its plasticity characteristics when air-drying. Whilst, the silt has no strength or negligible strength when air-drying. Another difference between the clay and silt is that when drawing the point of intersection between liquid limit and plasticity index of clay on the plasticity chart, that point would be located above the A-line and the plasticity index of clay is greater than four. Conversely, in case of silt, the point of intersection will be located below the A-line and the plasticity index of soil would be less than 4, ASTM D2487 [11].

1.3.2. Classification of Soils for Highway Construction Purposes. Classification of soils for highway construction purposes, ASTM D3282 [14], was used to classify the soil as this system describes the classification of mineral and organomineral soils for engineering uses. This system classifies the soil into two main groups depending on the percent passing sieve No. 200 (0.075 mm). These two groups are granular materials group and silt-clay materials group. The first group, granular material, is divided into 3 groups, A-1, A-3, and A-2. The other group, silt-clay materials group, divides the soil into 4-subaltern groups, A-4, A-5, A-6, and A-7. If the precision is in demand while classifying the soil for highway construction uses, particle size distribution characteristics and plasticity characteristics, liquid limit and plasticity index, are required.

2. Materials and Soil Testing Method
All the soil samples in this study were carried out from almost all over Iraq country. The soils have wide-ranging physical properties and initial vision classification, but all soil samples were classified as fine-grained soils. The grain size distribution curves, following the ASTM D6913 [15] procedure, of all the soil samples used in this study are presented in figure 2. The samples were collected from Baghdad, the capital of Iraq, and cities located at different distances from Baghdad. As the target of

this study is evaluating the effect of the liquid and plastic limits on the soil classification, some samples were eliminated because they did not show any plasticity properties, non-plastic soils. As a result, nine soil samples from different cities, Baghdad, Anbar, Karbala, Diwanya, Samawa, Nassirya, Rumaitha, Basrah, and Hillah, were chosen to evaluate their properties in this study.

The weight of each soil sample was 4 Kg. these samples were brought from nine cities and each sample was divided into two parts and prepared as mentioned in the soil preparation section. Each sample divided into two specimens. One of these two specimen was tested based on air-dried method and the other specimen was tested based on the oven-dried method.

All results of the samples were analysed manually and using a software program to make sure the accuracy of obtained results. The software that is called Soil Tester, which is specialized in analysing soil tests results, was used to analyse the results. The software program interface can be shown in figure 3. The test procedure that were performed in this study followed the ASTM recommendations, as it has been aforementioned.

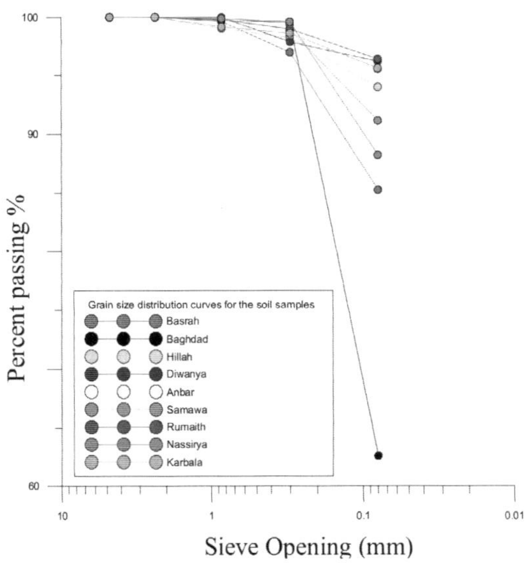

Figure 2. Grain size distribution curves for the soil samples used in this study.

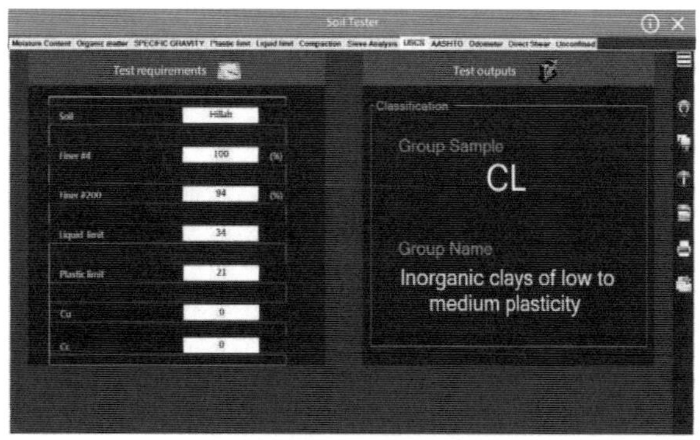

Figure 3. Soil tester program.

3. Results and Analysis

The effect of both soil-drying methods, oven dry and air dry, are discussed in following sections of this paper. Firstly, the influence of drying method on the soil consistency limits is analysed and then its effect on the soil classification.

3.1. Analysis of consistency limits for oven-dried and air-dried methods

All the Atterberg's test results were represented in table 1. The results showed that liquid limit of most oven-dried samples is higher than that of air-dried samples. Since, seven oven-dried samples have greater liquid limit and just two have less value of the liquid limit than air-dried samples. On the other hand, there is no Clear preference for the air or oven drying method regarding to the plastic limit results. Four oven-dried samples have greater plastic limit values. Whilst five air-dried samples have greater plastic limit values. All the results are presented in Radar charts to show the difference in the values. Figure 4 and 5 show the liquid and plastic limit results respectively based on the cities that have been taken from figure. 4 clarifies how the liquid limit values of oven-dried samples are greater than those of air-dried samples. Whilst figure 5 clarifies that, there is no clear correlation between plastic limit and the use of specific of the two methods of drying.

Table 1. Liquid and plastic limit for air and oven dried samples.

city	Coordinates		Liquid limit		plastic limit	
	latitude	longitude	oven dry	air dry	oven dry	air dry
Baghdad	33.340582	44.400876	26	25	15.81	19.38
Anbar	33.42056	43.307779	51.3	40	35.59	30.13
Hillah	32.463672	44.41963	34	28	21	23.8
Karbala	32.597902	44.016482	44.5	45.5	27.33	28.8
Diwanya	31.992886	44.925521	48	44	24.57	27.54
Samawa	31.305904	45.279884	38	36.5	28.5	25.51
Nassirya	31.057993	46.257262	26.8	27.2	22.51	22.7
Rumaitha	31.528454	45.203772	50	46	28.66	23.74
Basrah	30.533016	47.797466	35.2	34.2	30.64	27.64

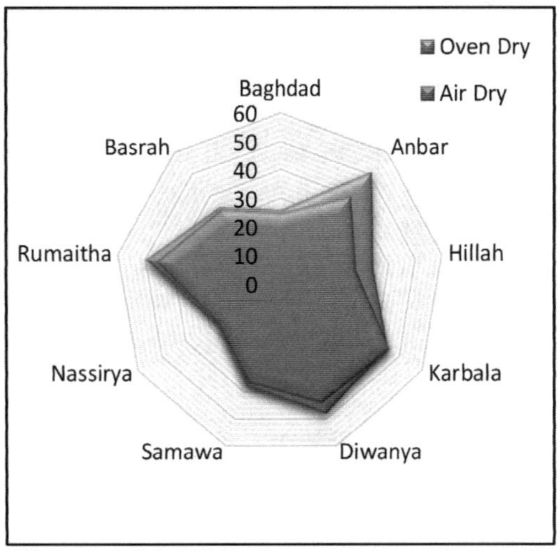

Figure 4. Liquid limit comparison of air and oven dried samples.

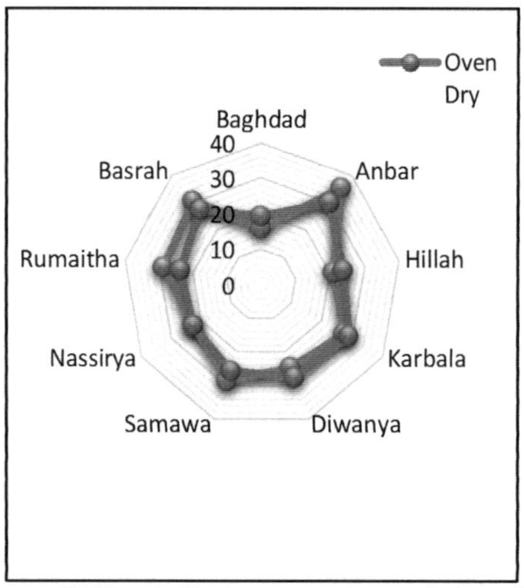

Figure 5. Plastic limit comparison of air and oven dried samples.

3.2. The effect of drying method on the soil classification

Soil classification due to unified soil classification system (USCS) and American Association of State Highway and Transportation Officials (AASHTO) system respectively were Presented in tables 2 and 3. Four of the samples exhibited similar classifications either oven-dried or air-dried. For example, the soil of Karbala, Diwanya, Samawa, and Nassirya were classified as CL, CL, CL, and CL-ML respectively for both methods of drying.

However, five samples showed different classifications due to the method of drying. For instance, Baghdad, Anbar, Rumaitha, Basrah, and Hillah have been classified as CL, MH, CH, ML, and CL, respectively if oven was used to dry the soils. On the other hand, these samples have been classified as CL-ML, CL, CL, CL-ML, and ML, respectively if the samples were exposed to weather temperature and dried naturally. Now it could be said that replacing the air-drying method with oven-drying method to speed up the drying process would give different classification results.

The classification results of the nine samples according to AASHTO classification system for both methods of drying were represented in table 3. The results showed that five of nine of samples have got different groups when dried by different methods. For example, Baghdad, Anbar, Karbala,

Samawa, and Hillah have been classified under the groups A-6, A-7-5, A-7-6, A-4, and A-6 respectively when samples were dried using oven. Whilst these samples have been classified as A-4, A-4, A-7-5, A-6, and A-4 respectively when the soils samples were dried naturally. These results confirm that there is no possibility to replace the natural drying method by weather temperature with the oven drying method without changing in the classification accuracy. Both the two ways of classification, USCS and AASHTO, exhibited a match in the classification groups regarding the two drying methods.

On the contrary, there is no any effect of the method of drying on the value of group index GI, but this is not a significant point to be focused on, as the group index is just a number to modify the classification obtained from the procedure specified by ASTM D3282 – 09. The group index in a number gives a prediction whether the soil is suitable to be used as a highway subgrade or not, Das 2013. Table 4 represents the values of the group index for all soil samples for both methods of drying. While table 5 clarifies how these numbers, classify the soil as a highway subgrade material.

Table 2. Classifying the sample due to USCS.

(a) oven dried

city	percent finer sieve # 4	percent finer sieve # 200	Oven Dry			
			L.L	P.L	P.I	USCS
Baghdad	100	62.56	26	15.81	10.19	Sandy Lean Clay CL
Anbar	100	95.54	51.3	35.59	15.71	Elastic Silt MH
Karbala	100	95.6	44.5	27.33	17.17	Lean Clay CL
Diwanya	100	96.2	48	24.57	23.43	Lean Clay CL
Samawa	100	91.15	38	28.5	9.5	Lean Clay CL
Nassirya	100	88.22	26.8	22.51	4.29	Silty Clay CL-ML
Rumaitha	100	96.37	50	28.66	21.34	Fat Clay CH
Basrah	100	85	35.2	31.64	3.56	Silt with Sand ML
Hillah	100	94	34	21	13	Lean Clay CL

(b) air dried

city	percent finer sieve # 4	percent finer sieve # 200	Air Dry			
			L.L	P.L	P.I	USCS
Baghdad	100	62.56	25	19.38	5.62	Sandy Silty Clay CL-ML
Anbar	100	95.54	40	30.13	9.87	Lean Clay CL
Karbala	100	95.6	45.5	28.8	16.7	Lean Clay CL
Diwanya	100	96.2	44	27.54	16.46	Lean Clay CL
Samawa	100	91.15	36.5	25.51	10.99	Lean Clay CL
Nassirya	100	88.22	27.2	22.7	4.5	Silty Clay CL-ML
Rumaitha	100	96.37	46	23.74	22.26	Lean Clay CL
Basrah	100	85	34.2	27.64	6.56	Silty Clay with Sand CL-ML
Hillah	100	94	28	24.1	3.9	Silt ML

Table 3. Classifying the sample due to AASHTO.

(a) oven dried

city	percent finer sieve # 10	percent finer sieve # 40	percent finer sieve # 200	Oven Dry			
				L.L	P.L	P.I	AASHTO
Baghdad	100	99.57	62.56	26	15.81	10.19	A-6
Anbar	100	99.35	95.54	51.3	35.59	15.71	A-7-5
Karbala	100	98.59	95.6	44.5	27.33	17.17	A-7-6
Diwanya	100	97.8	96.2	48	24.57	23.43	A-7-5
Samawa	100	98.86	91.15	38	28.5	9.5	A-4
Nassirya	100	99.59	88.22	26.8	22.51	4.29	A-4
Rumaitha	100	98.96	96.37	50	28.66	21.34	A-7-5
Basrah	100	96.97	85.23	35.2	30.64	4.56	A-4
Hillah	100	98.24	94	34	21	13	A-6

(a) air dried

city	percent finer sieve # 10	percent finer sieve # 40	percent finer sieve # 200	Air Dry			
				L.L	P.L	P.I	AASHTO
Baghdad	100	99.57	62.56	25	19.38	5.62	A-4
Anbar	100	99.35	95.54	40	30.13	9.87	A-4
Karbala	100	98.59	95.6	45.5	28.8	16.7	A-7-5
Diwanya	100	97.8	96.2	44	27.54	16.46	A-7-5
Samawa	100	98.86	91.15	36.5	25.51	10.99	A-6
Nassirya	100	99.59	88.22	27.2	22.7	4.5	A-4
Rumaitha	100	98.96	96.37	46	23.74	22.26	A-7-5
Basrah	100	96.97	85.23	34.2	27.64	6.56	A-4
Hillah	100	98.24	94	28	23.8	4.2	A-4

Table 4. Group Index (GI) values for all soil samples.

city	Oven Dry		Air Dry	
	G.I	Subgrade Value	G.I	Subgrade Value
Baghdad	5.588	Poor	5.512	Poor
Anbar	12.544	V. Poor	8	V. Poor
Karbala	11.768	V. Poor	11.78	V. Poor
Diwanya	14.972	V. Poor	11.384	V. Poor
Samawa	8	Poor	8.396	Poor
Nassirya	8	Poor	8	Poor
Rumaitha	14.536	V. Poor	14.104	V. Poor
Basrah	8	Poor	8	Poor
Hillah	9.2	Poor	8	Poor

Table 5. Group Index and subgrade values.

Group Index (GI)	0	0-1	2-4	5-9	10-20
Subgrade value	Excellent	good	fair	poor	v.poor

4. Conclusion

Liquid and plastic limit tests were conducted on nine fine-grained soils collected from nine Iraqi cities. The aim of the study is to evaluate how the method of drying, oven and air-drying methods, could affect the liquid and plastic limit and then the classification of soil. Depending on the results and the analysis of these tests, it can be concluded:

- Soil classification systems, USCS and AASHTO, of fine-grained soils depends entirely on liquid limit, plastic limit, plasticity index, and grins size in addition to the group index which is consider as a modification factor for AASHTO classification system.
- The results showed that liquid limit of most oven-dried samples is higher than that of air-dried samples. Since, seven oven-dried samples have greater liquid limit and just two have less values of the liquid limit than air-dried samples.
- There is no Clear effect of the drying method on the plastic limit results. Since, four oven-dried, samples have greater plastic limit values. Whilst five air-dried samples have greater plastic limit values.
- Four of the samples exhibit similar classifications due to USCS either oven-dried or air-dried. For example, the soil of Karbala, Diwanya, Samawa, and Nassirya are classified as CL, CL, CL, and CL-ML respectively for both methods of drying. Whilst five samples show different classifications respect to the method of drying.
- The results show that five of nine of samples have different groups when classified due to AASHTO classification system regarding to the method of drying and the other four samples get same classification groups whether oven or air-dried.
- Both USCS and AASHTO exhibit a match in the classification groups regarding the two drying method with a possibility of less than 45%. Whilst 55% possibility show mismatch between the classification when air-drying method is used and that of oven drying method.
- There is no any effect of the method of drying on the value of group index GI, but this is not a significant point to be focused on, as the group index is just a number to modify the classification obtained from the procedure specified by ASTM D3282 – 09.
- It is elicited that replacing the air-drying method with oven-drying method to speed up the drying process if strictly forbidden.

Acknowledgements
This study was established by laboratory of Civil Engineering Department at Al-Muthanna University. Words of appreciation are presented to the staff of the laboratory and the authors are grateful for this support.

References

[1] Odell R T, Thornburn T H, McKenzie L J 1960 Relationships of Atterberg Limits to Some Other Properties of Illinois Soils 1, *Soil Sci. Soci. America J.* **24(4)** 297-300.
[2] Haigh S K, Vardanega P J, Bolton M D 2013 The plastic limit of clays, *Geotech.* **63(6)** 435.
[3] Leroueil S, Le Bihan J P 1996 Liquid limits and fall cones, *Canadian. Geotech. J.* **33(5)** 793-98.
[4] Haigh S K 2012 Mechanics of the Casagrande liquid limit test, *Canadian. Geotech. J.* **49(9)** 1015-23.
[5] Das B M, Sobhan K 2013 Principles of geotechnical engineering, *Ceng. Learn.*
[6] Craig R F 2004 *Craig's Soil Mechan.* CRC press.
[7] Sridharan A, Rao S M, Murthy N S 1986 Liquid limit of montmorillonite soils, *Geotech. Test. J.* **9(3)** 156-9.
[8] ASTM D 2010 D4318-10 Standard Test Methods for Liquid Limit, Plastic Limit, and Plasticity Index of Soils.
[9] Dolinar B, Misic M, Trauner L 2007 Correlation between surface area and Atterberg limits of fine-grained soils, *Clays Clay Miner.* **55(5)** 519-23.

[10] Sridharan A, Prakash K 2000 Classification procedures for expansive soils, *Proc. Instit Civil Engineers-Geotech. Eng.* **143(4)** 235-40.
[11] ASTM D2487-11 Standard Practice for Classification of Soils for Engineering Purposes (Unified Soil Classification System).
[12] Budhu M 2008 SOIL MECHANICS AND FOUNDATIONS (With CD) John Wiley and Sons
[13] ASTM D421-85 Standard Practice for Dry Preparation of Soil Samples for Particle-Size Analysis and Determination of Soil Constants.
[14] ASTM D3282-09 Standard Practice for Classification of Soils and Soil-Aggregate Mixtures for Highway Construction Purposes.
[15] Standard, A. S. T. M 2009 D6913-04: Standard Test Methods for Particle-Size Distribution (Gradation) of Soils Using Sieve Analysis *ASTM Int.*

IOP Conf. Series: Materials Science and Engineering **739** (2020) 012045 doi:10.1088/1757-899X/739/1/012045

Bending Analysis of Nonlocal Functionally Graded Beams

F Garbin[1], A Levano[1], R Arciniega[1]

[1]Department of Civil Engineering, Universidad Peruana de Ciencias Aplicadas, Surco, Lima, Peru

roman.arciniega@upc.edu.pe

Abstract. In this paper, we study the nonlocal linear bending behavior of functionally graded beams subjected to distributed loads. A finite element formulation for an improved first-order shear deformation theory for beams with five independent variables is proposed. The formulation takes into consideration 3D constitutive equations. Eringen's nonlocal differential model is used to rewrite the nonlocal stress resultants in terms of displacements. The finite element formulation is derived by means of the principle of virtual work. High-order nodal-spectral interpolation functions were utilized to approximate the field variables, which minimizes the locking problem. Numerical results and comparisons of the present formulation with those found in the literature for typical benchmark problems involving nonlocal beams are found to be satisfactory and show the validity of the developed finite element model.

1. Introduction

Classical continuum mechanics theories commonly have a local approach, which assumes that the stress at a point depends on the strain at that same point. These theories are true enough for numerous cases of study. On the other hand, nonlocal continuum mechanics [1, 2] assumes that the stress at a point is a function of strains at all points in the continuum. This theory has been used to study lattice dispersion of waves, wave propagation [2], crack propagation, dislocations and surface tension fluids. An important application of nonlocal theories occurs in functionally graded micro beams where small-scale effects are required to be taken into account. Due to the smooth and continuous variation of the properties of the material from one surface to another, FGMs are usually superior to conventional composite materials.

Many works discussed the importance of nonlocal theories in the analysis of functionally graded beams. A comparison between different beam theories (varying the displacement fields) are presented by Aydogdu [3], Eltaher et al. [4] and Reddy [5] using nonlocal constitutive equations to analyze bending, buckling and vibration. Finite element solutions within nonlocal beam theories (Timoshenko and other higher order theories) have already been proposed by [4, 6-8]. However, these studies do not contemplate models that use 3D constitutive equations.

In the present work, a nonlocal linear bending behavior of functionally graded beams under distributed loads is investigated. An improved first-order shear deformation theory (IFSDT) for beams with five independent parameters is proposed and implemented. The finite element formulation is derived by the principle of virtual work. The verification results of the formulation show that the proposed computational model is satisfactory.

2. Non-local differential theory and finite element formulation

According to Eringen [1, 2], the stress at a point does not only depend on the strain at the same point, but also on the strains at the vicinity of the continuum body. Eringen describes this phenomenon based on the atomic theory of lattice dynamics and experimental observations. The nonlocal stress tensor σ at a point x is defined as:

$$\boldsymbol{\sigma} = \int_V K(|\boldsymbol{x}' - \boldsymbol{x}|, \tau)\overline{\boldsymbol{\sigma}} dV \tag{1}$$

where $\overline{\boldsymbol{\sigma}}$ is the local stress tensor and the Kernel function $K(|\boldsymbol{x}' - \boldsymbol{x}|, \tau)$ represents the nonlocal modulus, $|\boldsymbol{x}' - \boldsymbol{x}|$ being the distance in the Euclidian norm and τ is a material constant that depends on internal and external characteristic lengths (such as the lattice spacing and wavelength, respectively).

The local stress $\overline{\boldsymbol{\sigma}}$ is defined by the generalized Hooke's law as:

$$\overline{\boldsymbol{\sigma}} = \mathbb{C} \cdot \boldsymbol{\varepsilon} \tag{2}$$

where $\mathbb{C}$ is the fourth-order elasticity tensor. Although the nonlocal stress-strain relation is based on an integral constitutive equation, Eringen proposed an equivalent differential model:

$$\boldsymbol{\sigma} - \mu \nabla^2 \boldsymbol{\sigma} = \overline{\boldsymbol{\sigma}} \tag{3}$$

where $\mu = \tau^2 l^2$.

Next, we develop the linear beam formulation by using an improved first shear deformation theory [9]. Let $\{x^i\}$ be a set of Cartesian coordinates with orthonormal basis $\{e_i\}$. The neutral axis of the beam is defined by the coordinate x^1. The displacement vector is expanded through the thickness coordinate as follows:

$$\mathbf{v}(x^1, x^3) = \mathbf{u}(x^1) + x^3 \boldsymbol{\varphi}(x^1) + (x^3)^2 \boldsymbol{\psi}(x^1) \tag{4}$$

where $\mathbf{u} = u_i \boldsymbol{e}_i$ denotes the displacement vector of the neutral axis, $\boldsymbol{\varphi} = \varphi_i \boldsymbol{e}_i$ and $\boldsymbol{\psi} = \psi_3 \boldsymbol{e}_3$ are difference vectors ($i = 1{,}3$). Equation (4) contains five independent variables. The quadratic term ψ is included to avoid the Poisson locking; therefore, no enhanced methods are needed.

For the given displacement field, we define the linear part of the strain tensor as

$$\boldsymbol{\varepsilon} = \boldsymbol{\varepsilon}^{(0)} + x^3 \boldsymbol{\varepsilon}^{(1)} \tag{5}$$

where high-order terms are neglected.

The weak form can be constructed by applying the principle of virtual work. The solution of the beam is defined by the triple $\boldsymbol{\Phi} \equiv (\mathbf{u}, \boldsymbol{\varphi}, \boldsymbol{\psi})$. Thus, we obtain

$$\mathcal{G}(\boldsymbol{\Phi}, \delta\boldsymbol{\Phi}) = \int_X \left(\mathbf{N}^{(0)} \cdot \delta\boldsymbol{\varepsilon}^{(0)} + \mathbf{N}^{(1)} \cdot \delta\boldsymbol{\varepsilon}^{(1)}\right) dx^1 - \int_X \mathbf{p} \cdot \delta\mathbf{u} dx^1 = 0 \tag{6}$$

where $\delta\boldsymbol{\Phi} \equiv (\delta\mathbf{u}, \delta\boldsymbol{\varphi}, \delta\boldsymbol{\psi})$. For beam structures and linear cases, the non-local stress resultant $\mathbf{N}^{(i)}$ is obtained from equation (3) as:

$$\mathbf{N}^{(i)} - \mu \nabla^2 \mathbf{N}^{(i)} = \mathbb{B}^{(i)} \boldsymbol{\varepsilon}^{(0)} + \mathbb{B}^{(i+1)} \boldsymbol{\varepsilon}^{(1)}, \qquad i = 0{,}1. \tag{7}$$

In tensor notation:

$$\mathbf{N}^{(i)} = N_{11}^{(i)} \boldsymbol{e}_1 \otimes \boldsymbol{e}_1 + N_{33}^{(i)} \boldsymbol{e}_3 \otimes \boldsymbol{e}_3 + N_{13}^{(i)} \boldsymbol{e}_1 \otimes \boldsymbol{e}_3, \qquad i = 0{,}1, \tag{8}$$

where the components of the nonlocal stress resultants are:

$$\begin{aligned} N_{11}^{(0)} &= B_{1111}^{(0)} \varepsilon_{11}^{(0)} + B_{1133}^{(0)} \varepsilon_{33}^{(0)} + B_{1111}^{(1)} \varepsilon_{11}^{(1)} + B_{1133}^{(1)} \varepsilon_{33}^{(1)} - \mu \frac{df_1}{dx^1} \\ N_{11}^{(1)} &= B_{1111}^{(1)} \varepsilon_{11}^{(1)} + B_{1133}^{(1)} \varepsilon_{33}^{(1)} + B_{1111}^{(2)} \varepsilon_{11}^{(1)} + B_{1133}^{(2)} \varepsilon_{33}^{(1)} - \mu f_3 \\ N_{13}^{(0)} &= B_{1313}^{(0)} \varepsilon_{13}^{(0)} - \mu \frac{df_3}{dx^1} \end{aligned} \tag{9}$$

The components of the tensor $\mathbb{B}^{(i)}$ are the material stiffness coefficients (see [9]) and f_1, f_3 are the axial and transverse body forces.

In the two-phase functionally graded materials, the properties are assumed to vary through the thickness of the beam. The materials in the bottom and top surfaces are metal and ceramic respectively. Therefore, the components of the elasticity tensor are functions of the thickness x^3:

$$C_{ijkl}(x^3) = C_{ijkl}^{c} f_c + C_{ijkl}^{m} f_m \tag{10}$$

where f_c,f_m are the volume fractions of the ceramic and metal phases which are computed by the power law.

Let Ω^e be the domain of the neutral axis where the finite element domain lies in. Recall that $\widehat{\Omega}^e \equiv [-1,1]$ is a parent domain in ξ -space and $x^1(\xi): \widehat{\Omega}^e \in \mathbb{R} \rightarrow \Omega^e \in \mathbb{R}$. The finite element equations are obtained by interpolating the components of the field variables written in terms of the base vectors, namely:

$$\mathbf{u}^{hp}(x^1) = \left(\sum_{j=1}^{m} u_k{}^{(j)} \phi^{(j)}(\xi)\right)\boldsymbol{e}_k, \quad \boldsymbol{\varphi}^{hp}(x^1) = \left(\sum_{j=1}^{m} \varphi_k{}^{(j)} \phi^{(j)}(\xi)\right)\boldsymbol{e}_k,$$
$$\boldsymbol{\psi}^{hp}(x^1) = \left(\sum_{j=1}^{m} \psi_3{}^{(j)} \phi^{(j)}(\xi)\right)\boldsymbol{e}_3, \qquad k = 1,3 \tag{11}$$

The adopted basis functions $\phi^{(i)}$ are C^0 interpolant polynomials of Gauss–Lobatto-Legendre quadrature points, which are particularly suitable for high-order expansions. Explicitly, the one-dimensional basis functions of order $p = m - 1$ are expressed using the p-order Legendre polynomial P_{m-1}, as shown

$$\phi^{(i)}(\xi) = \frac{(1-\xi^2)P'_{m-1}(\xi)}{m(m-1)P_{m-1}(\xi_i)(\xi-\xi_i)} \tag{12}$$

High-order spectral elements, in contrast to low-order finite elements, do not exhibit locking problems.

3. Numerical Results

In the following section, numerical results are used to verify the proposed finite element model. Macro-beams and micro-beams are evaluated by varying the Eringen's nonlocal parameter μ and the power law index n. We study cases for simply supported and clamped-clamped boundary conditions with uniformly transverse load. A FGM beam with nonlocal effects is evaluated. A normalized center of deflection in the neutral axis is compared by using $\bar{w} = w \times \left(100 \frac{EI}{qL^4}\right)$.

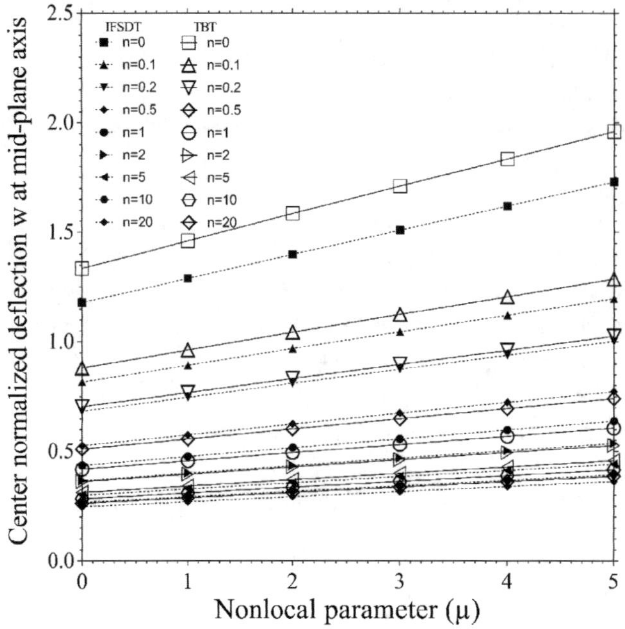

Figure 1. Nonlocal parameter μ vs dimensionless center of deflection for simply supported FG beam under a uniformly distributed load.

The present formulation is compared with Timoshenko beam (TBT) and the results are depicted in figure 1. As it was seen above, IFSDT formulation shows stiffer deflections for isotropic beams (n = 0). However, for FGM beam (n = 1 for example), IFSDT exhibits slightly more flexible results than TBT. Also, a FGM clamped-clamped beam under a uniformly distributed load is evaluated and compared with Reddy and El-Borgi [8]. Results are given in Table 1.

Table 1. Comparison of finite element results for the transverse center of deflection at for a clamped-clamped nanobeam under uniformly distributed load.

Case	Ref. [8]	Present IFSDT
C-C (n = 0)	1.0865×10^{-4}	0.9214×10^{-4}
C-C (n = 1)	2.5422×10^{-4}	2.1554×10^{-4}

For the following example, we will evaluate an FGM nonlocal beam under a uniformly distributed load of intensity $q_0 = 10$, with two different boundary conditions: cantilever beam and clamped-hinged-free beam. The material and geometric properties are:

$$L = 300, \quad b = 30, \quad H = 15$$
$$E_m = 3 \times 10^6, \quad E_c = 30 \times 10^6, \quad \nu = 0.3 \tag{13}$$

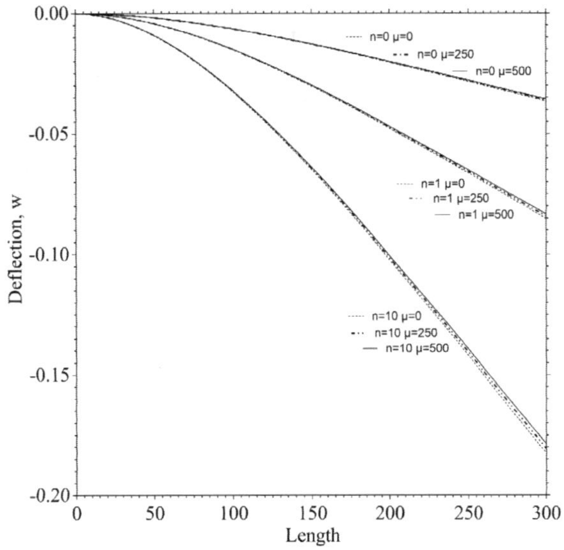

(a)

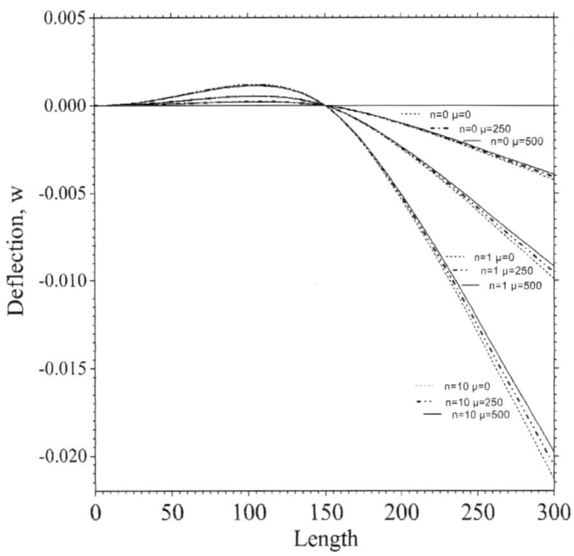

(b)

Figure 2. Length vs deflection of a non-local beam under uniformly distributed load and boundary conditions (a) clamped-free, (b) clamped-hinged-free.

It is noticeable boundary conditions effects when nonlocal approach is applied in structural analysis. In figure 2, the deflection at the free-end for a cantilever beam decreases as the nonlocal parameter increases. Likewise, the figure shows that the deflection in the middle-span for a clamped-hinged-free beam increases as the nonlocal parameter increase. Thus, for simply supported boundary conditions, an increment of μ means an increment of bending deflection, but for clamped boundary conditions, an increment of μ means a decrease of the deflection.

4. Conclusions

IFSDT and TBT linear nonlocal finite element models of FGM beams are presented with the aim of study its bending behavior. The formulation is derived by applying the principle of virtual work. Nonlocal constitutive equations of Eringen are used to develop the nonlocal finite element formulation. After showing the results, we came to the following conclusions: It is noticeable that the deflection increase at the rate of increase of the power law index *n* of FGM beams. That occurs because when the value of *n* increases the material approaches to metal. Metal materials have a smaller elasticity modulus, making the beam more flexible.

References

[1] Eringen A C 1972 Nonlocal polar elastic continua, *Int. J. Eng. Sci.* **10(1)** 1-16.
[2] Eringen A C 1983 On differential equations of nonlocal elasticity and solutions of screw dislocation and surface waves, *J. App. Phys*. **54(9)** 4703-10.
[3] Aydogdu M 2009 A general nonlocal beam theory: its application to nanobeam bending, buckling and vibration, *Physica E: Low-Dimen. Sys. Nanostruc.* **41(9)** 1651-5.
[4] Eltaher M A, Khairy A, Sadoun A M, Omar F 2014 Static and buckling analysis of functionally graded Timoshenko nanobeams, *Appl. Math. Compu.* **229** 283-95.
[5] Reddy J N 2007 Nonlocal theories for bending, buckling and vibration of beams, *Int. J. Eng. Sci.* **45(2-8)** 288-307.
[6] Alshorbagy A E, Eltaher M A, Mahmoud F F 2013 Static analysis of nanobeams using nonlocal FEM, *J. Mech. Sci. Technol*. **27(7)** 2035-41.
[7] Reddy J N 2010 Nonlocal nonlinear formulations for bending of classical and shear deformation theories of beams and plates, *Int. J. Eng. Sci.* **48(11)** 1507-18.
[8] Reddy J N, El-Borgi S 2014 Eringen's nonlocal theories of beams accounting for moderate rotations, *Int. J. Eng. Sci.* **82** 159-77.
[9] Soncco K, Jorge X, Arciniega R 2019 Postbuckling Analysis of Functionally Graded Beams, *IOP Conf. Ser: Mater. Sci Eng.* **473(1)** 012028.

IOP Conf. Series: Materials Science and Engineering **739** (2020) 012046 doi:10.1088/1757-899X/739/1/012046

Evaluation of the Evacuation of Essential Buildings: Interaction of Structural and Human Behaviour through Nonlinear Time-History Analysis and Agent-Based Modelling

M Delgado[1], A Rosales[1], V Arana[2]

[1]Student, Universidad Peruana de Ciencias Aplicadas
[2]Professor, Universidad Peruana de Ciencias Aplicadas

pccivara@upc.edu.pe

Abstract. In this article, a performance assessment of the evacuation system is established for educational buildings. Structural and geotechnical information of the building is collected and introduced into a database. A similar procedure was realized for the information related to the occupants. Using this information, a) the structural fragility and localized collapse were determined and b) the interaction of the person with the partial collapse was established. For the first aspect, nonlinear time history was used, and for the second, the agent-based modeling was applied to recreate the reaction of people that face the micro collapse. Therefore, the important results of this evaluation are: 1) To localize collapsed beans and columns that make inoperable evacuation routes, 2) to localize bottleneck areas that people concentration during evacuation, and 3) quantification of affected people, in terms of persons caught up in the building that cannot evacuate.

1. Introduction

In the period 1995-2015, more than 600,000 million people have died in natural disasters, economic losses are around 300,000 million dollars annually and 10% of natural disasters are of geophysical origin, all of the above in the five countries most affected by this type of disaster [1].

Now, the causes of the collapse of a building are: The breach of the specifications by the developers, corruption, inadequate design, defective construction methodology, inadequate supervision, change of use of buildings and low maintenance. Each cause is more recurrent according to the level of professionalism and development in each part of the world [2].

However, apart from structural concern, the researchers perform calculations to evaluate safety parameters in evacuation of establishments. Usually, these calculations are focused on architectural details, such as the location of the doors and reduce the time of arrival at an exit. However, they are not always reliable because evacuation drills are conducted in controlled environments; Here, for example, does not provide more complex factors such as emotions of people facing emergencies such as panic, fear and stress [3].

Currently there is commercial software for the evaluation of seismic performance of individual buildings that allow predicting the response of structural damage. The programs are applicable to new or existing buildings, and can be used to: (1) evaluate the probable performance of a building, (2) design new buildings providing the desired performance, or (3) improve the seismic design of buildings existing [4]. Likewise, software for modeling agents, which allow occupant profiles to be made, making it possible to simulate evacuations of buildings in case of emergencies [5].

This research has integrated the structural and human behavior of an educational institution during various seismic events. For the first, structural performance was evaluated by nonlinear time-history analysis. For the second, the existing capacity of the institution was taken to represent the behavior of people using the methodology of decision making. Finally, the efficiency of the evacuation system is questioned based on the results obtained.

2. Method

The evaluation of the evacuation system of a building has two important parts. First, structural behavior was analyzed by nonlinear time-history analysis. This allows us to evaluate the dynamics of the structure considering the inelastic behavior of the elements, through the use of standardized seismic registers. Second, user behavior was analyzed using agent-based modeling. This consists of modeling an agent, in which certain behaviors can occur in order to observe complex effects of the flow of people in a given environment. Finally, both structural and human behaviors interact together to obtain system results as a whole.

2.1. Data collection

A study of an establishment of interest that concentrates a large number of people is carried out. These can be hospitals, municipalities, educational institutions and others. From these categories an essential building is chosen, to begin with the compilation of building data; such as, structural plan, recognition of structural elements, detailed, etc. The building can be a structural system of reinforced concrete with structural walls and porches.

For the elaboration of the agents that will be used in the simulation, physical and psychological variables are taken into account. For the correct use of these variables it is proposed to make a recording of the evacuation drill of the building selected for the analysis. For physical variables is correct to take account the position of the agents, speed, vision, width and height. In the same way, for psychological variables, reaction time, acceleration time, persistence time and comfort distance are indispensable for the analysis.

With the aforementioned variables, the agents are programmed to make decisions during the simulation (decision-making methodology). This programming is done in commercial software, such as Pathfinder.

2.2. Definition of scenarios

The presence of people in the building was determined according to the number of classes that are being taught at certain times. It is proposed in this methodology 3 different schedules. For example: 7am, 12pm and 7pm. In addition, seismic records representative of the history for the area under study are identified. This data will be used for the Analysis of the response history, since a set of soil movement records selected and modified correctly is required as input.

2.3. Structural analysis

The purpose of the structural analysis is to obtain the distribution of the stresses and the corresponding displacements of the structure subjected to a given load. It can be calculated by using appropriate methods that express nonlinear analyzes such as Pushover, History Time, Incremental Analysis, etc. There are Software for the structural analysis of buildings with 3D modeling and visualization tools such as ETABS or SAP2000.

2.4. Human analysis

In addition to the way agents act, start and exit points must also be programmed for evacuation. Clearly this type of activity is not the cream of an agent who can make decisions according to situations, but it helps to keep the modeled evacuation under control. Agents will be generated by zones. The programmer can define whether these zones are interconnected with each other or the accesses are blocked at certain outputs. In this case the zones were created according to the distribution of halls, passageways and among other places they are delimited by zones.

3. Results

3.1. Evacuation drill

The recording made on May 31, 2019 provided important results to perform a more realistic evacuation simulation. These results were decided to separate them into physical and psychological.

On the part of the physical results, it is emphasized that the capacity of the people during the drill is the same as the one in the database of the study center. Also, the average speed of the people was 1.3 m/s. There were no people with any complications to move, however, during the drill it started to rain so some people decided to go slower because of the fear of slip of, preventing a faster speed than the one described. The turning radius of people was 0.65 m. The average height of the people in the drill was 1.70 m. This feature affects directly the range of vision, which it was found as 3 m.

On the other hand, in terms of the psychological, people had a reaction time of 0.8 s. facing any external stimulus they faced. These stimuli were mostly closed doors and accesses, imminent clash with other people and perception of great congestion of people. At the time of evacuation, people who realized that the congestion of people was high took 2 seconds to decide to change routes. In the same way, if they perceived that their evacuation route was clear, they took 2.5 seconds to increase their speed to 1.8m / s. In the bottleneck areas, it was determined that people took a comfort distance of 0.9 meters.

All these variables were taken into account for the preparation of the evacuation simulation, both for scenarios without collapses and with collapses.

3.2. Structural behavior

3.2.1. General structural features. This paper analyzes a seven-story reinforced concrete building located in the city of Lima. The predominant system is structural walls, it is a regular system, it has no irregularity in plan or height. As the mass and weight has a total of 696,453 tons. and 6832.208 tonf. respectively. Likewise, the fundamental period of greatest translational is determined in the direction of analysis (x). where the major structural period approaches in 0.698 seconds.

3.2.2. Nonlinear static analysis. The development of the graph that relates basal effort versus superior displacement is presented. The result of one of the axes is shown in the following Figure 1, where in the elastic range the maximum displacement of the building is 17,588 mm when the basal effort is 2931. 966 KN. No structural risk is recorded, so the building can be occupied immediately after a seismic event. In the inelastic range, the maximum displacement of the defined structure is 240.019 mm when the basal shear is 8635.636 KN.

Figure 1. Capacity curve xx direction.

3.2.3. Nonlinear Time history analysis. For response analysis over time, historical records are used. The forces, deformations and distortions are analyzed from the maximum values of the records used, in this case the earthquake of the year 1966, which is of greater magnitude, since it has a greater impact on our analysis because it seeks structural failures in the simulation. As soon as the seismic record to be used, three historical cases have been scaled up with the "SeismoMatch" software, providing us with a new scaled record. These results are indicated in Table 1.

Table 1. Historical records.

Seism	Latitude	Length	Depth (km)	Magnitude (Mb)	Intensity
1966	-10.70	-78.70	24	8.1 Mw	VIII
1970	-9.36	-78.87	64	7.8 Mw	VIII
1974	-12.50	-77.98	13	6.6 Mb	VII

Therefore, after analyzing the structure, the lateral design displacements for each floor are obtained. Table 2 shows the values obtained for each floor of the building in the analysis direction (X). Where you can see that the highest value is 2,632 cm.

Table 2. Maximum displacement of the building-1966 seism.

Story	Load (Case/combo)	Desplac.Max(cm)
Floor 7	DIN X-X	2.632
Floor 6	DIN X-X	2.1677
Floor 5	DIN X-X	1.6903
Floor 4	DIN X-X	1.219
Floor 3	DIN X-X	0.7781
Floor 2	DIN X-X	0.4002
Floor 1	DIN X-X	0.1228

Evaluating the results of the plastic hinges, it can be seen that the formation of the hinges was occurring progressively, which indicates that the analysis performed well as shown in Figure 2a.

This behavior is demonstrated in the moment-curvature graph of the analyzed elements. Figure 2b shows the beam graph. The elements most affected by the earthquake entered were the walls and beams. The columns did not suffer much stress because the plates withstood 82% of shear forces, which justifies that the columns did not incur a mostly plastic range.

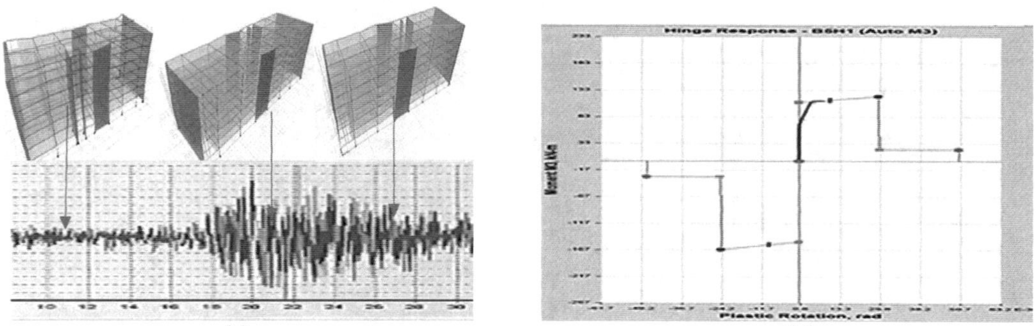

(a) (b)

Figure 2. (a) Plastic hinge formation,1966 seism. (b) Moment curvature graph of beam 34.

This structural behavior can be reflected in the modeling of buildings identifying the position and the moment in which the structural elements collapse. This type of processing of the results is essential to carry out the evacuation simulation later. Next, the collapsed structural elements, their place of collapse and instant of collapse are presented in Table 3.

IOP Conf. Series: Materials Science and Engineering **739** (2020) 012046 doi:10.1088/1757-899X/739/1/012046

Table 3. Collapsed structural elements.

Collapse of structural elements					
Element	Location: Passage	Intant of collapse (Seconds)	Element	Location: Passage	Intant of collapse (Seconds)
Beam	Floor 1	28	Beam	Floor 4	28
Beam	Floor 2	27	Beam	Floor 5	28
Beam	Floor 2	24	Beam	Floor 5	30
Beam	Floor 2	24	Beam	Floor 6	33
Beam	Floor 3	26	Beam	Floor 6	33
Beam	Floor 3	26	Beam	Floor 7	33
Beam	Floor 4	29	Beam	Floor 7	33

3.3. Agent behavior without collapses

As indicated in the methodology section, evacuation simulations were performed for 3 scenarios. For the specific case of the educational institution, the most significant schedules were 7 AM, 12 PM and 7 PM. The simulations will be divided by simulation without collapses and collapses at the same period of time.

Referring to drills without collapses, these took an average evacuation time of 430 seconds, similar time as the recorded drill that took 420 seconds People moved along evacuation routes in the same way as in the drill. Bottlenecks formed by people on the different evacuation routes were witnessed, specifically at the stairway entrances.

The control areas that were developed to simulate the collapse of structural elements, showed a large flow of people circulating. This is important because it is identified that these areas with high flow can be critical and important for the subsequent section of person-structure interaction. Figure 3a and 3b.

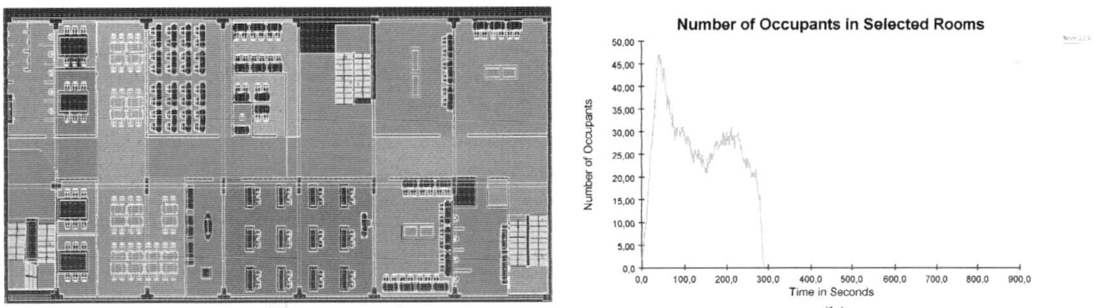

(a) (b)

Figure 3. (a)Agent control zones. (b)graphical flow of people in room 23.

3.4. Interaction of behaviors

Collapses were obtained according to the plastic hinges of the nonlinear history time analysis. Only the hinges that marked impending collapse were taken into account to ensure that entrances to different areas are affected. With this information it was possible to extract the trapped people due to the collapses.

The zones carried out for this simulation are divided according to the route of the structural beams of the building. This procedure was carried out to take into account the collapse of beams and columns in all possible floors and places.

In addition to taking new evacuation routes, the collapse led to the formation of larger bottlenecks. It also created greater congestion of people in areas where there was free movement. It was determined that evacuation routes are not designed for this sudden increase of people. Figure 4a.

The collapse of continuous or parallel structural elements indicates that the area of involvement is large. This could be seen in the area where the agents were trapped during the evacuation without being able to leave by another route. This is the case of the 4th floor where it can be seen that the collapse of beams in parallel meant that 32 people were trapped without the possibility of evacuation. Figure 4b.

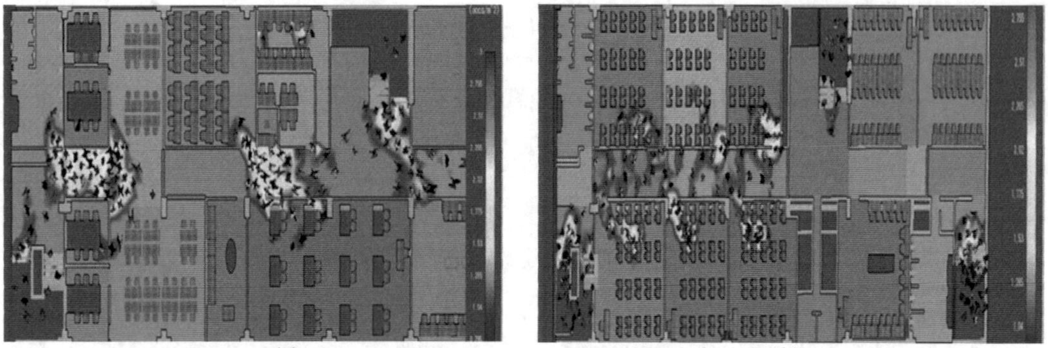

(a) (b)

Figure 4. (a)Landslide evacuation route change. (b) People trapped floor 4.

Evacuation times were 560, 490, 520 seconds for the 7 AM, 12 PM and 7 PM schedules respectively. These took longer than in simulations without landslides, due to alternate route taking and increased accumulation of people on evacuation routes. A total of 65 agents were trapped along the evacuation routes on the different floors for the morning schedule, 43 for the midday and 55 for the evening.

4. Conclusions

The evaluation of essential buildings through the interaction of structural and human behavior provides a more complete analysis of important aspects, such as evacuation during emergencies, finding structural, architectural flaws and possible human affectations.

The structural damage of the building per floor is moderate due to the presence of collapsing beams. 94% of structural elements of the building did not enter a plastic state, remaining in an elastic behavior.

Evacuation routes were directly affected by the collapse of beams. According to the location of the collapsed beams, they traveled along the evacuation routes making it difficult to evacuate the simulated agents. This situation is critical considering that evacuation routes should be the least affected in the building.

100% of the floors of the building presented bottlenecks around the emergency exit entrances. The largest bottlenecks appeared on the stairs at the ends because there are access doors before it, causing a greater delay in the start of movement of people in them.

5% of the total simulated agents were trapped inside the building during the simulation. The areas with the highest number of trapped agents were in the central sector before the central stairs. The flow of people or agents was greater in that area due to the easy access of the central stairs. In addition, it evidences the structural deficit presented by the building in that area.

The interpolation of the duration of the earthquakes with the evacuation time of the agents shows an alarming situation. 95% of people to evacuate are still in the building even when the earthquake is over. The duration of the selected earthquakes is part of 17% of the total evacuation time both in the simulation and in the simulation. The structure experiences maximum stress during the first 100 seconds of evacuation, leaving the greatest number of people in the building in danger. The occurrence of collapses during this short time interval means people unable to evacuate to the point of being physically and psychologically damaged.

References

[1] Liu X F, Lim S S 2016 Integration of spatial analysis and an agent-based model into evacuation

management for shelter assignment and routing, *J. Spatial Sci.* **61(2)** 283-98.
[2] Mohamed Marzouk Basma and Mohamed 2018 "Multi-classification criteria instrument for evacuation building evaluation in agent-based simulation function". Construction Research Congress.
[3] Ren C J, Yang C H, Jin S Y 2009 Agent-Based Modeling and Simulation on Emergency Evacuation, *Int. Conf. Complex Sci.* **5** 1451-61.
[4] José Ignacio García-Valdecasas Medina (2011). “Agent-based simulation: a new way to explore social phenomena” 91-110.
[5] Mehmet Erkan Yuksel (2018). "Evacuation model based on agents with multiple exits using NeuroEvolution". Advanced engineering computing 30-55.
[6] Gana H S, Richter K F, Shic M, Winterd S 2016 Integration of simulation and optimization for evacuation planning, *Simul. Model. Pract. Theory* **67** 59-73.
[7] Curtis S, Best A, Manocha D 2016 Menge: a modular framework for simulating crowd movement, *Collect. Dynamics* **1** 1-40.
[8] Pelechano N, Malkawi A 2008 Evacuation simulation models: challenges in modeling high rise building evacuation with cellular automata approaches, *Autom. Constr.* **17(4)** 377-85.

IOP Conf. Series: Materials Science and Engineering **739** (2020) 012047 doi:10.1088/1757-899X/739/1/012047

Artificial Intelligence of the Water Supply System of Plant Growing Objects

D Yu Polenov[1]

[1]Moscow State University of Civil Engineering, 26, Yaroslavskoye Shosse, Moscow, Russia

Polenoff.mitya@yandex.ru

Abstract. The article proposes a solution to the problem of achieving the maximum value of productivity of plant growing objects based on the introduction of artificial intelligence on the basis of neural networks in the water supply system of plants. Various factors affecting crop yields and plant growth are presented. Factors are considered as incoming signals for mathematical neurons of artificial intelligence. A diagram of the artificial intelligence neural network of the plant water supply system, its topology, and the corresponding mathematical model are constructed.

1. Introduction

AI is used in medicine, forensics, diagnostics of malfunctions of technical objects, etc. [1, 2, 3]. As for plant growing objects in terms of their water supply, it is necessary to make decisions of the following nature: turn on/turn off the water supply, reduce / increase the volume of its supply, etc. If such decisions are timely (based on the readings of soil moisture sensors, temperature and other properties of the water used, its consumption for each crop, etc.), the growth and productivity of plants will increase, which means their productivity.

Plant productivity is influenced by a large number of factors. At the same time, one set of factors can be easily regulated (water supply, aeration and soil moisture), another complex can be regulated with difficulty using high energy consumption (humidity, soil acidity, ambient temperature, light exposure), and the third is the presence and volume of snow cover, the duration of the frost-free period, the amount of precipitation, their distribution by months - to compensate as much as possible by regulating the first and second parts of the factors. It is known, for example, that productivity will not be affected if watering volume is reduced while the ambient temperature is lowered, etc. The aforesaid allows us to conclude that achieving the maximum value of the productivity of plant production objects, taking into account the influencing factors, is a rather difficult task.

It can be solved by introducing AI into the water supply system of plant growing objects (hereinafter - the system). In other words, it is recommended to make the so-called “smart” irrigation system. Due to the presence of AI, the system will be adapted to the characteristics of the environment (region, country) of the plants, it will follow year-round weather conditions, receive and process weather forecast information from a hydro meteorological center, etc. - the system will be based on the principle of watering plants with such a volume of water that they need. This can be done by:

1) the information stored in the memory when the system was put into operation and the system received experimental information over time (including from the sensors) regarding the required volume of water supply for each plant;

2) obtaining weather forecast information on the amount of precipitation in the form of rain, both in real time and in total for months, seasons, etc.;

3) dividing the site of the location of objects containing various crops in water supply zones;

4) processing the above data;

5) making decisions on the water supply of a separate plant area of the facility.

Table 1 provides indicative data for the analysis of AI, on the basis of which a decision will be made on the water supply of plant growing objects.

Table 1. The accumulated knowledge for the analysis of AI.

Information from sensors	Persistent memory	Information from a hydro meteorological center
1. The pressure	1. Dependences of soil acidity, ambient temperature, soil moisture, pressure, illumination	1. Weather forecast
2. Illumination		2. The direction and speed of the wind
3. Humidity		3. The distribution of precipitation by month
4. The presence of precipitation		
5. The temperature of the environment		
6. The acidity of the soil		
7. Soil moisture		
8. Water hardness		

Having explained the essence and need for the development and implementation of AI in the water supply system of plant growing objects, we proceed directly to the description of the principles of its organization. As you know, there are two main approaches used in creating systems containing AI: the creation of expert systems and the use of neural network technologies [4]. As a rule, a person directly deals with the water supply to plant growing objects - he either manually irrigates his plot or includes the available irrigation system. In this regard, the water supply system will consider the features of the use of AI based on neural network technologies - a system with AI in the image and likeness of the human brain.

The core of the AI of such a system will be considered a mathematical neuron - an object that is a mathematical model of the neuron of the human brain. It will have similar properties: contain inputs *i*, where will the input signals x_{ij} and one way out y_i.

Moreover, each signal arriving at the *j*-neuron signal x_{ij} will be multiplied by some corrective weighting coefficient b_{ij} (1):

$$S_j = \sum_{i=1}^{i} b_{ij} x_{ij} \tag{1}$$

The output signal y_i of a *j*-th mathematical neuron will take one of two values (2), (3):

$$y_i = 1, \text{ if } S_j \geq \theta_j \tag{2}$$

$$y_i = 0, \text{ if } S_j < \theta_j \tag{3}$$

where θ_j – sensitivity threshold of the j-th mathematical neuron.

If the resulting total value S_j j-th mathematical neuron of the water supply system does not reach a certain value of the sensitivity threshold θ, then the mathematical neuron will not be excited - its output will be zero, and vice versa [5, 6].

As signals entering the inputs x_{ij} j-th mathematical neuron of the water supply system, the output values of the sensors included in its composition will be used. Such sensors include: humidity, temperature, pressure sensors, etc. In addition to sensors, information from the Hydro Meteorological Center containing data on upcoming precipitation will be sent to the inputs of mathematical neurons: estimated time of precipitation, duration of exposure.

We list the main types of sensors, the output signals of which will arrive at the inputs of the mathematical neurons of the AI System, and briefly describe their purpose:

- pressure and temperature sensors monitor the current values of atmospheric pressure and ambient temperature (*T1*);
- the illuminance sensor of the plot shows the level of illumination of plants (*L1*);
- rainfall sensor detects rainfall / snowfall (*S1*);
- moisture sensors monitor current moisture values in the plant rhizome area (*W1*).

Let us present in figure 1 an image of the topology of the water supply system of plant growing objects, where *BUP* is the irrigation control unit, *G1* is the GSM-module for connecting the system to the Hydro meteorological Center website and receiving a weather forecast from there, *L1* is the light sensor, *S1* is the precipitation sensor, *T1*is the ambient temperature sensor, *W1* - *W5*are the soil moisture sensors, *K1* - *K5*are the valves for turning on / off the water supply to sections *Z1* - *Z5* of planted plants.

The AI of the System is located in the *BUP* unit, to which all “sensory organs” are connected (sensors *L1, S1, T1, W1* - *W5*) and output commands are generated for the electromagnetic valves (*K1* - *K5*) for switching on / off the water supply of the sections (*Z1* - *Z5*), where the plants are planted. Proceeding from this, the neural network of the AI System shown in figure 1 can be represented in accordance with figure 2.

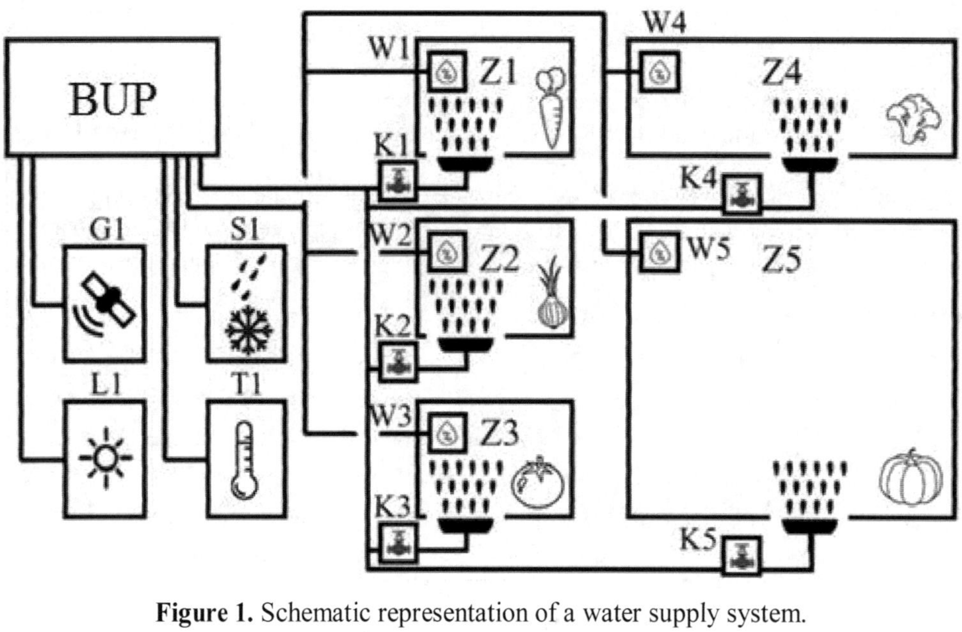

Figure 1. Schematic representation of a water supply system.

IOP Conf. Series: Materials Science and Engineering **739** (2020) 012047 doi:10.1088/1757-899X/739/1/012047

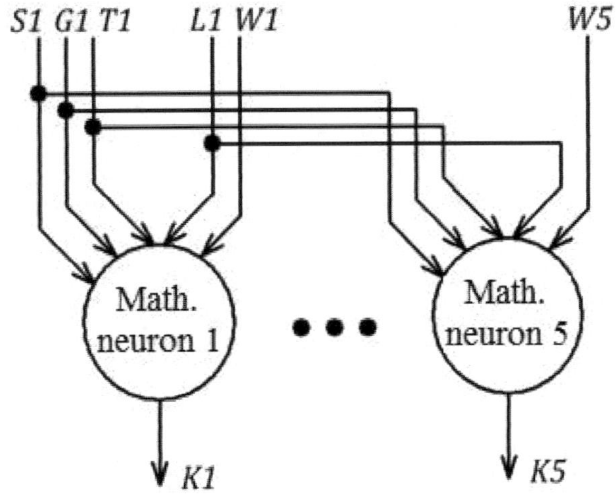

Figure 2. System Neural Network.

Marking $\overrightarrow{X_j}$, as a vector of input signals, and $\overrightarrow{Y_j}$, as a vector of output signals, based on figure 2 we get (4) and (5):

$$\overrightarrow{X_j} = \begin{pmatrix} S_1 \\ G_1 \\ T_1 \\ L_1 \\ Wj \end{pmatrix} \tag{4}$$

$$\overrightarrow{Y_j} = (K_j) \tag{5}$$

Moreover, the training vector $\overrightarrow{B_j}$ with correction factors $b_{ij} = \overline{0,1}$ is equal to (6):

$$\overrightarrow{B_j} = \begin{pmatrix} b_{11} \\ \dots \\ b_{ij} \end{pmatrix} \tag{6}$$

For the case shown in figure 2, according to expressions (1), (4) and (6) we have:

$$S_1 = b_{11}x_{11} + \dots + b_{51}x_{51} = b_{11}s_1 + b_{21}g_1 + b_{31}t_1 + b_{41}l_1 + b_{51}w_1 \tag{7}$$

$$\dots$$

$$S_5 = b_{15}x_{15} + \dots + b_{55}x_{55} = b_{15}s_1 + b_{25}g_1 + b_{35}t_1 + b_{45}l_1 + b_{55}w_5 \tag{8}$$

Based on (1) - (3), we can construct the threshold activation function [2] of the *j*-mathematical neuron for it output signal y_i(figure 3). The figure shows that after accepting the input signals x_{ij}, multiplied by correction factors b_{ij}, calculated weighted sum of input signals S_j, and based on its value, a decision is made on the output value y_i *j*-mathematical neuron. It is worth noting that the values of the correction factors will change over time (9):

$$b_{ij}(t+1) = b_{ij}(t) + \Delta b_{ij} \tag{9}$$

where $b_{ij}(t+1)$ – the new value of the correction factor, $b_{ij}(t)$ – the old value of the correction factor. Reason for changing odds Δb_{ij} will depend on the volume and quality of plant productivity obtained at the end of the reporting period.

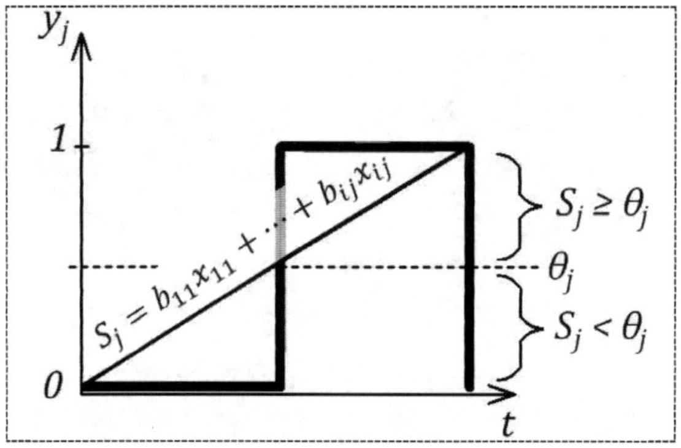

Figure 3. Threshold activation function.

As further research prospects, it is proposed to carry out work in the following areas:

- research and identification of additional input parameters of a mathematical neuron, identification of significant and insignificant parameters, exclusion (minimization) of insignificant ones. This should improve the quality of the neural network. However, it must be taken into account that too few input parameters may not be enough for the neural network to correctly determine the result y_i;
- the complexity of the algorithm for adjusting the weight coefficients b_{ij}.

As a result of the research, the following results were obtained:

1) a solution has been developed to achieve the maximum value of productivity of plant growing objects, taking into account various factors affecting the growth and productivity of plants. Its solution is proposed by introducing artificial intelligence into the water supply system of plant growing objects;

2) the organization of artificial intelligence has been developed for the water supply system of plant growing objects based on neural network technologies;

3) the organization of a water supply management system based on artificial intelligence has been developed; the topology of water supply for crop production facilities has been developed with the division of the entire irrigation area into zones;

4) a mathematical model of a water supply system with control based on artificial intelligence was developed.

References

[1] Holland J H 1975 Adaptation in Natural and Artificial Systems Ann Arbor,*The Uni. Michigan Press* **1** 975.

[2] Winston P H 1984 *Artific.Intellig.* 2nd Edit. Reading MA: Addison-Wesley.

[3] Yasnitsky L N2008 Introduction to artificial intelligence, *Book for learning schools 2nd Edition M.: publishing center Academy, p.* 10-14.

[4] Hopfield J J 1984 Neurons with graded response have collective computational properties like thous of two-state neurons,*Proc. Nation. Acade. Sci.* **81(10)** 3088-92.

[5] Minsky M, Seymour P2017 Perceptrons: An Introduction to Computational Geometry*MIT Press*.

[6] Rosenblatt F1958 The perseptron: a probalistic model for information storage and organization in the brain,*Psych. Rev.***65(6)** 386.

IOP Conf. Series: Materials Science and Engineering **739** (2020) 012048 doi:10.1088/1757-899X/739/1/012048

Exploration and Practice of Digital Factory of Complex Electronic Equipment

Hu Changming[1], Cao Yaqi[1], Zhu Zhaoyu[1]

[1]Nanjing Research Institute of Electronic Technology, Nanjing 210039, China

1169668672@qq.com

Abstract. For the characteristics of many varieties, small batch, short delivery cycle and high quality requirements of complex electronic equipment manufacturing, combined with a new generation of information and communication technologies and the research of advanced manufacturing technology, This paper puts forward the implementation framework of digital factory of complex electronic equipment, analyzes the connotation of digital factory, explores the application of digital twin in the manufacturing stage, discusses the process of data perception, fusion and analysis of digital factory, and establishes the factory production operation control platform. This paper verifies the feasibility of complex electronic equipment digital factory floor by the example of radar assembly digital factory, and provides theoretical basis and method reference for the comprehensive application of digital factory in complex electronic equipment field in the future.

1. Introduction

Complex electronic equipment is widely used in the fields of machinery, electric power, ships and aircraft, and is an important pillar of the national people's livelihood and national defense security. Complex electronic equipment involves many subjects, ranging from information science and technology such as electronics, communication, computer and control, to advanced manufacturing technologies such as additive manufacturing, micro-assembly and automation. It covers a wide range of knowledge and many fields, which means higher requirements for complex electronic equipment enterprises [3]. Its typical characteristics is strict manufacturing process standards, fast switching of mixed production line, strict production task planning, high product quality requirements. The traditional "equipment + artificial management" model has been difficult to meet the requirements of fine control in factory. Therefore, we must be with the aid of new technology and production mode to adapt to the intense competition of product performance, quality, cost and delivery of complex electronic equipment factory.

Generalized digital factory is geared to the needs of whole life cycle of products, based on the simulation technology and virtual reality technology, and that a physical entity of the whole production process is digital simulated, evaluated and optimized in virtual environment is a new production organization way. Digital information and the data flow as the main characteristics, its purpose is to achieve rapid, low cost and high quality manufacturing [7]. In the narrow sense, digital factory is oriented to the production and manufacturing process of products, and design, management, simulation, optimization and visualization of the production resources and production process of people, machine, material, method, environment and measurement, etc., which is also aimed at reducing cost, improving quality and increasing efficiency[6] [23]. This article discusses the complex digital electronic equipment factory in digital information and the data flow as the main characteristics, by means of digital, networked, intelligent, improving the production organization and management to realize workshop sophisticated controls, to satisfy flexible, efficient, high quality, transparent manufacturing needs of complex electronic equipment [5].

The world's most prestigious digital factory is the Siemens Amberg plant in Bavaria, Germany. The factory was founded in 1989. During 30 years of development, it experienced from the introduction of workshop management system, to RFID, to data optimization management, and then to the process line management system, to achieve a high degree of automation of the production process, 75% of the process work by the production equipment and computer processing. In recent years, the product information and physical entities in the virtual environment of Digital Twin system are mapped to each other to optimize and control the manufacturing process in real time. At present, the factory can produce one control device per second on average, and the product qualification rate is as high as 99.9988%[8].

GE aviation engine manufacturing center, located in Ohio, is a typical representative of another model of digital factory: "sensor + big data" industrial Internet model. Predix is GE's first cloud service platform developed specifically for industrial data analysis, enabling engine data acquisition and status monitoring through seamless integration with various applications and services deployed in the cloud [24]. At the same time, for the complex discrete manufacturing process of engines, GE has established a powerful production management and manufacturing execution system, which combines lean concepts with automation and digital methods to achieve comprehensive improvement of manufacturing performance [9, 10]

Qingdao Haier mold co., LTD. (hereinafter referred to as Haier mold) is the largest mold and fixture manufacturer in China, and is one of nine interconnected factory demonstration units. Aiming at the digital factory featuring automation and fewer people, Haier mold has built the leading CPS collaborative production management system in China since 2013, realizing the free flow of data through equipment interconnection. Through system integration, information sharing and collaborative manufacturing can be realized; Through data collection, analysis and display, to achieve scientific management; Through real-time monitoring of equipment status, predictive maintenance is realized. The successful implementation of the CPS system has improved the production efficiency and enterprise competitiveness of Haier mold, and achieves good economic and social benefits.

Ningxia sharing group, located in western China, which belongs to traditional manufacturing industry, is one of the first batch of 46 intelligent manufacturing pilot and demonstration projects of the ministry of industry and information technology. Through the information integration of ERP, PLM, MES, DNC and other systems, it realizes the digitalization and transparency of workshop production process. The free and orderly flow of data in the workshop can be realized through the integration between information systems, production equipment, physical world and virtual world.

Digital factory is the foundation of intelligent manufacturing and the only way for industry 4.0. Many domestic and foreign manufacturers have built or started to build their own digital factories [2]. In the face of rapid emerging technologies and increasingly fierce market competition, digital factories will integrate various new technologies such as artificial intelligence and data mining, so that factories can learn, think and make decisions just like people, and finally realize intelligent factories.

Compared with Siemens, GE and other international leading enterprises, the domestic digital factory started relatively late. At present, that is mainly used in casting, machining and other traditional industries. It is not yet mature for electronics industry whose characteristics are many varieties, small batch. And the construction of standard system, data fusion analysis, digital twin, control operating platform construction of research remains to be strengthened.

This paper, firstly, introduces the concept and implementation method of digital factory for complex electronic equipment, and then analyzes its connotation and frame, and discusses the key technology of digital factory practice. Finally, based on theoretical exploration, the paper combined digital factory with complex electronic equipment production and manufacturing, introduces the application cases of complex electronic equipment in digital factory, and provided a method reference for the implementation of digital factory in many-varieties and small-batch manufacturing enterprises.

2. Connotation of complex electronic equipment digital factory

Complex electronic equipment is the product of the integration of multiple subsystems through complex coupling relation. In addition to meeting the individual functions in their respective fields, it is also required to meet the overall functions of the system after coupling integration. Therefore, the manufacturing process of complex electronic equipment emphasizes more on reliability and integrity [4].

Digital factory of complex electronic equipment is the mutual mapping between physical entity operation and information flow in the assembly process of complex electronic equipment from parts to complete machine, which can truly reflect the integration process and meshing relationship of each part from single to whole. On the basis of operation control platform, it is targeted at refined production factors such as human, machine, material, method and environment, and combined with advanced scheduling, data collection and analysis, digital twining and other application functions to digitize physical entities and visualize plant operation, so as to realize the research and application of complex electronic equipment digital factory.

Digital factory, factors of production, operation process and product as the core, on the basis of whole life cycle of product data, applies simulation technology, virtual reality technology, data fusion and analysis technology, etc., digitize all the physical activities of products in the factory, and analyze these activities, optimization and feedback mode of an integrated organization [8].

Digital factory emphasizes the digitization of production factors, processes and products, which is the digitization record of each step in the real factory (such as process planning, manufacturing, quality inspection, process management, fault maintenance, etc.). Compared with traditional factories, what is more important is that digital factories can optimize and control the production process, through the perception, integration and analysis of data, real-time monitoring and adjustment of product quality, equipment status and worker operation, so as to keep the factory on a stable and reliable running track.

3. Digital factory framework for complex electronic equipment

Digital factory covers the whole process of manufacturing complex electronic equipment. Its framework is shown in figure 1, including execution layer, acquisition layer and equipment layer. In a general sense, digital factory also includes business layer.

The equipment layer of the digital factory includes various production resources necessary for the production and manufacturing of the digital factory, including manufacturing equipment, production personnel, tooling and so on. Among them, manufacturing equipment undertakes the tasks of material distribution, production execution, testing and inspection, and a large number of digital equipment is adopted to automatically collect information or execute instructions. A small part of equipment can be equipped with sensors to achieve information collection and transmission. The production staff undertakes the task of monitoring, guiding and correcting the production process, and digitizes the operation through human-computer interaction interface. Equipment tools assist manufacturing equipment to complete the production activities of the digital factory, but they also do not have the ability of digital communication. They can participate in the digital activities of the factory through bar code, scanning gun, RFID and other technologies.

The acquisition layer of complex electronic equipment includes data acquisition and feedback control on the equipment layer. The data collection is to perceive the production process data through RFID, WIFI, Ethernet and other communication technologies, and effectively utilize the data through multi-source heterogeneous fusion analysis technology to achieve predictive maintenance, visual display and other functions. Feedback control is to transfer the orders of equipment execution and correction issued by the execution layer to the equipment layer, so as to realize accurate control of production equipment.

Execution layer includes process management, schedule management, order management, material management, quality management, equipment management, and other six function module and a basic support module, for all kinds of business, activity in the production process and elements of digital management, realizing the ordering of factory production, lean and transparent.

Four layers framework of digital factory and two basic requirements, "digital" and "manufacturing", is considered from the production implementation, information collection and manufacturing resources three dimensions, covering the technology, production, logistics, quality, resources, connectivity, systems integration and other factors, to complete and accurate description of digital factory [26].

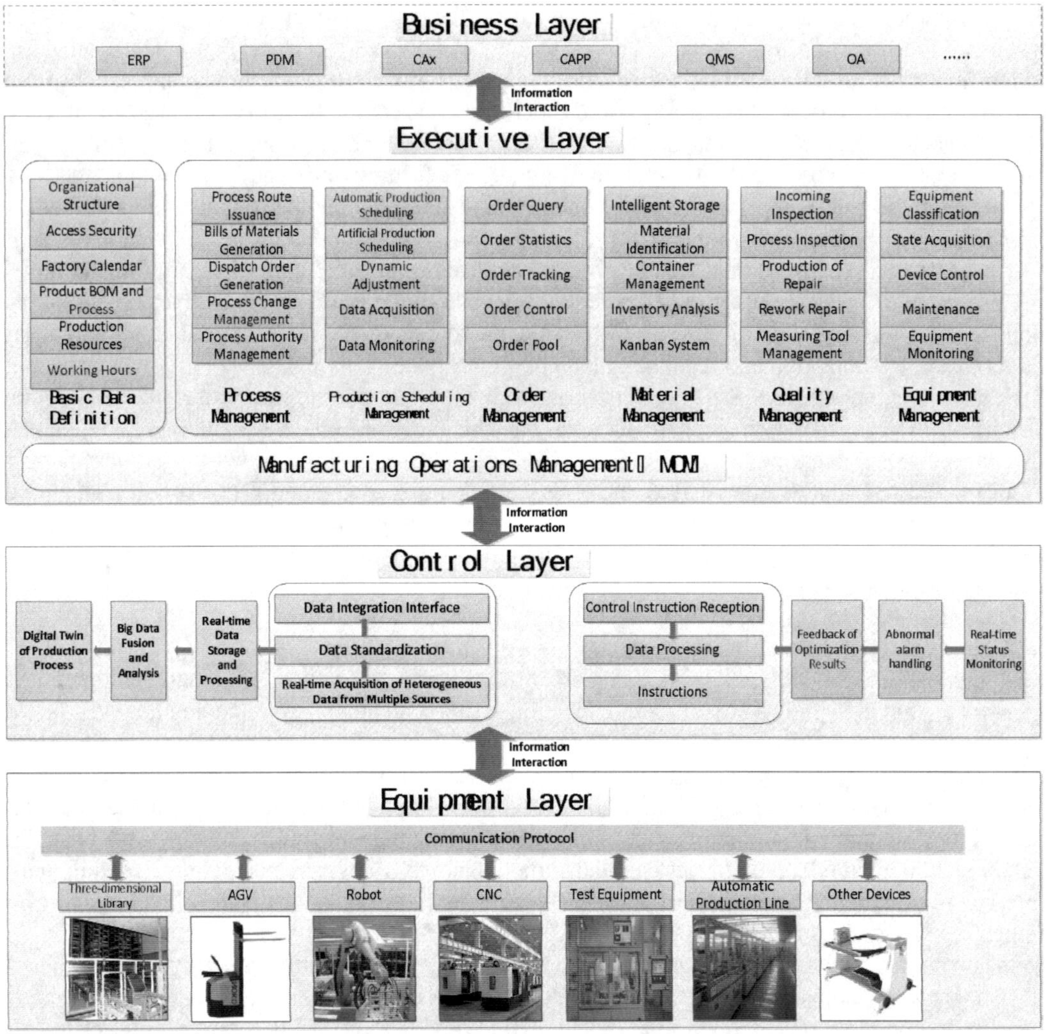

Figure 1. Digital factory structure of complex electronic equipment.

4. Key technology exploration

As the core unit of intelligent manufacturing, digital factory is a complex coupling of multiple fields and disciplines [25]. Digital factory framework and its technology have been widely used in casting, machining and other traditional manufacturing, but in complex electronic equipment manufacturing field, characteristics of many varieties, small batch, the application is still not in-depth enough. Based on the above mentioned digital factory architecture, and need to explore a series of key technologies, including production operation control technology, digital twin dynamic perception and fusion technology, and production data analysis technology.

4.1. Production data dynamic perception and fusion analysis technology

In the manufacturing process of complex electronic equipment, the data collected by various sensor devices are characterized by syntactical, semantic and structural heterogeneity. Analysis technology can achieve dynamic perception and production data fusion of different interface and communication protocol data conversion and packaging, and through the data classification, association and combination operation, integrating multi-source information for data mining, digital twin model needed by the simulation analysis, the computer can recognize the certainty of information, digital twin model for data mining, provide the underlying data support.

4.1.1. Production data dynamic perception and modelling. Complex electronic equipment manufacturing system is a dynamic system with continuous information generation and change. Its assembly site data has the characteristics of multi-source, real-time, massive and heterogeneous [20]. The field data of complex electronic equipment assembly can be roughly divided into product data, production process data and equipment data. Production process data mainly includes task information, quality information, personnel information, material information, environmental information and abnormal production information. Device data mainly refers to device state information, device operation parameters information, etc., classified as shown in figure 2.

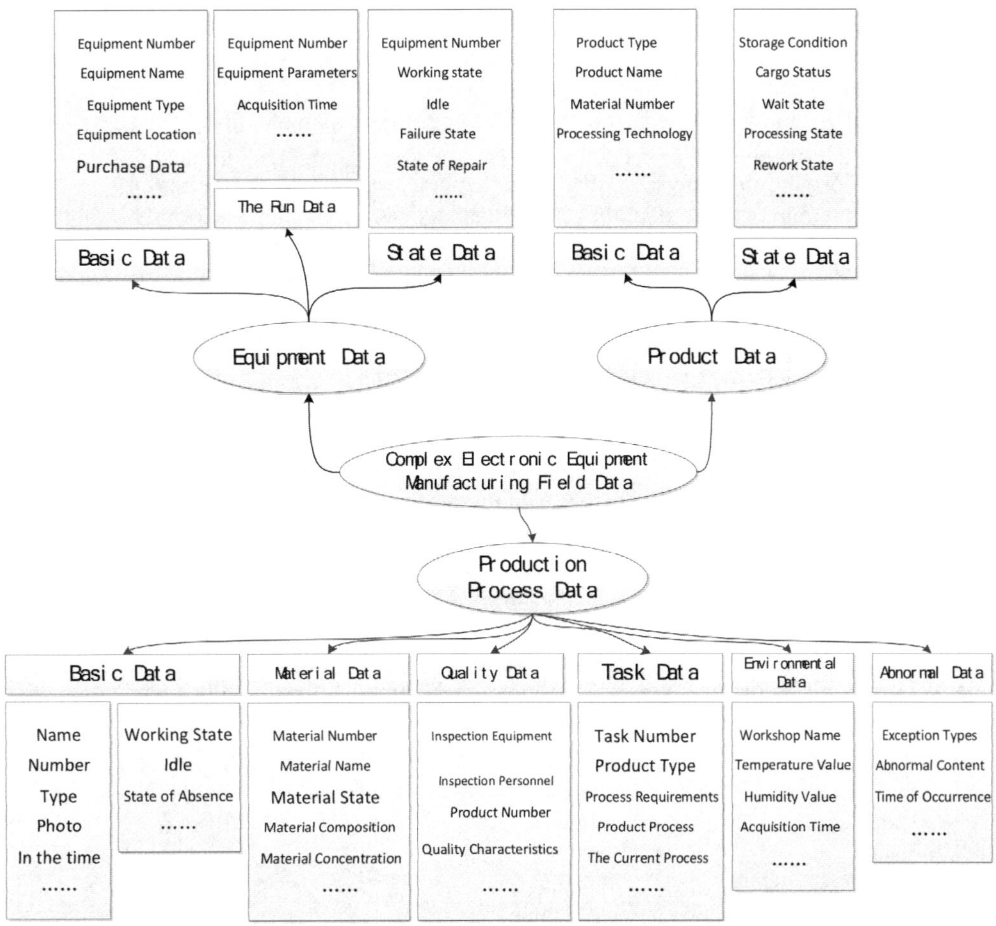

Figure 2. Data classification of complex electronic equipment manufacturing process.

On the basis of the analysis and classification of these data, the data collection objects are established according to various elements of personnel, equipment, materials, testing equipment and other elements involved in equipment manufacturing. For the monitoring of equipment, the objects to be collected are assembly equipment and the collected information is equipment status and other information. As for the production quality control, the production quality is affected by the operation parameters of the equipment, that is, the collection object is the assembly equipment. As the collected objects are different, some of the collected data are real-time, some are discrete, and differ in time dimension, and their data structure, format and collection method are also different. Therefore, it is necessary to establish a multi-source data collection object model with standard and formal methods, so as to realize the unified expression of the collected data. On the basis of the establishment of the acquisition model, the corresponding acquisition methods can be determined, such as sensor, bar code, device networking, human-machine interaction (HMI) and MES/ERP interface.

According to the assembly process data analysis and classification, combined with the data collection model, with modern sensor and the corresponding intelligent terminal as the main acquisition methods, and combining the industry establish a IOT workshop environment of Internet technology, using the unified and efficient data exchange protocol and the data interface, realize the complex electronic equipment manufacturing field of multi-source heterogeneous data real-time, accurate, reliable, acquisition and transmission.

4.1.2. Production data fusion and analysis. With the big data analysis method, combined with artificial intelligence as the main tool and the technology of data fusion, data feature extraction, data mining and visualization as the technical support, the production data of complex electronic equipment was analyzed and processed to provide data support for the intelligent decision of the management of the factory.

According to the business content in the process of equipment manufacturing, based on association rules, such as correlation coefficient data correlation analysis method, based on quantitative indicators, such as coverage, correlation in multi-source heterogeneous data based on the unified description of the process of extracting associated with strong business data, data on the original data cleaning, integration, reduction and transformation operations such as pretreatment, get consistent, accurate, high quality data, to construct a data warehouse of equipment manufacturing in the main body, finally realizes the data in the complex electronic equipment data integration platform, for equipment manufacturing process and dynamic optimization to provide reliable, reusable data resources.

By means of wavelet transform, Fourier transform, principal component analysis, mathematical statistics and other methods, the business characteristic data is extracted, and the business data model is simplified, which is convenient for subsequent data analysis and processing. Then on the basis of subject oriented data warehouse, using real-time stream processing combined with static batch hybrid parallel computing framework, through the behavior of the large data analysis, semantic analysis, statistical analysis, combined with the association rules, linear regression, forecasting, clustering data mining methods, to achieve implementation of the process parameters, equipment status, such as manufacturing data correlation analysis. According to the manufacturing data coupling mechanism of the business which is strong correlation between the data and its relevance as based on big data technologies, such as the depth of the intelligent technology such as neural network, the convolutional neural network input, to implement production scheduling, quality control, failure prediction and other business of pattern recognition, the results of prediction, workshop evolution rule knowledge acquisition etc., provide service for the intelligent decision support in the end.

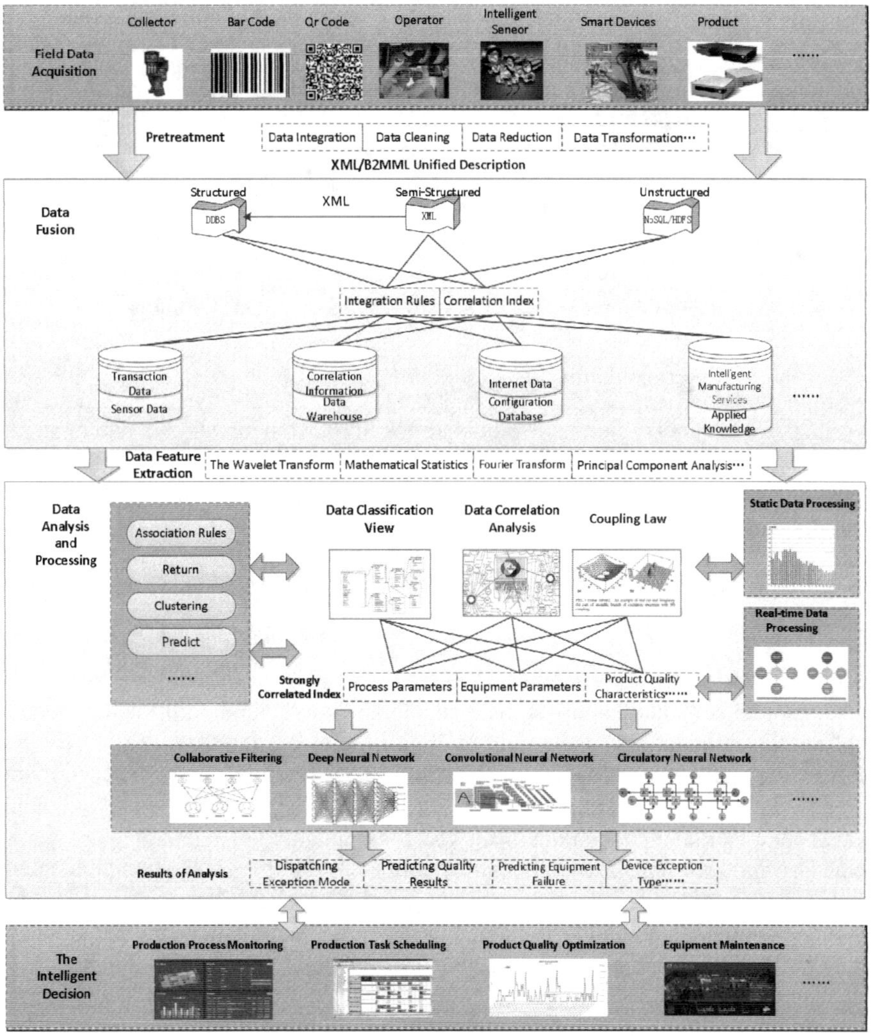

Figure 3. Technical framework of multi-source heterogeneous manufacturing data fusion and analysis.

4.2. Digital twin technology

Based on the digital twin, plant operation monitoring and prediction is on the basis of the connectivity of the whole process of equipment, collects total assembly factor information and mapped to a virtual environment, building digital twin model including assembly environment for assembly entity and equipment elements, by the deep fusion of assembling the physical process and the information world, constructs an extra realistic model of virtual assembly process, using intelligent methods of data analysis, deep learning to run the digital factory monitoring and prediction, realize the whole process of assembly production "model driven, real-time sensing, dynamic feedback, analysis, optimization", promote the transformation of the assembly mode from "open loop" to "closed loop" intelligent control mode, and improve the assembly quality and efficiency of complex electronic equipment [1].

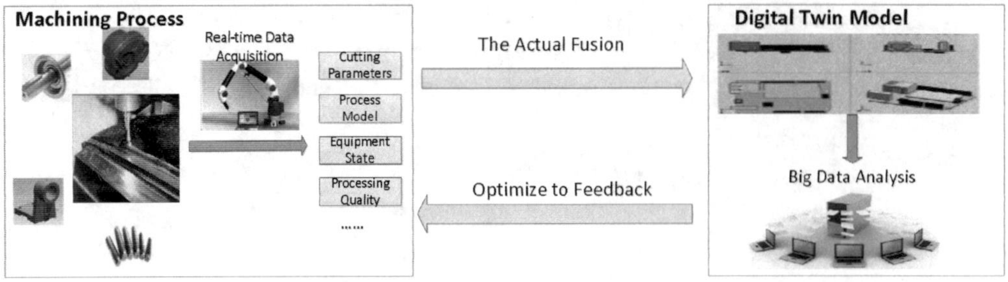

Figure 4. Complex parts processing based on digital twinning.

4.2.1. Workshop operation process modelling. To establish the digital twin model, a uniform structure should be constructed first. Facing different physical entity types and diversified functions, as well as the data generated by the entity, the twin model in the digital space should be constructed. In the whole machine assembly process, key factors involved in production include products/parts/materials, equipment, personnel and production environment, etc. Therefore, the whole process digital twin model is uniformly expressed as follows:

$$DT_{\mathrm{ws}} = DT_{equip} \cup DT_{prod} \cup DT_{pers} \cup DT_{env}$$

DT_{ws} DT_{equip} DT_{prod} DT_{pers} DT_{env} on behalf of digital twin model of workshop production process, equipment, product, personnel and environment respectively [27].

Digital factory equipment includes industrial robot, special process equipment, AGV car, etc. In order to complete the real mapping of the twin model to the physical entity, the 3d size and behavior of the model must be highly consistent with the entity [18]. At the same time, the twin model needs to establish the interface of virtual and real communication control, obtaining the entity data in real time, and defining the related virtual service to complete its behavior.

In the production process, human twin is mainly reflected in human movement and space position. Human body structure is mapped through 3d structure model, physical space location and joint motion data are obtained through location/motion data interface, and activity monitoring service is used to drive the update of movement and position of digital twin model.

For products/parts/materials, in different stages of assembly process, the product corresponding to different geometry, and accompanied by orders, coding and the quality of the whole life cycle information. This information may be retained by data interface in virtual label each product digital space according to its process data driven products/parts/material geometric state of evolution.

For the production environment, its twin model is identified by virtual label, and environmental data is acquired through sensors and displayed in quantitative formation.

4.2.2. Workshop running real-time mapping. The mapping of digital twin to physical space is the basis of virtual and real interaction of digital twin technology. After realizing the real-time mapping, the analysis and optimization of the real-time production situation in the digital twin space will be more real-time, more convenient and more multidimensional than the traditional way, which is also the basis of "virtual control real". The real-time mapping and interaction of digital space are mainly divided into five parts: product, equipment, personnel, system and environment.

1) The product is the core of the production line. The entire life cycle from the removal of parts from the warehouse to the storage of finished products is driven by the actual production data to complete the product evolution. The process data and quality data of the product are stored dynamically in the label of the virtual product and accompanied by the whole life cycle of the product.

2) Carry out real-time mapping of the movement, spatial position and running state of robots, AGVs, processing equipment and other equipment in the production line to complete the processing of each station.

3) Real-time mapping of personnel's identity, location and other information to complete visual management of personnel.

4) Real-time mapping of production schedule, operation schedule, process schedule and other information. Among them, the information such as inventory status, logistics situation, process of processing station and work-in-process data volume can be visualized and managed by digital twin space.

5) Display the current assembly environment parameter information of the factory and the changes of the factory environment parameter in a period of time in real time, and be able to give early warning of the current parameter as well as the assembly process requirements.

In order to realize the synchronous operation of digital space to each element of physical space, a large number of driving signals and data need to be obtained from the field to effectively drive data at all levels in the digital space. Multi-source data acquisition is the basis of realizing digital twin.

4.2.3. Monitor and forecast workshop operation. By constructing the digital twin of complex electronic equipment manufacturing factory, the production site situation can be understood in real time and the global information can be grasped [19]. In view of abnormal production conditions in manufacturing process, the sensitivity technology of collaborative perception is used to improve the sensitivity of important information closely related to production and reduce the sensitivity of interference information that is not important or averse to production. Combined with event processing function, processing workflow can be established for different alarm events. Realize the notice, judgment, processing, tracking, analysis, recovery and closure of the exception, form a closed loop of workshop exception management, improve the timeliness of exception handling.

Through real-time collection of quality data and correlation with production order, batch and other information, potential quality risks and problems can be found in time to realize rework and repair, quality visualization, and quality whole-process traceability management. In addition, analyzing the causes of quality problems in time provides analysis data and statements for the factory and quality department, improves the product qualification rate, and reduces the occurrence of similar problems.

Based on digital twin technology, the professional diagnosis of equipment in the factory is completed, including state analysis and fault diagnosis. Based on the equipment residual life evaluation technology, the evaluation model of equipment state deterioration trend was established, and the prediction model reflecting the correlation rule between variables was established with the quantitative analysis technology. The accumulated historical data are used for statistical analysis and data-driven equipment state modeling. According to different types of equipment objects, nonlinear prediction models, parameter optimization techniques and alarm strategies are established to accurately grasp the deterioration trend of equipment state, prevent abnormal equipment and ensure smooth production.

4.3. Production operation control technology

Through the construction of production operation control platform, the flow of production factors such as personnel, equipment, materials and energy in the factory is recorded and managed. Coordination and optimization are the most distinctive functional characteristics of the platform. The three major technologies for building a production operation control platform include dynamic scheduling and configuration technology, material identification and logistics tracking technology, and product quality control and traceability technology.

4.3.1. Production dynamic scheduling configuration technology. The production of complex electronic equipment is characterized by multi-variety, variable batch and mixed line production[11]. No matter how optimized the job plan generated by the traditional mixed line production scheduling algorithm is, it only ensures the rationality of the planning level, but cannot consider the disturbance factors such as production time/sequence change, equipment failure, production preparation and order change.

Dynamic scheduling configuration technology for mixed flow production, driven by production disturbance, is a planning and scheduling technology, mainly realizing two goals of intelligent scheduling and lean coordination: optimizing production scheduling of core equipment and site station, and lean coordination of main line/auxiliary line/complete set [13]. The overall process of dynamic scheduling configuration technology for mixed flow production is shown in figure 5.

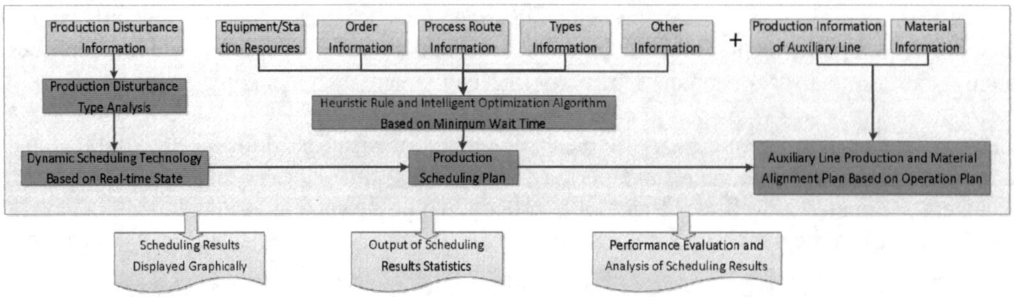

Figure 5. Overall research scheme of dynamic scheduling configuration technology for mixed flow production.

Dynamic scheduling configuration technology for mixed flow production is mainly reflected in the following aspects:

(1) On the basis of the scheduling target and constraint modeling: with complicated electronic equipment manufacturing process, the core equipment/position resource, process execution working time data as the basis, establishing process - resource map matching relationship, according to the difference of task to clear order priority, combining with the complex electronic equipment machine delivery date, resource utilization, form the coupling relation of constraint model [14].

(2) Focus on mainline production planning and scheduling: generate integrated production plans that are refined to specific process links and specific equipment/site resources based on the thought of combining heuristic scheduling rules and intelligent optimization algorithm.

(3) Disturbance driving the dynamic planning and scheduling of production as the core: adopting production disturbance events modeling and its classification modular processing mechanism, realizing timely response of production disturbance events such as the orders from task, resource allocation, equipment failure, field conflict, executive deviation, sub line/kitltems ill-prepared, to ensure the effectiveness of the production plan for the actual execution status.

(4) Lean coordination of the production plan of kitltems/sub line driven by the main line process plan: based on the key process nodes of kitltems/sub line, and guided by the process link plan of production, lean coordination and supply can be realized on time and according to the quantity.

4.3.2. Material identification and logistics tracking technology. The whole machine assembly of complex electronic equipment involves many processes and a long production process. The identification, storage and distribution of materials have a great impact on the production efficiency and quality. With the acceleration of batch production and research and development of complex electronic equipment, the characteristics of co-production and mixed production are more and more prominent, and the demand for fast introduction, fast switch, flexible production and accurate control of production lines is increasingly high. The material identification and logistics tracking technology of digital factory of complex electronic equipment focuses on solving the problems of insufficient material perception coverage, high material distribution error rate, low delivery accuracy and low efficiency.

The overall scheme of material identification and logistics tracking for digital factory of complex electronic equipment is shown in figure 6:

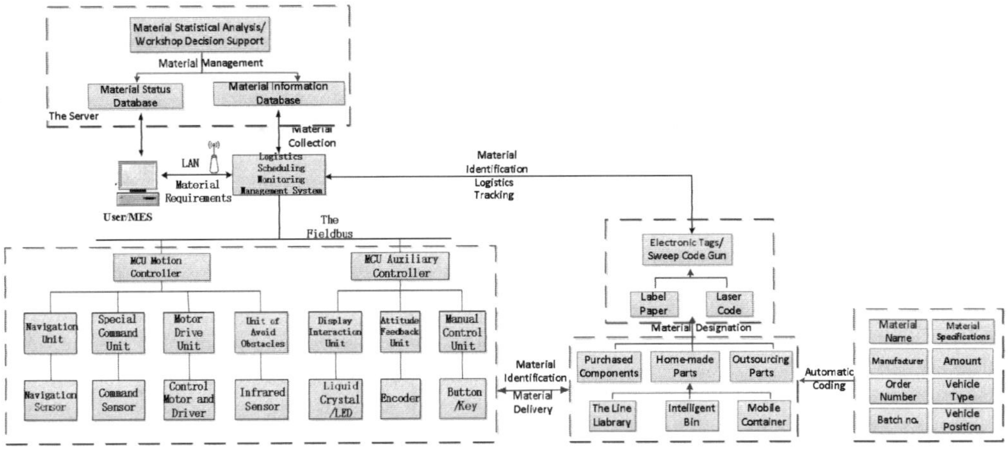

Figure 6. Material identification and logistics tracking technical scheme.

Material identification is the accurate definition and systematic classification of material type, and the unique code is given to each material according to its inherent properties. Through visual sensors, electronic tags, RFID and other technical forms to identify the material, material data real-time collection and tracking to provide technical support.

Logistics tracking is the collection and tracking management of the whole process of materials from storage, storage to delivery, circulation to finished product storage, so as to grasp the storage position, product in process position, process status, assembly queue, work station, operator and other information corresponding to materials in real time.

Material identification and logistics tracking are based on material data collection and management. Based on the management and maintenance of the basic material data and material knowledge base, apply the six sigma concept and SPC control analysis technology to conduct statistical analysis on material data, generate material inventory list, consumption list, quality list, material distribution matching degree, etc.

Material identification and logistics tracking technology through the acquisition of real-time data of the whole process of materials in the digital factory, make material management optimization plan, realize accurate and fast identification of materials, improve the high utilization rate of cargo space, eliminate the phenomenon of material shortage or accumulation; At the same time, the quantity and time of material distribution meet the production rhythm of the assembly line, to ensure continuous production without shortage of materials, accurate material distribution, avoid the occurrence of wrong materials, missing materials.

4.3.3. Product quality control and traceability technology. Users pay more and more attention to product quality, which requires enterprises to strictly control the quality of complex electronic equipment in the production process to prevent the production from unqualified products and avoid economic losses of enterprises and users [12]. In the digital factory, through real-time collection and analysis of the process parameters of manufacturing equipment and testing equipment in the production process, the fluctuation of product quality can be found timely and unqualified products can be reduced. After a period of production and data accumulation, a comprehensive analysis of process flow, process parameters and resource allocation will be conducted to provide scientific and quantitative reference data for managers to improve the factory, and to ensure the consistency and stability of products by controlling these parameters [15].

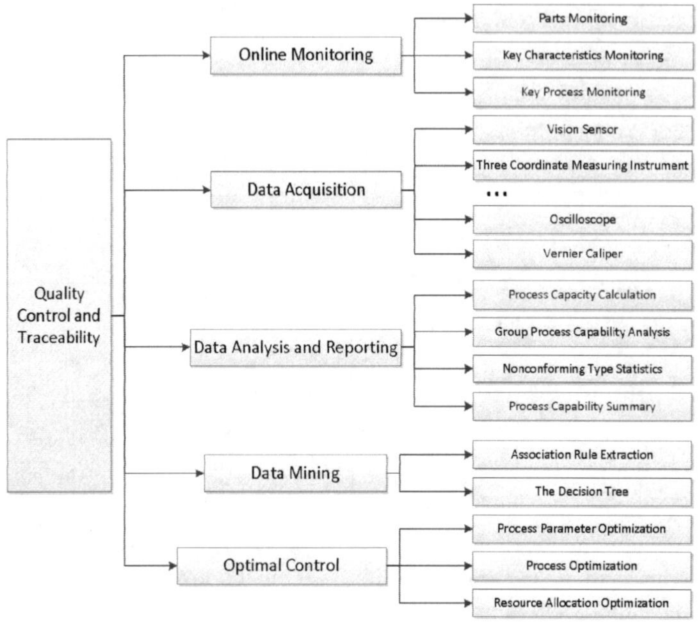

Figure 7. Quality control and traceability system.

The quality control and traceability system of digital factory of complex electronic equipment includes five functional modules: online monitoring, data collection, data analysis and report, data mining and optimization control [17].

Online monitoring module is mainly used to comprehensively monitor parts, key parameters and key processes in the production process through digital means [22].

The data acquisition module mainly faces all the detection equipment in the factory, realizes the collection of data in the inspection and measurement process, and stores the data into the database, providing support for data analysis and statements.

Based on the collected test data, the analysis and report module mainly realizes the functions of process capability calculation, key process capability analysis, unqualified type statistics and process capability summary.

The data mining module is a further study of the analysis and report module. By extracting association rules and establishing decision trees, it can find out the potential and valuable associations of each data item, and provides quantitative basis for optimal control.

Based on data mining, the optimal control module integrates and optimizes the process flow, process parameters and resource allocation [21]. In the digital factory, in order to ensure the stability and continuity of product quality, each process should carry out real-time optimization feedback according to the set value of process parameters and quality characteristics of the pre-process monitored online, so as to realize dynamic quality control and accurate manufacturing of complex electronic equipment.

5. Application example of radar assembly digital factory

Radar is a complex electromechanical coupling system and typical complex electronic equipment. The final assembly process has complex process flow and high requirements for controlling process parameters, and many factors that affect the stability of product quality and the precise execution of the production process in the manufacturing process. It is necessary to improve the overall production level of the radar final assembly workshop through the implementation of digital factory.

Based on the production requirements of many varieties, small batch and the beat of flexible, this example constructed the manufacturing operations management system based on big data and data acquisition and control system, enhanced the level of digital production process and assembly process equipment specialization level, implementation capacity up to that of origin can shorten the production waiting time by 30%or more, assembly a success rate is better than that of 99%, realize intelligent scheduling of production plan and production process of intelligent control.

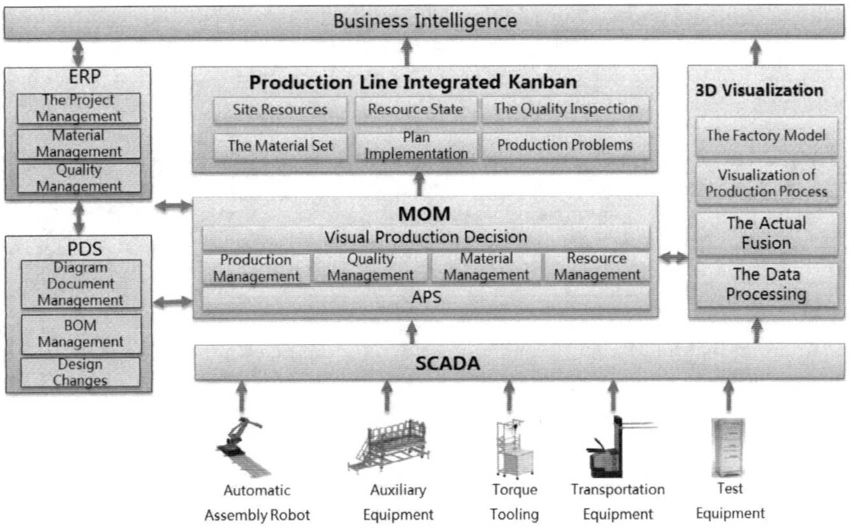

Figure 8. Overall framework of radar assembly digital factory.

The radar final assembly digital factory would build intelligent assembly lines characterized by the organic combination of intelligent units, human-machine hybrid lines and automatic production lines, and form a production mode of "three flows in one" of human flow, logistics and information flow, so as to enhance the production capacity and core competitiveness of the factory. The implementation idea of radar assembly digitization factory is as follows:

- Product Cluster Analysis: Design batch production line and develop production line according to product structure and process characteristics.
- Process Reengineering: Break through the traditional design idea, the whole life cycle analysis of the product assembly and general adjustment process, reconstructing the radar machine assembly and adjustment process.
- Optimize Station Layout: Analyze the product assembly process, integrate and optimize the operation content of each station.
- Production Line Means Upgrade: Configuration automation and professional equipment, planning and construction of production line intelligent logistics system.
- Improve Networking and Informatization: Build the production operation control platform to realize the interconnection of production and test equipment and data acquisition and application.

5.1. Process reengineering and layout optimization

Process reengineering and layout optimization are the premise of digital factory construction, including process reengineering and layout optimization.

Process reengineering mainly improves the overall assembly efficiency by optimizing the process route, balancing the beat of the station and improving the assembly capacity of the station. Firstly, the existing general assembly process was analysed in depth, and part of the mechanical assembly and electronic assembly were changed from serial work to parallel work, and the electrical performance test was changed from the test rack to the product loading, which greatly reduced the assembly and separation time between the array and the test rack. Then, the optimized process flow was divided into different parts to realize the parallel operation of multiple stations and to balance the operation time of each station, so as to achieve the pulsating production. This example set 8 stations in the process of product assembly and adjustment, including 6 mainline stations and 2 branch stations.

Plant layout is simulated and optimized based on process reengineering. Digital process simulation and verification system is used for digital simulation of dynamic assembly, man-machine cooperation, welding assembly, mechanical assembly and assembly accuracy, combining with the system for production line layout and simulation, workshop layout and simulation, logistics planning and simulation to realize the plant layout adjustment and the optimization of process equipment design.

Through the simulation process reengineering and factory layout optimization, shortening the period of the complicated electronic equipment assembly, assembly and tuned efficiency increased 28.7%.

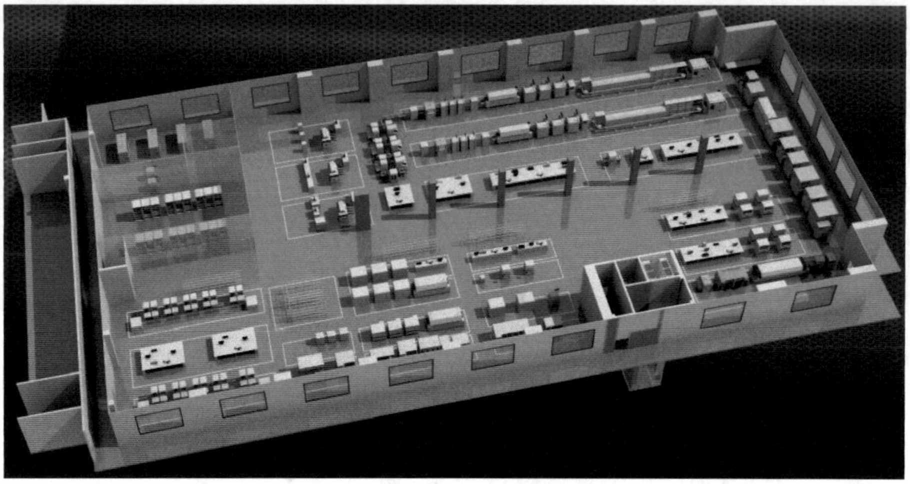

Figure 9. Layout of digital factory.

5.2. Digitalization of plant equipment

Digitalization of plant equipment is the premise and foundation of the construction of digital factory, which provides basic data support for design, research, production and other links.

The digital equipment accomplishes process management, network transmission, data acquisition, data analysis and visualization display with one or more servers to connectivity of factory equipment, through the Ethernet, RS232, RS485, Modus, OPC and other communication protocols. And also the digital factory realizes a more detailed and comprehensive management through the deep integration with the production and operation control platform.

Digital factory of complex electronic equipment includes not only traditional equipment such as CNC machine tools and PLC, but also advanced processing equipment such as robot, AGV and automatic three-dimensional warehouse. Therefore, digital transformation of complex electronic equipment factory includes DNC/MDC and other functions, as well as digital monitoring and intelligent control and other functions.

Taking the radar final assembly factory as an example, equipment digitalization mainly includes digitizing the existing equipment and developing the numerical control intelligent unit [16]. Digital transformation of existing equipment is mainly to embed sensors, integrated circuits, software or other digital components into existing equipment to add data acquisition interface, and realize the function of state monitoring. Intelligent unit mainly based on industrial robots, combined with the actual production demands, equipped with sensors, track, control software, etc., of intelligent equipment of non-standard custom development, assembly parameters can be controlled, process parameters can collect, and other functions, such as T/R component assembly robot, electric assembly robots, etc.

As shown in figure 10, with the support of the data collection system and the visual display function of the production line, the equipment status, production orders, maintenance records and task completion in the factory are displayed comprehensively to realize the connectivity rate of equipment 100%, achieve factory transport management.

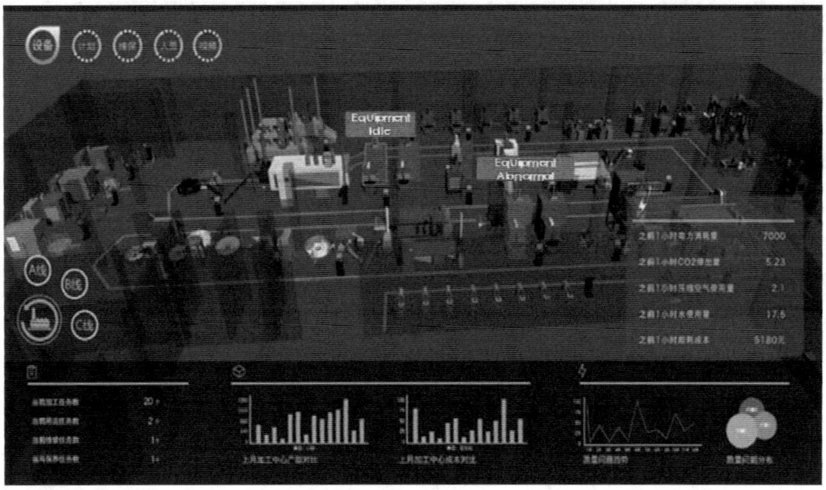

Figure 10. Digital display of factory production line and equipment

5.3. Digitalization of factory logistics

Logistics digitization is mainly to realize intelligent identification and bidirectional traceability of materials, including material identification, material data collection and material statistical analysis.

Firstly, the material identification module establishes coding rules for different types of materials and storage facilities, and endows manufacturing resources with unique identification. Then use the scanning gun, electronic label, RFID and other forms to quickly identify the material code, and bind the material information with processes, products and operators, so as to facilitate follow-up inquiries and traceability.

Material data acquisition module based on distributed collection technology, realizes real-time acquisition of multi-source material data. Automatic sensing of material storage location by means of electronic label; Identify the basic attribute information of the material by the scanning gun; Obtain material processing information from automatic equipment; In addition, MOM, ERP and other information systems are used to collect information such as subordinate orders, processes and quality feedback of materials. All collected material data is classified and stored in the remote database.

Material statistical analysis module adopts six sigma management idea and SPC and other data analysis tools to conduct statistical analysis on material status. Build the material query and traceability model, establish the connection between the operating system on application layer and the background material data, take the order name, material name, process, personnel and so on as the screening conditions, can query the consumption of a certain material; With the order name as the filter, you can query out the inventory list associated with the order. The forward and reverse traceability algorithm of material state is studied. To realize forward traceability, name, package, batch number and other information of the assembly material can be indexed according to number, batch number and other information of the order, while the assembly equipment information of the material can be inquired according to the name, batch number and other information of the material.

As shown in figure 11, the material management and logistics tracking module is constructed with the support of the production operation control platform, and the storage and transportation of materials in the factory are recorded digitally. The logistics operation in the factory is visualized by means of graphical modeling, and the on-time delivery rate of materials is 100%. The matching degree of material distribution and plan is better than 99%.

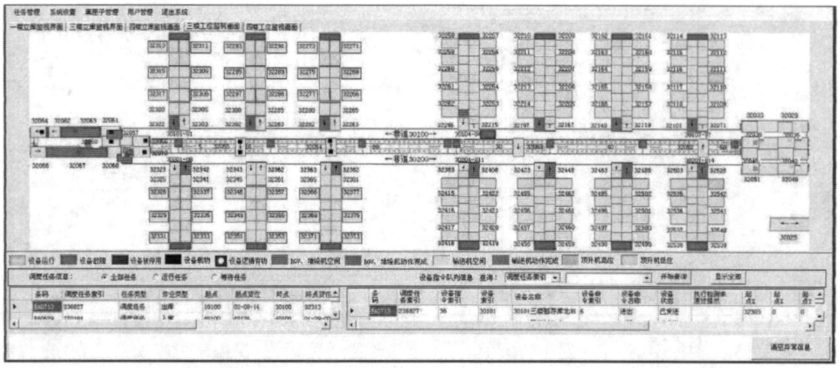

Figure 11. Plant logistics management and tracking.

5.4. Digitalization of production process

Digitalization of production process means that all production activities in the factory are recorded, managed and monitored through the information system, which mainly includes manufacturing execution system, data acquisition system and visual KANBAN.

Manufacturing execution system (MES) includes 6 functional modules including production twin, scheduling, production execution, quality management, material management and equipment management, covering manufacturing resources, planning control, production execution, quality management and other production total factors management. Record the plan, task, quality, material, equipment, personnel and other elements through the information system, form the production process resume, and provide support for follow-up inquiry and traceability.

Based on the Internet of things technology, the data collection system collects data of raw materials, processed components, equipment, personnel and other production factors, so as to realize the dynamic perception of production line equipment information, personnel information, logistics information and quality information. Taking radar as an example, the main data types collected include: data obtained by industrial sensors such as photoelectric sensors, thermal sensors and force sensors; 3 d drawings, processes and procedures of field assembly; Calibration pictures of key processes; Data generated by MES, QMS, WMS and other information systems.

Visual kanban is based on the manufacturing executive management system and data acquisition system to monitor the running status of the radar assembly plant in real time. Through transformation processing of the collected data, the LAYOUT module, rendering module, scene and UI editing module were imported, and the overall running state, assembly process and key process running state of the factory were visually managed from three dimensions of factory level, production line level and unit level.

As shown in figure 12, real-time display of information such as plans, tasks and capacity in the factory is conducted by using the perceived and collected data. The data acquisition coverage is greater than 99% to realize the transformation and upgrading of factory management ability.

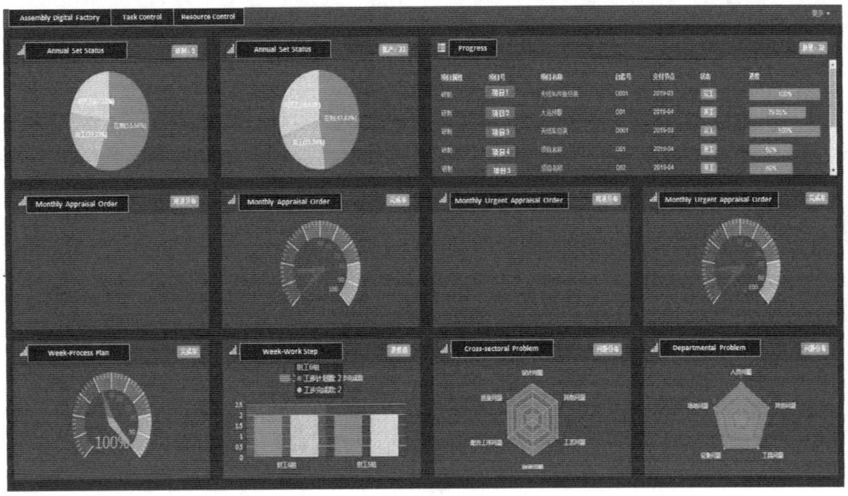

Figure 12. Digital display of production process.

The implementation of the digital factory of radar final assembly, through the functions of the intelligent material management system, such as early warning, pulling, etc., to improve the material evenness. Through the management and control of the manufacturing operation management system, reduce the stagnation of the final assembly process, realize the traceability of quality data, and improve the degree of informatization and digitalization of the factory; Through data collection and control system, we can improve the professional level of final assembly process equipment, improve the efficiency of factory management, and form a final assembly digital factory with batch production line and intelligent final assembly unit as the main body.

6. Conclusion

This paper studied the important technology of digital factory of complex electronic equipment including the perception and fusion of dynamic production data, digital twin and manufacturing operation management. Based on the method of the configuration of dynamic scheduling of production, the materials and logistics tracking and product quality control and trace, this paper developed the production operation control platform, realized the monitoring and prediction of digital factory running, and applied in the radar assembly production factory, which improved the efficiency and quality of radar assembly and the plant management efficiency, to form the radar digital factory of mixture of batch and research production lines. And this paper heralded the digital factory is mature, and can be extended to other fields of complex electronic equipment. Further research will be carried out on key technologies and operating mechanisms such as artificial intelligence, additive manufacturing, data mining, etc. that would help digital factory advances to intelligent factory.

References

[1] Tao F, et al. 2018 Digital Twin and Its Potential Application Exploration, *Comput. Integr. Manuf. Syst.* **24(1)** 1-18.

[2] Shoudian L I 2018 Discussion on Construction Plan of Digital Factory, *Manuf. Antom.* **40(4)** 109-14.

[3] Liu S 2010 Research into Complicated Electronic Equipment Design System Based on Knowledge Engineering, *Shipb. Elect. Counterm.* **33(5)** 111-14.

[4] Hu C, et al. 2017 Exploration and Practice Structure Digital Prototype of Complex Electronic Equipment, *Electro-Mechan. Eng* **33(6)** 1-9.

[5] State Administration for Market Regulation China Standardization Administration. GB/T 37393 2019 Digital Factory-General Technology Requirements *China Standard Press.*

[6] Zhu D, Zhao M 2018 Machine Intelligence: From Digital Workshop to Intelligent

Manufacturing, *Beijing: China Machine Press* **10**.
[7] Chen Y 2013 Research on Digital Factory Application Technology for Discrete Manufacturing *Electr. World* 49-50.
[8] Bo J 2016 Framework and Implementation of Digital Factory *China Industrial Review* **5** 28-33.
[9] Zhu Q 2018 Research and Development of Smart Jobshop for Commercial Aero-Engine Based on Industrial Internet, *Microcomp. App.* **34(8)** 108-11.
[10] Li T, Li S 2016 Digital Factory Path Exploration of Aircraft Complex System Assembly, *Avia. Manuf. Technol.* **10** 41-47.
[11] Li Z, Wang A, Long Z 2014 Dynamic Batch and Materials Coordination Technology of Complex Manufacturing Execution, *Compu. Integr. Manuf. Syst.* **20(4)** 836-46.
[12] Yang H, et al. 2016 Discussion on Quality Control and Supervision Mode and Method of Aircraft Manufacturing Based on MBD, *Avia. Manuf. Technol.* **5** 105-9.
[13] Li J, et al. 2012 Distributed Coordination Scheduling Technology Based on Dynamic Manufacturing Ability Service, *Compu. Integr. Manuf. Syst.* **18(7)** 1563-74.
[14] Lu Z, Wang A, Tang C 2017 Multi-Level Relevance Resources Coordinated Scheduling Based on Improved Genetic Algorithm, *Transac. Beijing Instit Technol.* **37(7)** 711-6.
[15] Xu G, et al. 2017 Online Monitoring and Control Method of Product Quality Based on Embedded Cyber-Physical System Models, *J. Mechan. Eng.* **53(12)** 94-100.
[16] Cai M, Wang T, Shang T 2016 Evaluation of Digital Factory for Manufacturing Enterprisem, *Sci. Technol. Manage. Res.* **15** 63-9.
[17] Mu P 2016 Research on Fine Quality Control Under Flexible Manufacturing Mode, *Jiangsu Sci. Technol. Info*. **10** 54-6.
[18] Chen H, et al. 2014 Data-driven Rapid Modeling of Manufacturing System *J. Hebei Uni. Sci. Technol.* **35(6)** 504-11.
[19] Yue J 2017 Discussion on Construction of Digital Factory, *Electr. Proc. Technol.* **38(4)** 193-6.
[20] Zhang G, Huang G 2013 Digital Factory: Its Application Situation and Trend, *Avia. Manuf. Technol.* **8** 34-7.
[21] Sun Y, et al. 2013 Digital Factory Application in Electronic Manufacturing, *J. CAEIT* **6** 551-6.
[22] Wang G 2016 Design and Implementation of Manufacturing Process Quality Control Software System, *Avia Standa. Quali.* **5** 20-3.
[23] Chan D S 2002 Digital Factory-A Technology Show Case for Teaching and Research in Digital Manufacturing, *Machi. Develop*. **31(6)** 109-11.
[24] GE, Predix 2018. Available: https://www.ge.com/cn/b2b/difital/predix.
[25] Kossiakoff A, Sweet W N, Seymour S J, Biemer S M 2011 Systems engineering principles and practice, *2nd Ed. John Wiley Sons* **83**.
[26] Pouly M, Berthold S 2010 Digital Factory for Collaborative Supply Chains, *Procee. Int. Conf. Compu. Support. Coope. Work Desg. (CSCWD)* 579-84.
[27] Liu L, et al. 2019 Construction and Application of Digital Twin System for Production Process in Workshop, *Compu. Integr. Manuf. Syst.* **25(6)** 1536-45.

IOP Conf. Series: Materials Science and Engineering **739** (2020) 012049 doi:10.1088/1757-899X/739/1/012049

Forecasting Change of the Roller Screw Mechanism Efficiency on the Basis of a Comparative Analysis of Coated Surfaces

L.V. Belyaev[1], A. V. Zhdanov [1], N. S. Dovbysh[1], V.V. Morozov[1]

[1]Department of the mechanical engineering, Vladimir State University, Gorky str. 87, Vladimir, 600000, Russia

Abstract. The paper presents the results of the influence of materials, lubricants and the type of coatings in friction roller screw mechanism on the values of the coefficient of friction and the coefficient of efficiency required to predict the efficiency of the drives. Theoretical studies were carried out for different types of roller screw mechanism coatings. The results can be used to predict the functional characteristics of the mechanisms during their design.

1. Introduction

During the computer-aided design of a roller screw mechanism it is necessary to predict its functional characteristics [1]. In some cases, especially for friction roller screw mechanism [2] (figure 1), a preliminary estimation of their efficiency is necessary if they are made of different materials, with different coatings and lubricants.

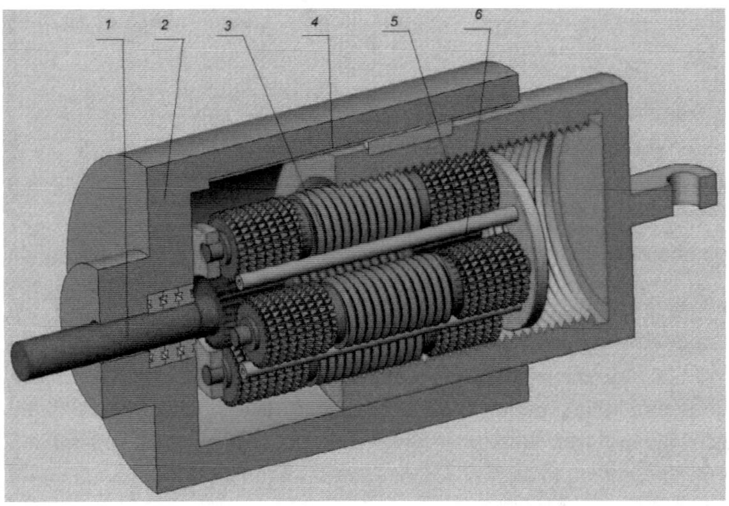

Figure 1. CAD–drawing of friction type roller screw mechanism:
1 - screw, 2 - housing, 3 - roller, 4 - nut, 5 - annular gear, 6 - bearing base.

In this paper, the influence degree of the material, the type of lubricant and the type of coating on the coefficient of friction and efficiency has been theoretically established.

2. Materials and methods

The friction coefficients for uncoated surfaces has been determined from the reference data. Materials and steels were taken from [3-5] depending on the roller screw mechanism. Those are steel 40Kh, ShKh15, U8, 08Kh17T and coatings MoS_2, (AlSiCr)C:H, AlN, TiN. At the first stage, the friction coefficients for uncoated steels were found during the dry interaction of the parts. The efficiency of the roller screw mechanism [3] was found by using the dependencies:

$$\eta_{\text{пх}} \approx \frac{1}{1 + \text{А}_{\text{тр.с.}}/(\text{А}_{\text{п}} + \text{А}_{\text{тр.к.}})/(\text{А}_{\text{п}} + \text{А}_{\text{тр.о.}})/\text{А}_{\text{п}}}$$

where $\text{А}_{\text{п}}$ – the effective work per one revolution of the screw $\text{А}_{\text{п}} = FS$;

$\text{А}_{\text{тр.с.}}$ – the work of the sliding friction forces beyond the pole connection of the threads per one screw revolution at a half profile angle of $\alpha=\pi/4$.

$$\frac{\text{А}_{\text{тр.с.}}}{\text{А}_{\text{п}}} = \frac{2f}{1-f}$$

where f – the coefficient of sliding friction;

$\text{А}_{\text{тр.к.}}$ – the loss due to the rolling friction of the connected threads in the transmission:

$$\frac{\text{А}_{\text{тр.к.}}}{\text{А}_{\text{п}}} = \mu \frac{2\pi}{s}(k+1)tg\alpha,,$$

where μ – the coefficient of rolling friction (shoulder of the rolling friction torque);

$\text{А}_{\text{тр.о.}}$ – the losses due to the rolling friction in the transmission supports.

3. Results

The obtained values of the efficiency of steels during dry friction are listed in table 1.

Table 1. Coefficients of friction *(f)* and efficiencies of the uncoated roller screw mechanism under dry friction and lubricated friction.

Steel name	Dry friction		Lubricated friction	
	f	Efficiency	*f*	Efficiency
40Kh	0.20	0.66	0.14	0.74
ShKh15	0.17	0.71	0.12	0.79
U8	0.15	0.74	0.10	0.81
08Kh17T	0.25	0.58	0.17	0.65

For roller screw mechanism with lubrication, the efficiency is 10.3...24.2% higher than for non-lubricated transmissionss, however, it still remains low for those types of friction mechanisms.

At the next stage, the coefficients of friction and efficiency were determined for roller screw mechanism with coatings. Type of coating: PVD; installation: UniCoat 600 SL+; thickness: up to 3 microns; type of coating: multilayer. The friction for the coating was obtained independently of the steels for each material. The measurement results are listed in table 2.

The analysis of the tables shows that during the use of coatings, the coefficient of efficiency can both increase and decrease. The coatings with MoS_2 and (AlSiCr) C:H structures have shown positive dynamics. During the use of traditional AlN and TiN coatings, the coefficient of efficiency can be reduced. This result applies only to the friction mechanisms of the roller screw mechanism and requires additional experimental confirmations.

Table 2. Coefficients of frictions *(f)* and efficiencies under dry friction and lubricated friction for surfaces with coatings.

Coating	Dry friction *f*	Dry friction Efficiency	Lubricated friction *f*	Lubricated friction Efficiency
MoS_2	0.1	0.82	0.05	0.91
(AlSiCr)C:H	0.2	0.66	0.10	0.82
AlN	0.4	0.43	0.28	0.56
TiN	0.6	0.25	0.52	0.32

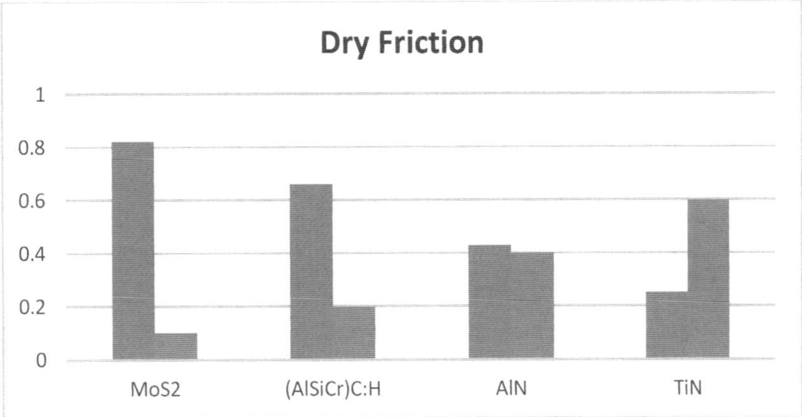

(a)

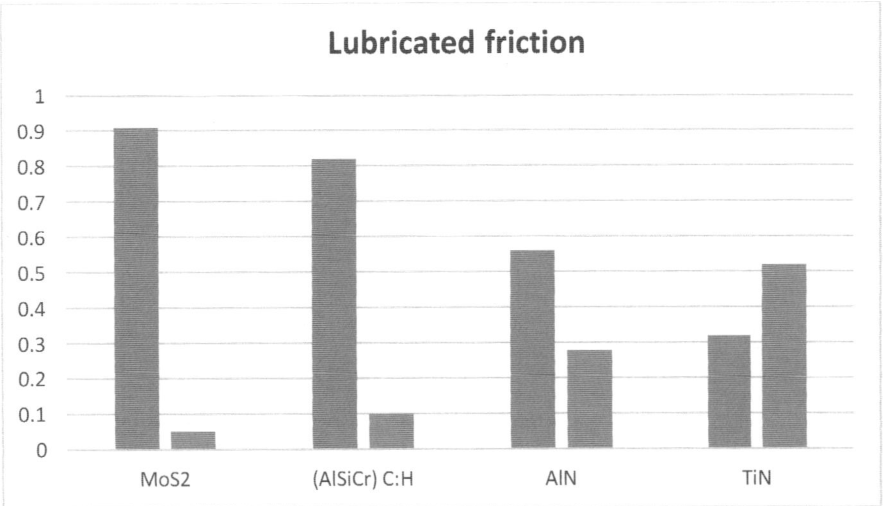

(b)

Figure 2. Comparison of the coefficient of efficiency (▪) and the coefficient of friction (▪) in coated RSM under dry friction conditions (a) and with lubrication (b).

At the third stage, the values of the coefficient of friction and efficiency have been compared for the friction-type roller screw mechanism with deposited coatings under dry friction conditions and with grease as the most common operation alternative (see figure 2).

4. Conclusions

The use of grease slightly increases the coefficient of efficiency by 10...13% in coated mechanisms. The use of grease can lead to a negative effect on the efficiency, which, obviously, also requires experimental verification.

Acknowledgments

This research was supported by the the Ministry of Science and Higher Education of the Russian Federation in the scope of scientific project No. 9.11716.2018/11.12.

References

[1] Zhukov I O, Ivanchenko A B, Zhdanov A V 2014 Singularities of 3D modelling of Roller Screw Mechanisms, *Sci. Techni. Bulle. Volg. Regi.* **4** 100-3.
[2] Morozov V V 2005 RSM: Kinematic characteristics, monograph *Vladimir State University* **78**.
[3] Morozov V V, Zhdanov A V 2015 *RSM: Reliability and durability* (Moscow, monograph) 152 p.
[4] Kononov D M, Zhdanov A V, Korolev A N 2011 Study of the properties of nanostructured PVD coatings based on carbon, *Mode. Prob. Sci. Edu.* **6** 130.
[5] Bukarev I M, Zhdanov A V 2011 Experimental researches of the physicomechanical properties of multilayer nanostructured coatings, *Sci. Techni. Bulle. Volg. Regi.* **6** 116-9.

Development of Fire Information Integration Application Platform

Ching-Shun Chen[1], Yu-Tsung Ho[1], Yi-Lin Chen[2], Chung-Chyi Chou[3]

[1]College of Mechatronics, Quanzhou University of Information Engineering
[2]Department of Environmental Engineering, Da-Yeh University
[3]Da-Yeh University of Department of Fire and Safety

dustin@mail.dyu.edu.tw

Abstract. The aging society is a challenge facing all countries in the world. The collection and analysis of fire information in elderly care institutions has become a key factor in seeking shelter time. This study will build a fire information platform and its reporting mechanism through HTML, PHP, MySQL and other software development tools, and through the source and distribution of fire information. Public announce technology to improve the fire response capacity of the elderly institutions. This paper quotes Marchant (1980), a British scholar, who regards fire and refuge action as the parameters of elderly people's refuge. The system of fire information platform consists of three parts: the "manager interface", "cloud database" and "fire information platform". The results show that the fire information reported after the confirmation of the management personnel can improve the residents' confidence in the source of the information. The test results in the experimental field show that the higher the integrity of fire information content and the more rigorous distribution process are, the more helpful it is to shorten the judgment time of pre-avoidance tragedy, and the visual fire information platform is helpful to the correct judgment of the refuge route. Adding a call button to the platform can also increase residents' sense of security. Fire information integration platform can help elderly care institutions take immediate shelter action in case of fire.

1. Introduction

France has experienced 115 years to come, "aged society" from "aging society", perhaps because of advances in medical science, people on average life expectancy, coupled with the low birth rate trend factors, Japan went to the "aged society" from "aging society" within 25 years. In response to the needs of an aging society, many governments enacted long-term care services regulations. A. Khitab et al. [1]. It is believed that nanotechnology is going to play an important role in the development of futuristic building materials. The innovations could be two-fold; one is the modification of classical materials and the other should cover the invention of novel materials. Disaster prevention also requires innovative technology to integrate fire information.

Many international long-term care facilities are converted from other uses of the building, even if the world building and fire codes conform to the conditions of time and upgrading, but the old building in which the establishment of long-term care institutions, only a refuge or safe stairs stairs. The fire information, shelter information, fire extinguishing information cannot be immediately regrouped, and the evacuation guide is carried out. As older people over the age of 65, auditory, visual and other physiological functions may deteriorate due to age, the ability to judge fire messages has not been as quick and agile as young people. Groner (2016) [2]. Point out that the user behavior of the usual place will be related to the architectural form, and the user and the user will also be related to each other. Besides the four correlations, the correlations of building types to evacuation behavior of the vulnerable, fire response of the vulnerable, fire response of the vulnerable, and between the vulnerable and the general users during the period of fire refuge become eight more complex correlations.

Deng et al (2012) [3] definition of the refuge of the weak: "refers to the refuge of the weak elderly patients (contains disabilities), children and other persons with reduced mobility or bed and so on, when a fire or disaster, unable to complete action on their own or evacuate personnel, to be assisted by the actions of others evacuation of vulnerable persons belong. " Elderly home or long-term care institutions not only to face the refuge of the weak elderly, but also consider the issue once the place of fire, whether the elderly person will immediately evacuate. Shen (1993) [4]. Put forward the content of Formula 1 according to the sequence of fire and refuge action, using the concept of time.

$$\frac{Tp+Tr+Ta}{Tf} \leq 1 \tag{1}$$

Research by D'Orazio et al. (2015) [5]. Shows that proper setting of reflective index, night light index, luminous index, dynamic guidance system and voice guidance system can effectively shorten the time of refuge. This study builds a fire information platform and reporting mechanism to enhance the trust mechanism of fire information in pre-evacuation process by improving the source, quality and distribution process of fire information. The development process of fire information platform adopts on-the-spot investigation and interview to understand the needs and past problems of the elderly residents and managers, and to serve as an important reference for the development of information system foreground and background. After the completion of the fire information platform, qualitative interviews with managers and residents were conducted at the site again.

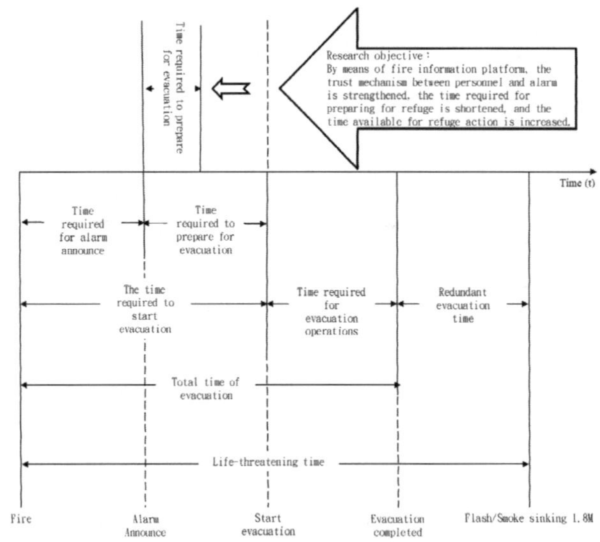

Figure 1-2. Research Concept Description.

2. Research Methods

Based on the relevant literature, this study infers that the behavioral decision-making of fire occurrence depends on the degree of risk perception. Risk perception is influenced by "source of information", "quality of information" and "mode of information announce". Therefore, this study intends to improve the trust mechanism of fire information in the process of avoiding distress by improving the source, quality and publishing process of fire information. Based on this, a research framework is proposed as shown in Figure 2-1.

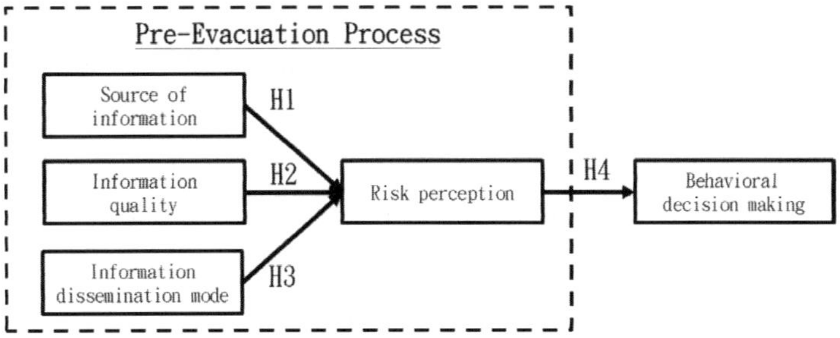

Figure 2-1. Research Architecture Diagram.

To ensure the reliability of fire information announces, information announce must be audited and confirmed. When the system collects fire detector signals, the system will automatically send messages to the mobile phones of authorized personnel above the level 2 supervisor. If the evacuation action is deemed necessary after verification and confirmation, the fire information platform can be announced immediately after entering the account password through the mobile phone, instead of the detector action. In the part of information quality, referring to Wiegman et al. (1992) [6], there are three psychological assessment processes before a decision is made: clue, situation and risk, while the three assessment processes are not sequential. Mileti and Sorensen (1990) [7] Humans undergo a specific process before responding to asylum, including hearing, understanding, belief and personalized warning. As Ramachandrn (1990) [8] pointed out, when people receive fire alarms, whether through fire, smoke, or alarm bells, they will try to understand the reasons and verify, if found to be false, they will choose not to change. Make the fire information platform conform to Mishra (1996) [9], "When people make decisions based on the information they trust, they are mainly based on four beliefs: Competent, Open, Concerned and Reliable". The research process can be divided into three stages:

1. Investigation stage: after exploring the behavior of the elderly, the behavior of pre-refuge, the behavior of fire refuge, the research of evacuation plan for the weak and the research of information trust, the feasible ways to improve the trust mechanism are summarized and the research process is established accordingly.
2. Construction stage: design fire information platform from three directions: information source, information quality, information announces mode, referring to the theory and results put forward by scholars, establish the fire information platform release mechanism, by whom, how to announce and what content should be included in the fire information announces.
3. Adjustment stage: After interviewing the elderly housing administrators and elderly resident through the use of survey questions, the paper analyses whether the administrative administrators and senior residents can enhance the trust mechanism between the fire information platform and the alarm, and then assists the senior citizens in decision-making of evacuation action in case of fire, so as to increase the time available for evacuation action.

3. Results and Discussions

3.1. System components and development tools

This study divides the system of fire information platform into three parts: "Management Interface", "Cloud Database" and "Fire Information Platform". Use 4G or Wifi network connection to enable administrators to immediately receive fire signals through mobile vehicles and evacuate immediately. The transmission format of the Web page as the "manager interface" should be consistent with the research intention. The administrator connects to the management webpage of Amazon Web Service through the action vehicle, edits and adjusts it, and the changed data will be linked with the cloud database written by MySQL inside AWS and update the data synchronously.

In the critical situation, when evacuation is necessary, the fire information platform is also released through this interface to force the pre-set display. The system uses Hypertext Preprocessor (PHP, Hypertext Preprocessor) as a development tool for the administrator's user interface. In order to make the fire information platform more convenient and convenient, and to avoid the obstacles that the use organization may be unfamiliar with the information program and difficult to replace or adjust the platform content, this study first uses the PHP hypertext preprocessor syntax to create the background interface. 3-1 cases. In the editing of each web page, according to the design requirements of the platform information, the attributes (text, picture, film) of Cascading Style Sheets (CSS, splicing style sheet) can be selected, and the input/output action logic or hyperlink can be given to other URL or file.

```
}elseif($mod type == "update"){
    $ic no = $ POST["ic no"];
    $ic name = $ POST["ic name"];
    $ic x = $ POST["ic x"];
    $ic y = $ POST["ic y"];

    if($ FILES["ic file"]["tmp name"] != ""){
      $tmp = explode(".",$ FILES['ic file']['name']);
      $subname = strtolower($tmp[sizeof($tmp)-1]);

      $ic file = $ic no.".".$subname;
      move uploaded file($ FILES["ic file"]["tmp name"], "./img/icon/".$ic file);
      $sql = "update control panel icon set
cp ic name='".$ic name."',cp ic x='".$ic x."',cp ic y='".$ic y."',cp ic img='".$ic file."' Where
cp ic no='".$ic no."'";
      $result = $mysqli->query($sql);
    }else{
      $sql = "update control panel icon set
cp ic name='".$ic name."',cp ic x='".$ic x."',cp ic y='".$ic y."' where cp ic no='".$ic no."'";
      $result = $mysqli->query($sql);
      //echo $sql;
    }
    redirect("./list icon.php");
```

Figure 3-1. Fire information platform PHP syntax example.

System foreground development uses Hyper Text Markup Language 5 (HTML, Hyper Document Markup Language) and jQuery: a cross-browser JavaScript library that can be used to simplify the operation between HTML and JavaScript. Figure 3-2 shows the partial syntax of HTML in the fire information platform, and integrates the functions of the Internet and remote monitoring of the smart networking function TV. It can be customized for individual TV broadcasts according to different situations.

```
<tr><td colspan='5'>[illegible]<br>
    <form name="main" method="post" action="./list icon.php" enctype="multipart/form-data">
    <input type="hidden" name="mod type" value="new">
    [illegible] : <input type="text" name="ic_name">
    [illegible] : <input type="file" name="ic_file">
    X:<input type="text" name="ic x">
    Y:<input type="text" name="ic y">
    <input type="submit"></td>
  </form>
  </tr>
  </table>
```

Figure 3-2. Example of HTML syntax for fire information platform.

```
$sql = "select b.cp ic no,b.cp ic name,a.cp bg no,a.cp bg name,b.cp ic x,cp ic y,cp ic img from
control panel background a,control panel icon b where a.cp bg no=b.cp bg no";
    $sql .= " order by b.cp ic no desc";
    //echo $sql;
    $result = $mysqli->query($sql);
    while($row = $result->fetch row()){
      print "<tr><td>".$row[0]."</td><td>".$row[1]."</td><td>".$row[4]."</td><td>".$row[5]."</td><td><a
href='./list_icon.php?mod_type=modify&ic_no=".$row[0]."'>[illegible]</a></td></tr>\n";
      $out html .= "<div
style='position:absolute;float:left;top:".$row[5]."px;left:".$row[4]."px;z-index:".$row[0].";'><img
src='./img/icon/".$row[6]."'></div>\n";


    }
    $result->close();
```

Figure 3-3. Fire Information Platform MySQL Syntax Example

The database is written by MySQL (My Structured Query Language), in which SQL is Structured Query Language, which is an open source relational database management system, which can be used together with PHP for database development [20], Figure 3-3 shows the MySQL syntax example of the fire information platform. This study stores the information about the Fire Information Platform on the Amazon Elastic Compute Cloud (Amazon EC2) cloud computing server. And uses Amazon Web Service (AWS) to adjust computing resources and program, store, query and update files.

3.2. Fire information platform

The programming of the research platform uses PHP + MySQL as the background webpage and database system. The database is set up on the Amazon EC2 cloud server, and the Amazon Web Service (AWS) system is used to convert the server data into a webpage operation interface for programming and storage., query and update files. The front page is used with HTML5 and jQuery. It is currently installed in a private network space. It is preset to automatically request updates from the server every 1 second. The fire information platform functions completed in this study include "Evacuation Location", "Floor Plan", "Escape Path", "Fire Location", and "Call for Help". The evacuation site is a text warning message. The administrator can edit the evacuation instruction or the collection location in the user interface. Can immediately edit any message that you want to inform the asylum seeker according to the fire situation or pre-edit the settings according to your own plan.

Mishra (1996) [12] and believes that the recipient of the message will be since the other party is "Qualified," "Confessed," "For Care," and "Depends on"; the message can only be Designated professional editors to meet the above four reasons. The platform provides a top-level plan view of the user's floor structure, allowing the user to quickly understand the floor structure and the stair position in the screen, and the dynamic escape path window is displayed in the upper right corner, so that the user can switch to the size screen according to the requirements.

The results of the interview conducted by the elderly care field showed that the higher the integrity of the fire message content and the more rigorous the management level of the message has a higher degree of trust; the user can shorten the judgment time of the pre-avoidance process in the event of a fire. The fire information is integrated with the imaged evacuation direction to the fire information platform to help the correct judgment of the evacuation route. At the same time, users also expect to be able to add a call for help button on the platform, which can increase the security of residents. The fire information integration platform can assist elderly care institutions to take refuge actions immediately in the event of a fire.

4. Conclusion and Recommendations

This study builds a fire information platform and its reporting mechanism through HTML, PHP, MySQL and other information tools to enhance the fire response capability of the elderly institutions through fire message sources and release technologies. The fire information platform system is divided into three parts: "Manager Interface", "Cloud Database" and "Fire Information Platform". The results of the study show that the fire message sent by the manager after confirmation can increase the trust of the residents in the source of the message. Test results in the experimental field show that the higher the integrity of the fire message content and the rigorous release process, it is helpful to shorten the judgment time of the pre-avoidance process. The visual fire information platform is helpful for the correct judgment of the evacuation route. Adding a call for help button on the platform can also increase the security of the residents. The fire information integration platform can assist elderly care institutions to take refuge actions immediately in the event of a fire.

References

[1] Khitab A, Anwar W, Mansouri I, Tariq M K 2015 Future of civil engineering materials: A review from recent developments, *Rev. Adv. Mater. Sci.* **42(1)** 20-7.

[2] Groner N E 2016 A decision model for recommending which building occupants should move where during fire emergencies, *Fire Safety J.* **80** 20-9.
[3] Deng Z, Zeng W, Shen Z, Cai Z, Yang S 2012 Research on the Impacts of Building Fire Asylum Weak Refuge and Human Factors Data, *J. Architec.* **79** 131-45.
[4] Shen Z 1996 Theoretical analysis of human asylum in fire, *Police Series* **26(6)**.
[5] D'Orazio M, Bernardini G, Tacconi S, Arteconi V 2016 Fire safety in Italian-style historical theatres: How photoluminescent wayfinding can improve occupants' evacuation with no architecture modifications, *J. Cultur. Herit.* **19** 492-501.
[6] Wiegman O, Komilis E, Cadet B, Boer H, Gutteling J M 1992 The Response of Local Residents to a Chemical Hazard Warning: Prediction of Behavioral Intentions in Greece, France and the Netherlands, *Int. J. Mass Emerg. Disas* **10(3)** 499-515.
[7] Mileti D S, Sorensen J H 1990 Communication of Emergency Public Warnings: A Social Science Perspective and State-of-the-Art Assessment, *Oak Ridge National Lab., TN(USA)* No. ORNL-6609.
[8] Ramachandrn G 1990 Human Behavior in Fire--A Review of Reserch in the United Kingdom, *Fire Technol.* **26(2)** 149-55.
[9] Mishra A K 1996 Organizational Responses to Crisis, *Trust Org. Fronti. Theor. Res*. **261**.

IOP Conf. Series: Materials Science and Engineering **739** (2020) 012051 doi:10.1088/1757-899X/739/1/012051

Determination of Changes in the Parameters of Electrical Machines after Repairs to Prevent an Unacceptable Increase in Vibration of the Building Structure

A Belyi[1]

[1]Department for Automation and Electric Supply, Moscow State University of Civil Engineering, 129337, 26 Yaroslavskoeshosse, Moscow, Russia

belyynmstu@gmail.com

Abstract. This article discusses the improvement of control methods for electrical machines after repair. These methods allow during testing of an electric machine to determine deviations of parameters that can lead to an increase in the level of vibration and subsequent failure of both the machine itself and the elements of building structures.

1. Introduction

One of the safety factors for building structures is the permissible level of vibration. For industrial buildings, the level of vibration is affected by the operation of mechanical equipment driven by electric machines. When designing such buildings, measures are taken to minimize the vibration load on building structures. Over time, the parameters of the electric machine and the mechanical transmission of movement can change, especially if the equipment is out of order and has been repaired. Testing of equipment put into operation after repair allows you to identify unacceptable changes in the parameters of electrical machines and prevent their commissioning.

2. Materials and methods

Methods of controlling changes in the parameters of electric machines are: vibration diagnostics using special devices and analysis of voltages and currents to exclude false diagnoses of previous studies.
Spectral analysis is the main procedure for diagnosing electrical machines. The purpose of the analysis is to identify vibration at frequencies characteristic of particular malfunctions. The operation of the machine is considered normal if the amplitude of all components does not exceed the specified limits. Typically, the quantity to be analyzed is the speed of vibration, but acceleration in the case of high speed machines or motion for low speed can also be used. To study the bearing assemblies, the spectral curve envelope method is often used. The bearing is a resonant system. It has three resonating parts. These are the outer, inner rings and rolling bodies. The outer ring, for example, has several intrinsic resonances: longitudinal, transverse, in thickness, and so on. Each of these resonances has another set of harmonics. That is, depending on the configuration, the ring has several orders of resonance [1].

Diagnostic signs of one or another deviation of the parameters of the mechanical and electrical parts of the electric machine are constantly being improved. Separately, we can single out the problem of improving test benches to control the parameters of electrical machines and mechanical equipment.

On the basis of the electrical repair workshop in the city of Magnitogorsk (Russia), studies were conducted to improve the diagnostic signs of changes in the parameters or defects of electrical

IOP Conf. Series: Materials Science and Engineering **739** (2020) 012051 doi:10.1088/1757-899X/739/1/012051

machines after repair. An example is the study of induction motors after repair from the point of view of changing parameters that can lead to an increase in the level of vibration.As a measuring tool, we used a vibration analysis instrument manufactured by VAST (Novosibirsk, Russia) with «DREAM" software. This program allows you to process vibration diagnostic signals by all known methods.

It is proposed at the initial stage of testing to identify electrical machines with uneven air gap between the rotor and stator. The unevenness of the gap indicates inaccuracies in the installation of the rotor or inaccuracies in the position of the electromagnetic system of the electric machine. It is known that the unevenness of the air gap can be detected by the spectrum of vibration acceleration.

3. Results

As an example in figure 1 shows the processing of the spectrum of vibration accelerations for an induction motor 1 after repair. Frequencies whose level exceeds the noise level are highlighted. The highest signal levels have frequencies f1, f2 and f3.

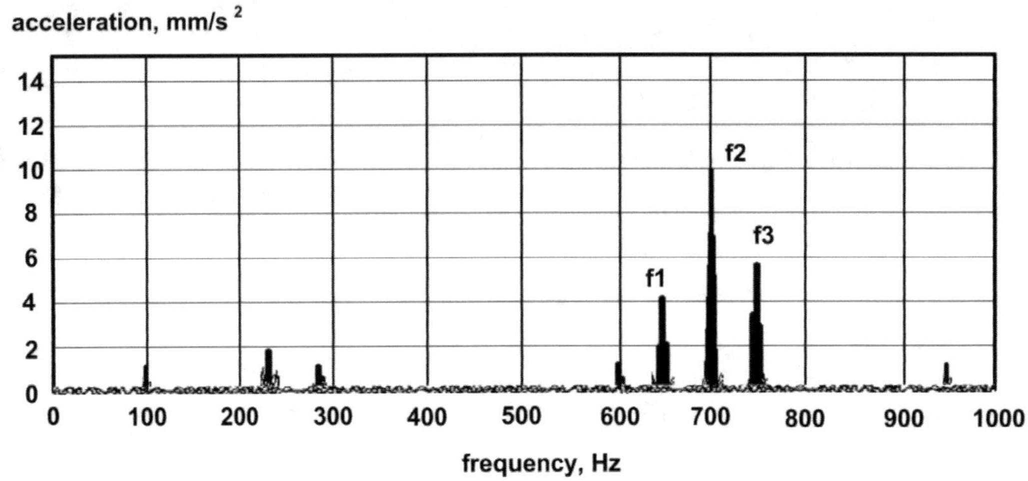

Figure 1. Spectrum of vibration accelerations with a predominance of frequencies of static rotor eccentricity.

The frequency f2 corresponds to the tooth frequency equal to the product of the rotor speed and the number of teeth of the rotor. The frequencies f1 and f3 differ from the frequency f2 by the level of the double frequency of the mains voltage. The predominance of such frequencies in the spectrum indicates a static eccentricity of the rotor. The frequencies f1 and f3 differ from the frequency f2 by the level of the double frequency of the mains voltage. The predominance of such frequencies in the spectrum indicates a static eccentricity of the rotor [1].

In figure 2 shows the processing of the vibration acceleration spectrum signal for an asynchronous electric motor 2. In figure 2, the prevailing frequencies f4, f5, f6, f7 are highlighted. The frequency f6 is the tooth frequency. The frequencies f5 and f7 differ from the frequency f7 by the value of the rotor speed. Separately, the frequency f4, equal to the rotor speed, is distinguished. The predominance of such frequencies in the spectrum indicates the dynamic eccentricity of the rotor [1].

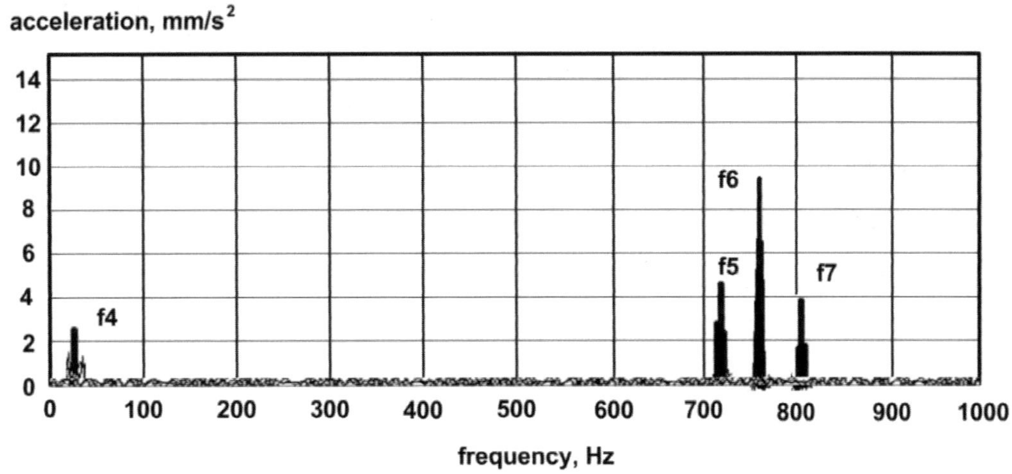

Figure 2. Vibration acceleration spectrum
with a predominance of rotor dynamic eccentricity frequencies.

It is further proposed to clarify the effect of defects in the electromagnetic system of an electric machine. If, when the electric machine is disconnected from the mains, the diagnostic indicators change very fast, then the electromagnetic system affects the diagnostic results. This requires additional research. If the diagnostic indicator does not change, the mechanical part should be tested. In the study of electric motors 1 and 2, the influence of the electromagnetic system on the results of the study was excluded. After this, it is proposed to use the diagnostic technique for bearing assemblies using the method of spectrum analysis of the envelope curve. The following results were obtained for the engines under study.

Figure 3 shows the processing of the envelope spectrum signal on a logarithmic scale for motor 1 (values above 5 dB are indicated). In figure 3, the frequency prevails, which corresponds to the frequency of the rolling elements of the bearing of the rotor of the electric motor [1, 2].

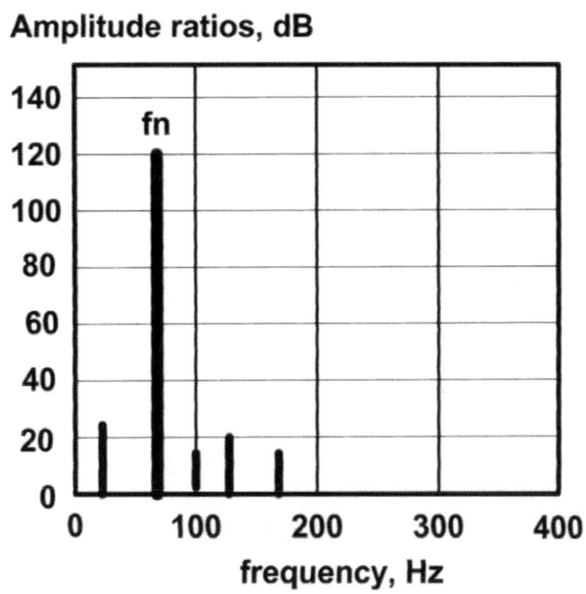

Figure 3. The envelope spectrum of the vibration acceleration curve
with a predominance of rolling frequency.

Standard software diagnoses this situation as deterioration on the outer ring of the bearing. In the case of motor 1, new bearings were installed during the repair that could not be subject to wear. When disassembling the motor, it was found that the bearing was installed with a weak landing, and the completion of this unit is necessary.

Figure 4 shows the processing of the envelope spectrum signal on a logarithmic scale for motor 2. In this figure, a predominance of several frequencies multiple of the rolling frequency of the rolling elements of the bearing is observed. Standard software diagnoses this situation as well as deterioration on the outer race of the bearing. In the case of a new bearing, this range turned out to be for a bearing with a slight misalignment [1-3].

Based on the research results, the diagnostic signs of deviation of the parameters of the electric machine from the nominal ones were expanded. This allowed us to more quickly identify and eliminate defects in electrical machines after repairs. These types of defects can lead to unacceptable levels of vibration, which lead to failure of mechanical and electrical equipment [4,5].

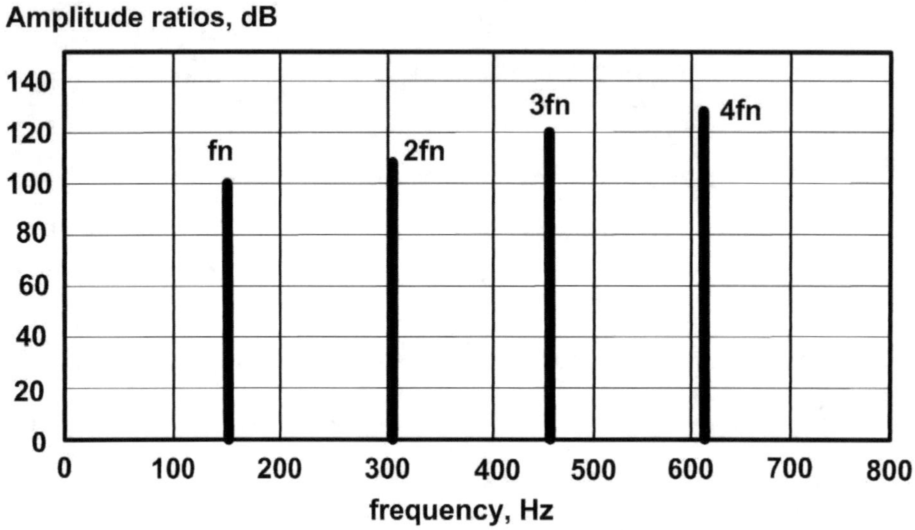

Figure 4. The envelope spectrum of the vibration acceleration curve with a predominance of frequencies that are multiples of the rolling frequency.

4. Conclusion

The paper provides examples of clarifying diagnostic signs of changes in the mechanical parameters of electrical machines after repair. According to the method of diagnosing changes in the parameters of electrical machines, further studies are carried out. The result of these studies is the quick elimination of defects in electrical machines. If these defects are not detected in a timely manner, they can lead to failure of mechanical and electrical equipment, as well as to an increase in the effect of mechanical vibrations on structural elements of industrial buildings.

References

[1] Barkov A, Barkova N, Azovtsev A 2000 *Monitoring and diagnostics of rotary machines by vibration* St. Petersburg 159.

[2] Ypma A 2001 *Learning methods for machine vibration analysis and health monitoring* Leeuwarden 217.

[3] Swam A, Mendel J M, Nikias C L 1998 Higher-order spectral analysis toolbox manual. The Mathworks 315.

[4] Buzdugan G, Mihailescu E, Rades M 1986 *Vibration Measurement* Dordrecht, The Netherlans 347.

[5] 1982 *Measuring Vibration* Bruel & Kjaer, Naerum, Denmark 40.

IOP Conf. Series: Materials Science and Engineering **739** (2020) 012052 doi:10.1088/1757-899X/739/1/012052

Use of Galactose Oxidase in Hydrazide-Capturing Technique to Isolate N-linked Glycoallergens from Olive Tree Pollen

A. Napoli[1]

[1]Dipartimento di Chimica e Tecnologie Chimiche. Università della Calabria, Arcavacata di Rende (Cs). I-87036 –Italy

amc.napoli@unical.it

Abstract. In this work, a simple method hydrazide-capturing technique followed by MALDI MS and MS/MS analysis for characterization of major pollen allergens of Olea europaea pollen was exploited. The strategy developed has different stages: devising of sample preparation protocol to sequester glycoproteins from a crude extract, and allowing the identification of antigenic peptides, glycosylation sites and modified glycans by MALDI MS and MS/MS.

1. Introduction

Air pollutants can directly interact with pollen and other plant materials [1-5], modify the allergen content of pollen further influencing its immunostimulatory properties. Many allergens from pollens are N-glycosylated and it is reported that the N-glycans themselves does not represent the major epitopes for the antibodies. The antibodies probably recognize a conformation constructed by the peptide portion around the glycosylation site and the N-glycan [6]. Considering that immunostimulatory properties of glycoallergens are essentially determined by the peptide portion around the glycosylation site, the application of hydrazide-capturing techniques followed by MALDI MS/MS analysis offers significant opportunities to obtain precious information [7-9]. Mass spectrometry is a well-established method for the high-throughput detection and quantitative analysis [10-12] of metabolites [13-15], amino acids [16] and their synthetic analogues [17], and proteins [18-22]. MS analysis of crude extracts [23] (i.e. pollen, fruit etc.) has been found to be a useful and robust approach for the chemical fingerprinting when a screening of matrix is wanted [24, 25]. MALDI-TOF-MS has been reported to be a very useful analytical technique for allergens detection [26, 27]. The specificity of MS and MS/MS experiments enable the identification of biomarkers through the elucidation of their molecular structures [28]. In this work, a simple method hydrazide-capturing technique followed by MALDI MS and MS/MS analysis for characterization of major pollen allergens of Olea europaea pollen was exploited. The strategy developed has different stages: devising of sample preparation protocol to sequester glycoproteins from a crude extract, and allowing the identification of antigenic peptides, glycosylation sites and modified glycans by MALDI MS and MS/MS.

2. Materials and Methods

2.1. Materials

Bovine serum albumin (BSA, P02769), Fetuin from fetal calf serum (F3385), Ribonuclease B from bovine pancreas (RNase B, R1153), PNGAse F Proteomics Grade (P7367), Trypsin from porcine pancreas (T6567), Galactose Oxidase from *Dactylium dendroides* (GAOX, G7400), Catalase from bovine liver (C1345) are purchased from Sigma Aldrich- Merck (Milano, Italy). Affi-Gel® Hz Hydrazide Gel is purchased from BIO RAD (Italy).

2.2. Preparation of standard protein mixture
Bovine serum albumin (P02769), Fetuin (F3385), and Ribonuclease B (R1153) are used as individual standards and also to prepare in-house proteins standard solution. Each protein is solubilized in a phosphate buffer (pH 6) at a concentration of 5ppm (1mg/200□L). Stock solution of proteins is prepared at concentrations of 5ppm (BSA: Fetuine: RNAse B/1:1:1 molar ratio) in the same phosphate buffer.

2.3. Oxidation of glycoproteins
The oxidation is performed by using Galactose Oxidase from *Dactylium dendroides* (GAOX), 4U/mg of protein, in presence of 10 μL di $CuSO_4$ (0.78 M) and a little amount of catalase. The oxidation is performed overnight in the dark glass vial, under magnetic stirring and at room temperature. Desalting step is subsequently performed by means of the SPE Strata C18-E column (55um, 70A, Phenomenex, USA). Condition step of the column is performed by washing twice with 4 ml of 0.1% TFA in 50% CH_3CN and then twice with 4 ml 0.1% TFA. Sample containing oxidized proteins is added with 3.5ml of TFA 0.1% and loaded onto conditioned SPE column. The device was washed tree times with 2 mL of 0.1% TFA, while the proteins were eluted with 4mL of 80% CH_3CN with 0.1% TFA, 4mL of 50% CH_3CN with 0.1% TFA and 4mL of 0.1% TFA in order to collect twelve protein fractions. The collected fractions are unified and completely dried by Concentrator Plus system (Eppendorf, Germany) and suspended in 200 μL of Affi-Gel coupling buffer, pH 5.5 (BIO RAD).

2.4. Coupling of oxidized glycoproteins to hydrazide resin
Affi-Gel Hz hydrazide gel is supplied in isopropanol, therefore 200 □l of gel/isopropanol slurry is transferred into a new centrifuge tube. When the gel settled, the supernatant is removed and the resin is washed two times with 800 □l of water and two times with 800 □l of coupling buffer, pH 5.5. The coupling is performed using 200 □l of oxidized proteins (pH 5.5) added with 10 □l of SDS (10%). The coupling reaction is conducted overnight on a rocking platform at room temperature. After that coupling reaction is completed, the supernatant containing the uncoupled protein is removed and the resin is washed several times to remove the unreacted material using three times the following series of solutions: a) 500 □l of NaCl (1.5 M); b) 500 □l of CH_3CN (80%), c) 500 □l of CH_3OH; d) 500 □l H_2O; e) 500 □l of NH_4HCO_3 (50mM). The last one step is performed with 200 □l of NH_4HCO_3 (50mM), ideal condition for tryptic digestion. The digestion is performed overnight at 37°C, by adding 160 pmol of trypsin. The tryptic peptides are recovered and analyzed by MALDI MS and MS/MS.

2.5. Recovery of glycopeptides and glycan moietys
The resin coupled with glycopeptides is suspended in 200 □l of NH_4HCO_3 (50mM) and added with 5□l of PNGase F (500U/□l). The de-glycosylation step is performed at 37 °C for 8h, the de-glycosylated peptides are recovered and analyzed by MALDI MS and MS/MS. Finally, the glycan moieties are removed from the resin with formic acid (0.1 M, 100 □l) at 80°C for 1 hour. The supernatant together with the wash solution (500 □l, CH_3CN: H_2O, 50:50, v: v) was concentrated in the Concentrator Plus system and analyzed by mass spectrometry.

2.6. Tree olea europaea pollen sample preparation
Tree Olea europaea pollen (2mg) are extracted with 400 □l of phosphate buffer (pH 6) for 20 min at room temperature, as previously reported. The protein mixture is then treated as reported above for the oxidation, coupling and digestion steps.

2.7. Mass spectrometry analysis

IOP Conf. Series: Materials Science and Engineering **739** (2020) 012052 doi:10.1088/1757-899X/739/1/012052

A 1 μL portion of a premixed solution of each sample and □-CHCA (H_2O/CH_3CN, 40:60, v/v, 0.3% in TFA) was spotted on the matrix target, dried at room temperature, and directly analyzed by MALDI mass spectrometry. MALDI-time-of-flight (TOF) analyses are performed using a 5800 Proteomics Analyzer mass spectrometer, from ABI SCIEX (AB SCIEX, Darmstadt, Germany) in reflectron positive mode with a mass accuracy of 5 ppm. At least 4000 laser shots are typically accumulated with a laser pulse rate of 400 Hz in the MS mode, whereas in the MS/MS mode spectra up to 5000 laser shots are acquired and averaged with a pulse rate of 1000 Hz. CID (Collision Induced Dissociation) MS/MS experiments are performed at a collision energy of 1 kV, ambient air is used as collision gas at a pressure of 10^{-6} Torr. Proteins are identified by searching a comprehensive protein database using Mascot programs (www.matrixscience.com). All MS/MS searches were performed using an initial mass tolerance of 50 ppm with enzyme cleavage specificity.

3. Results and discussions

A schematic illustration of the adopted approach is shown in Scheme 1.

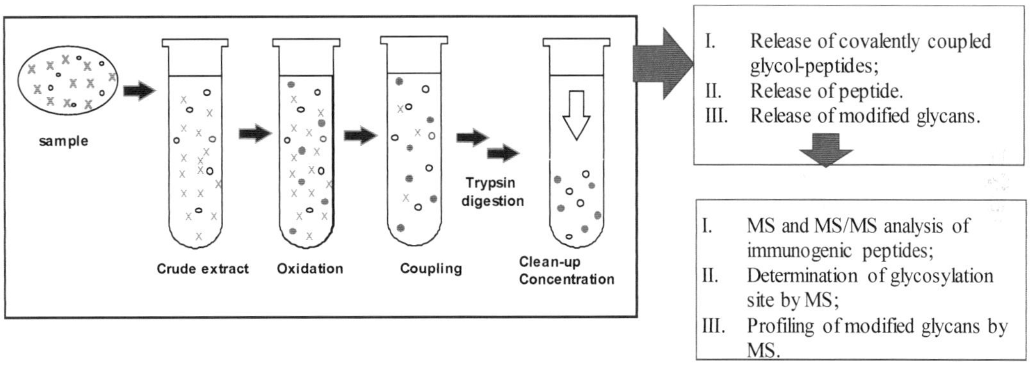

Scheme 1.

Hydrazide-capturing technique, consists in a chemical coupling between bead-immobilized hydrazide group and dialdehyde group derived from glycan moiety of glycoprotein. This approach requires that cis-diols on glycans are oxidized into aldehydes by sodium periodate, then covalently coupled with hydrazide immobilized to solid beads This approach is selective to *cis*-diols on monosaccharides consisting of glycans, but is not promising for selective capturing of protein glycoforms having a specific glycan structure. The extension of hydrazide-capturing techniques to glycoallergens, belonging to antigenic plant complex type structures, and consisting of □1–2 xylose and/or □1–3 fucose residues requires the use of an alternative oxidant reagent. Here, we describe the oxidation of the primary alcohol on glycans with galactose oxidase (GOAX) to the corresponding aldehyde derivatives, which can be subsequently covalently bound to hydrazide beads. A protein standard mixture was tested in order to study the chemical reactivity of GOAX reagents and the recovery. Bound glycoproteins were trypsin-digested in situ, the nonglycosylated peptides are removed by washing, peptides are released by PNGase F and finally derivatized glycans are released by acidic hydrolysis (figure 1).

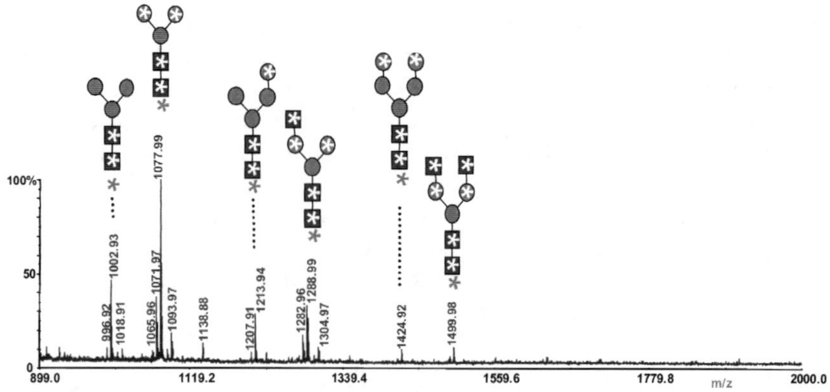

Figure 1. MALDI MS spectrum of derivatized glycans from standard protein mixture.

Covalently coupled glycopeptides were then released by PNGase F, and the resulting deglycopeptides are analyzed by mass spectrometry. The bound glycopeptides were released by acidic hydrolysis rather than enzymatic cleavage to analyze selectively glycopeptides. Finally, the methodology was extended to crude extract of olive tree pollen. The N-linked glycoallergens were selectively GOAX-oxidized, captured on hydrazide beads, trypsinized (figure 2A) and released by PNGase F (figure 2B). Tryptic peptides obtained by in situ digestion allowed the identification of the major pollen allergen (figure 2A).

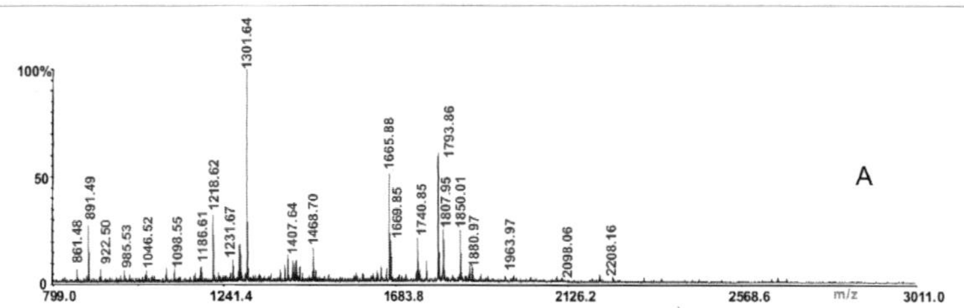

[M+H]+	Sequence	M.C.	Protein name
1468.70	NEFCEITLISSGR	0	Major pollen allergen (Allergen Ole e 1) (gi|14424429) Ole e 1.0102 protein (gi|2465127), Ole e 1.0103 (gi|2465129)
1665.88	FITELSEFIPGASVR	0	Ole e 1 protein (gi|2465131)
1669.85	FITEFSEFIPGAGVR	0	major allergen OLE1c (gi|1362131)
1793.86	AGFITELSEFIPGASVR	0	Ole e 1.0102 protein (gi|2465127), Ole e 1.0103 (gi|2465129), main olive allergen (gi|13196753)
1807.95	AGFITELSEFIPGASLR	0	Major pollen allergen (Allergen Ole e 1) (gi|14424429)

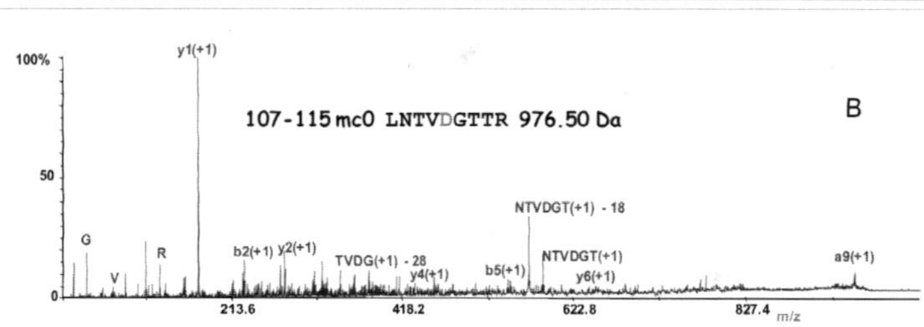

Figure 2. MALDI MS of tryptic peptides from captured major pollen allergen (A); MALDI MS/MS spectrum of peptide released by PNGases F (B).

The asparagine residue of the NX(S/T) consensus sequence of N-glycosides hydrazide-captured is converted into an aspartic acid (D) containing COOH functionality at side chain, as shown in figure 2B. Therefore, the PNGases F-mediated enzyme action allows the determination of glycation site.

IOP Conf. Series: Materials Science and Engineering **739** (2020) 012052 doi:10.1088/1757-899X/739/1/012052

4. Conclusions

In this paper, a hydrazide-capturing technique combined with GAOX strategy was developed to capture olive tree pollen glycoallergens, belonging to antigenic plant complex type structures. The developed strategy allowed the identification of antigen, the glycosilation site and glycans composition.

References

[1] De Luca P, Carbone I, B 2017 Green building materials: A review of state of the art studies of innovative materials, *J. Green Build.* **12(4)** 141-61.

[2] De Luca P, De Luca P, Candamano S, Macario A, Crea F, Nagy J 2018 Preparation and Characterization of Plasters with Photodegradative Action, *Build.* **8(9)** 122.

[3] De Luca P, Roberto B, Vuono D, Siciliano C, Nagy J B 2018 Preparation and optimization of natural glues based on Laricio pine resin, *IOP Conf. Ser. Mater. Sci. Eng.* **374(1)** p.012071.

[4] De Luca P, Bernaudo I, Elliani R, Tagarelli A, Nagy J B, Macario A 2018 Industrial Waste Treatment by ETS-10 Ion Exchanger Material, *Mater.* **11(11)** 2316.

[5] De Luca P, Pane L, Vuono D, Siciliano C, Candamano S, Nagy J B 2017 Preparation and Characterization of Natural Glues with Carbon Nanotubes, *Environ. Eng. Manag. J.* **16(8)** 1659-71.

[6] Jahouh F, Saksena R, Aiello D, Napoli A, Sindona G, Kovac P, Banoub J H 2010 Glycation sites in neoglycoglycoconjugates from the terminal monosaccharide antigen of the O-PS of Vibrio cholerae O1, serotype Ogawa, and BSA revealed by matrix-assisted laser desorption–ionization tandem mass spectrometry, *J. Mass. Spect.* **45(10)** 1148-59.

[7] Zhang H, Li X J, Martin D B, Aebersold R 2003 Identification and quantification of N-linked glycoproteins using hydrazide chemistry, stable isotope labeling and mass spectrometry, *Nat. Biotechnol.* **21(6)** 660-6.

[8] Liu T, et al. 2005 II.; Monroe, ME; Moore, RJ; Smith, RD Human Plasma N-glycoproteome analysis by immunoaffinity subtraction, hydrazide chemistry, and mass spectrometry, *J. Proteome Res.* **4(6)** 2070-80.

[9] Zeng X, Hood B L, Sun M, Conrads T P, Day R S, Weissfeld J L, Siegfried J M, Bigbee W L 2010 Lung cancer serum biomarker discovery using glycoprotein capture and liquid chromatography mass spectrometry, *J. Proteome Res.* **9(12)** 6440-9.

[10] Mazzotti F, Di Donna L, Napoli A, Aiello D, Siciliano C, Athanassopoulos C M, Sindona G 2014 N-hydroxysuccinimidyl P-methoxybenzoate as suitable derivative reagent for isotopic dilution assay of biogenic amines in food, *J. Mass Spectrom.* **49(9)** 802-10.

[11] Napoli A, Athanassopoulos C M, Moschidis P, Aiello D, Di Donna L, Mazzotti F, Sindona G 2010 Solid phase isobaric mass tag reagent for simultaneous protein identification and assay, *Analyt. Chem.* **82(13)** 5552-60.

[12] Di Donna L, Benabdelkamel H, Taverna D, Indelicato S, Aiello D, Napoli A, Sindona G, Mazzotti F 2015 Determination of ketosteroid hormones in meat by liquid chromatography tandem mass spectrometry and derivatization chemistry, *Analyt. Bioanal. Chem.* **407(19)** 5835-42.

[13] Aiello D, Furia E, Siciliano C, Bongiorno D, Napoli A 2018 Study of the coordination of ortho-tyrosine and trans-4-hydroxyproline with aluminum (III) and iron (III), *J. Mol. Liq.* **269** 387-97.

[14] Aiello D, Cardiano P, Cigala R. M, Gans P, Giacobello F, Giuffre O, Napoli A, Sammartano S 2017 Sequestering ability of oligophosphate ligands toward Al3+ in aqueous solution, *J. Chem. Eng. Data* **62(11)** 3981-90.

[15] Furia E, Aiello D, Di Donna L, Mazzotti F, Tagarelli A, Thangavel H, Napoli A, Sindona G 2014 Mass spectrometry and potentiometry studies of Pb (II)–, Cd (II)–and Zn (II)–cystine complexes, *Dalton Transa.* **43(3)** 1055-62.

[16] Antoniou A I, Pepe D A, Aiello D, Siciliano C, Athanassopoulos C M 2016 Chemoselective Protection of Glutathione in the Preparation of Bioconjugates: The Case of Trypanothione Disulfide, *J. Organ. Chem.* **81(10)** 4353-8.

[17] Siciliano C, De Marco R, Guidi L E, Spinella M, Liguori A 2012 A one-pot procedure for the preparation of N-9-Fluorenylmethyloxycarbonyl-α-amino diazoketones from α-Amino Acids, *J. Organ. Chem.* **77(23)** 10575-82.

[18] Aiello D, Casadonte F, Terracciano R, Damiano R, Savino R, Sindona G, Napoli A 2016 Targeted proteomic approach in prostatic tissue: a panel of potential biomarkers for cancer detection, *Oncoscience* **3(7-8)** 220-41.

[19] Aiello D, Giambona A, Leto F, Passarello C, Damiani G, Maggio A, Siciliano C, Napoli A 2018 Human coelomic fluid investigation: A MS-based analytical approach to prenatal screening, *Scitif. Rep.* **8(1)** 10973.

[20] Napoli A, Aiello D, Di Donna L, Prendushi H, Sindona G 2007 Exploitation of endogenous protease activity in raw mastitic milk by MALDI-TOF/TOF, *Analyt. Chem.* **79(15)** 5941-48.

[21] Napoli A, Aiello D, Aiello G, Cappello M S, Di Donna L, Mazzotti F, Materazzi S, Fiorillo M, Sindona G 2014 Mass spectrometry-based proteomic approach in Oenococcus oeni enological starter, *J. Prot. Res.* **13(6)** 2856-66.

[22] Li Y, Cappello A, Muto L, Martello E, Madeo M, Curcio R, Lunetti P, Raho S, Zaffino F, Frattaruolo L, Lappano R, Malivindi R, Maggiolini M, Aiello D, Piazzolla C, Capobianco L, Fiermonte G, Dolce V 2018 Functional characterization of the partially purified Sac1p independent adenine nucleotide transport system (ANTS) from yeast endoplasmic reticulum, *J. Biochem.* **164(4)** 313-22.

[23] Aiello D, Materazzi S, Risoluti R, Thangavel H, Di Donna L, Mazzotti F, Casadonte F, Siciliano C, Sindona G, Napoli A 2015 A major allergen in rainbow trout (Oncorhynchus mykiss): complete sequences of parvalbumin by MALDI tandem mass spectrometry, *Molecul. BioSyst.* **11(8)** 2373-82.

[24] Aiello D, Siciliano C, Mazzotti F, Di Donna L, Athanassopoulos C M, Napoli A 2018 Molecular species fingerprinting and quantitative analysis of saffron (Crocus sativus L.) for quality control by MALDI mass spectrometry, *RSC Adv.* **8(63)** 36104-13.

[25] Aiello D, Siciliano C, Mazzotti F, Di Donna L, Athanassopoulos C M, Napoli A 2019 A rapid MALDI MS/MS based method for assessing saffron (Crocus sativus L.) adulteration, *Food Chem.*125527.

[26] Napoli A, Aiello D, Di Donna L, Moschidis P, Sindona G 2008 Vegetable proteomics: the detection of Ole e 1 isoallergens by peptide matching of MALDI MS/MS spectra of underivatized and dansylated glycopeptides, *J. Prot. Res.* **7(7)** 2723-32.

[27] Napoli A, Aiello D, Di Donna L, Sajjad A, Perri E, Sindona G 2006 Profiling of hydrophilic proteins from Olea europaea olive pollen by MALDI TOF mass spectrometry, *Analyt. Chem.* **78(10)** 3434-43.

[28] Reddy C E, Albanito L, De Marco P, Aiello D, Maggiolini M, Napoli A, Musti A M 2013 Multisite phosphorylation of c-Jun at threonine 91/93/95 triggers the onset of c-Jun pro-apoptotic activity in cerebellar granule neurons, *Cell Death Disease* **4(10)** e852.

IOP Conf. Series: Materials Science and Engineering **739** (2020) 012053 doi:10.1088/1757-899X/739/1/012053

Use of Geopolymers in the Treatment of Water Contaminated by Industrial Waste

I Bernaudo[1], A Tagarelli[2], R Elliani[2], S Candamano[3], A Macario[1], P De Luca[3]

[1]Dipartimento di Ingegneria per l'Ambiente ed il Territorio ed Ingegneria Chimica. Università della Calabria, Arcavacata di Rende (Cs).I-87036 –Italy
[2]Dipartimento di Chimica e Tecnologie Chimiche. Università della Calabria, I-87030 Arcavacata di Rende (CS) – Italy
[3]Dipartimento di Ingegneria Meccanica, Energetica e Gestionale. Università della Calabria, Arcavacata di Rende (Cs). I-87036 Rende- Italy

pierantonio.deluca@unical.it

Abstract. The present research aimed to study the use of geopolymers in the removal of heavy metals in water contaminated by an industrial waste. The geopolymer was synthesized and characterized beforehand in the laboratory while the industrial waste is the so-called zinc ferrite that comes from a zinc processing industry in southern Italy. It is known that this waste contains within it a variety of elements that over time can be released if in contact with water. The study was carried out preparing different aqueous systems containing predetermined quantities of zinc ferrite and geopolymer. The systems were stirred for different reaction times and at room temperature. Subsequently, after the established time the systems were filtered and the resulting solution was analyzed by a mass spectrometer (ICP-Ms). The variation in the concentration of various elements (such as Fe, Zn, Pb, Mn, Ca, Mg, Na and K) was followed as a function of contact times. The obtained data allowed to evaluate the effectiveness of geopolymers in the removal of metals released by zinc ferrite in water. In particular, the geopolymers, in the systems studied, have shown particular propensity in the abatement of iron and zinc which has reached 100%, followed by an abatement of about 95% for Mn and Pb after a 1.5-hour contact time.

1. Introduction

Many researchers are oriented to study industrial processes that are as environmentally sustainable as possible through the use of natural raw materials, the reduction of by-products or their reuse for the preparation of new materials that respect the environment [1-14]. The industrial by-products that are of particular concern are those waste containing heavy elements. This type of slag includes that from the zinc extraction process from the blende, the so-called zinc ferrite [15, 16]. Some recent studies report the removal of some elements released from this slag in water through the use of microporous materials [17]. The titanosilicates belonging to the Englhard Titanium Silicate (ETS) family characterized by a porous structure and properties such as ion exchange, catalysis, molecular sieving, photocatalysis [18-20], have been shown to be advantageously used for the removal of heavy metals in water contaminated with zinc ferrite [21].

In this research the action of geopolymers in the removal of some elements released by zinc ferrite in water was studied. Geopolymers are inorganic materials with a chemical composition similar to that of zeolites, but unlike these, they have an amorphous structure. The prefix "geo" indicates that the present material is characterized by a mineralogical structure very similar to natural rocks, thus exhibiting similar properties, such as hardness, chemical stability and longevity. The term geopolymer aims to generally identify an amorphous or semi-crystalline alkaline-silicate alumino material, obtained by condensation polymerization, called geopolymerization or geosynthesis, or the reaction between an alumino-silicate powder with an aqueous solution of alkaline hydroxides or silicates [22-25].

2. Materials and methods

The zinc ferrite used appears as a material of intense brown color, grainy and characterized by a certain humidity since it comes from open-air storage [17]. Before testing, the latter was dried for two hours in an oven at 100 °C to eliminate the moisture present in it. Subsequently the dry material was finely ground in a mortar mill to ensure adequate homogeneity. The geopolymers that have been used to study their potential use in the removal of metals released from ferrite in water have been previously synthesized in the laboratory. The latter were obtained by mixing meta kaolin, supplied by Personal Factory Spa (Simbario, Italy) and an activating solution prepared from sodium silicate supplied by Sasol Italy SPA (Milan, Italy), sodium hydroxide (Sigma Aldrich) and water, in quantities sufficient to obtain a solid / water ratio of 0.51 in the finished product, a Na_2O/Al_2O_3 ratio of 1 and a SiO_2/Al_2O_3 ratio of 3.8. During the mixing phase a quantity of 5A zeolite equal to 5% by weight on the metakaolin was added to the system (figure 1a). In order to accelerate the geopolymerization reaction, the pastes were sealed and placed in the oven at 50 °C for an hour and then allowed to mature at room temperature for 7 days. The geopolymer was obtained as a monolithic piece (figure 1b), therefore before its use in experimentation, it was finely ground into a mortar mill in order to increase its active surface (figure 1c).

Figure 1. (a) Initial geopolymeric paste; (b) geopolymer; (c) pulverized geopolymer.

After pretreating the zinc ferrite and synthesizing the geopolymer, two aqueous systems were prepared, the first containing zinc ferrite and geopolymer e while the second containing only geopolymer. The latter was used as blanc during analytical measurements to take into account any releases of some metals by the geopolymer. The following table 1 shows the proportions by weight of the systems used.

Table 1. Proportions by weight of the aqueous system used.

System	Zinc ferrite (g)	Geopolymer (g)	H_2O (g)
Zinc Ferrite/Geoplymer	5	10	100
Blanc	----	10	100

In particular, the zinc ferrite was added directly in water, while the geopolymer was previously inserted in a filter paper bag and kept suspended in the solution (figure 2a, b), to allow at the end of the release tests an easy separation of the geopolymer from the system.

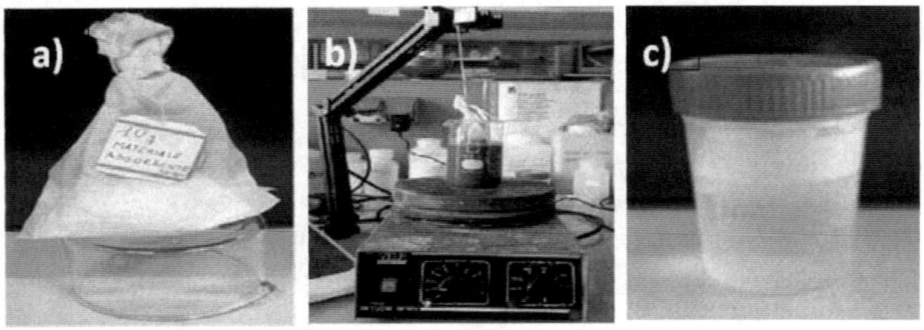

Figure 2. (a) Filter paper bag containing the geopolymer; (b) aqueous system: Geoplymer/Zinc ferrite; c) filtered solution after release tests.

All systems were kept stirring at room temperature. The tests were repeated varying the stirring times equal to: 0.5, 1, 1.5, 2 hours. At the end of the programmed times, each system was filtered and the resulting solution (figure 2c) was analyzed through a mass spectrometer (ICP-Ms) which allowed to follow the concentration variations of some elements such as: Fe, Zn, Pb, Mn, Mg and Na as contact times vary.

3. Results and discussion

The following graphs show the concentration of some elements released by zinc ferrite in the aqueous systems containing the geopolymer when the stirring times vary. In each graph, the data obtained are compared with those reported in a previous work [21], relating to the aqueous system containing only zinc ferrite. The data reported in figure 3a show how the geopolymer acts effectively on the abatement of the iron concentration.

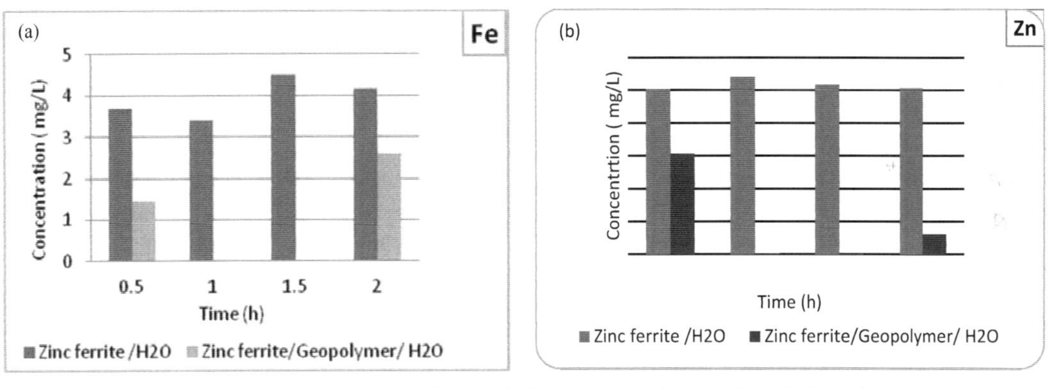

Figure 3. Variation of the concentration of Fe and Zn as a function of the stirring time for the aqueous system containing geopolymer/zinc ferrite and for the aqueous system containing only zinc ferrite [21].

After half an hour of contact there is a reduction of the Fe-concentration of over 60%, to reach an almost total reduction to an hour. The increase in concentration that occurs after two hours can be attributed to the particular and complex system of zinc ferrite, where for these times of agitation there is the probable descent into the field of other elements which are solubilized and which compete with iron on the exchange by the geopolymer. The data on Zinc (figure 3b) show a trend of variation of concentrations very similar to that of iron. After about half an hour of contact there is a reduction in the concentration of zinc of about 40% followed by a reduction close to 100% at an hour of agitation. At about two hours there is a slight increase in concentration, confirming in this case the thesis described above. The geopolymer also shows an effective role in the abatement of Pb (figure 4a) although for longer times than the two cases previously discussed. After an hour and a half there is a reduction of about 95%.

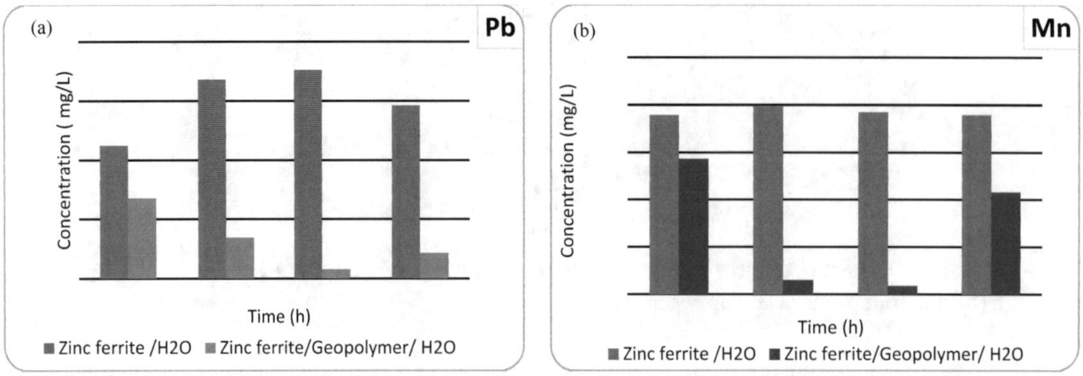

Figure 4. Variation of the concentration of Pb and Mn as a function of the stirring time for the aqueous system containing geopolymer/zinc ferrite and for the aqueous system containing only zinc ferrite [21].

An efficacy on the part of the geopolymer, very similar to that on the Pb, occurs in the case of Mn (Fig. 4b) where even for this element, at an hour and a half, the concentration is reduced by about 95%. In the case of magnesium (figure 5a), there is instead a lower efficacy of the geopolymer, in fact the best reduction does not exceed 60%, which occurs at about an hour and a half.

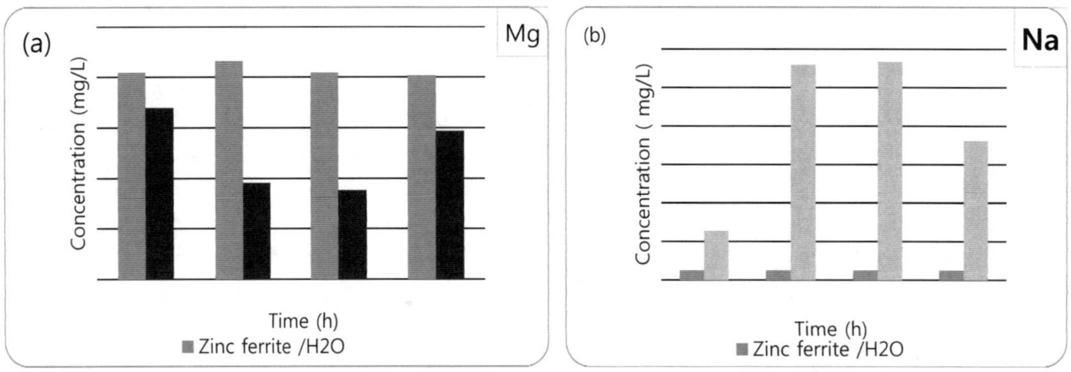

Figure 5. Variation of the concentration of the Mg and Na as a function of the stirring time for the aqueous system containing geopolymer/zinc ferrite and for the aqueous system containing only zinc ferrite [21].

The figure 5b shows the data related to the sodium concentration in the analyzed systems. It is possible to observe how the concentration of sodium increases with increasing agitation time and furthermore it is much higher than the zinc ferrite only system. All this evidently demonstrates the action of the geopolymer which exchanges the sodium ion in favour of other elements. At two hours, on the other hand, there is a slight decrease in the sodium concentration which is in perfect agreement with the decrease in the abatement of the elements analyzed, demonstrating that for these times of agitation the system becomes particularly detrimental to ion exchange.

From the set of data reported in all the previous graphs it is possible to deduce a common trend for all the analyzed elements: an hour and a half of agitation time is the optimal one to have the best reduction. After this time all the elements recorded a decrease in their concentration which confirms that after two hours of agitation there are important solubilization reactions by zinc ferrite components that create a crowding of elements released in the system giving rise to competitive phenomena between the elements and occlusion of the porous matrix of the geopolymer with reduction of the exchange action.

4. Conclusion

The set of results obtained showed that geopolymers can be advantageously used in the removal of elements released in water from zinc ferrite.

In particular, the geopolymer exhibits a constant behavior regardless of the metal analyzed. The highest felling has been recorded for treatment times of 1 and 1.5 hour. Iron and zinc were almost completely removed after an hour of treatment, followed by Mn and Pb with a 95% removal at an hour and a half. Lower abatement rates were recorded for Magnesium. Finally, it is possible to conclude that the geopolymer used has shown particular efficacy to remove Fe, Zn, Mn, and Pb.

Reference

[1] Venkataram Reddy B V. 2009 Sustainable materials for low carbon buildings, *Int. J. Low-Carbon Technol.* **4(3)** 175-81.

[2] Yen C, Tseng D, Tung T 2011 Characterization of eco-cement paste produced from sludge, *Chemo sphere* **84** 220-26.

[3] De Luca P, Carbone I, Nagy J. B 2017 Green building materials: A review of state of the art studies of innovative materials, *J. Green Build.* **12(4)** 141-61.

[4] Aiello D, Cardiano P, Cigala R. M, Gans P, Giacobello F, Giuffre O, Napoli A, Sammartano S 2017 Sequestering ability of oligophosphate ligands toward Al3+ in aqueous solution, *J. Chem. Eng. Data* **62(11)** 3981-90.

[5] De Luca P, De Luca P, Candamano S, Macario A, Crea F, B.Nagy J 2018 Preparation and Characterization of Plasters with Photodegradative Action, *Build.* **8(9)** 122.

[6] Aiello D, Furia E, Siciliano C, Bongiorno D, Napoli A 2018 Study of the coordination of ortho-tyrosine and trans-4-hydroxyproline with aluminum (III) and iron (III), *J. Molec. Liq.***269** 387-97.

[7] Lin Y, Zhou S, Li F, Lin Y 2012 Utilization of municipal sewage sludge as additives for the production of eco-cement, *J. Hazard. Mater.* **213** 457-65.

[8] Mohanty A. K, Misra M, Drzal L. T 2002 Sustainable bio-composites from renewable resources: opportunities and challenges in the green materials world, *J. Polym. Environ.* **10(1-2)** 19-26.

[9] Govindan K, Shankar K. M, Kannan D 2016 Sustainable material selection for construction industry–A hybrid multi criteria decision making approach, *Renew. Sustain. Energy Rev.* **55** 1274-88.

[10] Siciliano C, De Marco R, Guidi L. E, Spinella M, Liguori A 2012 A one-pot procedure for the preparation of N-9-Fluorenylmethyloxycarbonyl-α-amino diazoketones from α-Amino Acids, *J. Org. Chem.* 77(23) 10575-82.

[11] De Luca P, Roberto B, Vuono D, Siciliano C, Nagy J. B 2018 Preparation and optimization of natural glues based on Laricio pine resin, *IOP Conf. Ser. Mater. Sci. Eng.* **374(1)** 012071.

[12] De Luca P, Pane L, Vuono D, Siciliano C, Candamano S, Nagy J. B 2017 Preparation And Characterization Of Natural Glues With Carbon Nanotubes, *Environ. Eng. Manag. J.* **16(8)**.

[13] Sudin R, Swamy N. 2006 Bamboo and wood fibre cement composites for sustainable infrastructure regeneration, *J. Mater. Sci.* **41(21)** 6917-24.

[14] Intiya W, Thepsuwan U, Sirisinha C, Pongdhorn S 2017 Possible use of sludge ash as filler in natural rubber, *J Mater Cycles Waste Manage.***19(2)** 774-81.

[15] Zhao Y, Stanforth R 2000 Extraction of zinc from zinc ferrites by fusion with caustic soda, *Miner. Eng.* **13(13)** 1417-21.

[16] De Luca P, Bernaudo I, Tagarelli A, Candamano S, Nagy J. B 2018 Study On The Release Conditions Of The Metals Present In The Zinc Ferrite, *Proce. Envrio. Sci. Manage.* **5** 29-38.

[17] De Luca P, Bernaudo I, Candamano S, Siciliano C, Macario A 2019 Treatment Of Industrial Slag Zinc Ferrite By Zeolitic Sludge, *IOP Conf. Ser. Mater. Sci. Eng.* **572** 012046.

[18] Choi J. H, Kim S. D, Kwon Y. J, Kim W. J. 2006 Adsorption Behaviors of ETS-10 And Its Variant, ETAS-10 On The Removal of Heavy Metals, Cu2+, Co2+, Mn2+ And Zn2+ From A Waste Water, *Micropor. Mesopor. Mater.* **96(1-3)** 157-67.

[19] De Luca P, Mastroianni C, Nagy J. B 2018 Synthesis of self-bonded pellets of ETS-4 phase by new methodology of preparation, *IOP Conf. Ser. Mater. Sci. Eng.* **347(1)** 012003.

[20] De Raffele G, Aloise A, De Luca P, Vuono D, Tagarelli A, Nagy J. B 2016 Kinetic and thermodynamic effects during the adsorption of heavy metals on ETS-4 and ETS-10 microporous

materials, *J. Porous Mater* **23(2)** 389 400.
[21] De Luca P, Bernaudo I, Elliani R, Tagarelli A, Nagy J. B, Macario A 2018 Industrial Waste Treatment by ETS-10 Ion Exchanger Material, *Mater.* **11(11)** 2316.
[22] Davidovits J 1994 Geopolymers: man-made rock geosynthesis and the resulting development of very early high strength cement, *J. Mat. Edu.* **16** 91.
[23] Al-Harahsheh M, Al Zboon K, Al-Makhadmeh L, Hararah M, Mahasneh M 2015 Fly ash based geopolymer for heavy metal removal: A case study on copper removal, *J. Environ. Chem. Eng.* **3(3)** 1669-77.
[24] Candamano S, De Luca P, Frontera P, Crea F 2017 Production of geopolymeric mortars containing forest biomass ash as partial replacement of metakaolin, *Environ.* **4(4)** 74.
[25] Davidovits J *Geopolymer Chemistry & Application, Geopolymer Institute Saint-Quentin France* ISBN 2-951-14820-1-9.

IOP Conf. Series: Materials Science and Engineering **739** (2020) 012054 doi:10.1088/1757-899X/739/1/012054

Use of Zinc Ferrite for Photocatalytic Treatment of Water Contaminated with Organic Dye

A Cantiello[1], S Candamano[2], P De Luca[2]

[1]Dipartimento di Ingegneria per l'Ambiente ed il Territorio ed Ingegneria Chimica. Università della Calabria, Arcavacata di Rende (Cs). I-87036 –Italy
[2]Dipartimento di Ingegneria Meccanica, Energetica e Gestionale. Università della Calabria, Arcavacata di Rende (Cs). I-87036 Rende- Italy

pierantonio.deluca@unical.it

Abstract. The present work aims to test the use of zinc ferrite as a photocatalytic agent for the discoloration of water contaminated by organic dyes. The zinc ferrite used is an industrial waste that comes from an industrial plant destined for the extraction of zinc. The organic dye studied was the Blue Patent V. The photocatalytic tests were prepared using aqueous solutions contaminated by the organic dye to which predetermined quantities of zinc ferrite were added and were subsequently exposed to sunlight for programmed times. Furthermore, systems in the presence of hydrogen peroxide have also been studied. After the exposure time, the systems were filtered and the concentrations of the solutions were measured with UV spectrophotometry. The ferrite after the photocatalytic tests was characterized by thermal analysis (DSC). The data obtained were analyzed according to the experimental parameters used. Zinc ferrite showed photodegradative abilities respect to the analyzed dye.

1. Introduction

Water pollution is an increasingly current problem. There are many causes that contribute to threatening the state of rivers and seas, which are almost always the major receptors of pollutants that are released into the environment. Increased sensitivity is driving research to act on several fronts [1-3]. The first important action is surely that of avoiding the production of harmful substances itself by production of materials and life cycles eco-sustainable [4-9]. Unfortunately, this is not always possible and therefore it is fundamental to seek new and adequate methods of treatment for the abatement of pollutants. Particularly important is the purification of contaminated water which is addressed using new materials such as molecular sieves [10-16], carbon nanotubes [17-20], geopolymers [21, 22], photo catalytic materials [23-26]. Zinc ferrite is an industrial waste and comes from industrial plants destined to extract zinc. Many studies have shown that it contains many metals [15, 27, 28]. Precisely this great abundance of metals led us to suppose that zinc ferrite could be a potential photochemical agent. In this work, therefore, we wanted to consider ferrite not only as a problem but as a potentially useful waste, giving it a new function. In particular, in this work, zinc ferrite has been tested as a photocatalytic agent for the discoloration of water contaminated by the organic dye.

2. Materials and Methods

The materials used in the experiments were Blue Patent V dye, Zinc Ferrite and hydrogen peroxide. Blu Patent V, or E131, in the European code for food additives, is a dark blue synthetic dye. It is a salt that can be in the sodium (CAS: 20262-76-4) or calcium (CAS: 3536-49-0) form. The appearance is of a violet powder very soluble in water. In this research the sodium form was used.

Figure 1. (a) Structure and (b) powder of Blue Patent V;(c) Zinc ferrite.

The Blu Patent V raises concern when it is spilled in an abusive manner, and not treated according to the existing standards in the environmental field, in natural water courses from the various industries that use it both in the food industry and in the textile, tanning, etc. Zinc ferrite is an industrial waste derived from the treatment of zinc extraction from blende, a mineral that includes zinc sulfide. This waste is presented as a dark colored powder (figure 1c). The hydrogen peroxide solution used has a concentration of 30% (w/w). The experimental procedure initially involved the preparation of the aqueous solution of Blue Pantent V with a concentration of 0.1 g / L. Subsequently different systems were prepared by adding predetermined quantities of zinc ferrite to established volumes of the initial solution. Other systems have been prepared also by adding hydrogen peroxide. In order to have a reference, zinc ferrite-free systems have also been considered. The following table 1 summarizes the composition of the different systems considered.

Table 1. System composition.

System	Composition
0	15 mL Blu Patent solution [0.1g/L]
1	15 mL Blu Pantent solution [0.1g/L] + 0.1g Zinc Ferrite
2	15 mL Blu Patent solution [0.1g/L] + 1 mL H_2O_2 [30%w/w]
3	15 mL Blu Patent solution [0.1g/L] + 1 mL H_2O_2 [30%w/w] + 0.1g Zinc Ferrite

Subsequently all systems were exposed to sunlight for 4, 8, 16, 16, 20 days. The reason for exposing them only to sunlight and avoiding other types of radiation was motivated by the choice of studying the most natural and economic methods possible. Once the predetermined exposure times had elapsed, the systems were filtered by a vacuum pump and the solution obtained was analyzed by UV spectrophotometry to search for the residual concentration of the dye. Furthermore, on the solid residue, that is the zinc ferrite after the test, thermal analyzes (DSC) were carried out in order to verify also possible adsorption phenomena.

3. Results and discussions

The first phase of the research was to verify the photostability of Blu Patent V in the experimental conditions used. The following figure 2 shows the concentration of the dye solutions after the photocatalytic tests at different times of exposure to sunlight.

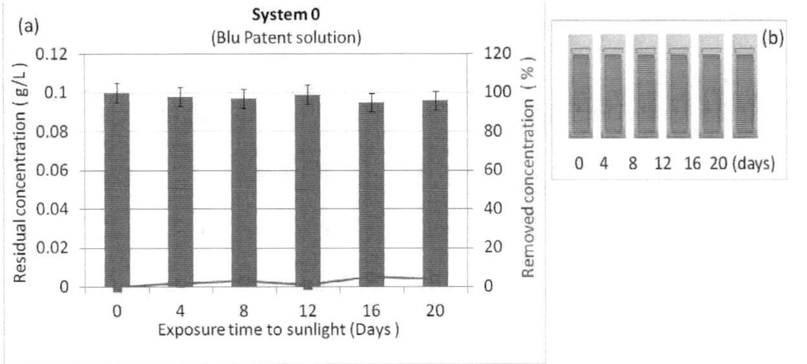

Figure 2. (a) Concentration and percentage of abatement of the system 0 and (b) respective images of the solutions as a function of the time of exposure to sunlight.

The data show that the Blue Patent V dye, in the experimental conditions used, presents a strong photostability in fact after 20 days of exposure to sunlight the concentration has remained almost unchanged. The following figure 3 shows the variation of the concentration of the dye solutions in the presence of zinc ferrite (system 1). The data obtained show that zinc ferrite accelerates dye degradation.

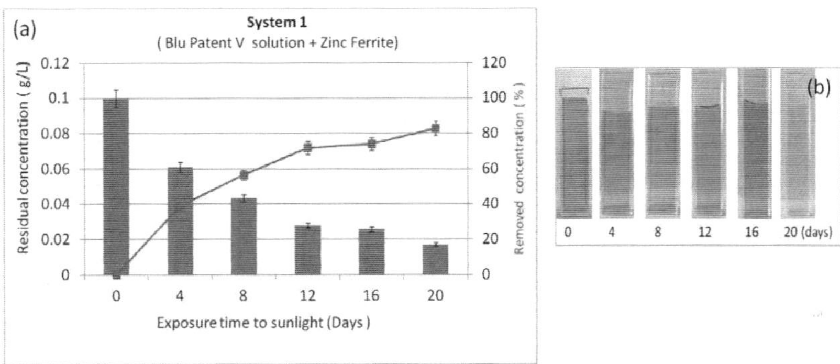

Figure 3. (a) Concentration and percentage of abatement of the system 1 and (b) respective images of the solutions the function of the time of exposure to sunlight.

The following figure 4 shows the variation of the concentration of the dye solutions in the presence of hydrogen peroxide but in the absence of zinc ferrite (system 2).

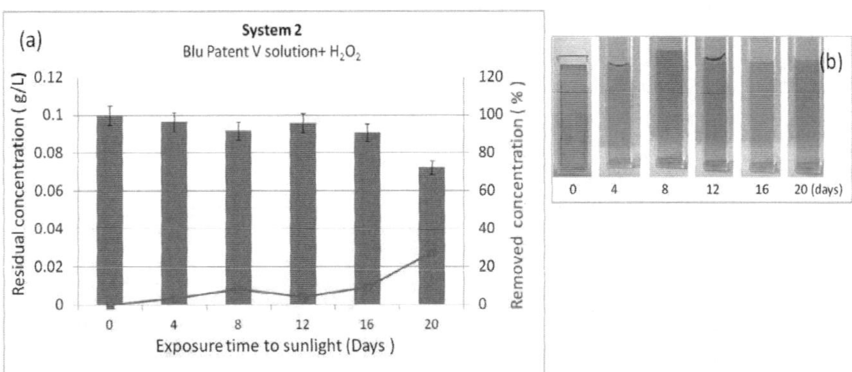

Figure 4. (a) Concentration and percentage of abatement of system 2 and (b) respective images of the solutions the function of exposure time to sunlight.

This test was carried out to verify the influence of hydrogen peroxide on the photostability of the dye. The data reported in figure 4 show that the dye is photostable even in the presence of hydrogen peroxide in fact a more significant decrease in concentrations is recorded only on the 20th day of exposure to sunlight. The following figure 5 shows the data obtained in the presence of zinc ferrite and hydrogen peroxide.

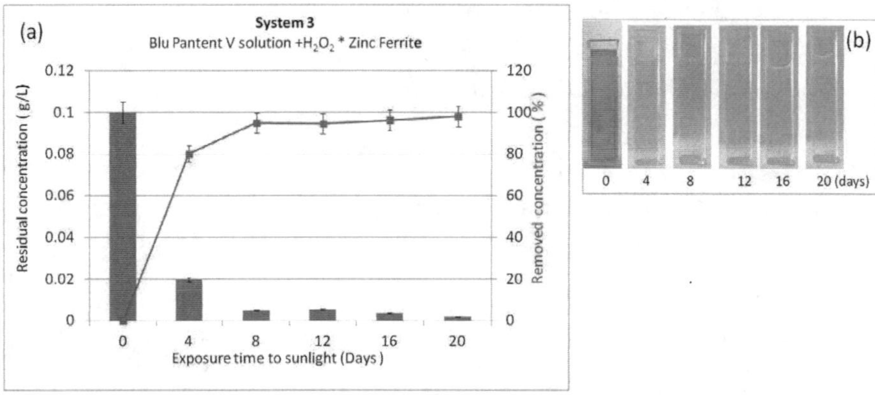

Figure 5. (a) Concentration and percentage of abatement of system 1 and (b) respective images of the solutions the function of exposure time to sunlight.

The addition of hydrogen peroxide to the system containing zinc ferrite increases the photodegradative action. In fact, even after only four days of exposure to sunlight there is a significant reduction in dye concentrations. From the thermal analysis (DSC) (figure 5) carried out it emerged that there was no adsorption action, in fact on the zinc ferrite samples (curves b and c) there are no peaks attributable to the dye. The degradation of the contaminant can be attributed exclusively to a photodegradative action.

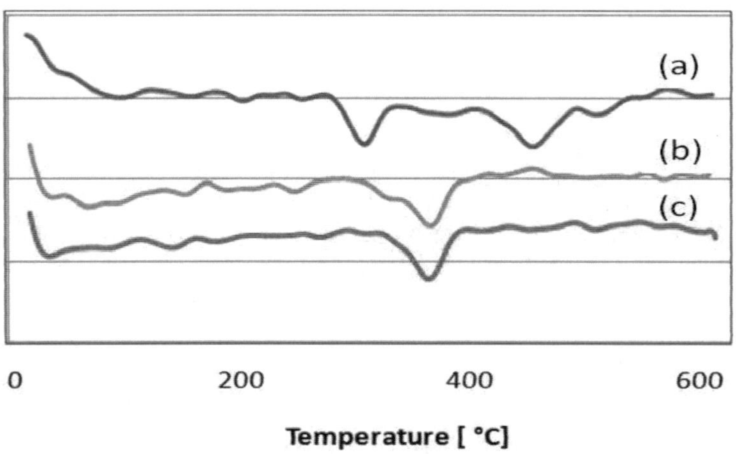

Figure 5. Thermal Analysis (DSC): (a) Blue Pantent V; Zinc Ferrite after test (b) of system 1 (b) and system 2 (c).

4. Conclusions

The results obtained showed that zinc ferrite showed photocatalytic capacities that are improved in the presence of hydrogen peroxide. Although this compound has metals in it, it has been used experimentally to see if it has a degradation action and surprisingly showed a high photodegradative behaviour.

Reference

[1] Aiello D, Cardiano P, Cigala R. M, Gans P, Giacobello F, Giuffrè, O, Napoli A, Sammartano S 2017 Sequestering ability of oligophosphate ligands toward Al3+ in aqueous solution *J. Chem. Eng. Data* 62(11) 3981-90.

[2] Aiello D, Furia E, Siciliano C, Bongiorno D, Napoli A 2018 Study of the coordination of ortho-tyrosine and trans-4-hydroxyproline with aluminum (III) and iron (III) *J. Molec. Liq.* **269** 387-97.

[3] Siciliano C, De Marco R, Guidi L. E, Spinella M, Liguori A 2012 A one-pot procedure for the preparation of N-9-Fluorenylmethyloxycarbonyl-α-amino diazoketones from α-Amino Acids *J. Org. Chem.***77(23)** 10575-82.

[4] Ljungberg L. Y. 2007 Materials selection and design for development of sustainable products *Mater. Desig.* **28(2)** 466-79.

[5] De Luca P; Carbone I, Nagy J. B 2017 Green building materials: A review of state of the art studies of innovative materials *J. Green Build.* **12(4)** 141-61.

[6] Guigo N, Mija A, Vincent L, Sbirrazzuoli N 2010 Eco-friendly composite resins based on renewable biomass resources: Polyfurfuryl alcohol/lignin thermosets *Eur. Polym. J.* **46(5)** 1016-23.

[7] De Luca P, Pane L, Vuono D, Siciliano C, Candamano S, Nagy J. B 2017 Preparation And Characterization of Natural Glues with Carbon Nanotubes *Environ. Eng. Manag. J.* **16(8)**.

[8] De Luca P, Roberto B, Vuono D, Siciliano C, Nagy J. B 2018 Preparation and optimization of natural glues based on Laricio pine resin *IOP conf. Ser. Mater. Sci. Eng.* **374(1)** 012071.

[9] Cheung H. Y, Ho M. P, Lau K. T, Cardona F, Hui D 2009 Natural fibre-reinforced composites for bioengineering and environmental engineering applications *Comp. Part B Eng.* 4(**7**) 655-63.

[10] Barakat M. A 2011 New trends in removing heavy metals from industrial wastewater *Arab. J. Chem.* **4(4)** 361-77.

[11] Lopes C B, Coimbra J, Otero M, Pereira E, Duarte A. C, Lin Z, Rocha J 2008 Uptake of Hg2+ from aqueous solutions by microporous titano-and zircono-silicates *Quimica Nova* **31(2)** 321-5.

[12] Motsi T, Rowson N. A, Simmons M. J. H 2009 Adsorption of heavy metals from acid mine drainage by natural zeolite *Int. J. Miner. Process.* **92(1-2)** 42-8.

[13] De Luca P, Mastroianni C, Nagy J. B 2018 Synthesis of self-bonded pellets of ETS-4 phase by new methodology of preparation *IOP conf. Ser. Mater. Sci. Eng.* **347(1)** 012003.

[14] Kang S. Y, Lee J. U, Moon S. H, Kim K. W 2004 Competitive adsorption characteristics of Co2+, Ni2+, and Cr3+ by IRN-77 cation exchange resin in synthesized wastewater *Chemosphere* **56(2)** 141-7.

[15] De Luca P, Bernaudo I, Elliani R, Tagarelli A, Nagy J. B, Macario A 2018 Industrial Waste Treatment by ETS-10 Ion Exchanger Material *Mater.* **11(11)** 2316.

[16] Huang C. P, Blankenship D. W 1984 The removal of mercury (II) from dilute aqueous solution by activated carbon *Water Res.* **18(1)** 37-46.

[17] Logar N. Z, Kaucic V 2006 Nanoporous materials: from catalysis and hydrogen storage to wastewater treatment *Acta Chim. Slov.* **53(2)** 117.

[18] De Luca P, Nappo G, Siciliano C, Nagy J. B 2018 *J.* The role of carbon nanotubes and cobalt in the synthesis of pellets of titanium silicates *Porous Mater.* **25(1)** 283-96.

[19] Lico D, Vuono D, Siciliano C, Nagy J. B, De Luca P 2019 Removal of unleaded gasoline from water by multi-walled carbon nanotubes *J. Environ. Manag.* **237** 636-43.

[20] Sebastiano C, De Luca P, Frontera P, Crea F 2017 Production of geopolymeric mortars containing forest biomass ash as partial replacement of metakaolin *Environ.* **4(4)** 74.

[21] Davidovits J 2015 *Geopolymer Chemistry & Application 4th Ed. Institute Geopolymere Saint-Quentin France* ISBN 9782951482098.

[22] Al-Harahsheh M. S, Al Zboon K, Al-Makhadmeh, L, Hararah M, Mahasneh M 2015 Fly ash based geopolymer for heavy metal removal: A case study on copper removal *J. Environ. Chem. Eng.* **3(3)** 1669-77.

[23] De Luca P, Foglia P, Siciliano C, Nagy J. B, Macario A 2019 Water Contaminated by Industrial Textile Dye: Study on Decolorization Process *Environ.* **6(9)** 101.

[24] De Luca P, Chiodo A, Nagy J. B 2011 Activated ceramic materials with deposition of photocatalytic titano-silicate micro-crystals *WIT Trans. Ecol. Environ.* **154** 155-65.

[25] Di Paola A, Garcia-Lopez E, Marci G, Palmisano L 2012 A survey of photocatalytic materials for environmental remediation *J. Hazard. Mater.* **211** 3-29.

[26] De Luca P, De Luca P, Candamano S, Macario A, Crea F, Nagy J. B 2018 Preparation and Characterization of Plasters with Photodegradative Action *Build.* **8(9)** 122.

[27] De Luca P, Bernaudo I, Candamano S, Siciliano C, Macario A. 2019 Treatment of industrial slag zinc ferrite by zeolitic sludge *IOP conf. Ser. Mater. Sci. Eng.* **572(1)** 012046.

[28] De Luca P, Bernaudo I, Tagarelli A, Candamano S, Nagy J. B 2018 Study on the Release Conditions of the Metals Present in the Zinc Ferrite *Proce. Environ. Sci. Eng. Manage.* **5(1)** 29.

IOP Conf. Series: Materials Science and Engineering **739** (2020) 012055 doi:10.1088/1757-899X/739/1/012055

Research on Aerial View Point Planning of Drone based on Multi-View

Zhu Yan-yan Shi Yun[1]

[1]West Anhui University, Lu'an, Anhui, 237012, China

710667612@qq.com

Abstract. For multi-view viewpoint planning research, Firstly, using the UAV aerial multi-view image, the viewpoint view is first to establish a candidate view area according to the shooting distance. Secondly, the sub-viewpoint set is filtered and the view points are complemented by the region where the candidate view cannot be photographed, so that the three-dimensional reconstruction data has better integrity. Thirdly, the influence of camera incident angle on image pixel error is analyzed, and the corresponding 3D model is established. Finally, the stereoscopic overlap is used to perform viewpoint screening under the premise of ensuring image quality and shooting coverage. The experimental results show that the proposed viewpoint planning method has high coverage and the reconstruction accuracy is below 0.1% of the target size.

1. Introduction

Multi-view 3D reconstruction has the advantages of simple shooting and high measurement accuracy. It is widely used in reverse engineering, virtual reality, online detection and other occasions. For the 3D reconstruction of outdoor scenes, there are traditional measurement methods such as laser scanning and laser radar, but it is difficult to obtain complete measurement data due to measurement limitations. In this paper, the multi-rotor UAV is equipped with four monocular cameras for omnidirectional shooting of outdoor objects, and a more complete three-dimensional model can be obtained through multiple views. In order to improve the reconstruction accuracy, it is necessary to plan the viewpoint of the aerial photography process of the drone.

Viewpoint planning is a non-deterministic issue and a key technology in 3D reconstruction. In [1-3] has applications in target recognition and tracking, surface defect extraction, robot vision and other occasions. In [4] summarizes the camera viewpoint planning research. In [5], based on the principle of cross ratio invariance, the transfer relationship between image pixel error and 3D object point error is established, and the multi-view 3D reconstruction accuracy is improved. The implementation method of the motion mechanism is given in [6], and the accuracy of 3D reconstruction is improved based on the reconstruction accuracy estimation function. In [7], based on the known rough three-dimensional model of the measured object, the viewpoint is arranged in the model and then expanded. In [8] studied the extraction of key frames and 3D reconstruction in images, but did not discuss the viewpoint planning problem.

In order to obtain a complete, high-quality image, the accuracy of the three-dimensional reconstruction is high. In this paper, aiming at the aerial vision point planning problem of UAV, by analyzing the parameter control in the aerial photography of UAV, based on the rough 3D model, the candidate viewpoint is first set and then the viewpoint planning is performed according to the constraint, thus ensuring the accuracy of 3D reconstruction.

2. Parameter Control of Camera Pose

The viewpoint represents six degrees of freedom in the camera pose in space, and its viewpoint pose can be expressed as $V=[x_v\ y_v\ z_v\ \alpha_v\ \beta_v\ \gamma_v]^T$.

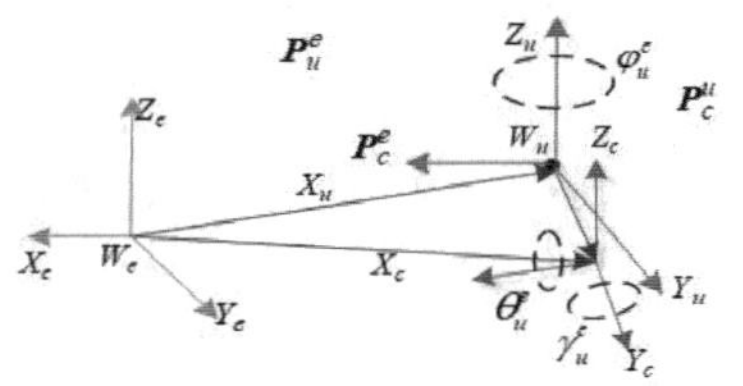

Figure 1. Relative pose of the earth, drone, camera.

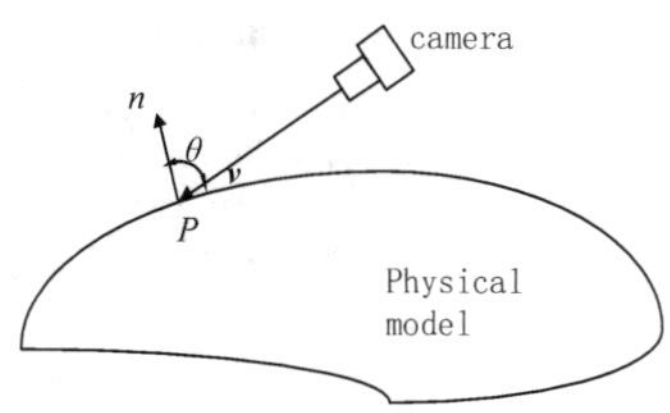

Figure 2. Schematic diagram of the viewing angle constraint.

Figure 1 shows the relative pose of the Earth, drone, and camera. The multi-rotor drone achieves flight control by setting P_u^e. When the drone is hovering to capture an image, $\theta_u^e=\gamma_u^e=0$ is required.

In order to meet the multi-degree of freedom requirements for shooting, this paper adds a two-dimensional pan/tilt. The camera is mounted on the gimbal, and the gimbal is fixed directly below the fuselage. The pan/tilt provides changes in the camera's pitch angle θ_c^u and roll angle γ_c^u.The camera is mounted close to the center of gravity of the drone, and the direction of the main axis of the lens is consistent with the direction of the longitudinal axis of the drone. The viewpoint of the drone aerial photography can be controlled by the parameters of equation (1):

$$V=[x_v\ y_v\ z_v\ \alpha_v\ \beta_v\ \gamma_v]^T=[x_u^e\ y_u^e\ z_u^e\ \varphi_u^e\ \theta_c^u\ \gamma_c^u]^T \tag{1}$$

3. Viewpoint Constraints

The target to be reconstructed needs to meet visibility constraints, overlap and coverage, motion platform constraints, and camera angle of incidence.

3.1. Visibility constraint

Visibility constraints include viewable angle, camera field of view, shooting distance, and occlusion.

(a)Viewable angle

P is a little on the surface of the object. ***v*** is the spindle vector of the camera. ***N*** is the normal vector at point *P*. θ is the angle between the two vectors. When $\theta<90^\circ$, *P* point is visible, as shown in figure 2.

(b) Camera field of view

The straight line OP should be in the longitudinal field of view angle α and the lateral direction β , and it is necessary to satisfy the formula (2).

$$\left.\begin{aligned} \boldsymbol{v}\cdot\boldsymbol{v}_{py}-\|\boldsymbol{v}\|\cdot\|\boldsymbol{v}_{py}\|\cos\frac{\alpha}{2}\geq 0 \\ \boldsymbol{v}\cdot\boldsymbol{v}_{px}-\|\boldsymbol{v}\|\cdot\|\boldsymbol{v}_{px}\|\cos\frac{\beta}{2}\geq 0 \end{aligned}\right\} \tag{2}$$

where $\boldsymbol{v}$ is the camera spindle vector. $\boldsymbol{v}_p=[x_p-x_o\ y_p-y_o\ z_p-z_o]^T$ is the vector of the OP . $\boldsymbol{v}_{px}$ and $\boldsymbol{v}_{py}$ denotes the projection loss of $\boldsymbol{v}_p$ in the plane transverse plane xO_pO and the longitudinal midplane yO_pO , respectively.

(c) Shooting distance

The shooting distance determines whether the feature can be recognized, and the shooting distance z from the lens to the surface of the object should satisfy the formula (3).

$$\left.\begin{aligned} z_{\min} &= \frac{afd_{\max}}{ad_{\max}-af+cf} \\ z_{\max} &= \frac{afd_{\min}}{ad_{\min}-af-cf} \end{aligned}\right\} \quad (f<d_{\min}<d_{\max}<2f) \tag{3}$$

Where a represents the aperture diameter of the camera, c represents the minimum acceptable resolution.

(d) Occlusion

As shown in figure 3, O is the center of the camera lens, P is the space object point, and triangle *efg* is *a* patch on the physical model. The parameter expression of the line *PO*:

$$p(t)=t\cdot P+(1-t)\cdot O\quad(0\leq t\leq 1) \tag{4}$$

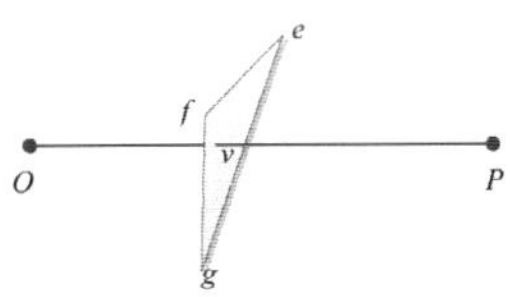

Figure 3. Blocking diagram.

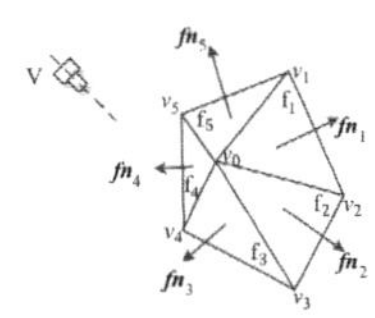

Figure 4. Supplementary view.

3.2. Overlap and coverage

The overlap between the areas captured by the camera can be calculated using equation (5).

$$\phi=S(C_{b1}C_{a2})/S(C_{a1}C_{a2})\times 100\% \tag{5}$$

where S represents the area of a certain area on the model.

3.3. Motion platform constraints

The camera is fixed directly below the fuselage body during aerial photography. The pan/tilt can provide angle changes of pitch and roll. The pitch angle of the viewpoint is $[-90^\circ\sim 90^\circ]$, and the roll angle is

IOP Conf. Series: Materials Science and Engineering **739** (2020) 012055 doi:10.1088/1757-899X/739/1/012055

$[-45 \sim 45]$.

3.4. Camera angle of incidence

In [9], the shooting angle of the camera is too large, which causes the extraction accuracy of the feature points to decrease. The spacing λ on the calibration plate should satisfy the formula (6).

$$\sigma_p = \sigma_k \lambda / \|\boldsymbol{J}_k\| = \sigma_k 3\sqrt{2}\lambda/4 \tag{6}$$

4. Viewpoint planning and screening

4.1. Candidate view area layout

The candidate viewpoints are distributed on the hemisphere with the center of the target model as the center of the sphere and the shooting distance as the radius. The shooting distance of the camera used in this paper is less than 7.5m.

4.2. Viewpoint Supplement

In order to have better integrity of the subsequent 3D reconstruction data, we use the expansion method to add visual points to the complex geometric regions that cannot be captured by candidate viewpoints, as shown in figure 4.

5. Experiment

5.1. Camera incident angle experiment and results

The screening viewpoint experiment uses a two-dimensional calibration plate that distributes 100 dot circular markers ($\lambda = 35mm$) in a range of 35 mm pitch in the 350×280 mm^2 range. In this experimental operation, the camera position is fixed, and the calibration plate attitude is changed from 0° to 75° with a minimum of 5°, and the experimental results are shown in table 1.

Table 1. Experimental Results.

Incident Angle	$\sigma_k(10^{-3})$ First group	Second	Third	Average	Error $\sigma_p(10^{-3})$
0°	1.456	2.194	2.462	2.037	7.971
5°	2.132	2.722	2.121	2.325	9.098
10°	2.413	2.763	2.474	2.55	9.978
15°	2.447	2.083	2.262	2.264	8.859
20°	2.954	2.980	2.93	2.955	11.56
25°	3.231	3.223	2.957	3.137	12.28
30°	3.243	3.378	3.15	3.257	12.75
35°	3.152	3.072	2.986	3.07	12.01
40°	3.391	4.433	3.395	3.74	14.64
45°	3.173	3.013	3.176	3.121	12.21
50°	4.343	3.755	4.431	4.176	16.34
55°	3.758	4.968	3.698	4.141	16.21
60°	6.045	7.733	5.955	6.578	25.75
65°	6.683	7.50	6.435	6.873	26.89
70°	8.521	8.236	8.325	8.361	32.72
75°	12.93	9.574	12.36	11.63	45.5

5.2. 3D Reconstruction Experiment

According to the above viewpoint planning and screening method, the viewpoint planning and reconstruction for the physical target.

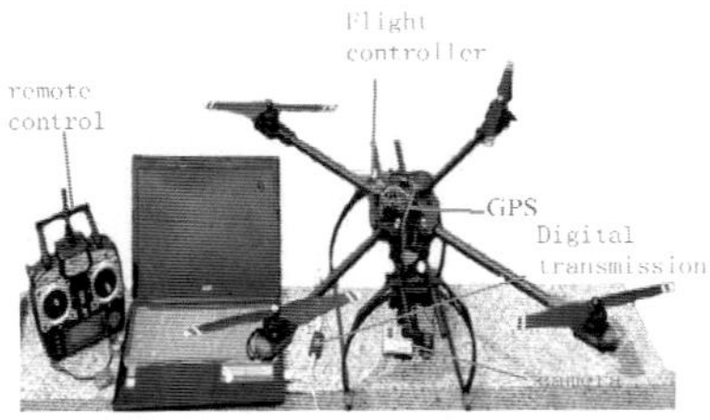

Figure 5. experiment system.

Figure 6. Planning viewpoint reconstruction results.

Figure 5 is an experimental system. The pose control software loads the planned waypoint file and transmits the waypoint command to the flight controller via the ground station using the wireless data transmitter. After receiving the command, the flight controller controls the drone and the camera to perform corresponding actions. When the planned specified viewpoint pose is reached, the pose control software sends a signal to trigger the camera to shoot. Cycle through the above process until all the waypoints are completed, and the shooting will automatically return to the landing.

The candidate viewpoint area is arranged, and the space arranged by the viewpoint is a hemisphere having a radius of a shooting distance (7.5 m). Using the latitude and longitude division method to obtain 500 candidate viewpoints, and then screening to obtain a set of 70 viewpoint sub-viewpoints, the final planning and reconstruction results are shown in figure 6 and table 2.

Table 2. Comparison of shooting effects.

	This paper	Traditional method
Number of viewpoints	70	70
Number of reconstructed feature points	212202	162320
Laser point cloud contrast deviation	7.08cm	7.92cm
Laser point cloud comparison standard deviation	55.57mm	62.05mm

Through the comparison of reconstruction results, the reconstruction method of the viewpoint layout proposed in this paper can be more complete and the reconstruction accuracy is higher.

6. Conclusion

In this paper, the viewpoint path planning problem of multi-view 3D reconstruction for aerial image is determined, and the parameters of camera pose control are determined. The experimental results show that the proposed method has high coverage and the reconstruction accuracy is less than 0.1%.

Acknowledgment

This work was supported by the Natural Science Research Project of West Anhui University (grant numbers WXZR201906), Key Teaching and Research Projects of West Anhui University (grant numbers wxxy2018020, wxxy2019065), National College Students Innovation and Entrepreneurship Training Program 2018 (grant numbers 201810376038).

References

[1] Cowan C K, Kovesi P D 1988 Automatic sensor placement from vision task requirements, *IEEE Transac. Pattern Anal. Mach. Intel.* **10(3)** 407-16.

[2] Satorres Martínez S, Gómez Ortega J, Gámez García J, Sánchez García A 2009 A sensor planning system for automated headlamp lens inspection, *Expert Syst. Appl.* **36** 8768-77.

[3] Abrams S, P. K. Allen, K. Tarabanis 1999 Computing Camera Viewpoints in an Active Robot Work-Cell, *Int. J. Robot. Res.* **18(3)**.

[4] W. R. Scott, G. Roth, View Planning for Automated Three-Dimensional Object Reconstruction and Inspection, ACM Comput. Surveys, **35(1)** 64-96.

[5] G. Olague 2000 Design and Simulation of Photogrammetric networks using Genetic algorithms, ASPRS 2000 Annual Conference Proceeding, Washington DC, USA.

[6] E. Dunn, G. Olague 2004 Milti-objective Sensor Planning for Efficient and Accurate Object Recinstruction, EvoWorkshops 2004, LNCS 3005, pp. 312-21.

[7] K. Schmid, H. Hirschmuller 2012 View Planning for Multi-View Stereo 3D Reconstruction Using an Autonomous Multicopter, *J. Intell. Robot. Syst.* **65** 309-23.

[8] X. zhen 2014 UAV aerial image-based three-dimensional reconstruction of outdoor scenes, Zhejiang University of Technology.

[9] G. Olague, R. Mohr 2002 Optimal camera placement for accurate reconstruction, Pattern Recongnition, **35(4)** 927-44.

IOP Conf. Series: Materials Science and Engineering **739** (2020) 012056 doi:10.1088/1757-899X/739/1/012056

Automated Platform to the Deposition of Thin-Films and Characterization Process in Fiber Optics Sensors

Adolfo Josué Rodríguez-Rodríguez[1, a], Wenceslao Eduardo Rodríguez-Rodríguez[1, b*], Ricardo Iván Luis-Morales[1, c], Víctor Alan Cisneros-Rojas[1, d], Rocío Zamorano Báez[1, e], Jose Isidro Acosta-Morales[1, f], Luis Antonio García-Garza[1, g]

[1]Computational Sciences and Technologies Department. Autonomous University of Tamaulipas. Reynosa - San Fernando Highway C.P. 88790. Reynosa, Tamaulipas, México

[a]arodriguez@uat.edu.mx, [b]wrodriguez@uat.edu.mx, [c]ricardo.ivan09@gmail.com, [d]victoralanciseneros@gmail.com, [e]rocio4679@hotmail.com, [f]isidro81@gmail.com, [g]lgarcia@uat.edu.mx

Abstract. In the recent years the instrumentation has focused its development in devices fabricated in fiber optic applied as sensors; these devices are useful to monitorate physical, chemical and biological variables. By the previous, it is necessary its deposition of a thin- film sensible to temperature, humidity, pH, glucose, etc. There are two technics to deposit this into a fiber optic sensor: dip-coating and layer-by-layer. In both cases, is mandatory the application of a automated plataform to manipulate the fiber optic structure when it is immersed into the films to its depositon and the subsequent characterization in different liquid mediums; applying rutines stablished by times, positions and repetitions. The previous has been developed into a *Robot X-Y,* controlled by *LabVIEW* and its hardware *MyRIO* with an electronic arquitecture of low cost devices; also was applied the 3D impresion based on "open software" to design utensils to manipulate the devices and materials.

1. Introduction

Nowadays the instrumentation has interest in the development of devices based on the optical properties; one of those is the sensors fabricated on fiber optics. The preference by these devices has been by their characeterisitcs properties, sucha as: inmmunity to electromagnetic interference, small dimensions, and large bandwidths. Exist differents configurations of fiber optic sensors due its specific properties: Based on Grattins (Long Period and Bragg), Interferometers, Based on pulished regions. The present research has focus its attention in the SMS structures; based on SingleMode-Multi Mode-SingleMode Fiber Optics. The SMS structures have an operation based on a sensible region that posse an operational wavelenght peak that monitorates variables: physical, chemical and biological. A requirment to achieve sensitive the SMS structures to the mentioned variables is necessary to fabricate a sensible thin-film capable to distinguish the presence of different samples of pH, temperature, glucose, refractive index, etc. By the previous, is required the deposition of the thin-films by automated plataforms to realize the tasks under parameters, such as: temporization, position and repetions. In the case of thin-films there are two techniches to deposition in fiber optics: Dip-Coating and Layer-by-Layer. By the mentioned, in the Section 2, the authors will expose the theorical fundamentals of SMS strucures and thin-films deposition techniques. In section 3, will be exposed the robot that has been conditioned to realize the deposition of thin-films. A fiber optic sensor is a device that possess a micro metric dimensions, so is necessary the application of utensilums to manipulate these devices at the moment for depositing sensible materials and conteiners to contain them; therefore, it will exposed the design of 3D impression utensiliums and conteiners In Section 4, it will described the embedded architecture design applied to control the conditioned robot; where is segmented by: electrical-electronic using low cost devices, mechanical based on steppers-motors and computer desing by MyRIO microprocessor and LabVIEW to the Graphical User Interface (GUI). Then, in Section 5 will be presented experimental tests when in a SMS structure is deposited a thin-film by dip-coating and layer-by-layer. After that, will be exposed the results obtained when a SMS is tested in to different IR samples to the characterization the behavior of the sensor. Finally, in Section 6, the authors will propose their conclusions and the improvements to future.

2. Theory Fundamentals

2.1. Fiber Optic Sensors: The SMS configuration

The fiber optic sensors are fabricated by a configuration SingleMode Fiber (SMF)-MultiMode (MMF)-SingleMode Fiber (SMF); where an incident light is transmited in the structure at the SMF, then it gets to the sensible región (MultiMode Fiber) the light will be propagating in different modes; the MMF used is a special one called "Coreless (MMF-C)", due its properties to interactuate with the surrounding media. Then the propagation modes are coupled into a spliced SMF. The theory fundamental of SMS structure configuration is based on the *Multi-Modal Effect*; it wich says, that due de presence of supported modes in MMF-Coreless (MMF-C) are excited, occurs an interference between each of them producing a replica image of the input at the end of the MMF-C. In the SMS configurations, this replica can be obtained as an operational wavelenght peak by the equation 1.

$$L = \rho\left[(n_{MMF-C}\, D_{MMF-C}{}^{2})/\lambda\right] \tag{1}$$

L means the length of *MM-C,* ρ is the replica's order (been the fourth), n_{MMF-C} is the refractive index in the core of *MM-C,* D_{MMF-C} is the diameter in the MMF-C and λ is the position of the operational wavelenght peak [1-3]. In section 6, will be exposed a SMS based on the parameters $D_{MMF-C} \approx 62.5\mu m, \rho = 4, n_{MMF-C} = 1.444, \lambda = 850nm$, where the length of MM-C $\approx 28mm$ is obtained by equation 1, where the diameter is reduced by etching [1].

2.2. Dip-Coating

Based on the wet chemical thin-films the deposition by dip-coating represents one of the most popular coating processes. The process is based on three stages: First is the *immersion*, the device is immersed (in a vertical movement) into the wet thin-film at a speed constantly untill it is complety inside the container with the material. Then, the second stage is the *deposition;* the thin-film is incrusted on the SMS during at certain time. After that, the third stage is the *extraction*. The SMS is taken out from the container with sensitive material at a constant speed. During the whole deposition is necesary monitoring the spectral wavelenght responce to verify the process, waiting at least 6 hours that the thin-film in the SMS is dried. [4].

2.3. Layer-by-Layer

Is requiered the application of two films the fabrication of this technique, these most to posses' charges oppositely, after of the deposition of each material is mandatory a stage of washed dips. The SMS strucutre is required be immersed into each wet material, during a definited temporization; sucha as the dip-coating must to be immersed in a vertical movement. The movements to each step of fabrication in the horizontal must to be with a speed controllaed (such as the whole experiment) [5].

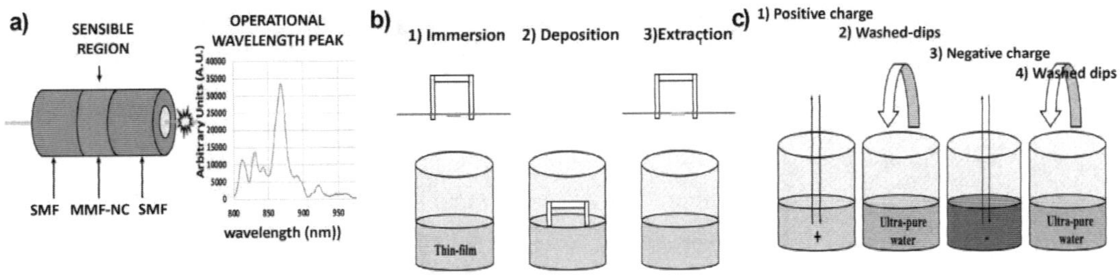

Figure 1. a) SMS structure with its operational wavelength peak.
b) Dip-coating process. c) Layer-by-Layer deposition.

3. Automated Platform and impressed utensils in 3D

A robot based on movement degrees on X-Y, has been adapted to realice such as an automated platform to deposite thin-films in SMS strucutres based on: dip-coating and LbL techiques; also to characterize the fabricated device on different liquid samples. To handle the device and the liqid samples, were fabricated by 3D desing and impression two utensils: a holder to manipulate de SMS embedded on the plataform and a container with 10 subdivisions to the diferent liquid samples. The previous ítems were designed on Sketchup and impressed on Anet A8.

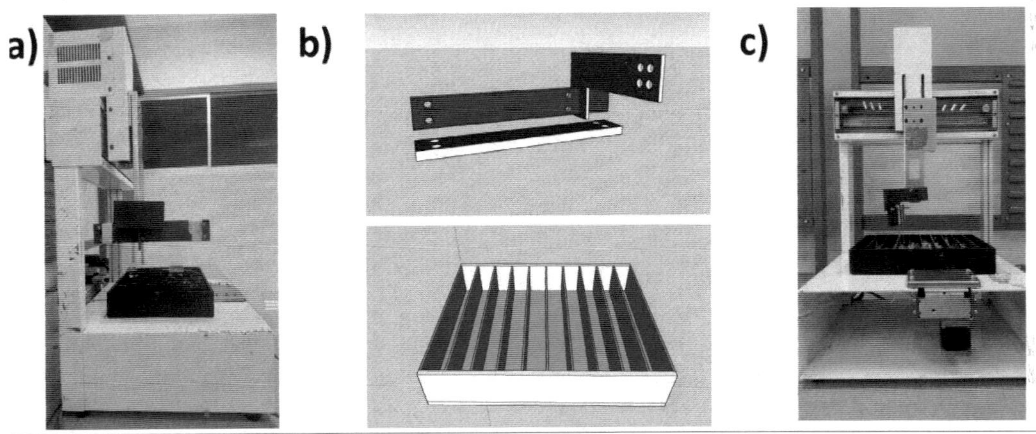

Figure 2. a) Lateral perspective of automated plataform.
b) Utensils based on 3D design and impression. c) Frontal perspective of automated platform.

4. Embedded architecture

In this section, the authors will explain the embedded architecture developed to manipulate the automated plataform. To communicate the instruccions established by the user, for the movements of the robot X-Z, has been used a National Instruments MyRIO; it wich, will controlate two steppers motors (by Sanyo Denki, 12V-1.4A and 2A). It is necessary, the utility for each stepper motor a Pololu A4988 Driver to assign the step direction and different microstep resolutions; also, an of power source (24V, 20A) to supply energy to the electric-mechanical instrumentation. In many ocassions, it is necessary to keep working the automated platform due the characterization process in the SMS devices; the MyRIO allows to stablish remote communications with the user without her/his prescence.

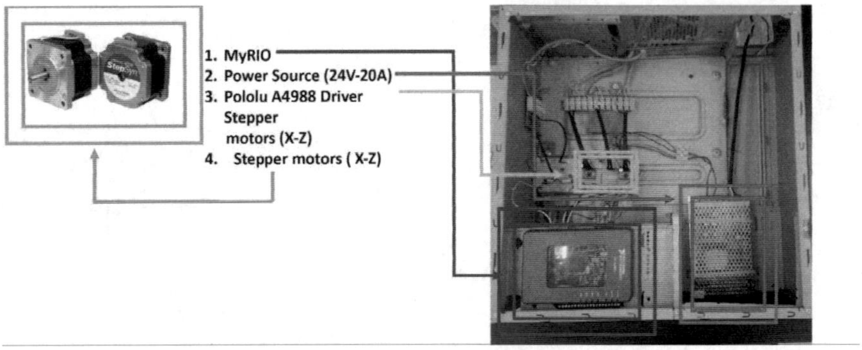

Figure 3. Embedded arquitechture mounted into a PC Desktop Case.

5. Graphical User Interface (GUI)

The Graphical User Interface (GUI), has been developed in National Instruments LabVIEW, due the user needs to establish a certain parameter to define the depositions routines; such as: position in X and Z, microstep resolutions, number of dips, times of steady state at up and down in X (inside or outside of the liquid sample), hystogram, number of samples and cycles. As most of all the automated machines is necessary to include virtual conditions of: Pause, Stop, Reset and Home; just in case of unforeseen scenarios.

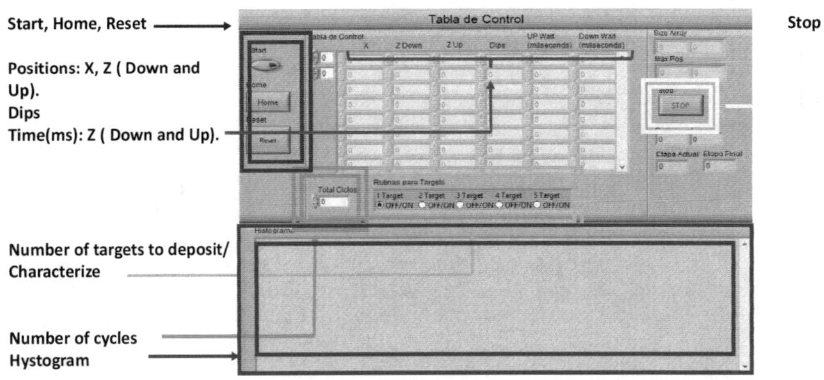

Figure 4. Graphical User interface of Automated Platform developed in LabVIEW.

6. Experimental tests

Using a SMS structure with $D_{MMF-C} \approx 62.5\mu m, L \approx 28mm$; the automated platform has been tested in two scenarios: The first one with a propuse to achieve in a SMS structure a Layer-by-Layer deposition using a Polycationic and a Polyanionic solutions at two layers, exposing a decreasing of power at the second layer; this deposition will allow to sense different liquid samples of physical, chemical and biological variables. The second application is based on a SMS device with the same diameter and length of $MMF - C$, but without any thin-film deposited into it; for a characterization based on different liquid samples of Refractive Index (RI) with the purpose for to know the absolute wavelength shift of the structure when the diameter is reduced at approximatly the half, Δ_{λ}= 19.63nm

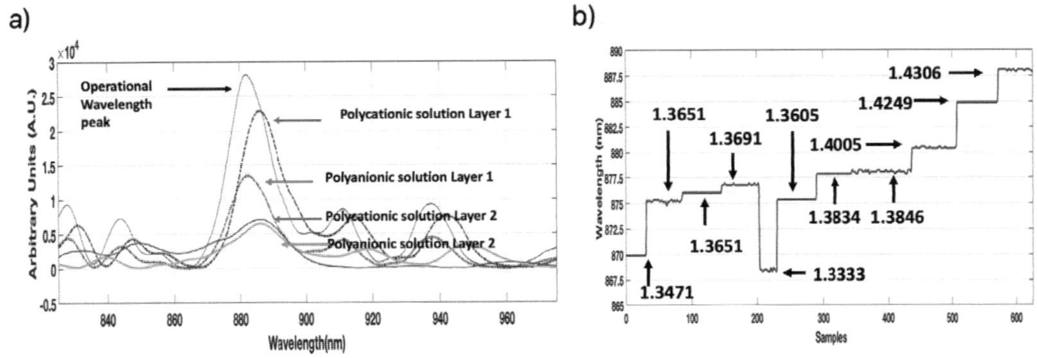

Figure 5. (a) Layer-by-Layer deposition. (b) Characterization of different RI samples.

7. Conclusions

The autors have developed an automated platform for deposition of thin-films to fiber optic sensores based SMS structures, where has exposed a Layer-by-Layer deposition and a characterization process by RI. The previous will be a great utility in the fabrication of fiber optic sensors, not only in SMS also in other structures; such as: D-shapes, MMF, Long Period and Bragg Grattings, etc. A future improvement will be the control by remote distance using a mobile device.

References

[1] Rodríguez-Rodríguez W E, et al. 2018 Sensitivity enhancement experimental demonstration using a low cutoff wavelength SMS modified structure coated with a pH sensitive film, *Sens. Actuat. B: Chemi.* **262** 696-702.

[2] Rodríguez A, et al. 2014 Gasohol Quality Control for Real Time Applications by Means of a Multimode Interference Fiber Sensor, *Sensors* **14(9)** 17817-28.

[3] Villareal-Jimenez L R, et al. 2018 An Efficient Optoelectronic System for Remote Salinity Water Sensing, *App. Mechan. Mater.*876 152-60.

[4] Schneller T, et al. 2013 Chemical Solution Depositon of Functional Oxide Thin Films, *Longdon: Springer* 233-61.

[5] Ariga K, Jonathan P H 2007 Layer-by-Layer (LbL) Assembly, A "Gentle Yet Flexible" Method Toward Functional Biomaterials, *Phys. Chem. Chem. Phys.* **9** 2319.

AUTHOR INDEX

Abdullah, Nur Athirah 100
Acosta-Morales, Jose Isidro 357
Adi, Tri Joko Wahyu 248
Al-Adhadh, Ahmed Raad 116, 274
Al-Rumaithi, Ayad 202
Al-Sherrawi, Mohannad 202
Alfonso, V. C. M. 35
Ali, M 208
Amalia, Aniendhita Rizki 248
Aminanda, Y 208
Arana, V 137, 289
Arciniega, R 283
Arias, C 137
Asim, Nilofar 100
Atabaev, Timur Sh. 243, 262
Atienza, A H 161, 166
Bagsit, M. B. 35
Baoyu, Xu 131
Baylon, M B 12
Baylon, M. B. 35
Belyaev, L. V. 257, 319
Belyi, A 329
Bernaudo, I 339
Betancourt, N 137
Bethary, Rindu Twidi 42
Báez, Rocío Zamorano 357
Candamano, S 339, 345
Cantiello, A 345
Chang, Yongjia 227
Changming, Hu 301
Chavez, B 177
Chen, Ching-Shun 323
Chen, L 237
Chen, Yi-Lin 323
Chou, Chung-Chyi 323
Chumchujan, Chaikit 172
Chung, S R 150
Chung, S. R 111
Chung, Shu-Ru 156, 172
Cisneros-Rojas, Víctor Alan 357
De Luca, P 339, 345
De Oliveira, Tiago 144
Delgado, M 289
Dovbysh, N. S. 319
Dukenbayev, Kanat 243
Durán, F 137
Duy, Vinh Nguyen 69
Elliani, R 339
Em, Svetlana 262
Embalzado, K A D 12
Embalzado, K A 59, 79
Eroles, R 166
Fajri, Dan 190
Fandiñola, E 166
Fauzi, A 190
Firsa, T 208
Garbin, F 283
Garcíagarza, Luis Antonio 357
Guodong, Wang 131
Gutlay, H 161
Guégan, Régis 144
Ho, Yu-Tsung 323
Hongliang, Yi 131
Huaco, G 177, 195
Huang, Shoudao 74
Huang, Y. X. 111
Husaini 84, 106
Ikbal, M 84
Iqbal, M 208
Iranata, Data 248
Irawan, Djoko 248
Ivanchenko, A. B. 257
Jiang, W 237
Jin, Xiaoguang 1
Jirsa, J 195
Kadhim, Zainab Jawad 116
Kadir, Nurul Huda Abd 100
Kang, W P 267
Khamkhash, Laura 243, 262
Kittikorn, C 64
Kochetov, D. O. 257
Kocich, R 48
Kuncická, L 48
Kwong, Kok Zee 90, 95
Lee, Foo Wei 90, 95
Levano, A 283
Li, Liangchao 215
Liew, Chi Hoe 90
Lim, Kok-Sing 90
Lim, Zhi Heng 95
Lin, M Y 150
Lin, Meng-Yen 156
Lin, Ronghui 215, 227
Lin, W X 150
Liu, Zihong 215, 227
Luis-Morales, Ricardo Iván 357

Macario, A 339
Majuar, E 190
Marquez, I 79
Matias, C 177
Mezeix, L 64
Minh, Dien Vu 69
Mo, Kim Hung 95
Mohammad, Masita 100
Molkenova, Anara 243, 262
Montes, K M S 12
Montes, K M 59, 79
Moolsarn, Kawinthida 156
Morozov, V. V. 319
Mulikova, Tomiris 243
Naeem, Zahraa Tawfiq 274
Napoli, A. 333
Nguyen, Duy Vinh 74
Nguyen, Khanh Duc 69
Nguyen, Thanh Viet 69
Novriandika, S 106
Pakpong, J 64
Payawal, J M 59
Peña, R Dela 166
Polenov, D Yu 296
Purith, P 64
Putra, T E 84, 106
Qi, X F 267
Qu, K 237
Rendong, Liu 131
Reyes, J. D. 35
Reza, M 190
Riyanto, Wahyu 248
Rizal, F 190
Rizov, V 25, 30
Rodrigo, P V 161
Rodríguez-Rodríguez, Adolfo Josué 357
Rodríguezrodríguez, Wenceslao Eduardo 357
Romanenko, Sofia A 126
Rosales, A 289
Sabur, Dhay Ali 116
Sainorudin, Muhammad Hanif 100
Sakban, Haider Kadhem 274
Sales, N M 59
Santos, G N C 183
Sayno, A J 59, 79
Stepenkin, A. V 257
Subagio, Bambang Sugeng 42
Syahyadi, R 190
Tagarelli, A 339
Tamayo, P 161
Tan, Chee Ghuan 90
Tani, L C 79
Thanh, Dinh Xuan 69
Torre, H M B Dela 183
Tran, Thi Diep 74
Trofimchuk, Oksana A 126
Turanov, Sergey B. 126
Uy, F A A 12, 59, 79
Uy, F. A. A. 35
Vu, Duc Ha 74
Wang, Q 267
Wang, Yuanqing 1
Xin, Xu 131
Xue, Zhenqian 54
Yaakob, Zahira 100
Yakovlev, Alexey N. 126
Yang, J F 237
Yang, J H 237
Yang, Qingting 1
Yang, Y 267
Yaqi, Cao 301
Yedigenov, Mussa 262
Yew, Ming Kun 95
You, Hui 215, 227
Yun, Zhu Yan-Yan Shi 351
Zhai, W 237
Zhang, Ping 215, 227
Zhao, G 267
Zhao, Xin 54
Zhaoyu, Zhu 301
Zhdanov, A. V. 257, 319